METEOROLOGY TODAY

AN INTRODUCTION TO WEATHER, CLIMATE, AND THE ENVIRONMENT

C. Donald Ahrens

SECOND EDITION

Modesto Junior College

West Publishing Company   St. Paul   New York   Los Angeles   San Francisco

Library of Congress Cataloging in Publication Data

Ahrens, C. Donald.
    Meteorology today.

    Bibliography: p.
    Includes index.
    1. Meteorology.   I. Title.
QC861.2.A3   1985   551.5   85–3214
ISBN 0–314–85212–3
2nd Reprint—1987

Design: Janet Bollow
Copy editing: Stuart Kenter
Illustrations: Joe Medeiros and Brenda Booth
Cover Photograph: Elizabeth Beaver Burnett
Composition: Janet Hansen & Associates

Credits

Chapter 1

Fig. 1.1 Palomar Observatory, California Institute of Technology. **Figs. 1.3, 1.4, 1.5** NASA photos. **Fig. 1.6** NOAA photo.

Chapter 2

Fig. 2.1 Adapted from *SMIC/Inadvertent Climate Modification*, The MIT Press, 1971, and other sources. **Fig. 2.2** U.S. Geological Survey photo.

Chapter 3

Fig. 3.7 Photo by author.

Chapter 4

Fig. 4.2 Gordon Newkirk, Jr., High Altitude Observatory, Boulder, Colorado. **Fig. 4.3** NASA photo. **Fig. 4** NOAA photo.

Chapter 5

Fig. 5.4 Courtesy of the American Museum of Natural History. **Fig. 5.10** Photo by Ross DePaola. **Figs. 4 and 5** Photos by author. **Figs. 6 and 7** From *Climates of the United States*, U.S. Department of Commerce, 1973. **Table 1** U.S. Department of Agriculture.

Chapter 6

Fig. 6.4 Photo by Berna Philbin. **Fig. 6.9** Photo by Ross DePaola. **Fig. 6.12** Photo by author.

Chapter 7

Fig. 7.12 Photo by Ross DePaola.

Chapter 8

Fig. 8.1 Photo by Elizabeth Beaver Burnett. **Figs. 8.3, 8.4, 8.5** Photos by author. **Figs. 8.6, 8.8** Photos by Ross DePaola. **Fig. 8.9** Photo by author. **Fig. 8.10** From *Climates of the United States*, U.S. Department of Commerce, 1973. **Fig. 2B** Courtesy of V. G. Plank of Air Force Geophysics Laboratory, Cloud Physics Branch.

Chapter 9

Figs. 9.1, 9.2, 9.3, 9.4, 9.5 Photos by author. **Fig. 9.6** Courtesy T. Ansel Toney and S. Horowitz. **Fig. 9.9** Courtesy of NASA. **Figs. 9.11a, b and 9.12** NOAA photos.

Chapter 10

Figs. 10.13, 10.14, 10.16 Photos by author.

Chapter 11

Figs. 11.8, 11.10 Photos by author. **Figs. 11.12, 11.15** NOAA photos. **Figs. 11.18, 11.20** NCAR photos.

Chapter 13

Fig. 13.6 U.S. Geological Survey, M. R. Campbell. **Fig. 13.8** NCAR photo. **Figs. 13.9, 13.14** Photos by author. **Fig. 13.17** Photo by Ross DePaola. **Fig. 2** U.S. Department of Energy. **Fig. 3** Sandia National Laboratories. **Fig. 4** Photo by Ross DePaola.

Chapter 14

Fig. 14.4 Photo by Jim Edwards, Riverside Press-Enterprise. **Figs. 14.10, 14.13** Photos by author. **Fig. 14.15** NOAA photo. **Figs. 14.17, 14.18** Courtesy Sherwood B. Idso.

Chapter 15

Figs. 15.12, 15.16 NOAA photos.

Chapter 16

Figs. 16.5, 16.13 NOAA photos.

Chapter 17

Fig. 17.17 NOAA photo.

Chapter 18

Fig. 18.3 Data taken from "Statistical Probabilities for a White Christmas," U.S. Department of Commerce, Washington, D.C., 1971. **Fig. 18.4** NOAA photo. **Fig. 18.11** Courtesy National Weather Service. **Figs. 18.14, 18.16** NOAA photos.

Chapter 19

Figs. 19.2, 19.3 Photos by author. **Fig. 19.6** Photo by Major Richard F. Picanso, USAF (Air Weather Service). **Fig.**

(Credits continued following Index.)

CONTENTS

CHAPTER 4

The Sun's Influence on the Upper Atmosphere   63

CHAPTER 5

Seasonal and Daily Temperatures   85

CHAPTER 6

Light, Color, and Atmospheric Optics   113

PREFACE

With satellites photographing the paths of hurricanes, computers forecasting the weather, aircraft seeding clouds with silver iodide, and radar tracking the precipitation inside violent thunderstorms, the science of meteorology (the study of the atmosphere) is appreciably more sophisticated today than it was just a few decades ago. Our technical knowledge of the atmosphere is expanding rapidly. Most people, though, become excited by meteorology simply by observing the atmosphere directly. This book, then, emphasizes watching the weather. In this way, meteorology becomes "alive" in that readers can immediately apply text material to the world around them.

About This Book

Meteorology Today is written for college-level students taking an introductory course on the atmospheric environment. The main emphasis of the text is on conveying meteorological concepts in a visual and practical manner, while at the same time providing students with a comprehensive background in basic meteorology. Written material covers topics directly related to one's everyday experiences and stresses the understanding and application of principles. As with the first edition, no special science prerequisites are necessary.

This second edition of *Meteorology Today* has been thoroughly revised and updated. In an attempt to stimulate interest, Chapter 1 now serves as a brief, but broad, introduction to the atmosphere. Atmospheric optics (Chapter 6) has been expanded to include topics on coronas, glories, and the *Fata Morgana*. Chapters 6 and 7 from the first edition have been combined and now form Chapter 5, "Seasonal and Daily Temperatures." Thunderstorms and tornadoes are now discussed within a single chapter (Chapter 19). Chapter 20

now covers hurricanes and tropical weather. There are two chapters on climate. One (Chapter 22, "Climate Modification and Change") explores such issues as urban climates, air pollution, acid deposition, and climatic change.

This new edition includes material on lightning suppression, Mesoscale Convective Complexes, microbursts, and the heat index (HI) initiated by the National Weather Service in 1984. Moreover, interesting new illustrations have been inserted and many existing ones redrawn to strengthen points and clarify more involved concepts. Also, totally new color photographs (as well as black-and-whites) of clouds and optical phenomena have been selected.

Another added feature is the "Problems and Exercises" sections that appear at the end of each chapter. The problems generally require mathematical calculations involving equations or formulas described in a focus section. The exercises can be used as lab work.

Each chapter still contains at least one focus section, which either extends textual material or explores a related topic. Focus sections are divided into five distinct categories: observations, applications, issues, instruments, and special topics. Some cover material not always found in introductory meteorology textbooks, such as topics on aviation weather, wind power, cloud seeding, and weather prediction. Others help to bridge theory and practice. Focus sections unique to this edition include such contemporary topics as "Climatic Change Induced by Nuclear War." Important equations, such as the geostrophic wind equation and the hydrostatic equation, are quantitatively described in new focus sections.

Each chapter incorporates a number of effective learning aids:

- A major topic outline at the beginning.
- The boldfacing of important terms (with definitions in the glossary or the text).

- Italicized key phrases.
- English equivalents of metric units given in parenthesis.
- An end-of-chapter summary, which reviews a chapter's key ideas.
- Questions for review, which check on students' assimilation of the material.
- Questions for thought, which require students to synthesize learning concepts for deeper understanding.
- Problems and exercises that require mathematical calculations.

The organization of the text into 22 chapters is designed to give instructors of atmospheric science courses maximum flexibility. Many of the chapters can be covered in any desired order. For example, Chapter 6 ("Atmospheric Optics") is self-contained and can be covered later, if so desired. Thus, instructors are free to tailor this text to their particular needs.

The book follows a traditional approach. After an introductory chapter, it covers the origin, composition, and behavior of air. It then discusses energy, temperature, moisture, precipitation, and winds. Then come chapters that deal with air masses, fronts, and middle-latitude cyclones. Weather prediction and severe storms are next. The final two chapters cover climate, its classification and its changing nature.

Eight appendices conclude the book. These are somewhat technical. One, "The Instant Weather Forecast Chart" (Appendix E), should be especially interesting to students. Also, at the end of the book, is a list of additional reading material and an extensive glossary. An instructor's manual is also available.

Acknowledgments

Thanks go to the many individuals who have contributed to the production of this second edition. Thanks to Ross DePaola who once again transformed my color transparencies into beautiful black-and-white photographs; and to Joe Medeiros and Brenda Booth for their care in drawing the illustrations. A special note of appreciation is due Janet Bollow, whose assistance and professional concern in designing the book is greatly appreciated;

and to Stuart Kenter, whose suggestions and conscientious editing has added greatly to this edition. I am especially grateful to my wife Suzy for her assistance during the preparation of the manuscript of the second edition and for her patience, encouragement, and understanding.

My special thanks to those individuals who provided me with slides and photographs and to those who took the time to read portions of the text. Thanks to Robert S. Robinson and Art Rangno, who made suggestions for an improved second edition. Finally, for their comments and suggestions, I am indebted to the professional reviewers, including:

Alan Robock, University of Maryland, College Park, Maryland

Michael F. Taugher, California State University, Northridge, California

Robert B. Batchelder, Boston University, Boston, Massachusetts

B. Ross Guest, Northern Illinois University, Dekalb, Illinois

Thomas T. Arny, University of Massachusetts, Amherst, Massachusetts

Gary D. Johnson, University of South Dakota, Vermillion, South Dakota

Jim Perry, University of Minnesota, St. Paul, Minnesota

Catherine M. Felton, San Francisco State University, San Francisco, California

To the Student

Learning about the atmosphere can be an enjoyable experience, especially if you become involved. This book is intended to give you some insight into the workings of the atmosphere. But, for a real appreciation of your atmospheric environment, you must go outside and observe. Mountains take millions of years to form, while a cumulus cloud can develop into a raging thunderstorm in less than an hour. All of the information in this book is out there—please, take the time to look.

Don Ahrens

METEOROLOGY TODAY

Earth viewed by *Apollo 8* astronauts, as their spacecraft circled the moon. (Photo: NASA)

CHAPTER 1

Introduction: The Earth and Its Atmosphere

The earth is surrounded by a life-giving blanket of air we call the *atmosphere*. In one way or another, our atmosphere influences everything we see and hear—it is intimately connected to our lives. Air is with us from birth; we cannot detach ourselves from its presence. In the open air, we can travel for many thousands of kilometers in any horizontal direction, but should we move a mere eight kilometers above the surface, we would suffocate. We may be able to survive without food for a few weeks, or without water for a few days, but, without our atmosphere, we would not survive more than a few minutes. Just as a fish is confined to an environment of water, so we are confined to our ocean of air. Anywhere we go, it must go with us.

The earth without an atmosphere would have no lakes or oceans. There would be no sounds, no clouds, no red sunsets. The beautiful pageantry of the sky would be absent. At night, it would be unimaginably cold, and, during the day, unbearably hot. All things on the earth would be at the mercy of an intense sun beating down upon a planet utterly parched.

Living on the surface of the earth, we have adapted so completely to our environment of air that we sometimes forget how truly remarkable this substance is. Even though air is tasteless, odorless, and (most of the time) invisible, it protects us from the scorching rays of the sun and provides us with a mixture of gases that allows life to flourish. Because you cannot see, smell, or taste air, you may have a hard time believing that between your eyes and the pages of this book are trillions of air molecules. Some of these may have been in a cloud only yesterday, or over another continent last week, or perhaps part of the life-giving breath of a person who lived hundreds of years ago.

Contents

To see how unique our earth's atmosphere really is, we will begin by comparing it with other planetary atmospheres. At the same time, we will examine a number of important concepts and ideas that will be expanded in subsequent chapters.

A Journey Through Space

Imagine—for a moment—that we have the power to travel through space at fantastic speeds. Suppose we are in a space capsule hurtling through the universe at nearly the speed of light, our destination a small planet called "Earth." What features might we see during our journey?

Peering outside, we would see bright specks by the hundreds of billions, surrounded by deep, dark space. As we approach one of these specks, it becomes larger and brighter, taking on the shape of an elongated saucer with spiraling arms. To our astonishment, it is not a single spot of light at all, but a cluster of one hundred billion stars—a *galaxy*. The width of the galaxy is enormous: Light takes nearly 100,000 years to travel from one side to the other.

The stars comprising a galaxy are hot, glowing balls of gas that generate energy by converting hydrogen into helium near their centers. The

Fig. 1.1   A spiral galaxy consisting of more than 100 billion stars.

brightness, color, and size of the stars vary. Some of the large stars are blue, whereas others are red. There are smaller yellow stars, white stars, and even faint red ones. The blue stars are hotter, "burn" their hydrogen quickly, and may last for only a few million years. The yellow and faint red stars, being cooler, "burn" more slowly and may exist for many billions of years. Between the stars are cloudlike regions of dust and gases, of which *hydrogen* is the most abundant element. Such cosmic "clouds" are known as *nebulae*. Occasionally, we see an exploding star—a *supernova*—which sends bits and pieces of stellar matter whirling through space in all directions, ultimately to become the material for the creation of new stars and planets.

Within this vast universe, there are billions of galaxies, each with hundreds of billions of stars of varying ages and sizes. Our task is to find a single planet that orbits an average size star known as the "sun," situated in a galaxy called the Milky Way. If we were able to locate the Milky Way galaxy, where would we expect to find the sun within it?

The sun is approximately 300 million billion kilometers (km) from the center of the galaxy. This extremely large number can be written in short form as 3×10^{17}, which means 3 followed by 17 zeros, or 300,000,000,000,000,000 km. (See Appendix A for further explanation of this system of notation.) Looking at the Milky Way galaxy, we see only stars, no planets. Planets are too small to be seen individually on this scale. They are also cooler objects and do not generate their own light. A planet, such as earth, is visible only in reflected light.

The Solar System

As we move toward the sun, it appears as a tiny speck of light surrounded by millions of other bright specks. Gradually, it appears larger and brighter, and soon we are able to make out the faint image of the largest planets that orbit it: Jupiter, Saturn, Uranus, and Neptune. At this distance (still many tens of billions of kilometers away), the smaller inner planets, including the earth, are lost in the sun's glare. Eventually, we move

Table 1.1 Data on Planets and the Sun

	DIAMETER		AVERAGE DISTANCE FROM SUN		AVERAGE SURFACE TEMPERATURE		MAIN ATMOSPHERIC COMPONENTS
	Kilometers	Miles	Millions of km	Millions of miles	°C	°F	
Sun	$1,392 \times 10^3$				5,800	10,500	—
Mercury	4,880	3,032	58	36	260*	500	—
Venus	12,112	7,526	108	67	480	900	CO_2
Earth	12,742	7,920	150	93	15	59	N_2, O_2
Mars	6,800	4,216	228	142	−60	−76	CO_2
Jupiter	143,000	88,700	778	483	−110	−166	H_2, He
Saturn	121,000	75,000	1,427	886	−190	−310	H_2, He
Uranus	51,800	32,187	2,869	1,780	−215	−355	H_2, CH_4, ?
Neptune	49,000	30,447	4,498	2,790	−225	−373	H_2, CH_4, ?
Pluto	3,100	1,926	5,900	3,670	−235	−391	CH_4, ?

*Sunlit side.

close enough to see the entire solar system, which includes the sun at the center, with nine orbiting planets: Mercury, Venus, Earth, Mars, Jupiter, Saturn, Uranus, Neptune, and Pluto. (See Table 1.1 for information about size, surface temperature, and atmospheric composition of the planets.)

From this vantage point, we can gather data about the planets. The sun, for instance, appears to be about 10 times the diameter of Jupiter, the largest planet. Jupiter itself is about 10 times the diameter of Earth. Since warmth for the planets is provided primarily by the sun's energy, those planets closer to the sun are generally warmer than those farther away. At an average distance of nearly 150 million kilometers (93 million miles) from the sun, the earth intercepts only a very small fraction of the sun's total energy output. However, it is this energy that drives the atmosphere into the patterns of everyday wind and weather, and allows the earth to maintain an average surface temperature of about 15°C (59°F).*

Although this temperature is mild, the earth experiences a wide range of temperature, as readings can drop to as low as −85°C (−121°F) during a frigid Antarctic night and climb, during the day, to almost 50°C (122°F) on the oppressively hot subtropical desert.

The most striking feature of the planets is the subtle nature of their colors. Mars appears red, Earth blue, Uranus green, and Jupiter everchanging shades of orange, brown, yellow, gray, and light blue.

Jupiter's colors are actually bands of clouds that spread laterally parallel to its equator. This spreading is probably due to Jupiter's rapid rate of rotation, which is about once every 10 hours.

*The abbreviation °C is used when measuring temperature in degrees Celsius, and °F is the abbreviation for degrees Fahrenheit. More information about temperature scales is given in Appendix A and in Chapter 2.

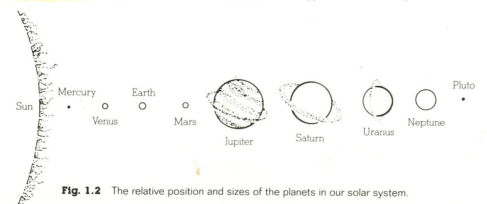

Fig. 1.2 The relative position and sizes of the planets in our solar system.

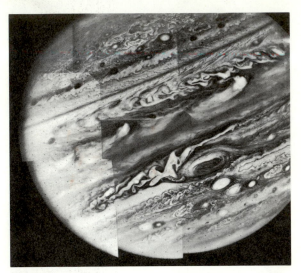

Fig. 1.3 A mosaic of Jupiter, assembled from nine individual photographs taken by *Voyager 1* on February 26, 1979. This shows Jupiter's great red spot (below and to the right of center) and the complex structure of its clouds. The great red spot, as well as the smaller ones, appear to be spinning eddies similar to the eddies (high and low pressure areas) that exist in the earth's atmosphere. (Distortion of the mosaic is caused by the rotation of the planet between picture frames.) (Photo: NASA)

A prominent feature on Jupiter is a great red spot about 3 times larger than the earth (Fig. 1.3). This spot appears to be a huge atmospheric storm that spins counterclockwise in Jupiter's southern hemisphere, completing one revolution every six days. Large, white ovals near the great red spot are probably similar, but smaller, storm systems.

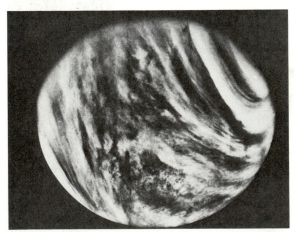

Fig. 1.4 Venus with its cloud cover. At the surface, beneath these clouds, the air temperature averages a scorching 480°C.

Studying the atmospheres of other planets may give us some insight into the behavior of the atmosphere on earth. Suppose, then, we briefly investigate the atmospheres of Jupiter, Venus, and Mars. Jupiter's atmosphere is composed primarily of hydrogen (H_2) and helium (He), with minor amounts of methane (CH_4) and ammonia (NH_3). Its opaque cloud cover is probably made of tiny, suspended flakes of frozen ammonia. The Venusian atmosphere is much different from that of Jupiter. It is mainly carbon dioxide (95 percent), with minor amounts of water vapor and nitrogen. An opaque cloud deck encircles the planet at about 50 km above the ground, preventing us from viewing its surface (Fig. 1.4). Since Venus is closer to the sun than is the earth, we would expect Venus to be slightly warmer. Instrument probes, however, indicate that surface air is not just slightly warmer but extremely hot, averaging 480°C (900°F).

Apparently, these exceptionally high temperatures result from the thick atmosphere of carbon dioxide. This gas allows the sun's energy to penetrate and heat the surface. As the surface warms, it must constantly lose energy, otherwise it will continually grow hotter. The carbon dioxide, however, temporarily prevents the surface heat from escaping into space. This trapped heat causes the high temperatures. This phenomenon is known as the *greenhouse effect* because the carbon dioxide behaves somewhat like the glass around a greenhouse—it allows sunlight to penetrate, but prevents heat inside from escaping and mixing with the outside air. (We will examine the earth's greenhouse effect, also called the *atmosphere effect*, in Chapter 3.)

The atmosphere of Mars, like that of Venus, is mostly carbon dioxide, with only small amounts of other gases. Unlike Venus, the Martian atmosphere is very thin, and heat escapes from the surface rapidly. Thus, surface temperatures on Mars are much lower, averaging around −60°C (−76°F). Because of its thin, cold atmosphere, there is no liquid water on Mars and virtually no cloud cover—only a barren desertlike landscape. (See Fig. 1.5.)

Occasionally, huge dust storms develop near the Martian surface. Such storms may be accompanied by winds of several hundreds of kilometers per hour. These winds carry fine dust around the

Fig. 1.5   Portion of Martian landscape photographed from the *Viking I* lander on July 20, 1976. Aside from an occasional dust storm, the weather on Mars could be very monotonous, with few clouds and no liquid water.

entire planet. The dust gradually settles out, coating the landscape with a thin reddish veneer. Studying the temperature changes that result from these storms may provide scientists with an understanding of the possible effects of dust in the earth's own atmosphere.

The Earth as a Planet

Our space capsule is now approaching the planet Earth. From space, the earth appears remarkably spherical, with a diameter of nearly 12,800 km or 8000 miles (mi). However, refined measurements reveal that the earth has "middle age spread," for it bulges slightly at its equator. Such a distorted shape is known as an *oblate spheroid*. The departure from a spheroid shape is small; the equatorial diameter is a mere 43 km (27 mi) greater than the polar diameter. The bulging is due to the prolonged stress put on its midsection as the earth spins.

Each day, the earth makes one complete rotation about its axis (an imaginary line running through the planet between the North and South Poles). As the earth rotates, it revolves around the sun, completing one orbit in a little more than 365 days, or one year. Water covers nearly three-fourths of the earth's surface, making it the only blue planet in the solar system. Its atmosphere is a thin, gaseous envelope comprised mostly of nitrogen (N_2)

and oxygen (O_2), with clouds of condensed water vapor and ice particles. Almost 99 percent of the atmosphere lies within 30 km (18 mi) of the surface. In fact, if the earth were to shrink to the size of a beach ball, its inhabitable atmosphere would be thinner than a piece of paper. Although thin, the atmosphere constantly shields the planet's surface and its inhabitants from the sun's dangerous radiation, as well as from the onslaught of material from interplanetary space.

Near the edge of the atmosphere, we occasionally see a sudden streak of light, a "shooting star," which remains visible for a second or two. These trails of light occur when small particles from space called *meteoroids* enter into the atmosphere, heat up due to friction with the air molecules, and vaporize. The daily bombardment of millions of these particles, many of which are no larger than a grain of sand, leaves a thin layer of dust in the upper atmosphere that slowly settles earthward.

A Satellite's View of the Earth   Suppose we orbit the earth at an altitude of 36,000 km (22,300 mi). At this elevation, if we travel at the same rate as the earth's rotation, we will remain positioned above the same spot; thus, we can continuously monitor what is taking place beneath us. Looking down upon the earth from this altitude, we obtain a visual picture of the earth and its atmosphere, such as that shown in Fig. 1.6.

Fig. 1.6 This satellite picture (taken in visible, reflected light) shows a variety of cloud patterns and storms in the earth's atmosphere.

In order to accurately locate and describe places and features on the earth's surface, we must construct a grid system. Imaginary lines running from pole to pole are called *meridians*. To have a reference line, we choose a zero meridian. Although the choice of a zero line is arbitrary, we will run ours through Greenwich, England, since that is the one used by the United States and other nations. This zero line is called the *prime meridian*. The longitude of any place on the earth is simply how far east or west, in degrees, it is from the prime meridian. North America is west of Great Britain, and, as we can see from Fig. 1.6, most of the United States lies between 75°W and 125°W longitude.

The lines that parallel the equator are called *parallels of latitude*. The latitude of any place is how far north or south, in degrees, it is from the equator. The latitude of the equator is 0°, while the latitude of the North Pole is 90°N and that of the South Pole is 90°S. From Fig. 1.6 it is evident that almost all of the United States is located between latitude 30°N and 50°N, a region commonly referred to as the *middle latitudes*.

Storms of All Sizes Probably the most dramatic spectacle in Fig. 1.6 is the whirling cloud masses of all shapes and sizes. The clouds appear white because sunlight is reflected back to space from their tops. The dark areas show where skies are clear. The largest of the organized cloud masses are the sprawling storms. One such storm shows as an extensive band of clouds, over 2000 km (1240 mi) long, west of the Great Lakes. This *middle latitude storm system* (or **extratropical cyclone**) forms outside the tropics and, in the Northern Hemisphere, has winds spinning counterclockwise about its center, which is presently over Minnesota.

A slightly smaller but more vigorous storm is located over the Pacific Ocean near latitude 12°N and longitude 116°W. This tropical storm system, with its swirling band of rotating clouds and surface winds in excess of 64 knots* is known as a **hurricane**. The diameter of the hurricane is about 800 km (500 mi). The tiny dot at its center is called the *eye*. In the eye, winds are light and

skies are generally clear. Around the eye, however, an extensive region of heavy rain and high surface winds are reaching peak gusts of 100 knots.

Smaller storms, less than 50 km (31 mi) in diameter, are seen as bright spots over the Gulf of Mexico near latitude 27°N and longitude 88°W. These spots represent clusters of towering clouds that have grown into thunderstorms. If you look closely at Fig. 1.6 you will see similar cloud forms in many regions. There were probably thousands of thunderstorms occurring throughout the world at that very moment. Although they cannot be seen individually, there are even some thunderstorms embedded in the cloud mass west of the Great Lakes. Later in the day on which this photograph was taken, a few of these storms spawned the most violent disturbance in the atmosphere—the **tornado**. Although normally less than a kilometer wide, these small, rapidly whirling funnels can have winds exceeding 200 knots.

A Look at a Weather Map We can obtain a better picture of the middle latitude storm system by examining a simplified surface weather map for the same day that the satellite picture was taken. In Fig. 1.7, the letter "L" on the map indicates a region of low atmospheric pressure, often called a **low**, which marks the center of the middle latitude storm. The two letters "H" on the map represent regions of high atmospheric pressure, called **highs**, or *anticyclones*. The *wind direction*—the direction from which the wind is blowing—is given by arrows.

Notice how the wind blows around the highs and the lows. The horizontal pressure differences create a force that starts the air moving from higher pressure toward lower pressure. Because of the earth's rotation, the winds are deflected toward the right in the Northern Hemisphere.* This causes the winds to blow *clockwise* and *outward* from the center of the highs, and *counterclockwise* and *inward* toward the center of the low.

As the surface air spins into the low, it converges (flows together) and rises, much like toothpaste does when its open tube is squeezed. The rising air cools, and the moisture in the air con-

---

*A knot is a nautical mile per hour. One knot is equal to 1.15 miles per hour (mi/hr), or 1.9 kilometers per hour (km/hr).

*This deflecting force, known as the *Coriolis force*, is discussed more completely in Chapter 12, as are the winds.

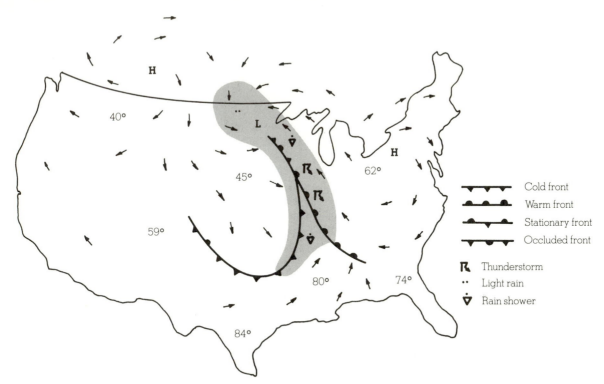

Fig. 1.7 Simplified surface weather map that correlates with the satellite picture shown in Fig. 1.6. Arrows on the map indicate wind direction. The shaded area represents precipitation. Air temperatures are in °F.

denses into clouds. Notice in Fig. 1.7 that the area of precipitation (the shaded area) in the vicinity of the low corresponds to an extensive cloudy region in the satellite photo (Fig. 1.6).

Also notice by comparing Figs. 1.6 and 1.7 that, in the regions of high pressure, skies are generally clear. As the surface air flows outward (diverges) away from the center of a high, air sinking from above must replace the laterally spreading air. Since sinking air does not usually produce clouds, we find generally clear skies and fair weather associated with the regions of high pressure.

Look at the middle latitude storm and the surface temperatures in Fig. 1.7 and notice that, to the southeast of the storm, warm southerly winds from the Gulf of Mexico are bringing warm, humid air northward over much of the southeastern portion of the nation. On the storm's western side, cool, dry northerly breezes combine with sinking air to create good weather over the Rocky Mountains. The boundary that separates the warm and cool air appears as a heavy, dark line on the map— a *front*, across which there is a sharp change in temperature, humidity, and wind direction.

Where the cool air from Canada replaces the warmer air from the Gulf of Mexico, a *cold front* is drawn, with arrowheads showing its direction of movement. Where the warm Gulf air is replacing cooler air to the north, a *warm front* is drawn, with half circles showing its direction of movement. Where the faster-moving cold front has overtaken the slower-moving warm front, an *occluded front* is drawn, with alternating arrowheads and half circles to show how it is moving. Along each of the fronts, warm air is rising, producing clouds and precipitation. In the satellite photo (Fig. 1.6), the occluded front and the cold front appear as an elongated, curling cloud band that stretches from the low pressure area over Minnesota into the northern part of Texas.

The Everchanging Weather and Climate
Observing storm systems, we see that not only do they move, but they constantly change. Steered by the upper-level westerly winds, the middle latitude storm in Fig. 1.7 intensifies into a larger storm, which moves eastward, carrying its clouds and weather with it. In advance of this system, a

FOCUS ON A SPECIAL TOPIC:

Meteorology—The Science of the Atmosphere

Meteorology is the study of the atmosphere and its phenomena. The term itself goes back to the Greek philosopher Aristotle who, about 340 B.C., wrote a book on natural philosophy entitled *Meteorologica*. This work represented the sum of knowledge on weather and climate at that time, as well as material on astronomy, geography, and chemistry. Some of the topics covered included clouds, rain, snow, wind, hail, thunder, and hurricanes. In those days, all substances that fell from the sky, and anything seen in the air, were called meteors, hence the term "meteorology." Today, we differentiate between those meteors that come from extraterrestrial sources outside our atmosphere (meteoroids) and particles of water and ice observed in the atmosphere (hydrometeors).

In *Meteorologica*, Aristotle attempted to explain atmospheric phenomena in a philosophical and speculative manner. Even though many of his speculations were found to be erroneous, Aristotle's ideas were accepted without reservation for almost 2000 years. In fact, the birth of meteorology as a genuine natural science did not take place until the invention of weather instruments, such as the thermometer at the end of the sixteenth century, the barometer in 1643, and the hygrometer in the late 1700s. With observations from instruments available, attempts were then made to explain certain weather phenomena using scientific experimentation and the physical laws that were being developed at the time.

As more and better instruments were developed in the 1800s, the science of meteorology progressed. Ideas about winds and storms became, at least, partially understood, and crude weather maps were drawn. Around 1920, the concepts of air masses and weather fronts were formulated in Norway. By the 1940s, daily upper-air balloon observations of temperature, humidity, and pressure gave a three-dimensional view of the atmosphere.

Meteorology took another step forward in the 1950s, when high-speed computers were developed to solve the equations that describe the behavior of the atmosphere. Along with drawing current weather maps, the computers were used to predict the state of the atmosphere at some desired time in the future. In 1960, *Tiros I*, the first weather satellite, was launched, ushering in space-age meteorology. Subsequent satellites provided a wide range of useful information, ranging from day and night time-lapse photographs of clouds and storms to pictures that depict swirling ribbons of water vapor flowing around the globe. Even today, more sophisticated satellites are being developed to supply computers with a far greater network of data, so that more accurate forecasts—perhaps up to a week or more—will be available in the future.

---

sunny day in Ohio will gradually cloud over and yield heavy showers and thunderstorms by nightfall. Behind the storm, cool, dry northerly winds rushing into eastern Colorado cause an overcast sky to give way to clearing conditions. Further south, the thunderstorms presently over the Gulf of Mexico (Fig. 1.6) expand a little, then dissipate as new storms appear over water and land areas. To the west, the hurricane over the Pacific Ocean drifts northwestward and encounters cooler water. Here, away from its warm energy source, it loses its punch; winds taper off, and the storm soon turns into an unorganized mass of clouds and tropical moisture.

As we can see, the **weather**—the condition of the atmosphere at any particular time and place—is always changing. Weather is comprised of the elements of *air temperature*, *air pressure*, *humidity*, *clouds*, *precipitation*, *visibility*, and *wind*. If we were to measure and observe these elements over a specified interval of time, say, for many years, we would obtain the "average weather" or the **climate** of a particular region. Climate, therefore, represents the accumulation of daily and seasonal weather events over a long period of time. The concept of climate also includes the extremes of weather—the heat waves of summer and the cold spells of winter—that occur in a particular region. The *frequency* of these extremes is what helps us distinguish among climates that have similar averages.

If we were able to watch the earth for many thousands of years, even the climate would change. We would see rivers of ice moving down stream-cut valleys and huge glaciers—sheets of moving snow and ice—spreading their icy fingers over large portions of North America. Advancing slowly from Canada, a single glacier might extend as far south as Kansas and Illinois, with ice many thousands of kilometers thick covering the region now occupied by Chicago. Over an interval of 2 million years or so, we would see the ice advance and retreat several times. Of course, for this phenomenon to happen, the average temperature of North America would have to decrease and then rise in a cyclic manner.

Suppose we could photograph the earth once every 1000 years for many hundreds of millions of years. In time-lapse film sequence these photos would show that not only is the climate altering, but the whole earth itself is changing as well: Mountains would rise up only to be torn down by erosion; isolated puffs of smoke and steam would appear as volcanoes spew hot gases and fine dust into the atmosphere; and the entire surface of the earth would undergo a gradual transformation as some ocean basins widen and others shrink.*

In summary, the earth and its atmosphere are dynamic systems that are constantly changing. While major transformations of the earth's surface are completed only after long spans of time, the state of the atmosphere can change in a matter of minutes. Hence, a watchful eye turned skyward will be able to observe many of these changes.

Summary

This chapter has given us a brief overview of the earth and its atmosphere. We have seen that the earth, one of nine planets that circle the sun, is the only planet with an atmosphere rich in nitrogen and oxygen. Dispersed throughout the earth's atmosphere are storms and clouds of all sizes and shapes. The largest of the storms are the middle latitude (extratropical) cyclones with their sweeping weather fronts, and the smallest are the violent tornadoes. The movement, intensification, and weakening of these systems, as well as the dynamic nature of air itself, produce a variety of weather events that we describe in terms of weather elements. The average of the elements is what we call climate. Although sudden changes in weather may occur in a moment, climatic changes take place gradually over many years. The study of the atmosphere and all of its related phenomena is called meteorology, a term whose origin dates back to the time of Aristotle.

Questions for Review

1. What is the primary source of energy for the earth's atmosphere?

2. Compare the composition of Earth's atmosphere to the atmosphere of Jupiter, of Mars, and of Venus.

3. Why is the surface temperature on Venus so much warmer than that on Earth or on Mars?

4. Explain how the atmosphere "protects" inhabitants at the earth's surface.

5. What are two other names often applied to an extratropical cyclone?

6. Rank the following storms in size from largest to smallest: hurricane, tornado, extratropical cyclone, thunderstorm.

7. In what direction do the winds generally blow around a surface high pressure area and a surface low pressure area in the Northern Hemisphere?

8. Distinguish between a warm front and a cold front.

9. Weather in the middle latitudes tends to move in what general direction?

10. Describe some of the features observed on a surface weather map.

11. List the common weather elements.

12. How does weather differ from climate?

13. Define *meteorology* and discuss the origin of this word.

*The movement of the ocean floor and continents is explained in the widely acclaimed theory of *plate tectonics*, formerly called the theory of continental drift.

Questions for Thought

1. Which of the following statements relate more to weather and which relate more to climate?

(a) The summers here are warm and humid.

(b) Cumulus clouds presently cover the entire sky.

(c) Our lowest temperature last winter was $-28°C$ ($-18°F$).

(d) The air temperature outside is $22°C$ ($72°F$).

(e) December is our foggiest month.

(f) The highest temperature ever recorded in Phoenixville, Pennsylvania, was $44°C$ ($111°F$) on July 10, 1936.

(g) Snow is falling at the rate of 5 cm (2 in.) per hour.

(h) The average temperature for the month of January in Chicago, Illinois, is $-3°C$ ($26°F$).

Problems and Exercises

1. Keep track of the weather. On an outline map of North America, mark the daily position of fronts and pressure systems for a period of several weeks or more. (This information can be obtained from newspapers or TV news broadcasts.) If upper-level winds are available, plot the general upper-level flow pattern on the map. Observe how the surface systems move. Relate this information to the material covered in later chapters.

The earth's atmosphere is a rich mixture of many gases, with clouds of condensed water vapor and ice particles. Here, rising air currents push a lazy cloud layer upward, as the sun spreads warmth over the surface. (Photo: Ross DePaola)

CHAPTER 2

The Earth's Atmosphere: Origin, Composition, and Behavior

The atmosphere that surrounds the earth is a mixture of many gases, as well as suspended particles of solid and liquid matter. The most abundant gases in our atmosphere are nitrogen and oxygen, which may seem surprising since neither the atmosphere of Venus nor Mars, our closest planetary neighbors, contains either of these gases in very large quantities. How, then, did the earth's atmosphere evolve into one rich in nitrogen and oxygen? What are other important atmospheric constituents? And how do these atmospheric gases behave? These are some of the questions we will address in this chapter.

The Origin of the Earth's Atmosphere

A variety of astronomical evidence and insight suggests that the earth's original atmosphere probably developed around the time the solar system formed, about 4.6 billion years ago. Some of the ideas about the formation of the atmosphere are speculative. However, scientists, with the aid of orbiting telescopes and theoretical computer models, have amassed a good deal of information about stars and planets that helps us to draw conclusions about our earth's early atmosphere. Because the earth and its earliest atmosphere are believed to have formed from the remnants of exploding stars, we will examine stars first.

Stars are composed mostly of hydrogen. In the star's core, where temperatures are several million degrees and pressures are enormous, hydrogen nuclei are able to fuse into a new element:

Contents

13

helium. This process, called *nuclear fusion*, may continue inside the star for billions of years. In older stars, the core gradually becomes depleted of hydrogen, and helium becomes the primary constituent. If temperatures and pressures in the core are high enough, the helium may fuse into the heavier element carbon. Likewise, in more massive stars, it is hypothesized that if temperatures and pressures continue to be sufficiently high, the carbon is converted into still heavier elements. In extremely massive stars, the production of elements goes on and on and, eventually, comes to an abrupt end in a great explosion—a *supernova*—that hurls the elements manufactured over thousands of millions of years out into space.

Suppose the remains of many exploding stars, as well as interstellar dust and gas (mostly hydrogen), form an enormous cloud—a *nebula*. Under the influence of its own gravity, the cloud slowly contracts toward its center. As it shrinks, it spins. The spinning increases as the cloud grows smaller, just as an ice skater spins faster as the arms are brought in closer to the body. Gradually, the spinning nebula takes on the shape of a disk. Near the center, the compressed gas and dust increase in temperature many millions of degrees, nuclear reactions begin, and a new star is born. Around the star are rings of matter left behind as the cloud contracted. As these smaller masses slowly coalesce, they do not become hot enough to produce nuclear reactions. Hence, they become planets—planets that may have atmospheres.

If, indeed, the earth formed from a contracting mass of dust and gas, its earliest atmosphere must have been comprised of the same gases from which it formed. These gases included hydrogen and helium (the two most abundant gases in the universe), as well as hydrogen compounds, such as methane (CH_4) and ammonia (NH_3). What has happened in the 4.6 billion years since the earth formed? Where, for example, has all the hydrogen gone?

Hydrogen and helium are extremely light molecules that could simply have escaped the pull of the earth's gravity. The remaining gases may have been blown off into space by heat generated from either the bombardment of rocky material from space or from an outpouring of extra energy from the sun. In any case, it appears that the earth was stripped of its primitive atmosphere early in its development. However, we do have an atmosphere today. Where did it come from?

Billions of years ago, gases trapped in the earth's hot interior during its formation were probably expelled at the surface by large numbers of volcanoes, fissures (giant cracks), and fumaroles (steam vents). We assume that volcanoes spewed out the same gases then as they do today: mostly water vapor (about 85 percent), carbon dioxide (about 10 percent), and up to a few percent nitrogen. These gases probably created the earth's second atmosphere.

As hundreds of millions of years passed, our atmosphere gradually cooled. The constant outpouring of gases from the hot interior—known as **degassing** or *outgassing*—provided a rich supply of water vapor, which formed into clouds. Rains fell upon the earth for many thousands of years, forming the rivers, lakes, and oceans of the world. During this time, large amounts of carbon dioxide were dissolved in the oceans. (Even today, the oceans act as a reservoir for carbon dioxide.) Through chemical and biological processes, much of the carbon dioxide became locked up in carbonate sedimentary rocks, such as limestone. With much of the water vapor already condensed and the concentration of carbon dioxide dwindling, the atmosphere gradually became rich in nitrogen (N_2), which is chemically inactive.

Where does oxygen, the second most abundant gas in today's atmosphere, fit into the picture? In the early 1980s, astronomical measurements made with an orbiting telescope revealed that young stars less than 1 billion years old give off many times more ultraviolet radiation than does the sun today. If the sun gave off such energetic rays in its early years, they would have been sufficient to split water vapor into molecules of hydrogen and oxygen. The hydrogen, being lighter, probably rose and escaped into space. The oxygen remained in the atmosphere, gradually increasing in concentration. In fact, analysis of the oldest known rocks (estimated to be 3.5 billion years old) reveal oxidized iron in amounts that suggest the early atmosphere's oxygen level was much higher than once thought. Eventually, the splitting of water vapor by the sun's energy—called *photodissociation*—provided enough oxygen for primitive plants to evolve, perhaps 2 to

FOCUS ON A SPECIAL TOPIC

A Breath of Fresh Air

If we could examine a breath of air, we would see that air (like everything else in the universe) is composed of incredibly tiny particles called **atoms**. We cannot see atoms individually, even with our most powerful microscopes. Yet, if we could see one, we would find electrons whirling at fantastic speeds about an extremely dense center, somewhat like hummingbirds darting and circling about a flower. At this center, or nucleus, are the protons and neutrons. Almost all of the atom's mass is concentrated here, in a trillionth of the atom's entire volume. In the nucleus, the proton carries a positive charge, whereas the neutron is electrically neutral. The circling electron carries a negative charge. As long as the total number of protons in the nucleus equals the number of orbiting electrons, the atom as a whole is electrically neutral (Fig. 1).

Most of the air particles are **molecules**, combinations of two or more atoms, and most of the molecules are electrically neutral. A few, however, are electrically charged, having lost or gained electrons. These charged atoms and molecules are called **ions**.

An average breath of fresh air contains a tremendous number of molecules. With every deep breath, trillions of molecules from the atmosphere enter your body. Some of these inhaled gases become a part of you, and others are exhaled.

The volume of an average size breath of air is about a li-

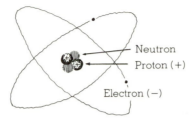

Fig. 1 An electrically neutral atom has the same number of protons in its nucleus as it has orbiting electrons.

ter.* Near sea level, there are roughly ten thousand million million million (10^{22}) air molecules in a liter. So,

1 breath of air $\approx 10^{22}$ molecules.

We can appreciate how large this number is when we compare it to the number of stars in the universe.

Astronomers have estimated that there are about 100 billion (10^{11}) stars in an average size galaxy and that there may be as many as 10^{11} galaxies in the universe. To determine the total number of stars in the universe, we multiply the number of stars in a galaxy by the total number of galaxies and obtain:

$10^{11} \times 10^{11} = 10^{22}$ stars in universe.

Therefore, each breath of air contains about as many molecules as there are stars in the known universe.

*One cubic centimeter is about the size of a sugar cube, and there are a thousand cubic centimeters in a liter.

In the entire atmosphere, there are nearly 10^{44} molecules. Since the number 10^{44} is 10^{22} squared, as

$10^{22} \times 10^{22} = 10^{44}$ molecules in the atmosphere,

we conclude that there are about 10^{22} breaths of air in the entire atmosphere. In other words, there are as many molecules in a single breath as there are breaths in the atmosphere.

Each time we breathe, the molecules we exhale enter the turbulent atmosphere. If we wait a long time, those molecules will eventually become thoroughly mixed with all of the other air molecules. If none of the molecules were consumed in other processes, eventually there would be a molecule from that single breath in every breath that is out there. So, considering the many breaths people exhale in their lifetimes, it is possible that in our lungs are molecules that were once in the lungs of people who lived hundreds or even thousands of years ago. Thus, in a very real way, we all share the same atmosphere.

Table 2.1 Composition of the Atmosphere Near the Earth's Surface

PERMANENT GASES			VARIABLE GASES		
Gas	Symbol	Percent (by volume)	Gas	Symbol	Percent (by volume)
Nitrogen	N_2	78.08	Water vapor	H_2O	0 to 4
Oxygen	O_2	20.95	Carbon dioxide	CO_2	0.034
Argon	Ar	0.93	Ozone	O_3	0.000004*
Neon	Ne	0.0018	Carbon monoxide	CO	0.00002*
Helium	He	0.0005	Sulfur dioxide	SO_2	0.000001*
Methane	CH_4	0.0001	Nitrogen dioxide	NO_2	0.000001*
Hydrogen	H_2	0.00005	Particles (dust, soot, etc.)		0.00001*
Xenon	Xe	0.000009			

*Average value in polluted air.

3 billion years ago. Because plants consume carbon dioxide and release oxygen, the atmospheric oxygen content increased more rapidly, probably reaching its present composition about several hundred million years ago.

Present Composition of the Atmosphere

Our atmosphere has gradually, yet delicately, evolved over hundreds of millions of years to the point where, today, we unthinkingly inhale and exhale about 12 times each minute a mixture of gases consisting mainly of nitrogen (78 percent) and oxygen (21 percent). Table 2.1 shows the percent by volume of the various gases in the earth's atmosphere.

Gases and Particles The percent of nitrogen and oxygen in the atmosphere remains essentially constant from the surface up to about 80 km (50 mi). Hence, near the surface, there is a balance between destruction (output) and production (input) of these gases. For example, nitrogen is removed from the atmosphere primarily by biological processes that involve soil bacteria. It is returned to the atmosphere mainly through the decaying of plant and animal matter. Oxygen, on the other hand, is removed from the atmosphere when organic matter decays and when oxygen combines with other substances, producing oxides. It is also taken from the atmosphere during breathing, as the lungs take in oxygen and release carbon diox-

ide. The addition of oxygen to the atmosphere occurs during photosynthesis, as plants, in the presence of sunlight, combine carbon dioxide and water to produce sugar and oxygen.

Carbon dioxide (CO_2) is a natural component of the atmosphere. It occupies a small (but important) percent of a volume of air, about 0.034 percent. Carbon dioxide enters the atmosphere mainly from the decay of vegetation, but it also comes from volcanic eruptions, the exhalations of animal life, and from the burning of fossil fuels, such as coal, oil, and natural gas. The removal of CO_2 from the atmosphere takes place during photosynthesis, as plants consume CO_2 to produce green matter. The oceans act as a huge reservoir for CO_2, as phytoplankton (tiny, drifting plants) in surface water fix CO_2 into organic tissues. Estimates are that the ocean holds more than 50 times the total atmospheric carbon dioxide content.

Figure 2.1 shows that the concentration of CO_2 in the air has been rising since the early part of this century, mostly because of the burning of fossil fuels and deforestation. Notice that the annual average values near the turn of this century were about 290 parts per million (ppm)* by volume, while the annual average in 1984 was about 340 ppm. Some scientists estimate that this value will increase to 600 ppm sometime within the next 100 years. Because CO_2 traps a portion of the earth's outgoing infrared heat energy (see Chapter 3), an increase in the concentration of CO_2

*Two hundred and ninety parts per million means that for every million air molecules, 290 were CO_2 molecules. This is the same as 0.029 percent.

should result in an increase in average global surface air temperatures.

Most of the mathematical model experiments that take into account the many complex physical and chemical atmospheric processes to predict future atmospheric conditions, estimate that a doubling of CO_2 will result in a global warming of surface air between 1.5°C and 4.5°C (about 2.5°F to 8°F). Such a warming could severely reduce the quantity and quality of water resources in the western United States, as well as raise sea level by about half a meter or so, flooding coastal lowlands, as polar ice begins to disintegrate and melt. Other consequences might include a shifting of the global air currents that guide the major storm systems across the earth. However, because of the complex nature of the atmosphere, much information has yet to be obtained before a definitive answer can be made to the question of what effect increasing CO_2 concentrations will have on surface air temperatures and global climate.

Impurities from both natural and human sources are also present in the atmosphere. Wind picks up dust from the earth's surface and carries it into the atmosphere. Small saltwater drops from ocean waves are swept into the air. Upon evaporating, these drops leave microscopic salt particles suspended in the atmosphere. Smoke from forest fires is often carried high above the earth, and volcanoes spew many tons of fine ash particles and gases into our air. For example, on several occasions during 1980, the volcano Mount St. Helens erupted in the Pacific Northwest, sending many tons of ash and rock into the atmosphere. The largest eruption (Fig. 2.2) was a lateral explosion that pulverized a portion of the volcano's north slope. The ensuing dust and ash cloud was confined mostly to the lower atmosphere (within 10 km of the surface) and fell out quite rapidly over a large area of the northwestern United States.

A more important eruption, in terms of its ash veil, occurred during April, 1982, when the volcano El Chichón erupted in Mexico. The summit-type eruption sent a plume of ash and gas, rich in sulfur, billowing into the upper atmosphere. Over a period of several months, the sulfur gases reacted with water vapor in the presence of sunlight, producing tiny, bright sulfuric acid particles that gradually joined together and grew in size, forming

a dense haze. By 1983, the height of the sulfuric acid cloud was detected at 30 km (18 mi) above Wyoming. The cloud particles, which resided in the upper atmosphere for several years, blocked and diffused sunlight. Hence, they prevented a small portion of the sun's energy from reaching the earth's surface. The effect of this may have been a slight lowering of surface air temperature (probably less than 0.3°C or 0.5°F) over portions of the Northern Hemisphere in 1984. The exact amount of cooling is difficult to determine because the average temperature varies slightly from year to year anyway.

Some natural impurities found in the atmosphere are quite beneficial. Small, floating particles, for instance, act as surfaces on which water vapor condenses to produce clouds. However, most human-induced impurities (and some natural ones) are a nuisance, as well as a health hazard. We generally refer to these as **pollutants**. For example, the burning of coal and oil releases the gas *sulfur dioxide* (SO_2) into the air. If the air also contains enough water vapor, droplets of sulfuric acid form. Rain containing sulfuric acid corrodes metals and painted surfaces and turns freshwater lakes acidic. *Acid rain* is a major environmental

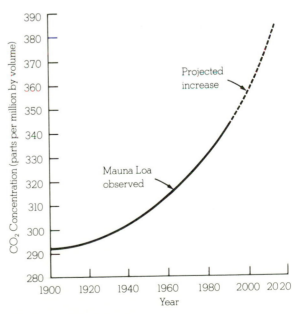

Fig. 2.1 Calculations of the increase in carbon dioxide in the atmosphere since about 1900, due mainly to the combustion of fossil fuels.

Fig. 2.2   On May 18, 1980, the eruption of Mount St. Helens sent many millions of tons of particles into the atmosphere, along with copious amounts of water vapor and carbon dioxide. However, unlike the sulfur-rich ash cloud that came from the eruption of El Chichón in April, 1982, the gases from Mount St. Helens that reached the upper atmosphere were deficient in sulfur, and did not essentially affect global temperatures.

problem, especially in areas downwind from major industrial areas.*

Automobile engines emit clouds of *nitrogen dioxide* (NO_2), *carbon monoxide* (CO), and *hydrocarbons*. Nitrogen dioxide gas often gives the atmosphere a dirty, light-brown color. In the presence of sunlight, it reacts with hydrocarbons and other gases to produce *ozone* (O_3), the primary ingredient of *photochemical smog*, which irritates the eyes and throat and damages vegetation.

The majority of atmospheric ozone is found in the upper atmosphere—in the *stratosphere*—where it is produced naturally, as oxygen atoms combine with oxygen molecules. Here, the concentration of ozone averages less than 0.002 percent by volume. This small quantity is important, however, because it shields plants, animals, and

*We will examine the acid rain problem and its effect on the environment in more detail in Chapter 22.

humans from the sun's harmful ultraviolet rays. In recent years, concern has arisen that certain human activities (e.g., gases emitted from high-flying aircraft and fluorocarbons from spray cans) may be upsetting the ozone balance and that the ozone shield may be gradually decreasing. There is speculation that a depletion of ozone will bring about an increase in the number of cases of skin cancer, as well as have an adverse effect on animals and plants. So, it is a paradox that ozone, which damages plant life in a polluted environment, provides a natural protective shield in the upper atmosphere, so that these same plants may survive.

Water in the Atmosphere   The concentration of the invisible gas water vapor varies greatly from place to place, and from time to time. In warm tropical locations, close to the surface, it may account for up to 4 percent of the atmospheric gases, while in cold polar regions, its concentration may dwindle to a mere trace. The measure of the air's water vapor content is called *humidity*.*

Water vapor is extremely important in the atmosphere. Not only does it transform into both liquid and solid cloud particles that grow in size and fall to earth as precipitation, but it also releases large amounts of heat—called **latent heat**—when it changes from vapor into liquid water or ice. Latent heat is an important source of atmospheric energy, especially for storms, such as thunderstorms and hurricanes. Because water vapor strongly absorbs infrared radiation (see Chapter 3), it is important in the heat-energy balance of the earth.

In the lower atmosphere, water is everywhere. In fact, it is the only substance that exists as a gas, a liquid, and a solid at those temperatures and pressures normally found near the earth's surface. As a gas, water vapor molecules move about quite freely, mixing well with neighboring atoms and molecules. As a liquid, the water molecules are closer together, and so they constantly jostle and bump each other. In the solid state (ice), the molecules arrange themselves into an orderly pattern, with each molecule more or less locked into a rigid position, able to vibrate, but not able to move about freely.

Water vapor molecules are, of course, invisible. They become visible, however, when they transform into larger liquid or solid particles, such as cloud droplets and ice particles. In this process—known as a *change of state* or, simply, a *phase change*—water only changes its disguise, it does not change its identity.

To better understand some of the important roles that water plays in the atmosphere, consider this picture: You catch a falling snowflake in the palm of a glove. If we could magnify that tiny crystal of ice about a billion times, we would see H_2O molecules in the shape of tiny heads that resemble Mickey Mouse (see Fig. 2.3). The bulk of the ''head'' of the molecule is the oxygen atom. The ''mouth'' is a region where an excess of negative charge is concentrated. The ''ears'' are par-

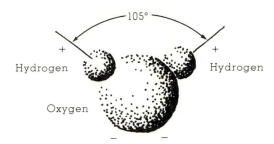

Fig. 2.3 The water molecule.

tially exposed protons of the hydrogen atom, which are regions where an excess of positive charge is concentrated.

When we look at many molecules (Fig. 2.4), we can see that they are locked into specific positions and arranged as a six-sided (hexagonal) crystal form we call *ice*. As we mentioned, the molecules are unable to move about freely, but they do vibrate and jiggle. As we observe the ice crystal in freezing air, we see an occasional molecule gain enough energy to break away from its neighbors and enter into the air above. The molecule changes from an ice molecule directly into a vapor molecule without passing through the liquid state. This ice-to-vapor phase change is called **sublimation**. If a water vapor molecule should attach itself to the

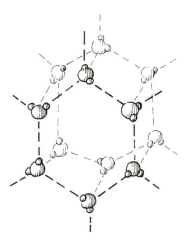

Fig. 2.4 The partially exposed hydrogen atom of one molecule attracts itself to the negative oxygen atom of another molecule. Because the hydrogen atoms of each water molecule are separated by an angle of 105°, the joining of many billions of molecules produces a hexagonal-shaped ice crystal. In the atmosphere, many ice crystals may join together to form a snowflake.

---

*The *relative humidity* is actually the amount of water vapor in the air compared to how much water vapor the air can hold. In Chapter 7, we will examine the various ways of describing the air's water vapor content.

Is Moist Air Less Dense Than Dry Air?

The masses of atoms and molecules, and the amount of space between them, determine the **density** of a substance. In other words, density tells us how much matter is in a given space (that is, volume):

$$Density = \frac{amount\ of\ matter}{volume}.$$

There are several ways of expressing density. The *molecular density* of a gas is the number of molecules in a given volume. The most common way to use density is to determine the mass of air in a given volume:

$$Density = \frac{mass}{volume}.$$

When mass is expressed in grams (g) or kilograms (kg), volume is given in cubic centimeters (cm^3) or cubic meters (m^3).

Air is surprisingly heavy. Near sea level, its density is about 1.2 kg/m^3 (nearly 1.2 ounces per cubic foot). Consequently, the air in

an average size room 6 m long, 6 m wide, and 2.5 m high (20 ft × 20 ft × 8 ft) weighs about 108 kg (237 lb). The weight of all of the air around the earth is a staggering 5600 trillion tons!

Even more interesting is the fact that air does not always weigh the same. The weight of the air on a warm, moist day is different than on a warm, dry day. Answer the following question. True or false: A given volume of warm, humid air will be heavier than the same volume of warm, dry air. Did you answer true? Most people do; but the answer is false! At the same temperature and at the same level, moist air weighs *less* than dry air. (Keep in mind that we are referring strictly to water vapor—a gas—and not suspended liquid droplets.) To understand why air containing a lot of water vapor is lighter, we must first see what determines the weight of atoms and molecules.

Almost all of the weight of an atom is concentrated in its nu-

cleus, where the protons and neutrons are found. Neutrons weigh nearly the same as protons. To get some idea of how heavy an atom is, we simply add up the number of protons and neutrons in the nucleus. (Electrons are so light that we ignore them in comparing weights.) The larger this total, the heavier the atom. Now, we can compare one atom's weight with another's. For example, hydrogen, the lightest known atom, has only one proton in its center (no neutrons). Thus, it has an *atomic weight* of 1. Nitrogen, with 7 protons and 7 neutrons in its nucleus, has an atomic weight of 14. Oxygen, with 8 protons and 8 neutrons, weighs in at 16.

A molecule's weight is the sum of the atomic weights of its atoms. For example, molecular oxygen, with two oxygen atoms (O_2), has a molecular weight of 32. The most abundant atmospheric gas, molecular nitrogen (N_2), has a molecular weight of 28.

---

ice crystal, the vapor-to-ice phase change is called **deposition**. (In meteorology, the vapor-to-ice transformation is sometimes called *sublimation* as well.)

As heat is added to our ice crystal, we would observe that the molecules start to vibrate faster. In fact, some of the molecules actually vibrate out of their rigid crystal pattern into a disorderly condition—the ice melts. If we could watch this magnified tiny water drop closely, we would see that the molecules are not all moving at the same speed—some are moving much faster than others. The energy associated with this motion is called **kinetic energy**. The temperature of the water (or any substance) is a measure of its average

kinetic energy. Simply stated, **temperature** *is a measure of the average speed of the molecules*. If heat is applied to the water, its molecules jiggle faster, and its temperature goes up. Conversely, if heat is removed, the molecules move more slowly, and the temperature goes down.

A dynamic process occurs continuously at the surface of water. Molecules with enough speed (and traveling in the right direction) break away from the liquid surface and enter the atmosphere. These molecules are changing from the liquid state into the vapor state in a process called **evaporation**. While some water molecules are leaving the liquid, others are returning. Those returning are going from the vapor state to the liquid state. This

FOCUS ON A SPECIAL TOPIC, continued

When we determine the weight of air, we are dealing with the weight of a mixture. As you might expect, a mixture's weight is a little more complex. We cannot just add the weights of all its atoms and molecules because the mixture might contain more of one kind than another. Air, for example, has far more nitrogen (78 percent) than oxygen (21 percent). We allow for this by multiplying the molecule's weight by its share in the mixture. Since dry air is essentially composed of N_2 and O_2 (99 percent), we ignore the other parts of air for the rough average shown in Table 1.

The double wiggley line ≈ means "is approximately equal to." Therefore, dry air has a molecular weight of about 29. How does this compare with moist air?

Water vapor is composed of two atoms of hydrogen and one atom of oxygen (H_2O). It is an invisible gas, just as oxygen and nitrogen are invisible. It has a molecular weight; its 2 atoms of hydrogen (each with atomic weight 1) and 1 atom of oxygen (atomic weight 16) gives water vapor a molecular weight of 18. Obviously, air, at nearly 29, weighs appreciably more than water vapor.

Suppose we take a given volume of completely dry air and weigh it, then take exactly the same amount of water vapor at the same temperature and weigh it. We will find that the dry air weighs slightly more. If we replace dry air molecules one for one with water vapor molecules, the total number of molecules remains the same, but the total weight of the drier air decreases. Earlier, we saw that

density is mass (weight) per unit volume. Therefore, *moist air is less dense than dry air.*

Think about what this fact means for our weather. The lighter the air becomes, the more likely it is to rise. All other factors being equal, warm, moist (less dense) air will rise more readily than warm, dry (more dense) air. It is the water vapor in rising air that changes into liquid cloud droplets and ice crystals. These, in turn, grow large enough to fall to the earth as precipitation in the form of rain or snow.

Table 1

GAS	WEIGHT		NUMBER OF ATOMS		MOLECULAR WEIGHT		PERCENT BY VOLUME	
Oxygen	16	×	2	=	32	×	21% ≈	7
Nitrogen	14	×	2	=	28	×	78% ≈	22
					Molecular weight of dry air ≈ 29			

process is called **condensation**. Hence, at the surface of the liquid, we always find some molecules evaporating (escaping) and others condensing (returning).

When a cover is placed just above the exposed water surface, we observe that the total number of molecules escaping from the liquid is quickly balanced by the number returning. At this point, the air above the water contains the maximum number of water vapor molecules that it is capable of holding. When this condition exists, the air is said to be **saturated** with water vapor. For every molecule that evaporates, one must condense, and no net loss of liquid or vapor molecules results.

If we remove the cover and blow across the top

of the water, some of the vapor molecules already in the air above would be blown away, creating a difference between the actual number of vapor molecules and the total number required for saturation. This would help prevent saturation from occurring and would allow for a greater amount of evaporation. Wind, therefore, enhances evaporation.

Water temperature also influences evaporation. All else being equal, warm water will evaporate more rapidly than cool water. Heating water increases its average molecular speed. This means that a larger fraction of the molecules have high enough speeds to break through the surface tension of the water and zip off into the air above.

Hence, the warmer the water, the greater the rate of evaporation, provided the air above does not become saturated.

The air temperature above the water has an additional effect on the rate of evaporation. To comprehend this effect, we need to see how condensation occurs in the atmosphere.

In the atmosphere, water vapor molecules are constantly moving around, bumping into each other as well as other molecules. When these gas molecules collide, they tend to bounce off one another, constantly changing in speed and direction. However, the speed lost by one molecule is gained by another, and so the average speed of all the molecules does not change. Consequently, the temperature of the air does not change. Mixed in with all of the air molecules are microscopic bits of dust, smoke, and other particles called *condensation nuclei* (so-called because water vapor condenses on them). In warm air, fast-moving vapor molecules strike the nuclei with such impact that they simply bounce away. In cold air, however, the slower-moving vapor molecules are more apt to stick to the nuclei, much like flies tend to cling to sticky flypaper. When many billions of these molecules condense on the floating particle, a liquid droplet forms. (See Fig. 2.5.)

Since condensation occurs when the air is saturated, we can see that condensation is likely to happen as the air cools and the speed of the vapor molecules decreases. Therefore, with the same

number of water vapor molecules, saturation is more likely to occur in cool air than in warm air. Put another way: *Warm air can hold more water vapor molecules before becoming saturated than can cold air.* (This concept is very important. We will use it throughout this book.)

In summary, water vapor is a small, yet important component of air. Under certain conditions, it transforms into liquid droplets and delicate ice crystals—two processes that remove water vapor from the air. Water vapor returns to the atmosphere during evaporation (and sublimation). The main factors that influence evaporation are: wind, degree of saturation of the air, temperature of the water, and air temperature above the water. Because water plays such an important role in atmospheric processes, we will study its influence in many sections of this book.

Up to this point, we have examined the various individual components of air. The following section describes how all of these gases behave collectively, producing the variables of air temperature, pressure, and density.

The Behavior of Air

We know that air is a mixture of countless billions of air molecules. If they could be seen, they would appear to be moving about in all directions, freely

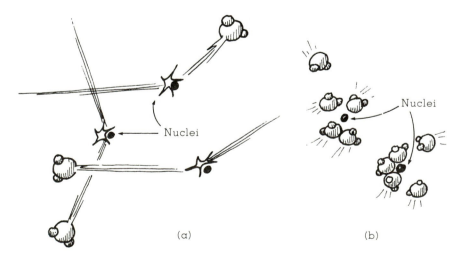

(a)   (b)

Fig. 2.5 (a) Fast-moving H_2O vapor molecules tend to bounce away after colliding with nuclei. (b) Slow-moving vapor molecules are more likely to join together on nuclei. The condensing of many billions of water molecules produces tiny liquid water droplets.

FOCUS ON AN APPLICATION

Circulation of Water in the Atmosphere

Within the atmosphere an unending circulation of water begins as energy from the sun evaporates enormous quantities of water from the oceans. Winds transport the moist air to other regions, where it condenses into clouds, some of which produce rain and snow. If the precipitation falls into an ocean, the water is ready to begin its cycle again. If, on the other hand, the precipitation falls on a continent, a great deal of the water makes its way back to the ocean in a complex journey. This cycle of moving and transforming water molecules from liquid to vapor and back to liquid again is called the **hydrologic** (water) **cycle**. In the form with which we are most concerned, water molecules travel from ocean to atmosphere to land and then back to the ocean.

Figure 2 illustrates some of the complexities of the hydrologic cycle. For example, before falling rain ever reaches the ground, a portion of it evaporates back into the air. Some of the precipitation may be intercepted by vegetation, where it evaporates or drips to the ground long after a storm has ended. Once on the surface, a portion of the water soaks into the ground by percolating downward through small openings in the soil and rock, forming ground water that can be tapped by wells. What does not soak in collects in puddles of standing water, or runs off into streams and rivers, which find their way back to the ocean. Even the underground water moves slowly and eventually surfaces, only to evaporate or be carried seaward by rivers.

Over land, a considerable amount of vapor is added to the atmosphere through evaporation from the soil, lakes, and streams. Even plants give up moisture by a process called *transpiration*. The water absorbed by a plant's root system moves upward through the stem and emerges from the plant through numerous small openings on the underside of the leaf. In all, evaporation and transpiration from continental areas amount to only about 15 percent of the nearly 1.5 billion billion gallons of water vapor that annually evaporate into the atmosphere; the remaining 85 percent evaporates from the oceans. If all of this vapor were to suddenly condense and fall as rain, it would be enough to cover the entire globe with 2.5 centimeters (cm) or 1 inch (in.) of water. The total mass of water vapor stored in the atmosphere at any moment adds up to only a little over a week's supply of the world's precipitation. Since this amount varies only slightly from day to day, the hydrologic cycle of evaporation, condensation, and precipitation is exceedingly efficient in circulating water in the atmosphere.

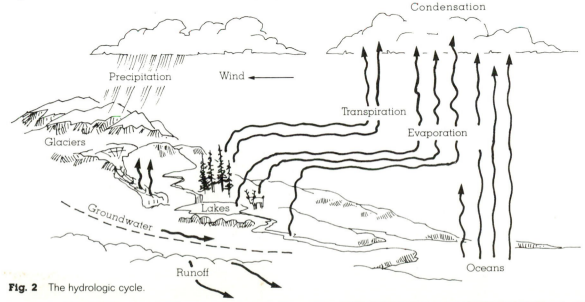

Fig. 2 The hydrologic cycle.

darting, twisting, spinning, and colliding with one another like an angry swarm of bees. Close to the earth's surface, each individual molecule would travel about a thousand times its diameter (nearly $\frac{1}{10,000}$ cm) before colliding with another molecule. Molecules would also collide with any object around them. As we study their average behavior, we will begin to see how these motions relate to temperature, pressure, and density.

Temperature, Pressure, and Density   Earlier in this chapter we learned that the temperature of a substance is determined by the average speed (average kinetic energy) of its molecules. At room temperature, the average speed of all the molecules is about 450 meters per second (m/sec) or about 1000 miles per hour. If the temperature were raised, the average speed of the molecules would increase; if lowered, the average speed would decrease.* If the air is cooled, molecules would continuously slow down until they reached $-273°C$ ($-459°F$), the coldest possible temperature, which is called **absolute zero**. At absolute zero, the molecules would possess a minimum amount of energy and, theoretically, no thermal motion.

As long as the air temperature is considerably above absolute zero, molecules are continually darting about and colliding. On a mild spring day, an average air molecule will collide about 10 billion times each second with other air molecules. It will also bump against objects around it—houses, trees, flowers, the ground, and even people. Each time an air molecule bounces against a person, it gives a tiny push. This small force (push) divided by the area on which it pushes is called *pressure*:

$$\text{Pressure} = \frac{\text{force}}{\text{area}}.$$

Billions of air molecules constantly push on the human body. At sea level, the molecules exert an average force of about 14.7 pounds (lb) on each square inch of area. This force is exerted equally in all directions. For instance, air pushing in on the back of the hand exerts the same pressure as air pushing in on the palm. We are not crushed by

*Average molecular speed also depends on the weight of the molecule. Lighter molecules move more rapidly than heavier ones at the same temperature.

the force because billions of molecules inside the body push outward just as hard.

Even though we do not actually feel the constant bombardment of air, we can detect quick changes in it. For example, if we climb rapidly in a small airplane, our ears may "pop." This experience happens because air collisions outside the eardrum lessen. The popping comes about as air collisions between the inside and outside of the ear equalize. The drop in the number of collisions informs us that the pressure exerted by the air molecules decreases with height above the earth. But why are there fewer molecules bouncing against us at higher atmospheric levels?

Air molecules, like everything else, are held near the earth by gravity. This strong, invisible force squeezes air molecules close together at the earth's surface, making them much more compressed here than at higher elevations. (See Fig. 2.6.) Because of gravity, nearly 25 billion molecules occupy a cubic centimeter (cm³) of air near sea level. In contrast, at an elevation of 600 km (370 mi), only about 10 million molecules occupy the same amount of space. Since there are appreciably more molecules within the same volume of air near the earth's surface than at higher levels, air density must decrease with height. Because of the air's compressibility, air density usually decreases rapidly at first, then more slowly as we move farther away from the earth.

We know from an earlier section that air molecules have weight. The weight of the air molecules exerts a force upon the earth. The amount of force exerted over an area of surface is called *atmospheric pressure* or, simply, **air pressure**. (The concept of air pressure is just another way of describing the pressure of air molecules bombarding a surface.) The pressure at any level in the atmosphere may be measured in terms of the total weight of the air above any point. As we climb in elevation, fewer air molecules are above us; hence *atmospheric pressure always decreases with increasing height*. Just as the air density, air pressure decreases rapidly at first, then more slowly at higher levels. (See Fig. 2.6.)

If we weigh a column of air 1 square inch in cross section, extending from the average height of the ocean surface (sea level) to the "top" of the atmosphere, it would weigh very nearly 14.7 pounds. Thus, normal atmospheric pressure near sea level is close to 14.7 pounds per square inch. If more molecules are packed into the column, it becomes more dense, the air weighs more, and the surface pressure goes up. On the other hand, when fewer molecules are in the column, the air weighs less, and the surface pressure goes down. So, a change in air density can bring about a change in air pressure.

Pounds per square inch is, of course, one way of expressing air pressure. However, the most common unit presently found on surface weather maps is the **millibar** (mb). If you are unfamiliar with this unit, or with the Kelvin or Celsius temperature scales, read the focus section on temperature and pressure units (p. 26) before reading the next section.

A Rapid Drop in Pressure and Temperature Figure 2.7 illustrates how rapidly air pressure decreases with height. Near sea level, atmospheric pressure is usually close to 1000 millibars (mb). Normally, just above sea level, atmospheric pressure decreases by about 10 mb for every 100 meters (m) increase in elevation (about 1 inch of mercury for every 1000 foot of rise). At higher levels, air pressure decreases much more slowly with height. For example, atmospheric pressure decreases by about 900 mb in the first 16 km (10 mi) above the surface, while only about 90 mb in the next 16 km. With a sea level pressure near 1000 mb, we can see in Fig. 2.7 that, at an altitude of only 5.5 km (3.5 mi), the air pressure is about 500 mb, or half of the sea level pressure, meaning that here we are above one-half of all the molecules in the atmosphere at a mere 18,000 feet (ft) above the surface.

At an elevation approaching the summit of Mount Everest (about 9 km—the highest mountain peak on earth), the air pressure would be about 300 mb. The summit is above nearly 70 percent of all the molecules in the atmosphere. At an altitude of about 50 km (30 mi), the air pressure is about 1 mb, which means that 99.9 percent of all the molecules are below this level. Yet the atmosphere extends upwards for many hundreds of kilometers—imagine how "thin" this air must be.

Air temperature normally decreases from the surface up to an elevation of about 10 km (6 mi). This decrease in temperature is due to the fact that sunlight warms the earth's surface, and the

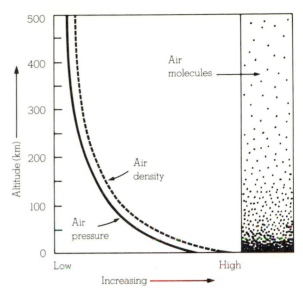

Fig. 2.6 Both air pressure and air density decrease with increasing altitude.

surface, in turn, warms the air from below (Chapter 3). The average drop in air temperature—called the *lapse rate*—is about 6.5°C per 1000 m (3.6°F per 1000 ft). Consequently, air only 2 km (about 6500 ft) above the surface is typically 13°C (23°F)

(*Text continues on p. 29.*)

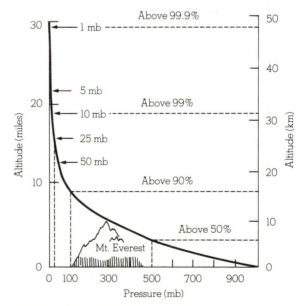

Fig. 2.7 Atmospheric pressure decreases rapidly with height. Climbing to an altitude of only 5.5 km, where the pressure is 500 mb, would put you above one-half of the atmosphere's molecules.

FOCUS ON INSTRUMENTS

Temperature Units, Pressure Units, and Barometers

Temperature Units

Most scientists use a temperature scale called the *absolute* or **Kelvin scale**, after the famous British scientist Lord Kelvin (1824–1907), who first introduced it. This scale starts at absolute zero. Since it contains no negative numbers, it is quite convenient for scientific calculations.

Two other temperature scales commonly used today are the Fahrenheit and Celsius (formerly centigrade). The **Fahrenheit scale** was developed in the early 1700s by the physicist G. Daniel Fahrenheit. On this scale, he assigned the number 32 to the temperature at which water freezes, and the number 212 to the temperature at which water boils. The zero point was simply the lowest temperature that he obtained with a mixture of ice, water, and salt. Between the freezing and boiling points are 180 equal divisions, each of which is called a *degree*. A thermometer calibrated with this scale is referred to as a Fahrenheit thermometer, for it measures the hotness or coldness of things in degrees Fahrenheit (°F).

The **Celsius scale** was introduced later in the eighteenth century. The number 0 (zero) on this scale is assigned to the temperature at which water freezes, and the number 100 to the temperature at which water boils. The space between freezing and boiling is divided into 100 equal degrees. Therefore, each degree Celsius (°C) is 180/100 or 1.8 times bigger than a degree Fahrenheit. Put another way, an increase in temperature of 1°C equals an increase of 1.8°F.

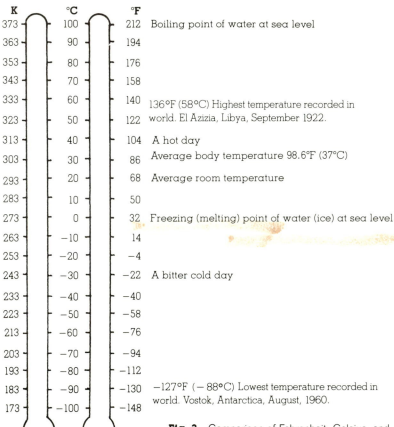

Fig. 3 Comparison of Fahrenheit, Celsius, and Kelvin scales.

On the Kelvin scale, degrees Kelvin are called **Kelvins** and they are abbreviated K. Each degree on the Kelvin scale is exactly the same size as a degree Celsius, and a temperature of 0 K is equal to −273°C.

Figure 3 compares the Fahrenheit, Celsius, and Kelvin scales. Converting a temperature from one scale to another can be done by simply reading the corresponding temperature from the adjacent scale. Thus, 303 on the Kelvin scale is the equivalent of 30°C and 86°F.

In most of the world, temperature readings are taken in °C. In the United States, however, temperatures above the surface are taken in °C, while temperatures at the surface are typically read in °F.[*] Hence, present temperatures on upper-level maps are plotted in °C, while, on surface weather maps, they are in °F. Since both scales are in use, most temperature readings in this book will be given in °C followed by their equivalent in °F.

[*]By the early 1990s, all surface temperature readings will likely be made in degrees Celsius.

FOCUS ON INSTRUMENTS, continued

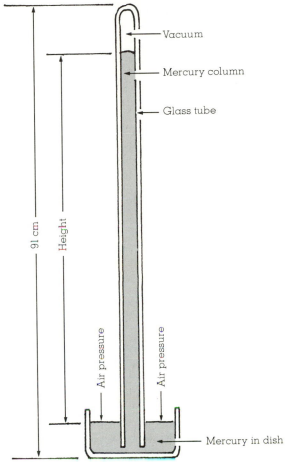

Vacuum

Mercury column

Glass tube

91 cm

Height

Air pressure

Air pressure

Mercury in dish

Fig. 4 The mercury barometer. The height of the mercury column is a measure of atmospheric pressure.

Pressure Units

In meteorology, the bar is a unit of pressure. By definition, a bar is a force of 100,000 Newtons (N) acting on a surface area of 1 square meter (m²). A Newton is the amount of force required to move an object with a mass of 1 kilogram (kg) so that it increases its speed at a rate of 1 meter per second (m/sec) each second.

Because the bar is a relatively large unit, and because surface pressure changes are usually small, the unit of pressure most commonly found on surface weather maps is, as we have mentioned, the millibar, where 1 mb = 1/1000 bar, or

1 bar = 1000 mb.

The unit of pressure designated by the International System (SI) of measurement is the *pascal*, named in honor of Blaise Pascal (1632–1662), whose experiments on atmospheric pressure greatly increased our knowledge of the atmosphere. A pascal (Pa) is the force of 1 Newton acting on a

surface area of one square meter. One hundred pascals equals one millibar; therefore,

1 bar = 1000 mb = 100,000 Pa.

At sea level, the average or *standard* value for atmospheric pressure is

1013.25 mb = 101,325 Pa.

Currently, pressure readings on all surface weather maps are expressed in millibars. However, when the National Weather Service converts entirely to the metric system (perhaps by the early 1990s), the pascal will probably be the unit of pressure displayed on surface charts.

Barometers

We measure the atmospheric pressure with an instrument called a **barometer**. Thus, atmospheric pressure is also referred to as *barometric pressure*. Evangelista Torricelli, a student of Galileo, invented the **mercury barometer** in 1643. His barometer, which is similar to those in use today, consisted of a glass tube over 91 cm (36 in.) long, open at one end, and closed at the other. Removing air from the tube and covering the open end, Torricelli immersed the lower portion into a dish of mercury. He removed the cover, and the mercury rose up the tube to nearly 76 cm (30 in.) above the level of the dish. Torricelli correctly concluded that the column of mercury in the tube was balancing the weight of the air above the dish, and, hence, its height was a measure of atmospheric pressure (Fig. 4).

Figure 5 compares pressure readings in millibars and in

FOCUS ON INSTRUMENTS, continued

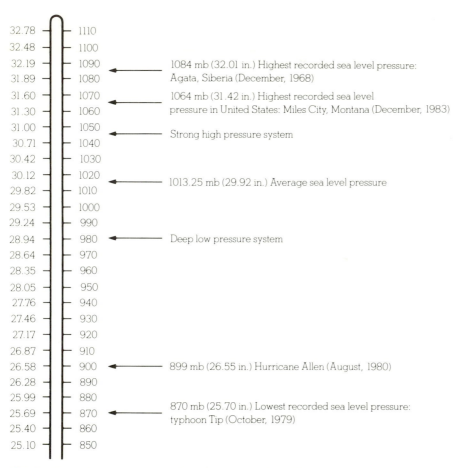

inches	millibars	
32.78	1110	
32.48	1100	
32.19	1090	← 1084 mb (32.01 in.) Highest recorded sea level pressure:
31.89	1080	Agata, Siberia (December, 1968)
31.60	1070	← 1064 mb (31.42 in.) Highest recorded sea level
31.30	1060	pressure in United States: Miles City, Montana (December, 1983)
31.00	1050	← Strong high pressure system
30.71	1040	
30.42	1030	
30.12	1020	← 1013.25 mb (29.92 in.) Average sea level pressure
29.82	1010	
29.53	1000	
29.24	990	
28.94	980	← Deep low pressure system
28.64	970	
28.35	960	
28.05	950	
27.76	940	
27.46	930	
27.17	920	
26.87	910	
26.58	900	← 899 mb (26.55 in.) Hurricane Allen (August, 1980)
26.28	890	
25.99	880	
25.69	870	← 870 mb (25.70 in.) Lowest recorded sea level pressure:
25.40	860	typhoon Tip (October, 1979)
25.10	850	

Fig. 5 Atmospheric pressure in inches of mercury and in millibars.

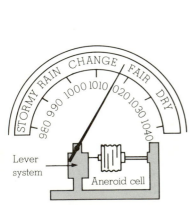

Fig. 6 The aneroid barometer.

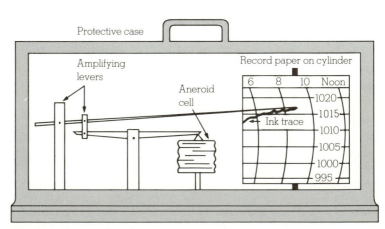

Fig. 7 A recording barograph.

FOCUS ON INSTRUMENTS, continued

inches of mercury (Hg). A few of the more useful conversions are:

1 mb	=	0.02953 in. Hg
1 in. Hg	=	33.865 mb
1 mm Hg	=	1.3332 mb
1 Pa	=	0.01 mb.

(Appendix A presents a more complete list of conversions.)

You may wonder why mercury rather than water is used in the barometer. The primary reason is convenience. (Also, water evaporating in the tube presents a problem.) Mercury seldom rises to a height above 79 cm (31 in.). Water, however, is 13.6 times less dense than mercury. Thus, an atmospheric pressure of 76 cm (30 in.) of mercury would be equivalent to 1034 cm (408 in.) of water. Consequently, a water barometer sitting on the ground would have to be read from a third story window.

The most common type of home barometer—the **aneroid barometer**—contains no fluid. Inside this instrument is a small, flexible metal box called an aneroid cell. Before the cell is tightly sealed, air is partially removed, so that small changes in external air pressure cause the cell to expand or contract. The size of the cell is calibrated to represent different pressures, and any change in its size is amplified by levers and transmitted to an indicating arm, which points to the current atmospheric pressure (Fig. 6).

Notice that the aneroid barometer often has descriptive weather-related words printed above specific pressure values. These adjectives indicate the most likely weather conditions when the needle is pointing to that particular pressure reading.

Generally, the higher the reading, the more likely good weather will occur, and the lower the reading, the better are the chances for inclement weather. This is because surface high pressure areas are associated with sinking air aloft, while surface low pressure areas are associated with rising air.

Two types of aneroid barometers are the **altimeter** and the **barograph**. Altimeters are aneroid barometers that measure pressure, but are calibrated to indicate altitude. Barographs are recording aneroid barometers. Basically, the barograph consists of a pen attached to an indicating arm that marks a continuous record of pressures on chart paper. The chart paper is attached to a drum rotated slowly by an internal mechanical clock (Fig. 7).

---

colder than air at the surface. Higher up, at a cruising altitude of 10 km (about 33,000 ft), the air surrounding a jet airliner would be about 65°C (117°F) colder than the air at the surface. With a surface air temperature near 15°C (59°F), the actual air temperature outside the aircraft would be near −50°C (−58°F).

Why Does Rising Air Cool and Sinking Air Warm? The decrease of air pressure with height ensures that rising air always cools. To see why, let's examine some rising air. First, we place air in an imaginary thin, elastic wrap about the size of a large balloon. We will call this "invisible" balloonlike container a **parcel**. The parcel of air can expand and contract freely, but neither external air nor heat is able to mix with the air inside. As the parcel moves, it does not break apart, but remains a single unit. The space occupied by the molecules within the parcel defines the air density (mass per unit volume). The average speed of the molecules is directly related to the air temperature, and the molecules colliding against the sides of the parcel determine the air pressure inside.

At the earth's surface, the parcel has the same temperature as the surrounding air. If it is carried up into the atmosphere, it enters a region where the air pressure is lower. The lower pressure outside allows the molecules inside to push the parcel walls outward, expanding it. Because there is no other energy source, the gas molecules inside must use some of their own energy to expand the parcel. Just as a Ping-Pong ball moves more slowly after rebounding from a paddle that is moving away from it, air molecules in an expanding parcel move more slowly after the parcel expands. This decrease in average molecular speed results in a lower parcel temperature. Therefore, air that rises always expands and cools.

If the parcel is lowered to the earth, it returns

to a region where the air pressure is higher. The higher pressure squeezes the parcel back to its original (smaller) volume (Fig. 2.8). This squeezing increases the average speed of the molecules inside because molecules have a faster rebound velocity after colliding with the sides of a collapsing parcel. (A Ping-Pong ball moves faster after striking a paddle that is moving toward it.) Because an increase in molecular speed results in a higher temperature, air that sinks (subsides) warms. *Just as rising air always cools by expanding, subsiding air always warms by compression.**

As air subsides and warms, its capacity for water vapor increases. Because more water vapor molecules are required to saturate a warmer parcel, sinking air inhibits the formation of clouds. On the other hand, as rising air cools, its capacity for holding water vapor decreases. Because less water vapor is required to saturate colder air, rising air enhances cloud formation. Consequently, cloudy skies are often due to rising (cooling) air, while clear skies may be the result of sinking (warming) air.

The Atmosphere Obeys the Gas Law

Air pressure, density, and temperature are all interrelated; thus, if one of these atmospheric variables changes, the other two usually change as well. For example, when we lifted the parcel of air, the air expanded and cooled. The number of molecules in the parcel remained the same, while the volume of the parcel grew larger. This situation indicates that, after expanding, the air density had decreased. We know that expanding air cools, and that molecules within the container move at slower speeds and strike the sides with less force. Since there is more surface area for the same number of molecules to hit, proportionately fewer molecules would bump against the sides. For these reasons, the pressure of expanding air also decreases. So when air rises or descends, air pressure, temperature, and density all change.

*When a parcel of air expands and cools, or compresses and warms, with no interchange of heat with its surroundings, the temperature change is said to be *adiabatic*. Chapter 10 examines adiabatic processes and how they relate to rising and sinking air.

The relationship among the pressure, temperature, and density of air can be expressed by

Pressure = temperature × density × constant.

This simple relationship, often referred to as the **gas law** (or *equation of state*), tells us that the pressure of a gas is equal to its temperature times its density times a constant. When we ignore the constant and look at the gas law in symbolic form, it becomes

$p \sim T \times \rho$

where, of course, p is pressure, T is temperature, and ρ (the Greek letter rho, pronounced "row") represents air density.* The line $\sim$ is a symbol meaning "is proportional to." A change in one variable causes a corresponding change in the other two variables. Thus, it will be easier to understand the behavior of a gas if we keep one variable from changing and observe the behavior of the other two.

Suppose, for example, we hold the temperature constant. The relationship then becomes

$p \sim \rho$ (temperature constant).

This expression says that the pressure of the gas is proportional to its density, as long as its temperature does not change. In other words, at a constant temperature, an increase in pressure results in an increase in density, and vice versa.

We can visualize this phenomenon by imagining air within a container that has rigid walls (Fig. 2.9). The air inside is at a certain temperature, pressure, and density. If more air molecules with the same average motion are stuffed into the container, the inside air temperature does not change. Yet, there are more molecules bouncing against the inside walls, which results in a greater air pressure (Fig. 2.9b). Adding more molecules increases the air density. The air inside is now at a higher pressure and at a greater density than it was originally. Thus, *at the same temperature, air at a higher pressure is more dense than air at a lower pressure.*

We can apply this concept directly to the atmosphere. In Fig. 2.10 the height, elevation, and

*The gas law may also be written as $p \times v = T \times$ constant. Consequently, pressure and temperature changes are also related to changes in volume.

average temperature of both columns are the same. However, there are more molecules in column H than in column L. Since atmospheric pressure is essentially the weight of the air in a column above an area, the surface air pressure at H is higher than at L. In other words, with nearly the same temperature and elevation, air above a region of surface high pressure is more dense than air above a region of surface low pressure.

Now, let's look at what happens to the gas law when the pressure of a gas remains constant. In shorthand notation, the law becomes

$$(\text{Constant pressure}) \times \text{constant} = T \times \rho.$$

This relationship tells us that when the pressure of a gas is held constant, the gas becomes less dense as the temperature goes up.

To see this phenomenon more clearly, let's observe the behavior of air in a container once again. In Fig. 2.11a, the air in the container is at some average temperature, density, and pressure. This time, we want to see how the temperature and density change as pressure is held constant.

The gas law states that, to keep the pressure constant, an increase in temperature must be offset by a decrease in density. Suppose we remove all the molecules from the container and replace them with fewer, faster-moving molecules, just enough to equal the original air pressure. This feat can be accomplished because the faster-moving molecules collide with the walls more frequently, and with greater force. We now have faster-moving molecules but fewer of them (Fig. 2.11b). Faster-moving molecules represent warmer air; fewer molecules indicate a lower air density. So, at higher temperatures, the air becomes less dense.

What would happen if slower-moving molecules were added to the original empty container? Figure 2.11c demonstrates that, in order to keep the pressure from varying, we have to add a greater number of molecules. These slower-moving, more tightly packed molecules represent a lower temperature and a higher density. Therefore, *at a given atmospheric pressure, cold air is more dense than warm air*.

If the air is usually cold aloft, why doesn't this cold air always sink? The idea that cold air is more dense than warm air applies only when we compare volumes of air at the same level. It does *not* apply when we compare air parcels, which are

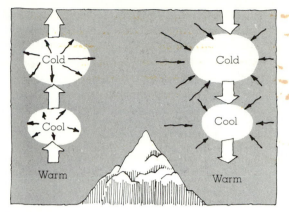

Fig. 2.8 Rising air expands and cools; sinking air is compressed and warms.

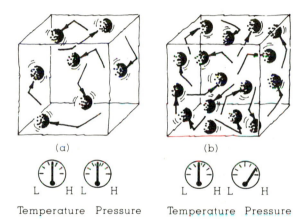

Fig. 2.9 As the pressure of a gas increases (at a constant temperature), the density also increases.

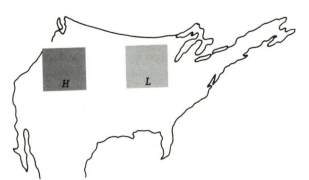

Fig. 2.10 Air above a region of surface high pressure is more dense than air above a region of surface low pressure (at the same temperature).

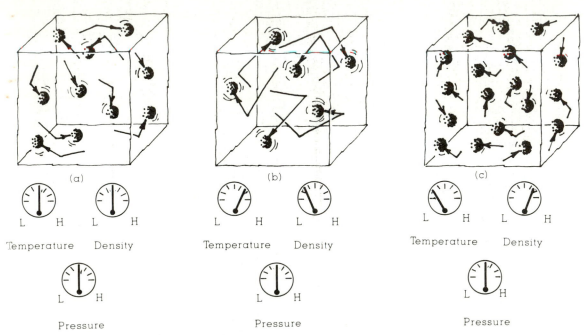

Fig. 2.11   If (a) the pressure of a gas is held constant, then (b) an increase in temperature results in a decrease in density, and (c) a decrease in temperature results in an increase in density.

located at different elevations and, hence, at different pressures. Because air density decreases rapidly with height, an air parcel many kilometers above the earth's surface may be colder than a parcel near the ground. It will not, however, be more dense. *The air parcel aloft will sink or rise depending on the temperature and density of the air surrounding it at that level.*

Pressure Changes Aloft   In the previous section, we examined the behavior of air molecules in a confined container. But the atmosphere is not confined—it is free to move from one region to another. This situation presents a problem when we examine air temperature, pressure, and density. For example, we already know that a change in one variable usually causes a change in the other two. Suppose, then, that the surface air is warmed over a field. As the air temperature increases, the air density decreases. This condition causes a change in pressure. The pressure change might start the air moving over the field which, in turn, might create a further change in density and pressure. Consequently, to simplify some of the atmosphere's complex behavior, we construct *models* of the atmosphere.

Figure 2.12a shows our simple model—two columns of air, each extending well up into the atmosphere. In each column, the dots represent air molecules. To better understand the air's behavior, our simplified model assumes the following three things about the atmosphere:

1. That air density remains constant from the surface up to the top of each column (this eliminates the variation of density with increasing height)
2. That the walls of the air columns act as invisible barriers preventing molecules from leaving or entering each column (this means the air pressure at the surface will not change)
3. That the width of each column does not change

With these assumptions in mind, we are now ready to examine the air inside the columns.

The two cities in Fig. 2.12 are located at the same elevation and have identical surface air pressure. These facts mean, of course, that there is the same mass of air in each column above both cities. Since the dots represent air molecules, each column must have an equal number of them. Suppose the air above city 1 cools, while the air above city 2 warms.

We know that cold air is more dense than warm air, as long as the pressure remains constant. To keep the surface pressure from varying, the total number of molecules above each city must remain the same. The cooling air above city 1 becomes denser as the molecules crowd closer together. Because the air is cooling and becoming more dense, air column 1 shrinks (Fig. 2.12b). In air column 2, the warm air expands and becomes less dense. (Notice that the total mass of air above each column is the same, so that the surface pressure of each column has not changed.) We now have a cold, shorter column of air above city 1 and a warm, taller air column above city 2. From this picture, we can conclude that *it takes a shorter column of cold, dense air to exert the same surface pressure as a taller column of warm, less dense air.* This concept has great meteorological significance.

Atmospheric pressure decreases more rapidly with elevation in a cold column of air. In the cold air above city 1 (Fig. 2.12b), move up the column and observe how quickly you pass through the densely packed molecules. This indicates a rapid change in pressure. In the warmer, less dense air, the pressure does not decrease as rapidly with height, simply because you climb above fewer molecules in the same distance.

Move up the cold column until you are above half of all the molecules. Mark this level. Now move up the same distance in the warm column and stop. Observe that there are more molecules (a greater pressure) above this level in the warm column than in the cold column. The fact that the number of molecules above any level is a measure of the atmospheric pressure leads to an extremely important concept: *Warm air aloft is normally associated with high atmospheric pressure, and cold air aloft with low atmospheric pressure.* *

Suppose we use our atmospheric model with its assumptions to explore the two air columns in Fig. 2.13a.

Pressure Changes at the Surface   Why does surface air pressure vary from one place to another? We know that changes in elevation cause pressure variations, but what about stations located at the same level? We have already learned that atmospheric pressure is the weight of air

*"Aloft" means at least a kilometer or more above the surface.

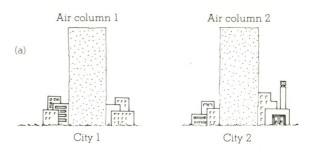

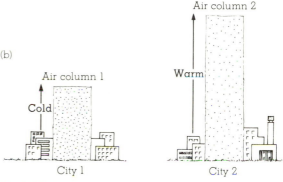

Fig. 2.12   It takes a shorter column of cold air to exert the same surface pressure as a taller column of warm air. Because of this, cold air aloft is associated with low pressure, and warm air aloft with high pressure.

above an area and that a change in air density can bring about a change in air pressure. One way to alter the density is to change the air temperature.

The two columns in Fig. 2.13a represent identical columns of air extending upward from the earth's surface to an altitude of 10 km. (About 75 percent of all the air molecules are found below this level.) The density, temperature, and surface pressure are the same in each column. The dashed line represents the level where the air pressure is 500 mb—exactly half the surface value.

Suppose the atmospheric pressure at the bottom of both columns remains constant as the air cools in one column and warms in the other (Fig. 2.13b). Notice that, in the warm column, a pressure of 500 mb is found at a higher elevation, while, in the cold column, an air pressure of 500 mb is found at a lower elevation. In the warm column, descend from the pressure reading of 500 mb to the letter H. Because air pressure increases as we move downward, the pressure at position H must be greater than 500 mb. Hence, at position H, the air pressure is greater than the pressure in the cold air column at the same level.

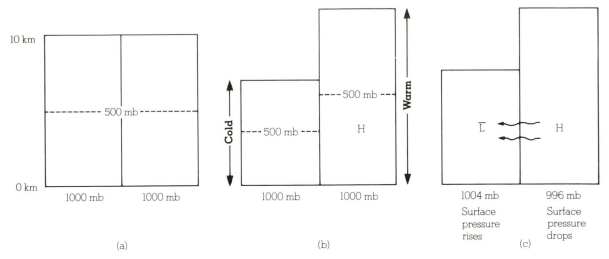

Fig. 2.13   The heating of one air column and the cooling of another set up pressure differences that force the air aloft to move from the warmer to the cooler column. The removal of air from the warm column causes its surface pressure to drop, while the addition of air into the cold column causes its surface pressure to rise.

This horizontal difference in pressure causes air to move from higher toward lower pressure. (The force that causes the air to move, the *pressure gradient force*, is discussed in depth in Chapter 12.)

Suppose we remove the invisible barrier between the two columns and allow the air aloft to move horizontally from the region of higher pressure toward the region of lower pressure. As air aloft leaves the warm column, the weight of the air in the column decreases, and so the surface pressure drops (Fig. 2.13c). Meanwhile, the accumulation of air in the cold column causes the surface pressure to rise.

In summary, heating or cooling an air column can influence surface air pressure. When either very cold or very warm air covers an extensive region, huge areas of either surface high or low pressure may form. These surface pressure systems are typically quite shallow and often disappear at a height of several kilometers above the surface.

From what we have seen so far, we might expect surface air pressure to decrease as the air temperatures increase and reach a minimum during the warmest part of the afternoon. After this time, pressure should increase as the air cools until a maximum is reached during the cool, early morning hours. In some areas of the world, this event actually happens. In other regions, large migrating pressure systems greatly influence daily pressure variations. In still other locations, heating of the air in the upper atmosphere appears to have a strong influence on surface air pressure.

In the tropics, for example, pressure rises and falls in a regular pattern twice a day. (See Fig. 2.14.) Maximum pressures occur around 10:00 A.M. and 10:00 P.M., minimum near 4:00 A.M. and 4:00 P.M.. The largest pressure difference, about 2.5 mb, occurs near the equator. This daily (*diurnal*) fluctuation of pressure appears to be due primarily to the sun's heating of the upper atmosphere, where, at altitudes above 100 km (60 mi), changes in daily temperatures of over 500°C (930°F) may occur. This rapid warming and cooling of the upper air creates density oscillations known as **thermal** (or *atmospheric*) **tides**. But because the air is very thin at such high altitudes, the tidal effect only shows up as small pressure changes near the earth's surface.

Movement of air aloft can also influence surface air pressure. Upper-level winds do not always blow at a constant speed from the same direction. In some places, the winds aloft will cause the air to funnel into an area, causing the air to crowd together in the manner of cars entering a congested highway. This piling up of air is called **convergence**. Because upper-level convergence increases the mass of air above the surface, it can

FOCUS ON A SPECIAL TOPIC

The Gas Law and Surface Pressure Changes in Cold and Warm Weather

Any person who has experienced a bitter, cold, winter day knows that, at the surface, very cold weather is normally associated with high atmospheric pressure. On the other hand, anyone who has experienced a clear, hot, summer day in a southwestern desert knows that, at the surface, very warm weather is normally associated with low atmospheric pressure. We can use the gas law (commonly called the *ideal gas law*) to see why this happens.

Recall that the gas law is written as

$$p = T \times \rho \times C.$$

With the pressure (p) in millibars, the temperature (T) in Kelvins, and the density (ρ) in kilograms per cubic meter (kg/m^3), the numerical value for the constant (C) is about 2.87.*

*The constant is usually expressed as 2.87 × 10^6 erg/g K, or, in the SI system, as 287 J/kg K. (See Appendix A for information regarding the units used here.)

Suppose that, first, we calculate the surface air pressure (near sea level) on a typical day that is neither too cold nor too warm. With an air temperature of 288 K (15°C, 59°F) and an air density of 1.226 kg/m^3, the air pressure becomes

$$p = 288 \times 1.226 \times 2.87$$
$$p = 1013 \text{ mb.}$$

Earlier in this chapter, we learned that, when the air temperature changes, it causes a corresponding change in the air pressure and air density. In very cold air, near the surface as the temperature drops, the air molecules slow down and crowd closer together. This causes an increase in air density. To see what happens to the surface air pressure, let's consider this: Suppose the air cools to a bone-chilling 253 K (−20°C, −4°F) and the air density increases to 1.450 kg/m^3. The pressure then becomes

$$p = 253 \times 1.450 \times 2.87$$
$$p = 1053 \text{ mb.}$$

Hence, at the surface in very cold air, as the air density increases, so does the air pressure.

What happens to the surface pressure on really hot days? Suppose the air temperature increases to a sizzling 314 K (41°C, 106°F). As the air warms, the molecules move faster, spread farther apart, and the air density drops. If the density lowers to 1.110 kg/m^3, the surface air pressure becomes

$$p = 314 \times 1.110 \times 2.87$$
$$p = 1000 \text{ mb.}$$

Therefore, at the surface in hot weather, as the air density decreases, so does the air pressure.

In summary, we can see with the aid of the gas law that, near the surface as the air temperature decreases to very low readings, the air density and air pressure both increase. Where the air temperature increases to very high readings, the air density and air pressure both decrease.

cause the surface air pressure to increase. In other regions, winds aloft can force air out of an area in many directions, causing the air to spread apart. This outflow of air is called **divergence**. Because upper-level divergence can decrease the mass of air above the surface, it can cause the surface air pressure to decrease.

Figure 2.15 illustrates how upper-level convergence and divergence can influence air pressure at the surface. Beneath the region of upper-level convergence, the surface air pressure is rising. As the surface pressure increases, surface winds begin to blow outward (diverge) away from the center of highest pressure. This causes the air

above the region of high pressure to slowly sink (subside), producing generally clear skies. Beneath the region of upper-level divergence, the surface air pressure is decreasing. This causes the surface winds to blow inward (converge) toward the center of lowest surface pressure. As the surface air converges, it rises, expands, and cools, often condensing into clouds that produce precipitation.

The areas of high and low atmospheric pressure that are created by the flow of air aloft are usually deep pressure systems that intensify with altitude. They can be observed on surface weather maps, and on upper-level maps as well. Deep

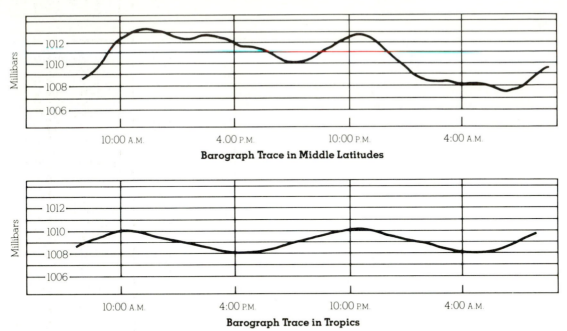

Fig. 2.14 Daily sea level pressure variations in middle and low latitudes.

areas of low pressure are also referred to as *depressions* and *middle latitude cyclones*. Elongated regions of low pressure are called **troughs**. As we learned in Chapter 1, areas of high pressure are called *anticyclones*. Elongated high pressure areas are known as **ridges**. Once these surface pressure systems form, they are often steered by the westerly flow of winds aloft, and tend to move eastward. When a surface low approaches a station, surface air pressure will normally decrease. When a surface high approaches, surface air pressure will normally increase. Hence,

the movement of these pressure systems with diameters measuring hundreds of kilometers can influence the surface air pressure over a vast area.

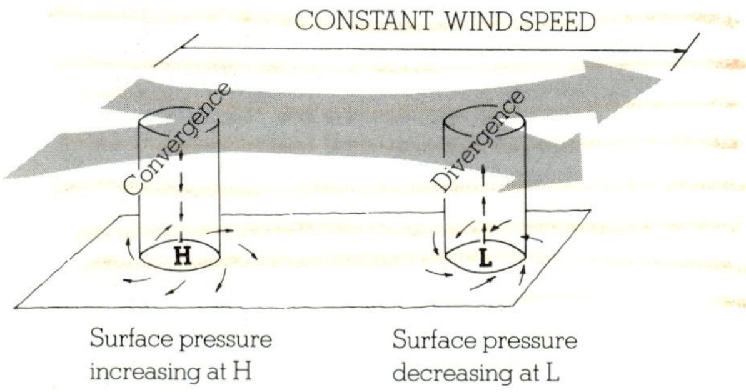

In summary, changes in surface air pressure can be brought on by changes in the air density above the surface. Such density changes may be due to:

1. Changes in air temperature in an air column above the surface
2. Sunlight warming the upper portion of the atmosphere, creating thermal tides
3. Convergence and divergence of air brought on by wind flow patterns

Once areas of surface high and low pressure form, they often move, causing surface pressure to either increase or decrease as they approach a particular area.

Summary

The earth's atmosphere has evolved over hundreds of millions of years from one composed mostly of hydrogen and helium into one rich in nitrogen and oxygen. Today's air also consists of many other

CONSTANT WIND SPEED

Convergence

Divergence

H

L

Surface pressure increasing at H

Surface pressure decreasing at L

Fig. 2.15 Horizontal convergence and divergence of air, created by upper-level winds of uniform speed, cause changes in air pressure at the earth's surface.

Wind Is the Movement of Air

Wind is the horizontal movement of air. If we imagine air molecules as being a swarm of bees, the wind may be seen as the movement of the entire swarm. This analogy can be carried a little further: On a calm day, the swarm will remain in one spot, with each bee randomly darting about; while, on a windy day, the entire swarm will move quickly from one place to another. The swarm's speed would be the rate at which it moves past you. In like manner, wind speed is the rate at which air moves by a stationary observer. This movement can be expressed as the distance in nautical miles traveled in one hour (knots) or as the number of meters traveled in one second (m/sec).

We know from Chapter 1 that wind direction is given as the *direction from which* the wind is blowing: A north wind blows from the north, an east wind

blows from the east, and so on. Unlike a swarm of bees, air is invisible; we cannot really see it. Rather, we see things being moved by it. Therefore, we can determine wind direction by watching the movement of objects as air passes them. For example, the rustling of small leaves, smoke drifting near the ground, and flags waving on a pole all indicate wind direction.

In a light breeze, a tried and true method of determining wind direction is to raise a wet finger into the air. The dampness quickly evaporates on the windward side, cooling the skin. Traffic sounds carried from nearby railroads or airports can be used to help figure out the direction of the wind. Even your nose can alert you to the wind direction as the smell of fried chicken or broiled hamburgers drifts with the wind from a local restaurant.

Wind direction may be diffi-

cult to determine in regions where the wind flow is disrupted into random swirls called **eddies**. You may feel eddies are a nuisance because they stir up small particles that may well end up in your eyes. Actually, though, eddies are quite beneficial. Their turbulent mixing disperses concentrated pollutants and carries heat and moisture away from the surface. Also, on cold nights, eddies help prevent damagingly low surface temperatures by mixing colder surface air with the warmer air above.

But it is the giant eddies—the high and low pressure areas of the middle latitudes—that are the major weather producers. Throughout this book we will examine these systems, looking at their winds, their vertical air motions, and the weather they bring.

gases and particles, some of which are extremely important. Even though the concentration of water vapor is normally less than 4 percent by volume, under certain conditions it transforms into liquid water droplets and delicate ice crystals that can grow in size and fall to earth as precipitation. Increasing levels of carbon dioxide may result in raising average surface air temperatures, while volcanic eruptions rich in sulfur may have the effect of lowering surface temperatures.

The temperature, pressure, and density of the air are all interrelated. The relationship among these variables is expressed by the gas law. We define air temperature in terms of the average kinetic energy of its molecules. Air pressure is defined in terms of the total weight of air above any point. Air density is the mass of air in a given column.

Changes in the density of an air column can bring about corresponding changes in surface air pressure. Such density changes can be brought on by either temperature changes or by winds producing mass convergence or divergence of air.

Questions for Review

1. How has the composition of the atmosphere apparently changed since the formation of the earth?

2. What are the four most abundant gases in today's atmosphere?

3. What is an ion?

4. Briefly explain the production and natural destruction of carbon dioxide near the earth's surface.

5. Why has the eruption of the volcano El Chichón in April, 1982, had a greater impact on surface temperature than the volcanic eruption of Mount St. Helens in 1980?

6. List some of the pollutants found in the atmosphere, and briefly describe their effect on the environment.

7. Basically, how do the three states of water differ?

8. What are the primary factors that influence evaporation?

9. At the same temperature and level in the atmosphere, why is moist air lighter than dry air?

10. When the air temperature increases, what happens to its capacity to hold water vapor?

11. Define: (a) evaporation; (b) condensation; (c) sublimation; (d) deposition; (e) saturation.

12. Describe how the average speed of air molecules relates to the temperature of air.

13. Explain the concept of air pressure in terms of: (a) molecular bombardment; (b) weight of air.

14. Why does air pressure always decrease with height?

15. With the aid of a diagram, describe how a mercury barometer works.

16. Explain why rising air always cools and sinking air always warms.

17. What is meant by the statement, "At the same pressure, cold air is more dense than warm air"?

18. Cold air aloft is normally associated with low atmospheric pressure. Explain why.

19. Why are regions of low atmospheric pressure frequently associated with clouds and precipitation? Explain why the opposite is generally true in a region of high pressure.

20. List and then explain four possible reasons that could account for a decrease in air pressure at the surface. Now list four reasons that could account for an increase in surface air pressure.

Questions for Thought

1. List several ways the composition of today's atmosphere may be changing.

2. Would you expect water in a glass to evaporate more quickly on a windy, warm, dry summer day or on a calm, cold, dry winter day? Explain.

3. Explain how frozen clothes can "dry" outside in subfreezing weather.

4. Why does the air pressure decrease with height more rapidly in cold air than in warm air?

5. Based on surface air temperature only, how would you expect the surface atmospheric pressure to change from early morning to evening on a clear, calm, summer day?

6. Use the gas law to explain why a car with tightly closed windows will occasionally have a window "blow out" or crack when exposed to the sun on a hot day.

7. Explain why, on a sunny day, an aneroid barometer would indicate "stormy" weather when carried to the top of a hill or mountain.

8. In Fig. 2.15, explain how the surface pressure would change if the outflow of air around the surface high were greater than the convergence of air aloft. How would the surface pressure change if the inflow of air around the surface low were greater than the divergence of air aloft?

9. Suppose the air column above city Q is completely saturated with water vapor, and the air column above city T is completely dry. If the temperature of the air in both columns is the same, which column will have the highest atmospheric pressure at the surface? Explain.

10. The total pressure of a mixture of gases is equal to the sum of the pressures of the individual gases. This is known as *Dalton's law of partial pressure.* Suppose, then, that water vapor occupies 4 percent of the air above city Q. If the total surface pressure is 1000 mb, what would be the pressure exerted only by the water vapor molecules?

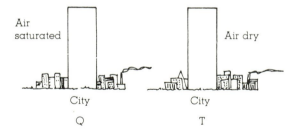

Air saturated   Air dry

City   City
Q   T

Problems and Exercises

1. A standard pressure of 1013.25 mb is also known as one atmosphere (1 ATM). (a) At approximately what elevation would you record a pressure of 0.5 ATM? 0.1 ATM? (b) The surface air pressure on the planet Mars is almost 0.005 ATM. If you were standing on Mars, the surface air pressure would be equivalent to a pressure observed at approximately what elevation in the earth's atmosphere?

2. The surface atmospheric pressure on the planet Venus is 90 times greater than that on Earth. (a) What would be the average surface pressure of Venus in millibars? (b) Because water is nearly incompressible, the pressure of sea water increases by about 1000 mb (1 ATM) for every 10 m (33 ft) of depth. About how deep in the ocean would one dive to experience the same pressure as that on the surface of Venus?

3. Use the gas law in the focus section on p. 35 to calculate the air pressure in millibars when the air temperature is −23°C and the air density is 0.700 kg/m³. (Hint: be sure to use the Kelvin temperature.) At approximately what elevation would you expect to observe this pressure?

4. Suppose air in a closed container has a pressure of 1000 mb and a temperature of 20°C. (a) Use the gas law to determine the air density in the container. (b) If the density in the container remains constant, but the pressure doubles, what would be the new temperature?

5. A large balloon is filled with air so that the air pressure inside just equals the air pressure outside. The volume of the filled balloon is 3 cubic meters, the mass of air inside is 3.6 kg, and the temperature inside is 20°C. What is the air pressure? (Hint: density = mass/volume.)

As the sun's radiant energy warms the wet ground and trees, convection currents carry the heated air upward. The water vapor in the rising air condenses into tiny water droplets, giving the appearance of smoke rising from the heated surfaces. (Photo: J. Medeiros)

CHAPTER 3

Energy: Warming the Earth and the Atmosphere

In Chapter 2 we investigated some of the applications of energy without really discussing energy itself. Energy is one of the most fundamental concepts in all of science. For this reason, defining "energy" adequately is a difficult (if not impossible) task. We know that it can come in various forms: Energy powers a car, warms a home, melts ice, and evaporates water. It also drives the atmosphere, producing the everyday weather events that we observe. Energy is something we buy when we pay our electric bill and something we feel when we stand close to a fire. Our supply of some forms of energy is dangerously low—we are all familiar with the "energy crisis." What, then, is this common, yet mysterious, quantity we call "energy"? What is its primary source? How does it warm our earth and provide the driving force for our atmosphere?

Energy

By definition, energy is the ability or capacity to do *work* on some form of matter. (Matter is anything that has mass and occupies space.) Work is done on matter when matter is either pushed, pulled, or lifted over some distance. When we lift a brick, for example, we exert a force against the pull of gravity—we "do work" on the brick. The higher we lift the brick, the more work we do. So, by doing work on something, we give it "energy," which it can, in turn, use to do work on other things. The brick that we lifted, for instance, can now do work on your toe—by falling on it.

The total amount of energy stored in any object (internal energy) determines how much work that object is capable of doing. A lake behind a dam, for instance, contains energy by virtue of its position. This is called *gravitational potential energy*

Contents

or simply **potential energy** because it represents the potential to do work. (Imagine the destructive work such energy would do if the dam were to break!) A volume of air aloft has more potential energy than the same size volume of air just above the surface. This is so because the air aloft has the potential to sink and warm through a greater depth of atmosphere. A substance also possesses potential energy if it can do work when a chemical change takes place. Thus, coal, natural gas, and food all contain chemical potential energy.

Any moving substance possesses energy of motion, or *kinetic energy*. The faster something moves, the greater its kinetic energy; hence, a strong wind possesses more kinetic energy than a light breeze. Kinetic energy also depends on the object's mass. A volume of water and an equal volume of air may be moving at the same speed, but, because the water has greater mass, it has more kinetic energy. The atoms and molecules that comprise all matter have kinetic energy due to their motion. This form of kinetic energy is often referred to as *heat energy*. Probably the most important form of energy in terms of weather and climate is the energy we receive from the sun—*radiant energy*.

Energy, therefore, takes on many forms, and it can change from one form into another. But the total amount of energy in the universe remains constant. Energy cannot be created nor can it be destroyed. It merely changes from one form to another in any ordinary physical or chemical process. In other words, the energy lost during one process must equal the energy gained during another. This is what we mean when we say that energy is conserved. This statement is known as the *law of conservation of energy*, and is also called the *first law of thermodynamics*.

Fig. 3.1 Even though the temperature of the beverage in both mugs is the same, the mug on the left contains more internal energy because it contains more molecules.

Up to now, we have examined the idea of energy without giving much attention to other important factors, such as temperature and heat. Since the concepts of energy, temperature, and heat are often incorrectly used interchangeably, let's see how they differ.

Energy, Temperature, and Heat

The atmosphere and oceans contain *internal energy*, which is the total energy (potential and kinetic) stored in their molecules. The temperature of air and water is determined only by the *average* kinetic energy (average speed) of *all* their molecules. For a *particular* molecule, greater speeds represent higher temperatures. Since temperature only indicates how "hot" or "cold" something is relative to some set standard value, it does not always tell us how much internal energy that something possesses. For example, two identical mugs, each half-filled with water and each with the same temperature, contain the same internal energy. If the water from one mug is poured into the other, the total internal energy of the filled mug has doubled because its mass has doubled. Its temperature, however, has not changed, since the average speed of all of the molecules is still the same. (See Fig. 3.1.)

Now, imagine that you are sipping a hot cup of tea on a small raft in the middle of a lake. The tea has a much higher temperature than the lake, yet the lake contains more internal energy because it is composed of many more molecules. If the cup of tea is allowed to float on top of the water, the tea would cool rapidly. The energy that would be transferred from the hot tea to the cool water (because of their temperature difference) is called *heat*.

In essence, **heat** *is energy in the process of being transferred from one object to another because of the temperature difference between them.* After heat is transferred, it is stored as internal energy. How is this energy transfer process accomplished? In the atmosphere, heat is transferred by *conduction*, *convection*, and *radiation*. We will examine these mechanisms of energy transfer, along with some of their implications, in the following sections.

FOCUS ON A SPECIAL TOPIC

Specific Heat

A watched pot never boils, or so it seems. The reason for this is that water requires a relatively large amount of heat to bring about a small temperature change. The **heat capacity** of a substance is the ratio of the amount of heat absorbed by that substance to its corresponding temperature rise. The heat capacity of a unit mass of a substance is called **specific heat**. In other words, specific heat is the amount of heat needed to raise the temperature of one gram (g) of a substance one degree Celsius.

If we heat 1 g of liquid water on a stove, it would take about 1 calorie* (cal) to raise its temperature by 1°C. So water has a specific heat of 1. If, however, we

*By definition, a calorie is the amount of heat required to raise the temperature of 1 g of water from 14.5°C to 15.5°C.

put the same amount of compact dry soil on the flame, we would see that it would take about one-fifth the heat (about 0.2 cal) to raise its temperature by 1°C. The specific heat of water is therefore 5 times greater than that of soil. In other words, water must absorb 5 times as much heat as the same quantity of soil in order to raise its temperature by the same amount. The specific heat of various substances is given in Table 1.

Not only does water heat slowly, it cools slowly as well. It has a much higher capacity for storing energy than other common substances, such as soil and air. A given volume of water can store a large amount of energy while undergoing only a small temperature change. Because of this attribute, water has a

Table 1 Specific Heat of Various Substances

SUBSTANCE	SPECIFIC HEAT (cal/g × °C)
Water	1.00
Wet mud	0.60
Ice (0°C)	0.50
Sandy clay	0.33
Dry air (sea level)	0.24
Quartz sand	0.19
Granite	0.19

strong modifying effect on weather and climate. Near large bodies of water, for example, winters usually remain warmer and summers cooler than nearby inland regions—a fact well known to people who live adjacent to oceans or large lakes.

Heat Transfer in the Atmosphere

Conduction   The transfer of heat from molecule to molecule within a substance is called **conduction**. Hold one end of a metal straight pin between your fingers and place a flaming candle under the other end. (See Fig. 3.2.) Because of the energy they absorb from the flame, the molecules in the pin vibrate faster. These molecules collide with neighboring molecules, causing them to move faster. These, in turn, collide with their neighbors, and so on until the molecules at the finger-held end of the pin begin to vibrate rapidly. These fast-moving molecules eventually cause the molecules of your finger to vibrate more quickly. Heat is now being transferred from the pin to your finger, and both the pin and your finger feel hot. If enough

heat is transferred, you will drop the pin. The transmission of heat from one end of the pin to the other, and from the pin to your finger, occurs

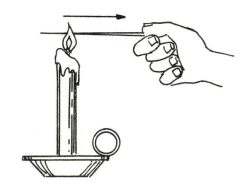

Fig. 3.2 The transfer of heat from the hot end of the metal pin to the cool end by molecular contact is called conduction.

by conduction. Heat transferred in this fashion always flows from *warmer* to *colder* regions. Generally, the greater the temperature difference, the more rapid the heat transfer.

When materials can easily pass energy from one molecule to another, they are considered to be good conductors of heat. How well they conduct heat depends upon how their molecules are structurally bonded together. Table 3.1 shows that solids, such as metals, are good heat conductors. It is often difficult, therefore, to judge the temperature of metal objects. For example, if you grab a metal pipe at room temperature, it will give you the sensation of being much colder than it actually is because the metal conducts heat away from the hand quite rapidly. Conversely, air is an extremely poor conductor of heat, which is why most insulating materials have a large number of air spaces trapped within them. Air is such a poor heat conductor that, in calm weather, the hot ground only warms a shallow layer of air a few centimeters thick by conduction. Yet, air can carry this energy rapidly from one region to another. How then does this phenomenon happen?

Convection The transfer of heat by the mass movement of a fluid is called **convection**. This type of heat transfer takes place in liquids and gases because they can move freely and it is possible to set up currents within them.

Convection happens naturally in the atmosphere. On a warm, sunny day certain areas of the earth's surface absorb more heat from the sun than others; as a result, the air near the earth's surface is heated somewhat unevenly. Air molecules adjacent to these hot surfaces bounce against them, thereby gaining some extra energy by conduction. The heated air expands and becomes less dense than the surrounding cooler air. The expanded warm air is buoyed upward, and rises. In this manner, large bubbles of warm air rise and transfer heat upward. Cooler, heavier air flows toward the surface to replace the rising air. This cooler air becomes heated in turn, rises, and the cycle is repeated. In meteorology, this vertical exchange of heat is called *convection* and the rising air bubbles are known as **thermals** (Fig. 3.3).

The rising air expands, cools, and gradually spreads outward. It then slowly begins to sink. Near the surface, it moves back into the heated region, replacing the rising air. In this way, a *convective circulation*, or thermal "cell," is produced in the atmosphere.

Although the entire process of heated air rising, spreading out, sinking, and finally flowing back toward its original location is known as a convective circulation, meteorologists usually restrict the term *convection* to the process of the rising and sinking part of the circulation.

The horizontally moving part of the circulation (called *wind*) carries properties of the air in that particular area with it. The transfer of these properties by horizontally moving air is called **advection**. For example, wind blowing across a body of water will "pick up" water vapor from the evaporating surface and transport it elsewhere in the atmosphere. If the air cools, the water vapor may condense into cloud droplets and release latent heat. In a sense, then, heat is advected (carried) by the water vapor as it is swept along with the wind. In the focus section on latent heat you will see that this is an important way to redistribute heat in the atmosphere.

In summary, convection is an important mechanism of heat transfer, as it represents the vertical movement of heated air upward and cooler air downward.

There is yet another mechanism for the transfer of energy—radiation, or *radiant energy*, which is what we receive from the sun. In this method,

Table 3.1 Heat Conductivity of Various Substances

MATERIAL	HEAT CONDUCTIVITY (Cal per sec per cm per °C)
Still air	0.0000614 (at 20°C)
Dry soil	0.0006
Water	0.00143
Snow	0.0015 (density 0.5 g/cm³)
Wet soil	0.0050
Ice	0.0053 (at 0°C)
Sandstone	0.0062
Granite	0.0065
Iron	0.161
Copper	0.918
Silver	1.006

FOCUS ON AN OBSERVATION

Watching for Convection

Since air is invisible, we cannot actually see convection and rising thermals. Often, however, there are signs that tell us where the air is rising. On a calm day, you can watch a hawk circle and climb high above level ground while its wings remain motionless. The rising air carries the hawk upward as it scans the terrain for prey. In a similar way, sailplanes remain airborne as glider pilots "look" for thermals, ride them as high as they can, then quickly search for another. A rising thermal that has picked up surface dust and debris in a cone of rotating motion is called a *dust devil.* Although dust devils are usually small, they occasionally become quite large, and can extend upward into the atmosphere for hundreds of meters. Thermals can form above just about any heated surface, such as volcanoes, fires, fields, and cities.

If the air in a rising thermal is sufficiently moist, puffy clouds with flat bases called *cumulus clouds* may form within them. A small aircraft flying below a de-veloping cumulus cloud may be jostled around by the rising and sinking air associated with convection. This jostling can be severe enough to cause passengers inside the aircraft to experience air sickness. Consequently, pilots would rather fly above cumulus clouds than below them. So, the next time you see a sky spotted with cumulus clouds growing in size, imagine how many thermals must be rising from the earth's surface.

---

energy may be transferred from one object to another without the space between them necessarily being heated.

Radiation

Have you ever noticed how warm and flushed your face feels as you stand in front of a huge bonfire on a bitterly cold evening, while the surrounding air remains quite cold? Somehow, energy from the fire is being transferred through the air with little effect upon the air itself. Your face, however, absorbs this energy and converts it to heat energy. Thus, you feel warm. You probably have experienced the same sensation of warmth while lying outdoors in the summer sun. The energy transferred from the fire and the sun to your face is called **radiant energy**, or **radiation**. It travels in the form of waves that release energy when they

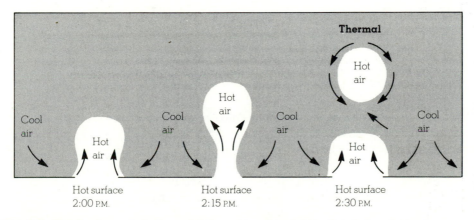

Fig. 3.3   The development of a thermal. A thermal is a rising bubble of air that carries heat upward by convection.

Latent Heat—The Hidden Warmth

The heat energy required to change a substance from one state to another is called **latent heat**. But why is this heat referred to as "latent"? To answer this question, we will begin with something familiar to most of us—the cooling produced by evaporating water.

Suppose we microscopically examine a small drop of pure water. We know from Chapter 2 that molecules are constantly evaporating at the drop's surface. Because the more energetic, faster-moving molecules escape most easily, the average motion of all the molecules left behind decreases as each additional molecule evaporates. Since temperature is a measure of average molecular motion, the slower motion suggests a lower water temperature. Evaporation is, therefore, a cooling process.

In the everyday world, we experience evaporational cooling as we step out of a shower or swimming pool into a dry area. Because some of the energy used to evaporate the water comes from our skin, we may experience a rapid drop in skin temperature, even to the point

where goose bumps form. In fact, on a hot, dry, windy day in Tucson, Arizona, cooling may be so rapid that we begin to shiver even though the air temperature is hovering around 38°C (100°F).

The energy lost by liquid water during evaporation can be thought of as carried away by, and "locked up" within, the water vapor molecule. The energy is thus in a "stored" or "hidden" condition and is, therefore, called *latent heat*. It is latent (hidden) in that the temperature of the substance changing from liquid to vapor is still the same. However, the heat energy will reappear as **sensible heat** (the heat we can feel and measure with a thermometer) when the vapor condenses back into liquid water. Therefore, condensation (the opposite of evaporation) is a warming process.

The heat energy released when water vapor condenses to form liquid droplets is called *latent heat of condensation*. Conversely, the heat energy used to change liquid into vapor at the same temperature is called *latent heat of evaporation* (vapor-

ization). Nearly 600 calories (cal) are required to evaporate a single gram of water at room temperature. This is enough heat to raise the temperature of two and one-half teaspoons of water from freezing to boiling. With many hundreds of grams of water evaporating from the body, it is no wonder that after a shower we feel cold before drying off.

In a way, latent heat is responsible for keeping a cold drink with ice colder than one without ice. As ice melts, its temperature does not change. The reason for this is that the heat added to the ice only breaks down the rigid crystal pattern, changing the ice to a liquid without changing its temperature. The energy used in this process is called *latent heat of fusion* (melting). Roughly 80 cal are required to melt a single gram of ice. Consequently, heat added to a cold drink with ice primarily melts the ice, while heat added to a cold drink without ice warms the beverage. If a gram of water at 0°C changes back into ice at 0°C, this same amount of heat (80 cal) would be released as sensible heat to the environment.

are absorbed by an object. Because these waves have magnetic and electrical properties, we call them **electromagnetic waves**. Electromagnetic waves do not need molecules to propagate them. In a vacuum, they travel at a constant speed of nearly 300,000 km (186,000 mi) per second—the speed of light. Let's first examine some characteristics and properties of electromagnetic waves. Then, we will see how they transfer energy.

Wave Characteristics   Some wave characteristics are shown in Fig. 3.4. The low points (valleys) of the waves are called *troughs*. The high points (*ridges*) of the waves are the *crests*. The horizontal distance between two successive crests is the **wavelength**, which is usually expressed by the Greek letter lambda (λ). Electromagnetic waves vary greatly in wavelength. Exceedingly short waves include gamma rays and X-rays (whose wave-

FOCUS ON A SPECIAL TOPIC, continued

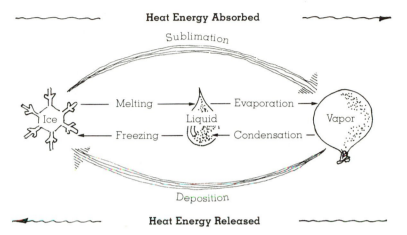

Fig. 1 Heat energy absorbed and released.

Therefore, when ice melts, heat is taken in; when water freezes, heat is liberated.

The heat required to change ice into vapor is called *latent heat of sublimation*. For a single gram of ice to transform completely into vapor at 0°C requires nearly 680 cal—80 cal for the latent heat of fusion plus 600 cal for the latent heat of evaporation. If this same vapor transforms back into ice, approximately 680 cal would be released.

Figure 1 summarizes the concepts examined so far. When the change of state is from left to right, heat is absorbed by the substance and taken away from the environment. The processes of sublimation, melting, and evaporation all remove heat from their surroundings. When the change of state is from right to left, heat is given up by the substance and added to the environment. The processes of deposition, freezing, and condensation all release heat into their surroundings.

Latent heat is an important source of atmospheric energy. Once vapor molecules become separated from the earth's surface, they are swept away by the wind, like dust before a broom. Rising to high altitudes where the air is cold, the vapor changes into liquid and ice cloud particles. During these processes, a tremendous amount of heat is released into the environment. For each kilogram of cloud particles that forms, more than 120,000 cal are given off. This heat provides energy for storms, such as hurricanes, middle latitude cyclones, and thunderstorms. In fact, the amount of heat released inside a huge thunderstorm is greater than the amount of energy unleashed by several small atom bombs! Water vapor evaporated from warm, tropical water can be carried into polar regions, where it condenses and gives up its heat. Thus, as we will see throughout the text, evaporation—transportation—condensation is an extremely important mechanism for the relocation of heat (as well as water) in the atmosphere.

lengths are shorter than one-hundredth of one-millionth of a meter). Other waves, such as television and radio waves, have lengths of more than one meter. Radiation that we can see (visible light) has a wavelength less than one-millionth of a meter (0.000001 m)—a distance nearly one-hundredth the diameter of a human hair.

Just as we measure the length of everyday household items in units of meters, centimeters,

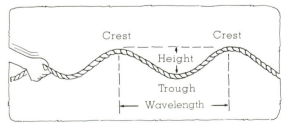

Fig. 3.4 Wave characteristics.

The Fate of a Sunbeam

Consider sunlight in the form of radiant energy striking a large lake. Part of the incoming energy heats the water, causing greater molecular motion and, hence, an increase in the water's kinetic energy. This greater kinetic energy allows more water molecules to evaporate from the surface. As each molecule escapes, work is done to break it away from the remaining water molecules. This energy becomes the latent heat energy that is carried with the water vapor.

Above the lake, a parcel of warm, moist air rises and expands. In order for this expansion to take place, the gas molecules inside the parcel must use some of their kinetic energy to do work against the parcel's sides. This results in a slower molecular speed and a lower temperature. Well above the surface, the water vapor in the rising, cooling parcel of moist air condenses into clouds. The condensation of water vapor releases latent heat energy into the atmosphere, warming the air. The tiny suspended cloud droplets possess potential energy, which becomes kinetic energy when these droplets grow into raindrops that fall earthward.

When the drops reach the surface, their kinetic energy erodes the land. As rain-swollen streams flow into a lake behind a dam, there is a buildup of potential energy, which can be transformed into kinetic energy as water is harnessed to flow down a chute. If the moving water drives a generator, kinetic energy is converted into electrical energy, which is sent to cities. There, it heats, cools, and lights the buildings in which people work and live. Meanwhile, some of the water in the lake behind the dam evaporates and is free to repeat the cycle. Hence, the energy from the sunlight on a lake can undergo many transformations and help provide the moving force for many natural and human-made processes.

---

or inches, we measure the wavelength of radiation in terms of **micrometers** (represented by the symbol μm). One micrometer is equal to a millionth of a meter:

$$1 \text{ micrometer } (\mu m) = 0.000001 \text{ m} = 10^{-6} \text{ m}.$$

The average wavelength of visible light is about 0.5 μm. To give you a common object for comparison, the average height of a letter on this page is about 2000 μm, or 2 millimeters.

Radiation and Temperature *All things, no matter how big or small, emit radiation.* This book, your body, flowers, trees, air, the earth, the stars are all radiating a wide range of electromagnetic waves. The energy originates from rapidly vibrating electrons, billions of which exist in every object.

The wavelengths that each object emits depend primarily on the object's temperature. The higher the temperature, the faster the electrons vibrate, and the shorter are the wavelengths of the emitted radiation. This can be visualized by attaching one end of a rope to a post and holding the other end. If the rope is shaken rapidly (high temperature), numerous short waves travel along the rope; if the rope is shaken slowly (lower temperature), longer waves appear on the rope. Although objects at a temperature of about 500°C (950°F) radiate many waves, some of them are short enough to stimulate the sensation of vision. We actually see these objects glow red. Objects cooler than this radiate at wavelengths that are too long for us to see. The page of this book, for example, is radiating electromagnetic waves. But because its temperature is only around 20°C (68°F), the waves emitted are much too long to stimulate vision. Yet, we can see the paper! Where do these visible waves originate? What we see are light waves that have *bounced off* the paper. These waves emanate from glowing or incandescent objects. Such sources include light bulbs, lamps, and the sun. If this book were carried into a completely dark room, it would continue to radiate, but the pages would appear black because there are no visible light waves in the room to reflect off the pages.

Objects that have a very high temperature emit energy at a greater rate or intensity than objects at a lower temperature. Thus, as the temperature of an object increases, more total radiation is emitted each second. This can be expressed mathematically as

$$R \sim T^4 \quad \text{(Stefan-Boltzmann law)*}$$

where R is the maximum rate of radiation emitted by each square centimeter of surface area of the object and T is the object's surface temperature in degrees Kelvin. This relationship, called the *Stefan-Boltzmann law* after Josef Stefan (1835–1893) and Ludwig Boltzmann (1844–1906), who derived it, states that all objects with temperatures above absolute zero (0 K or $-273°C$) emit radiation at a rate proportional to the fourth power of their absolute temperature. Consequently, a small increase in temperature results in a large increase in the amount of radiation emitted because doubling the absolute temperature of an object increases the maximum energy output by a factor of 16, which is 2^4.

Radiation of the Sun and Earth   Most of the sun's energy is emitted from its surface, where the temperature is nearly 6000 K (10,500°F). The earth, on the other hand, has an average surface temperature of 288 K (15°C, 59°F). The sun, therefore, radiates a great deal more energy than does the earth (see Fig. 3.5). At what wavelengths do the sun and the earth radiate most of their energy? Fortunately, the sun and the earth both have characteristics (discussed in a later section) that enable us to use the following relationship discovered by the German physicist Wilhelm Wien (1864–1928):

$$\lambda_{max} = \frac{\text{constant}}{T} \quad \text{(Wien's law)},$$

where λ_{max} is the wavelength in micrometers at which maximum energy emission occurs, T is the object's temperature in kelvins, and the constant is 2897 µm K. To make the numbers easy to deal with, we will round off the constant to the number 3000.

For the sun, with a surface temperature of 6000 K (10,500°F), the equation becomes

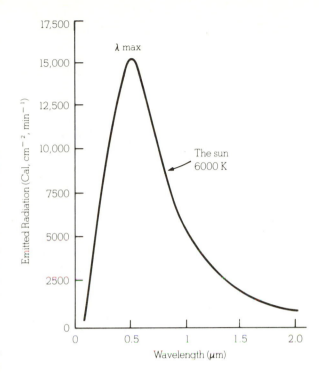

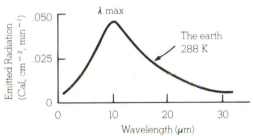

Fig. 3.5   Radiant energy emitted from a square centimeter of area of the surface of the sun (6000 K) and the earth (288 K). The hotter sun not only radiates much more energy than the cooler earth, it does so at shorter wavelengths.

$$\lambda_{max} = \frac{3000 \text{ µm K}}{6000 \text{ K}} = 0.5 \text{ µm}.$$

Thus, the sun emits a maximum amount of radiation at wavelengths near 0.5 µm. The cooler earth, with a surface temperature near 300 K (81°F), emits maximum radiation near wavelengths of 10 µm, since

$$\lambda_{max} = \frac{3000 \text{ µm K}}{300 \text{ K}} = 10 \text{ µm}.$$

Thus, the earth emits most of its radiation at longer wavelengths than does the sun. For this reason,

*The exact law is written as $R = \sigma T^4$, where σ is the Stefan-Boltzmann constant and is equal to 5.67×10^{-5} ergs/cm² K⁴ sec. In SI units, the constant is 5.67×10^{-8} W/m² K⁴.

Table 3.2 Colors that Correspond to Various Wavelengths of Radiation

COLOR*	WAVELENGTH RANGE (μm)	TYPICAL WAVELENGTH (μm)
Violet	0.40 to 0.44	0.42
Blue	0.45 to 0.49	0.48
Green	0.50 to 0.53	0.52
Yellow	0.54 to 0.58	0.56
Orange	0.59 to 0.64	0.60
Red	0.65 to 0.70	0.68

*Notice that violet is the shortest wavelength of visible light and red is the longest.

the earth's radiation is often called **longwave radiation**, whereas the sun's energy is referred to as **shortwave radiation**.

Wien's law demonstrates that, as the temperature of an object increases, the wavelength at which maximum emission occurs is shifted toward shorter values. For example, if the sun's surface temperature were to double to 12,000 K, its wavelength of maximum emission would be halved to about 0.25 μm. If, on the other hand, the sun's surface cooled to 3000 K, it would emit its maximum amount of radiation near 1.0 μm.

Even though the sun radiates at a maximum rate at one wavelength, it nonetheless emits some radiation at almost all other wavelengths. If we look at the amount of radiation given off by the sun at each wavelength, we obtain the sun's *elec-*

tromagnetic spectrum. A portion of this spectrum is shown in Fig. 3.6.

Our eyes are sensitive to radiation between 0.4 and 0.7 μm. These waves reach the eye and stimulate the sensation of color (see Table 3.2). This portion of the spectrum is referred to as the **visible region**. The sun emits nearly 44 percent of its radiation in this zone, with the peak of energy output found at the wavelength corresponding to the color blue-green. The color violet is the shortest wavelength of visible light. Wavelengths shorter than violet (0.4 μm) are **ultraviolet**. X-rays and gamma rays with exceedingly short wavelengths fall into this category. The sun emits only about 7 percent of its total energy at ultraviolet wavelengths.

The longest wavelengths of visible light correspond to the color red. Wavelengths longer than red (0.7 μm) are **infrared**. These waves cannot be seen by humans, but sometimes can be felt as heat. Nearly 37 percent of the sun's energy is radiated between 0.7 μm and 1.5 μm, with only 12 percent radiated at wavelengths longer than 1.5 μm.

Whereas the hot sun emits only about half of its energy in the infrared portion of the spectrum, the relatively cool earth emits almost all of its energy at infrared wavelengths. Although we cannot see infrared radiation, there are instruments called *infrared sensors*, that can. Weather satellites that orbit the globe use these sensors to observe radiation emitted by the earth, the clouds,

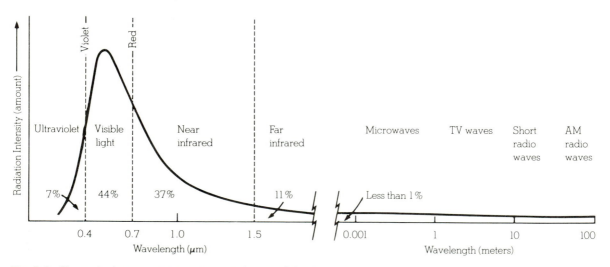

Fig. 3.6 The sun's electromagnetic spectrum and some of the descriptive names of each region. The numbers underneath the curve represent the percent of energy the sun radiates in various regions.

and the atmosphere. Since objects of different temperatures radiate their maximum energy at different wavelengths, infrared photographs can distinguish between objects of different temperatures. Clouds always radiate infrared energy. Thus, cloud pictures using infrared sensors can be taken during both day and night.

In summary, both the sun and earth emit radiation. The hot sun (6000 K) radiates nearly 88 percent of its energy at wavelengths less than 1.5 μm, with maximum emission near 0.5 μm. The cooler earth (288 K) radiates nearly all its energy between 1 and 40 μm with a peak intensity near 10 μm (see Fig. 3.5). The sun's surface is nearly 20 times hotter than the earth's surface. From the Stefan-Boltzmann relationship, this fact means that a unit area on the sun emits nearly 160,000 (20^4) times more energy during a given time period than the same size area on the earth. And since the sun has such a huge surface area from which to radiate, the total energy emitted by the sun each minute amounts to a staggering 6 billion, billion, billion calories!

Balancing Act—Absorption, Emission, and Equilibrium

If the earth and all things on it are continually radiating energy, why doesn't everything get progressively colder? The answer is that all objects not only radiate energy, they absorb it as well. If an object radiates more energy than it absorbs, it gets colder; if it absorbs more energy than it emits, it gets warmer. On a sunny day, the earth's surface warms by absorbing more energy from the sun and the atmosphere than it radiates, while at night the earth cools by radiating more energy than it absorbs from its surroundings. When an object emits and absorbs energy at equal rates, its temperature remains constant.

The rate at which something radiates and absorbs energy depends strongly on its surface characteristics, such as color, texture, and moisture, as well as temperature. For example, a black object in direct sunlight is a good absorber of radiation. It converts energy from the sun into internal energy, and its temperature ordinarily increases. You need only walk barefoot on a black asphalt road on a summer afternoon to experience this. The asphalt road can be used to illustrate an important general principle: *Good absorbers are also good emitters of radiation.* At night, the blacktop road will cool quickly and, by early morning, will be cooler than surrounding surfaces.

Any object that is a perfect absorber (that is, absorbs all the radiation that strikes it) and a perfect emitter (emits all possible radiation) is called a **black body**. Black bodies do not have to be colored black, they simply must absorb and emit all possible radiation. Since the earth's surface and the sun absorb and radiate with nearly 100 percent efficiency for their respective temperatures, they both behave as black bodies. This is the reason we were able to use Wien's law and the Stefan-Boltzmann law to determine the characteristics of radiation emitted from the sun and the earth.

The atmosphere, on the other hand, *does not* behave like a black body, as it absorbs some wavelengths of radiation and is transparent to others. Because the atmosphere both selectively absorbs and emits radiation, it is known as a **selective absorber**. Let's examine selective absorbers more closely.

Selective Absorbers   A substance that is a good absorber of *certain* wavelengths may not be a good absorber of *all* wavelengths. Just as some people are selective eaters of certain foods, most substances in our environment are selective absorbers; that is, they absorb only certain wavelengths of radiation. Glass is a good example of a selective absorber in that it absorbs some of the infrared and ultraviolet radiation it receives, but not the visible radiation that is transmitted through the glass. As a result, it is difficult to get a sunburn through the windshield of your car, although you can see through it.

Objects that selectively absorb radiation also selectively emit radiation at the same wavelength. This phenomenon is called **Kirchhoff's law**. This law states that *good absorbers are good emitters at a particular wavelength, and poor absorbers are poor emitters at the same wavelength.*

Snow is a good absorber as well as a good emitter of infrared energy. The bark of a tree absorbs sunlight and emits infrared energy, which the snow around it absorbs. During the absorption

Wave Energy

Standing close to a fire makes us feel warmer than we do when we stand at a distance from it. Does this mean that, as we move away from a hot object, the waves carry less energy and are, therefore, weaker? Not really. The intensity of radiation decreases as we move away from a hot object because radiating energy spreads outward in all directions. Figure 2 illustrates that, as the distance from a radiating object increases, a given amount of energy is distributed over a larger area, so that the energy received per unit of area and per unit of time decreases. In fact, at twice the distance from the source, the radiation is spread over 4 times the area.

This phenomenon raises another question: Do all wavelengths of radiation carry the same amount of energy? The answer is emphatically NO! Shorter waves carry much more energy than do longer waves. When comparing the energy carried by various waves, it is useful to give electromagnetic radiation characteristics of particles in order to "explain" some of the wave's behavior. We might think of radiation as streams of particles, or **photons**, which are discrete packets of energy.

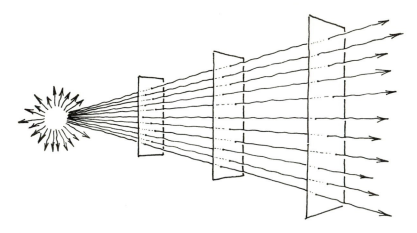

Fig. 2 The intensity, or amount, of radiant energy transported by electromagnetic waves decreases as we move away from a radiating object because the same amount of energy is spread over a larger area.

Short wavelength photons carry a great deal more energy than long wavelength photons. Short ultraviolet wavelengths, less than 0.32 μm, have enough energy per photon to produce sunburns and penetrate skin tissues, sometimes causing skin cancer. Oddly enough, these same wavelengths activate provitamin D in the skin and convert it into vitamin D, which is essential to health. Longer ultraviolet wavelengths, between 0.33 and 0.4 μm, are less energetic, but can still tan the skin. Fortunately for human existence, virtually all the ultraviolet radiation at wavelengths below about 0.3 μm is absorbed in the upper atmosphere. Very little is able to reach the earth's surface. Photons with wavelengths near 0.26 μm carry enough energy to cause chromosome mutations. If these waves were able to reach the earth's surface, the effects on plant and animal life could be drastic. In Chapter 4, we will see that it is primarily the gases oxygen and ozone in the upper atmosphere that protect us from these harmful rays.

process, the infrared radiation is converted into heat energy, and the snow melts outward away from the tree trunk, producing a small depression that encircles the tree (see Fig. 3.7).

At night, a snow surface usually emits much more infrared energy than it absorbs from its surroundings. This large loss of infrared energy (coupled with the insulating qualities of snow) cause the air above a snow surface on a clear, winter night to become extremely cold. This also explains why the temperature drops so suddenly after the season's first snowfall.

Gases are also selective absorbers. Ozone, for example, selectively absorbs ultraviolet radiation, especially at wavelengths between 0.2 and 0.3 μm, as well as infrared radiation at 9.6 μm. Mo-

Fig. 3.7 The melting of snow outward from the trees causes small depressions to form. The melting is caused mainly by the snow's absorption of the infrared energy being emitted from the trees.

lecular oxygen absorbs ultraviolet energy below a wavelength of 0.2 μm. Both of these gases absorb this radiation in the upper part of the atmosphere at levels above 10 km (6 mi). Below this level, water vapor and carbon dioxide (CO_2) are strong selective absorbers of infrared radiation. In Fig. 3.8, observe that water vapor is the most important absorber in that it strongly absorbs infrared energy at wavelengths between 1 and 8 μm, and also at wavelengths longer than about 12 μm. The gas CO_2 absorbs infrared radiation at 4 μm and between about 13 and 17 μm. Neither water vapor nor CO_2 readily absorb wavelengths between 8 and 11 μm. Because these wavelengths of emitted energy pass upward through the atmosphere and out into space, this wavelength range (between 8 and 11 μm) is known as the **atmospheric window.***

*There is some absorption of radiation at 9.6 μm by ozone in the upper atmosphere.

We know that the earth radiates most of its energy at wavelengths between 1 and 40 μm. Therefore, a large portion of this emitted energy is absorbed by water vapor and CO_2, which are abundant in the lower atmosphere. As these gases absorb this radiation, they gain kinetic energy (energy of motion). The gas molecules share this energy by colliding with neighboring air molecules, such as oxygen and nitrogen. These collisions increase the average kinetic energy of the air, which results in an increase in air temperature. So some of the infrared energy emitted from the earth's surface warms the lower atmosphere.

Besides being selective absorbers, water vapor and CO_2 selectively emit radiation at infrared wavelengths. This radiation travels away from these gases in all directions. A portion of this energy is radiated toward the earth's surface and absorbed, thus heating the ground. The earth, in turn, reradiates energy upward, where it is absorbed and warms the lower atmosphere. In this way, water

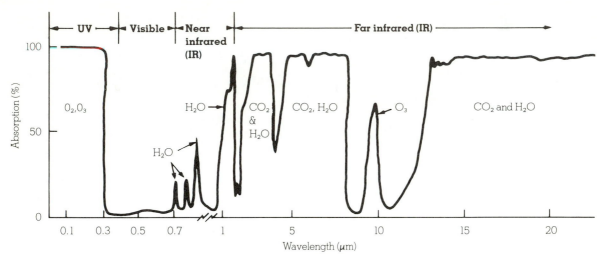

Fig. 3.8 Absorption of radiation by gases in the atmosphere.

vapor and CO_2 absorb and reradiate energy and act as an insulating layer around the earth, keeping part of the earth's infrared radiation from escaping rapidly into space. Consequently, the earth's surface and the lower atmosphere are much warmer than they would be if these selective absorbers were not present (Fig. 3.9a). In fact, the mean global temperature without CO_2 and water vapor would be around $-20°C$ ($-4°F$), or about $35°C$ ($63°F$) lower than at present.

The absorption characteristics of water vapor and CO_2 in the air were, at one time, thought to be similar to the glass of a florist's greenhouse. In a greenhouse, the glass allows visible radiation to come in, but inhibits to some degree the passage of outgoing infrared radiation. For this reason, the behavior of the water vapor and CO_2 in the atmosphere is commonly called the **greenhouse effect**. However, studies have shown that the warm air inside a greenhouse is probably caused more by the air's inability to circulate and mix with the cooler outside air, rather than by the entrapment of infrared energy. Because of these findings, the greenhouse effect is now called the **atmosphere effect** by meteorologists.

Tiny liquid cloud droplets are also selective absorbers. They are good absorbers of most wavelengths of infrared radiation, but poor absorbers of visible radiation. Low clouds with a thickness of over 100 m (300 ft) are effective absorbers of outgoing infrared radiation. They even absorb the wavelengths between 8 and 11 μm, which are

otherwise "passed up" by water vapor and CO_2. Thus, they have the effect of closing the atmospheric window.

So, the next time you are outside on a cloudy night, try to picture the billions of tiny cloud droplets absorbing much of the radiation that is emitted by the earth and everything on it. Even your body's radiation is being absorbed. Because these droplets are good emitters of infrared energy, they radiate energy back to earth where it is reabsorbed and, in a sense, reradiated back to the clouds. This process keeps calm, cloudy nights warmer than calm, clear ones (Fig. 3.9b). If the clouds remain into the next day, they prevent much of the sunlight from reaching the ground. Since the ground does not heat up as much as it would in full sunshine, cloudy, calm days are normally cooler than clear, calm days. In summary, the presence of clouds tends to keep nighttime temperatures higher and daytime temperatures lower.

On a cloudy day, what happens to the sunlight that is not absorbed by the clouds? Well, some of the radiation may bounce off the cloud's surface. This process is called **reflection**.

Reflected Radiation White or shiny objects reflect sunlight from their surfaces. An object that reflects a great deal of sunlight absorbs very little. This can be illustrated by placing thermometers inside two identical houses, one painted white, the other black. If both houses are in direct sunlight, the white house will heat up more slowly because it is a better reflector (and a poorer absorber) of solar radiation.

Clean, white snow reflects as much as 95 percent of the solar radiation that strikes it. Most of this energy is in the visible and ultraviolet wavelengths. Reflected radiation, coupled with direct sunlight, can produce severe sunburns on the exposed skin of unwary snow skiers. Excessive exposure to light reflected from snow can also cause *snow blindness*, a condition of temporary visual impairment.

Notice in Fig. 3.10 that water surfaces reflect only a small amount of solar energy. For an entire day, a smooth water surface will reflect an average of only 10 percent of the incident sunlight. Calm water reflects sunlight least (absorbs best) around noon when the sun is high in the sky. At this time calm water reflects only about 2 percent of the incoming radiation. Water reflects sunlight best (absorbs least) when the sun is low on the horizon and the water is a little choppy. This may explain why people who wear brimmed hats while fishing from a boat in choppy water on a sunny day can still get sunburned during the midmorning or midafternoon.

When sunlight strikes very small objects, such as air molecules and dust particles, the light itself is deflected in all directions—forward, sideways, and backwards. The distribution of light in this manner is called **scattering**. Because air molecules are much smaller than the wavelengths of visible light, they are more effective scatterers of the shorter (blue) wavelengths than the longer (red) wavelengths. Hence, when we look away from the direct beam of sunlight, blue light strikes our eyes from all directions, turning the daytime sky blue. At midday, all the wavelengths of visible light from the sun strike our eyes, and the sun is perceived as white. At sunrise and sunset, when the white beam of sunlight must pass through a thick portion of the atmosphere, scattering by air molecules and fine particles of dust removes the blue light, leaving the longer wavelengths of red, orange, and yellow to pass on through, creating the image of a ruddy or yellowish sun.

Albedo   The percent of radiation returning from a surface compared to that which strikes it is called the **albedo** of the surface. Albedo then represents the reflectivity of the surface. We have seen that clean, white snow can reflect up to 95 percent of the solar radiation that reaches it. Such snow, therefore, can have an albedo as high as 95 percent. The albedo of a water surface depends upon the angle at which the sunlight strikes it and whether the surface is smooth or rough.

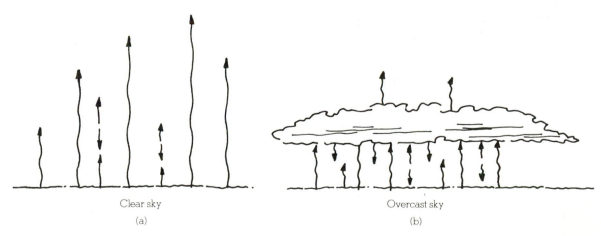

Clear sky
(a)

Overcast sky
(b)

Fig. 3.9 The atmosphere (greenhouse) effect. (a) In clear weather, water vapor and CO_2 absorb a portion of the earth's infrared energy and reradiate it back to the surface. This helps to keep air temperatures near the surface higher than they otherwise would be. (b) In cloudy weather, water vapor, CO_2, and liquid cloud droplets absorb the majority of the earth's infrared energy and reradiate a great deal of it back to the surface. In the absence of wind, this keeps cloudy nights warmer than clear nights.

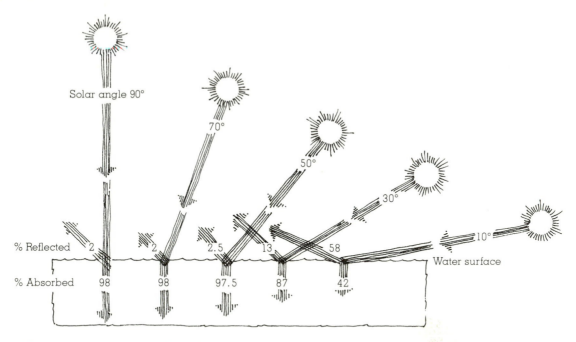

Fig. 3.10   The amount of sunlight reflected from water decreases as the sun increases in elevation above the horizon.

Most grassy fields, plowed fields, and rocky areas have albedos of between 10 and 30 percent. Thick clouds have a higher albedo than thin clouds. On the average, the albedo of clouds is near 60 percent. The albedo of the moon is about 7 percent, which means that out of all the solar radiation that hits the moon's surface, only 7 percent is reflected. What does this fact tell us about the surface color of most moon rocks? It indicates that, because moon rocks reflect so little sunlight, they must be good absorbers; hence, they tend to be dark in color.

Out of 100 percent of the solar radiation reaching the top of the atmosphere each year, nearly 20 percent is reflected upward by clouds and 6 percent is scattered back to space by the atmosphere. Only about 4 percent is reflected from the earth's relatively dark surface. Therefore, on an annual global average, the earth and its atmosphere redirect about 30 percent of the sun's incoming energy back to space. This gives the earth and its atmosphere a combined albedo of 30 percent. Table 3.3 summarizes the idea of reflectivity by showing the albedo of various surfaces.

In the last several sections, we have explored examples of some of the ways solar radiation is absorbed, emitted, and reflected by various objects. Before reading the next section, review a few important facts and principles:

1. An object that is a good absorber at some wavelength is also a good emitter at that same wavelength.
2. Water vapor, CO_2, and cloud droplets are selective absorbers that strongly absorb and emit infrared energy.
3. Because the earth's surface behaves as a black body, it is a much better absorber and emitter of radiation than is the atmosphere.
4. Opaque surfaces with a high albedo (high reflectivity) at some wavelength are generally poor absorbers at that same wavelength.
5. The earth and its atmosphere have a combined albedo, which averages near 30 percent.

Warming the Air from Below   On a clear day, solar energy—which passes through the lower atmosphere—is absorbed by the ground, warming it (Fig. 3.11). Air molecules in contact with the heated surface bounce against it, gain energy by conduction, then shoot upward like freshly popped kernels of corn, carrying their energy with them.

Table 3.3 Typical Albedo of Various Surfaces

SURFACE OR OBJECT	ALBEDO (percent)
Fresh snow	75 to 95
Clouds (thick)	60 to 90
Clouds (thin)	30 to 50
Venus	78
Ice	30 to 40
Sand	15 to 45
Earth and atmosphere	30
Mars	17
Grassy field	10 to 30
Dry, plowed field	5 to 20
Water	10*
Forest	3 to 10
Moon	7

*Daily average.

Because the air near the ground is very dense, these molecules only travel a short distance (about 10^{-7} m) before they collide with other molecules. During the collision, these more rapidly moving molecules share their energy with less energetic molecules, raising the average temperature of the air. But air is such a poor heat conductor that this process is only important within a few centimeters of the ground.

As the surface air warms, it actually becomes less dense than the air directly above it. The warmer air rises and the cooler air sinks, setting up thermals, or *free convection cells* that transfer heat upward and distribute it through a deeper layer of air. The rising air expands and cools, and, if sufficiently moist, the water vapor condenses into cloud droplets, releasing latent heat, which warms the air. Meanwhile, the earth constantly emits infrared energy, which is being absorbed and re-emitted by water vapor and carbon dioxide (atmosphere, or greenhouse, effect). Since the concentration of these gases decreases rapidly above the earth, most of the absorption occurs in a layer near the surface. Hence, the lower atmosphere is heated from below.

The Earth's Annual Energy Balance

Although the average temperature at any one place may vary considerably from year to year, the earth's overall average temperature changes little from one year to the next. This fact indicates that, each year, the earth and its atmosphere combined must send off into space just as much energy as they receive from the sun. The same type of balance must exist between the earth's surface and the atmosphere. That is, each year, the earth's surface must return to the atmosphere the same

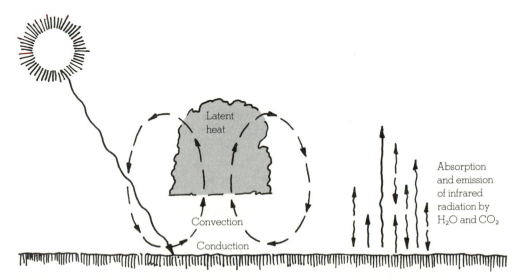

Fig. 3.11 Air in the lower atmosphere is heated from below. Sunlight warms the ground, and the air above is warmed by conduction, convection, and radiation. Further heating occurs during condensation as latent heat is given up to the atmosphere.

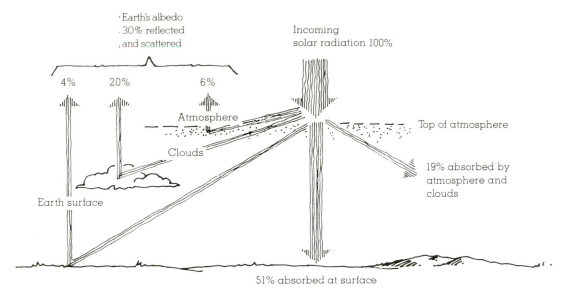

Fig. 3.12   On the average, of all the solar energy that reaches the earth's atmosphere, about 30 percent is reflected and scattered back to space, 19 percent is absorbed by the atmosphere and clouds, and 51 percent is absorbed at the surface.

amount of energy that it absorbs. If this did not occur, the earth's average surface temperature would change. How does the earth and its atmosphere maintain this yearly energy balance?

Suppose 100 units of solar energy reach the top of the earth's atmosphere each year, as shown in Fig. 3.12. We already know that clouds, the earth, and the atmosphere reflect and scatter 30 units back to space. The figure shows that the atmosphere and clouds together absorb 19 units, leaving 51 units to be absorbed by the earth's surface. Let's see what other energy the atmosphere and earth absorb. (You will find it helpful to refer continually to Fig. 3.13 as you read this section. Also, keep in mind that the numbers used in Figs. 3.12 and 3.13 are approximations based on satellite data and ground observations. While the actual value of each process may vary by several percent, it is the relative size of the numbers that is important in each diagram.)

Since water covers nearly three-fourths of the earth, it is estimated that as much as 23 units of the solar energy absorbed at the surface are used to evaporate water. These 23 energy units lost by the surface as latent heat are later gained by the atmosphere when the water vapor condenses into liquid (or ice) and the latent heat is released. Hence,

nearly 45 percent of the solar energy reaching the earth's surface is used to evaporate water. Over land, the warm ground heats the air directly above it by conduction. The heated air rises (convection) and redistributes heat to the atmosphere. The earth's surface loses about 7 units of energy through conduction and convection, while these same 7 units are gained by the atmosphere. So far, then, we have the atmosphere gaining 30 units of energy from the earth's surface.

The earth's surface continually radiates infrared energy upward (117 units).* Only a small fraction of this earth radiation (6 units) passes through the atmosphere and goes into space. The remaining infrared energy (111 units) is absorbed mainly by water vapor, CO_2, and clouds; and much of it (96 units) is, in effect, reradiated back to earth. We now have a complex situation. The earth's surface not only absorbs energy from the sun during the day, but it absorbs infrared energy from its own atmosphere both day and night. Up to this point, the total amount of energy gained by the atmosphere amounts to:

*Keep in mind that the earth's surface receives more than 100 units of energy because it receives solar energy from the sun, plus infrared energy from the atmosphere.

19 units of absorbed solar radiation
23 units by the release of latent heat
7 units by conduction and convection
111 units by absorbed infrared earth radiation
_____
Total 160 units of energy gained by the atmosphere

The atmosphere absorbs a total of 160 units of energy. Of these, the atmosphere radiates 96 units back to earth and 64 units into space. As we can see, the amount of radiation lost by the atmosphere is also 160 units.

At the top of the earth's atmosphere, the earth-atmosphere system is losing a total of 70 units of energy: 64 units radiated by the atmosphere and 6 units radiated from the earth's surface. This amount of energy (70 units) is the same gained from the sun each year by the earth-atmosphere system (51 units absorbed at the earth's surface and 19 units absorbed by the atmosphere). Averaged for a year over the entire earth, incoming energy from the sun exactly balances the outgoing energy at the top of the atmosphere.

There must be a similar yearly balance occurring

between incoming and outgoing energy at the earth's surface. The earth's surface gains 51 units of solar energy and 96 units of reradiated infrared energy from its atmosphere and clouds. The earth's surface absorbs a yearly total of 147 units. This same surface loses:

7 units by conduction and convection
23 units by evaporation
117 units by infrared radiation
_____
Total 147 units lost at surface

Therefore, at the earth's surface, the amount of energy gained each year (147 units) exactly balances the energy lost (147 units).

At first, it may seem odd that the earth has to radiate away more energy than it receives from the sun in order to maintain a heat balance. But this process makes sense when we consider that the earth receives a great deal of infrared energy from its own atmosphere. This energy is received 24 hours a day, while solar energy is only received at a given area during the daylight hours.

And so, the earth and the atmosphere absorb energy from the sun, as well as from each other.

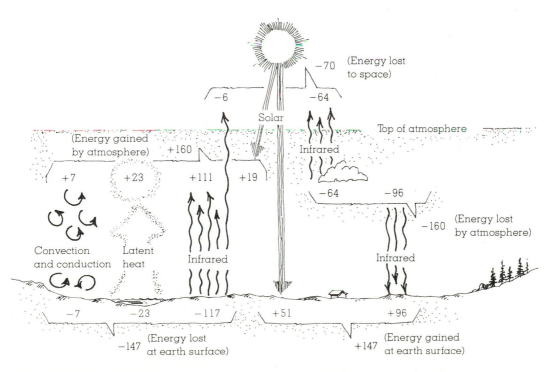

Fig. 3.13 The earth-atmosphere energy balance. (Numbers represent approximations based on surface observations and satellite data.)

In all of the energy exchanges, a delicate balance is maintained. Essentially, there is no yearly gain or loss of total energy, and the average temperature of the earth and the atmosphere remains fairly constant from one year to the next. This equilibrium does not imply that the earth's average temperature does not change. It probably does. But the changes are small from year to year—usually less than one-tenth of a degree Celsius—and become significant only when measured over many years.

Summary

In this chapter, we have seen how the concepts of heat and temperature differ, and how heat is transferred in our environment. We learned that conduction is the transfer of heat by molecular collisions and is most effective in solids. Because air is a poor heat conductor, conduction in the atmosphere is only important in the shallow layer of air in contact with the earth's surface. A more important process of atmospheric heat transfer is convection, which involves the mass movement of air (or any fluid) with its energy from one region to another. Another significant heat transfer process is radiation—the transfer of energy by means of electromagnetic waves.

The hot sun emits most of its radiation as short-wave radiation. A portion of this energy heats the earth, and the earth, in turn, warms the air above. The cool earth emits most of its radiation as long-wave infrared radiation. Selective absorbers in the atmosphere, such as water vapor and carbon dioxide, absorb some of the earth's radiation and reradiate a portion of it back to the surface, where it warms the surface, producing the atmosphere (greenhouse) effect. Because clouds are both good absorbers and good emitters of infrared radiation, they enhance the atmosphere effect and keep calm, cloudy nights warmer than calm, clear nights. The average temperature of the earth and the atmosphere remains fairly constant from one year to the next because the amount of energy they absorb each year is equal to the amount of energy they lose.

Questions for Review

1. Describe some of the transformations energy goes through as water evaporates from the ocean.

2. Distinguish between temperature and heat.

3. List and then explain the various ways heat is transferred in the atmosphere.

4. In the study of meteorology how does advection differ from convection?

5. How is latent heat an important source of atmospheric energy?

6. How does the temperature of an object influence the radiation that it emits?

7. How does the amount of radiation emitted by the earth differ from that emitted by the sun?

8. How do the wavelengths of most of the radiation emitted by the sun differ from those emitted by the surface of the earth?

9. Which photon carries the most energy—infrared, visible, or ultraviolet?

10. Explain how the atmosphere near the earth's surface is heated from below.

11. If the earth's surface continually radiates energy, why doesn't it become colder and colder?

12. What are some of the selective absorbers of radiation in the atmosphere?

13. Explain why snow usually melts around and away from tree trunks.

14. Explain how the earth's atmosphere (greenhouse) effect works on a(a) clear night, (b) cloudy night.

15. Why is the average albedo of the earth's surface only about 4 percent?

16. How is the earth able to maintain its average temperature from one year to the next?

Questions for Thought

1. Explain why the bridge in the diagram is the first to become icy.

2. Explain why the first snowfall of the winter usually "sticks" better to tree branches than to bare ground.

3. At night, why do materials that are poor heat conductors cool to temperatures less than the surrounding air?

4. Explain how ice can form on puddles (in shaded areas) when the temperature above and below the puddle is slightly above freezing.

5. In northern latitudes, the oceans are warmer in summer than they are in winter. In which season do the oceans lose heat most rapidly to the atmosphere by conduction? Explain.

6. How is heat transferred away from the surface of the moon? (Hint: The moon has no atmosphere.)

7. Why is ultraviolet radiation more successful in dislodging electrons from air atoms and molecules than visible radiation?

8. Why must you stand closer to a small fire to experience the same warmth you get when standing further away from a large fire?

9. Imagine you are a photon of infrared radiation that has just been emitted from the earth. Describe your journey as you travel into the earth's atmosphere on a cloudy night.

10. Which will show the greatest increase in temperature when illuminated with direct sunlight: a plowed field or a blanket of snow? Explain.

11. Why does the surface temperature often increase on a clear, calm night as a low cloud moves overhead?

12. If the amount of cloud cover around the earth were to increase sharply, what effect would this have on the earth–atmosphere energy balance?

13. Explain why an increase in cloud cover surrounding the earth would increase the earth's albedo, yet not necessarily lead to a cooler earth surface temperature.

14. If all of the water vapor, CO_2, and ozone were removed from the earth's atmosphere, how would this influence the earth's average temperature? Explain.

Problems and Exercises

1. Suppose that 500 g of water vapor condense to make a cloud about the size of an average room. If we assume that the latent heat of condensation is 600 cal/g, how much heat would be released to the air? If the total mass of air before condensation is 100 kg, how much warmer would the air be after condensation? Assume that the air is not undergoing any pressure changes. (Hint: Use the specific heat of air in Table 1, p. 43.)

2. Suppose planet A is exactly twice the size (in surface area) of planet B. If both planets have the same exact surface temperature (1500 K), which planet would be emitting the most radiation? Determine the wavelength of maximum energy emission of both planets, using Wien's law.

3. Suppose, in question 2, the temperature of planet B doubles. (a) What would be its wavelength of maximum energy emission? (b) In what region of the electromagnetic spectrum would this wavelength be found? (c) If the temperature of planet A remained the same, determine which planet (A or B) would now be emitting the most radiation (use the Stefan-Boltzmann relationship). Explain your answer.

The shimmering bands of the aurora borealis illuminate the night sky against a background of stars in northern Alaska. (Photo: National Oceanic and Atmospheric Administration)

CHAPTER 4

The Sun's Influence on the Upper Atmosphere

At high latitudes after darkness has fallen, a faint, white glow may often be seen near the horizon. Sometimes, this light moves across the sky as a yellow-green arc much wider than a rainbow. On other occasions, the sky becomes decorated with draperies of blue, green, and even purple light, which softly sweep across the sky, constantly changing in form and location, as if blown by a gentle breeze. In the Northern Hemisphere, this light show is the *aurora borealis*, or northern lights. What causes this atmospheric spectacle and why is it usually seen in northern latitudes?

In this chapter, we will address other questions as well. Why, for example, is the maximum concentration of ozone found in the stratosphere? What prevents ultraviolet radiation with wavelengths shorter than 0.3 μm from reaching the earth's surface? And what effect does the absorption of high-energy solar radiation have on the temperature structure and gaseous composition of the atmosphere?

The answers to these questions lie with the sun and its influence on the upper part of the atmosphere.

The Sun

The sun is our nearest star. It is some 150 million km (93 million mi) from earth. The next star, Alpha Centauri, is more than 250,000 times further away. Even though the earth only receives about one two-billionths of the sun's total energy output, it is this energy that allows life to flourish. Sunlight determines the rate of photosynthesis in plants and strongly regulates the amount of evaporation from the oceans. It warms this planet and drives the atmosphere into the dynamic patterns we experience as everyday wind and weather. Without

Contents

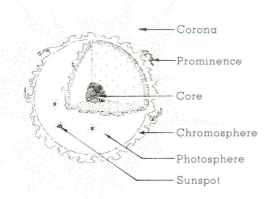

Fig. 4.1 Various regions of the sun.

the sun's radiant energy, the earth would gradually cool, in time becoming encased in a layer of ice! Evidence of life on the cold, dark, and barren surface would be found only in fossils. Fortunately, the sun has been shining for billions of years, and it is likely to shine for at least several billion more.

The sun is a giant celestial furnace. Its core is extremely hot, with a temperature estimated to be near 15 million degrees Celsius. In the core, hydrogen nuclei (protons) collide at such fantastically high speeds that they fuse together to form helium nuclei. This thermonuclear process generates an enormous amount of energy, which gradually works its way to the sun's outer luminous surface—the *photosphere* ("sphere of light"). Temperatures here are much cooler than in the interior, generally near 6000°C. We have noted already that a body with this surface temperature emits radiation at a maximum rate in the visible region of the spectrum. The sun is, therefore, a shining example of such an object.

Dark blemishes on the photosphere called *sunspots* are huge, cooler regions that typically average more than 5 times the diameter of the earth. Although sunspots are not well understood, they are known to be regions of strong magnetic fields. They are cyclic, with the maximum number of spots occurring approximately every 11 years. Recent measurements from satellites reveal that during a sunspot maximum, less radiation actually leaves the photosphere. Sunspots, apparently repress the convective outflow of energy from inside the sun.

Above the photosphere are the *chromosphere* and the *corona*. The chromosphere ("color sphere") acts as a boundary between the relatively cool (6000°C) photosphere and the much hotter (2,000,000°C) corona, the outermost envelope of the solar atmosphere. During a solar eclipse, the corona is visible. It appears as a pale, milky cloud encircling the sun (Fig. 4.2). Although

Fig. 4.2 The solar corona during a solar eclipse at Pulacoya, Bolivia.

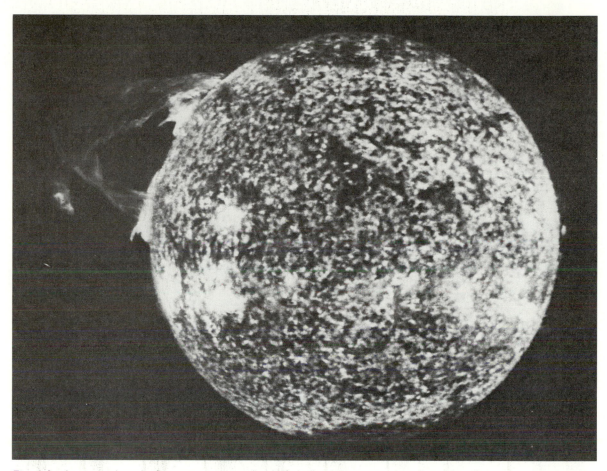

Fig. 4.3 A spectacular solar flare spanning more than half a million kilometers across the solar surface.

much hotter than the photosphere, the corona radiates much less energy because its density is extremely low. This very thin solar atmosphere extends into space for many millions of kilometers.*

Violent solar activity occasionally occurs in the regions of sunspots. The most dramatic of these events are *prominences* and *flares*. Prominences are huge cloudlike jets of gas that often shoot up into the corona in the form of an arch. Solar flares are tremendous, but brief, eruptions. They emit large quantities of high-energy ultraviolet radia-

---

*During a solar eclipse or at any other time, you should *not* look at the sun's corona either with sunglasses or through exposed negatives. Take this warning seriously. Much of the energy from this hot region is ultraviolet radiation. Viewing the corona directly permits large amounts of this radiation to enter the eye, causing serious and permanent damage to the retina. View the sun by projecting its image onto a sheet of paper, using a telescope or pinhole camera.

tion, as well as energized charged particles, mainly protons and electrons, which stream outward away from the sun at extremely high speeds. A huge flare on May 21, 1980, lasted only 40 minutes but covered more than 5.2 billion km² (2 billion mi²) of the sun's surface.

Solar Wind and the Magnetosphere

In the outermost region of the corona, there is a continuous discharge of particles. With temperatures millions of degrees Celsius, gases become stripped of electrons by violent collisions and acquire enough speed to escape the gravitational pull of the sun. As these charged particles (ions and electrons) travel through space, they are known as *plasma*, or **solar wind**. When the solar wind

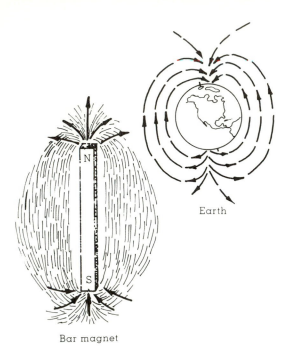

Fig. 4.4 Magnetic field about a bar magnet and about the earth.

magnetic north pole and leave near the magnetic south pole. Most scientists believe that an electric current coupled with fluid motions deep in the earth's hot, molten core are responsible for its magnetic field. This field protects the earth, to some degree, from the onslaught of the solar wind.

Observe in Fig. 4.5 that, when the solar wind encounters the earth's magnetic field, it severely deforms it into a teardrop-shaped cavity known as the **magnetosphere**. On the side facing the sun, the pressure of the solar wind compresses the field lines. On the opposite side, the magnetosphere stretches out into a long tail—the *magnetotail*—which reaches far beyond the moon's orbit. In a way, the magnetosphere acts as an obstacle to the solar wind by causing some of its particles to flow around the earth.

Inside the earth's magnetosphere are ionized gases. Some of these gases are solar wind particles, while others are ions from the earth's upper atmosphere that have moved upward along electric field lines into the magnetosphere. (How this process happens is not fully understood.)

Normally, the solar wind approaches the earth at an average speed of 400 km (250 mi) per second. However, during periods of high solar activity (many sunspots and flares), the solar wind is more dense, travels much faster, and carries more energy. When these energized solar particles reach the earth, they cause a variety of ef-

moves close enough to the earth, it interacts with the earth's magnetic field.

The magnetic field that surrounds the earth is much like the field around an ordinary bar magnet (Fig. 4.4). Both have north and south magnetic poles, and both have invisible lines of force (field lines) that link the poles. On the earth, these field lines form closed loops as they enter near the

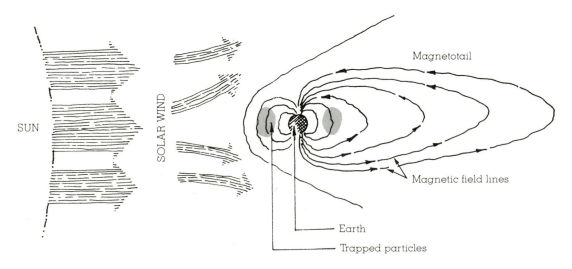

Fig. 4.5 The solar wind distorts the shape of the earth's magnetic field and produces the magnetosphere.

fects, such as changing the shape of the magnetosphere and producing auroral displays. (For information on how the aurora forms, read the focus section on p. 68.)

The Vertical Structure of the Atmosphere

Solar radiation travels through space at a speed close to 300,000 km (186,000 mi) per second. Because the earth's average distance from the sun is 150 million km, it takes sunlight about 8 minutes to reach our planet's tenuous outer atmosphere. Once there, the very short high energy waves are able to ionize certain atmospheric gases, namely, nitric oxide and atomic oxygen. As we can see in Fig. 4.6, above 80 km (50 mi), oxygen absorbs radiation having wavelengths shorter than 0.2 µm. At elevations between 20 and 50 km (12 and 31 mi), ozone selectively absorbs solar energy with wavelengths between 0.2 and 0.3 µm. Consequently, by the time the sun's radiation reaches the earth's surface, practically all of it with wavelengths below about 0.3 µm has been absorbed by gases in the upper atmosphere. To find out what effect this absorbed energy has on the temperature of these upper levels, we will journey from the earth's surface up through the atmosphere.

The Troposphere We begin our imaginary journey at the earth's surface. We climb inside a pressurized gondola attached to an enormous helium-filled balloon that will carry us all the way to the upper limits of the atmosphere. Outside our window are two instruments—a thermometer and a barometer. The air temperature near the ground is a rather pleasant 15°C (59°F), and the air pressure is 1000 mb. The weight holding the balloon is released, and up we go.

As we ascend, the air temperature outside constantly drops. The air grows colder as we climb higher because the lower atmosphere is being warmed from below (Chapter 3). Most of the solar radiation not absorbed in the upper atmosphere penetrates the lower atmosphere to warm the surface. The surface, in turn, acts like a heater, warming the air in contact with it. Right next to the ground, the air warms by conduction. The

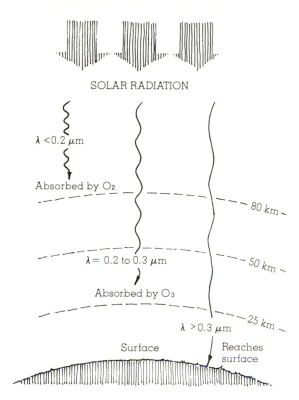

SOLAR RADIATION

$\lambda < 0.2\ \mu m$

Absorbed by O_2

80 km

$\lambda = 0.2$ to $0.3\ \mu m$

50 km

Absorbed by O_3

25 km

$\lambda > 0.3\ \mu m$

Surface

Reaches surface

Fig. 4.6 Because of the absorbing gases in the upper atmosphere, only solar radiation with wavelengths longer than 0.3 µm is able to reach the earth's surface.

heated air then rises (convection) and distributes its heat through a deeper layer of air. However, rising air expands and cools, which means that, at every level, the rising air will be cooler than the air directly beneath it. Meanwhile, the earth constantly emits infrared energy, which is being absorbed and re-emitted by water vapor and carbon dioxide (atmosphere or greenhouse effect). Since the concentration of these gases decreases rapidly above the earth, most of the absorption occurs in a layer near the surface. Hence, the lower atmosphere, being heated from below, usually is warmest at the surface and gradually cools as we climb.

Fortunately, our gondola is heated because, at an altitude of 5500 m (18,000 ft), the temperature has fallen to −20°C (−4°F). The barometer indicates 500 mb, only one-half of its surface reading of 1000 mb. We have traveled through half of the mass of the earth's atmosphere in approximately 5.5 km (3.5 mi).

(Text continues on p. 70.)

The Aurora—A Dazzling Light Show

The aurora is not reflected light from the polar ice fields, nor is it light from demons' lanterns as they search for lost souls (as once believed by the Eskimos). The aurora is produced by the solar wind disturbing the magnetosphere. The disturbance involves high-energy particles within the magnetosphere being ejected into the earth's upper atmosphere, where they excite atoms and molecules. The excited atmospheric gases emit visible radiation, which causes the sky to glow like a neon light. Let's examine this process more closely.

A high-energy particle from the magnetosphere will, upon colliding with an air molecule (or atom), transfer some of its energy to the molecule. The molecule then becomes excited (Fig. 1). Just as excited football fans leap up when their favorite team scores the winning touchdown, electrons in an excited molecule jump into a higher energy level as they orbit its center. As the fans sit down after all the excitement is over, so electrons quickly return to their lower level. When molecules de-excite, they release the energy originally received from the energetic particle, either all at once (one big jump), or in steps (several smaller jumps). This emitted energy is given up as radiation. If its wavelength is in the visible range, we see it as visible light. In the Northern Hemisphere, we call this light show the **aurora borealis**, or *northern lights*; its counterpart in the Southern Hemisphere is

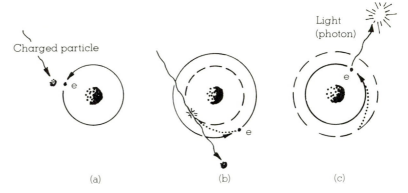

Fig. 1 When an excited atom, ion, or molecule de-excites, it can emit visible light. The electron in its normal orbit (a) becomes excited by a charged particle (b) and jumps into a higher energy level. When the electron returns to its normal orbit (c) it emits a photon of light.

the **aurora australis**, or *southern lights*.

Since each atmospheric gas has its own set of energy levels, each gas has its own characteristic color. For example, the de-excitation of atomic oxygen at altitudes between 100 to 150 km emits green light. At higher altitudes near 250 km, atomic oxygen produces red light. Molecular nitrogen gives off red and violet light. The shades of these colors can be spectacular as they brighten and fade, sometimes in the form of waving draperies, sometimes as unmoving, yet flickering, arcs and soft coronas. On a clear, quiet night the aurora is an eerie yet beautiful spectacle.

The aurora is most frequently seen in polar latitudes. Energetic particles trapped in the magnetosphere move along the earth's magnetic field lines. Because these lines emerge from the

earth near the magnetic poles, it is here that the particles interact with atmospheric gases to produce an aurora. Notice in Figs. 2 and 3 that the zones of most frequent auroral sightings (aurora belts) are not at the magnetic poles, but equatorward from them, where the field lines emerge from the earth's surface. At lower latitudes, where the field lines are oriented almost horizontal to the earth's surface, the chances of seeing an aurora diminish rapidly.

On rare occasions, however, the aurora is seen in the southern United States. Such sightings happen only when the sun is very active—as giant flares hurl electrons and protons earthward at a fantastic rate. These particles move so fast that some of them penetrate unusually deep into the earth's magnetic field before they are trapped by it. In a process not fully understood,

FOCUS ON A SPECIAL TOPIC, continued

particles from the magnetosphere are accelerated toward
the earth along electrical field
lines that parallel the magnetic
field lines. The acceleration of
these particles gives them sufficient energy, so that when they
enter the upper atmosphere they
are capable of producing an
auroral display much farther
south than usual.

How high above the earth is
the aurora? The exact height
appears to vary. The base of an
aurora is rarely lower than 80
km (50 mi), and it averages
about 105 km (65 mi). Since the
light of an aurora gradually
fades, it is difficult to define an
exact upper limit. Most auroras,
however, are observed below
200 km (124 mi), with a very
few extending upward to 950
km (600 mi).

In summary, energy for the
aurora comes from the solar
wind, which disturbs the earth's
magnetosphere. This disturbance causes energetic particles
to enter the upper atmosphere,
where they collide with atoms
and molecules. The atmospheric
gases become excited and emit
energy in the form of visible
light.

But there is other light coming
from the atmosphere—a faint
glow at night much weaker than
the aurora. This feeble lumines-
cence, called **airglow**, is detected
at all latitudes, and shows no
correlation with solar wind activity. Apparently, this light comes
from oxygen and nitrogen and
other gases that have been excited by solar radiation (not solar wind particles).

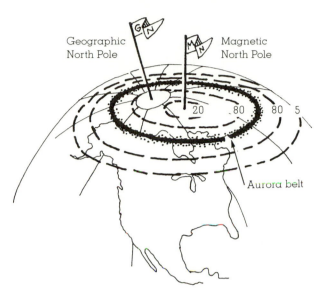

Fig. 2 The aurora belt represents the region where you would be able to see an aurora on most clear nights. (The dashed lines represent the average number of nights per year on which you might see an aurora if the sky were clear.)

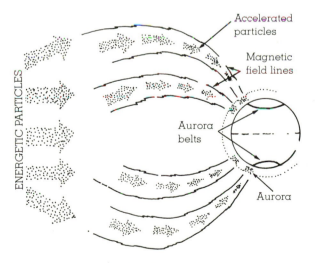

Fig. 3 As energetic particles from the magnetosphere accelerate toward the earth, they reach the upper atmosphere with enough energy to produce an aurora.

We notice that the wind speed is now much stronger compared to the light breezes on the surface. Rapidly, we pass through light and dark colored clouds. Some appear thin, wispy, and translucent, while others are much larger and thicker. Some extend from near the earth's surface to well over 8500 m (28,000 ft). Within these clouds, we see an occasional lightning flash. As we keep rising, the temperature keeps dropping, and the wind becomes even stronger, constantly blowing from the west. Near 11 km (36,000 ft), we find ourselves in a narrow flowing "river" of air—a *jet stream*. Slightly higher, the wind diminishes, and the air temperature suddenly stops decreasing. The thermometer now reads a chilling −57°C (−71°F).

From the surface to an altitude of 11 km, the air has cooled by 72°C (130°F). If we average this change for every 1000 m, we find that the air temperature drops by about 6.5°C for every 1000 m, or 3.6°F for every 1000 ft. You may recall from Chapter 2 that the rate at which the air temperature decreases with height is called the **lapse rate**. The lapse rate we calculated—6.5°C/

1000 m (3.6°F/1000 ft)—represents an *average* (or *standard*) *lapse rate*. If we made this ascent day after day, and year after year, then averaged the many readings, we would come up with a value close to that which we obtained here. Some days, the air would become cooler more quickly as we move upward. This would increase the lapse rate (see Fig. 4.7). On other days, the air temperature would not drop as fast, and the lapse rate would be less. So the lapse rate fluctuates, varying from day to day and season to season.

The part of the atmosphere we have just passed through contains all of the weather we are familiar with on earth: lightning, thunder, hail, hurricanes, dust, and so forth. Also, this portion of the atmosphere is kept well stirred by rising and descending air currents. Here it is common for air molecules to circulate through a depth of more than 10 km in just a few days. This region of circulating air extending upward from the earth's surface to where the air stops getting colder with height is called the **troposphere** (from the Greek *tropein*, meaning to turn or to change). The troposphere (changing sphere) is much different from what lies ahead.

The Tropopause—A Boundary Still rising into the atmosphere, we are now at an elevation of 17 km (56,000 ft). To our surprise the thermometer registers −57°C, the same as it did at 11 km (36,000 ft). Since there is no temperature change with height, the lapse rate in this region must be zero. A region such as this, where the air temperature remains constant with height, is referred to as an *isothermal* (equal temperature) *zone*.

The bottom of the isothermal zone marks the top of the troposphere and the beginning of another layer. This upper layer is the **stratosphere**. The boundary separating the troposphere and the stratosphere also has a name: the **tropopause**. Note in Fig. 4.8 that the altitude at which we find the tropopause depends on our latitude. The tropopause is normally found at higher elevations over equatorial regions, and it decreases in elevation as we travel poleward. Generally, it is higher in summer and lower in winter at all latitudes. Also, we can see in Fig. 4.8 that the height of the tropopause can be determined by plotting a vertical view (vertical profile) of air temperature above

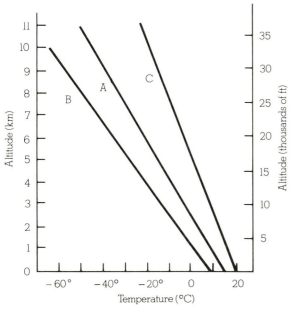

Fig. 4.7 The lapse rate is the rate at which the air temperature decreases with height. Line A represents a normal or average lapse rate for the troposphere. Line B represents the lapse rate when the air is cold aloft, and line C when the air is warm aloft.

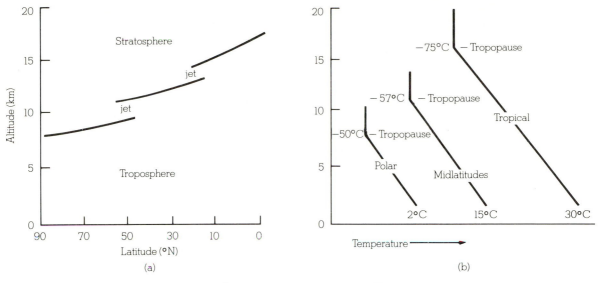

Fig. 4.8 (a) The average variation in height of the tropopause (heavy line). (b) The average vertical profile of temperature for three latitude regions.

the surface. The beginning of the isothermal zone marks the position of the tropopause.

Look at Fig. 4.8 again and notice that the tropopause occasionally "breaks" and is difficult to locate. In these regions, scientists have discovered stratospheric air mixing with tropospheric air and vice versa. These breaks also mark the position of jet streams—high winds that meander in a narrow channel, like an old river, often at speeds topping 100 knots. (Chapter 15, which deals with global wind systems, discusses jet streams further.)

The Stratosphere In our imaginary ascent, we have climbed higher into the stratosphere. At about 20 km (66,000 ft), the air temperature unexpectedly begins to increase. An increase in air temperature with height is called an **inversion**. This inversion, as well as the lower isothermal layer, keeps the vertical currents of the troposphere from spreading into the stratosphere. The inversion also tends to reduce the amount of vertical motion within the stratosphere. For this reason, the stratosphere is known as the *layered sphere.*

But what causes the inversion? A sensitive nose would give us the first clue as it detects the scent of ozone. Although we might smell ozone, its actual concentration is small. Even near 25 km (15 mi), where ozone is most dense, there are only

about 12 ozone molecules for every million air molecules. (Here the composition of air is still about the same as near the earth's surface, mainly nitrogen and oxygen.)

Despite its small concentration, ozone plays a major role in heating the air. It strongly absorbs ultraviolet solar radiation at wavelengths between 0.2 and 0.3 μm. Some of the absorbed energy increases the motion (kinetic energy) of the ozone molecules. They pass some of this energy onto other molecules as they collide with them. The increased motion of these gases represents a higher temperature, which explains why there is an inversion in the stratosphere. If the ozone were not present, the air probably would become colder with height, as it does in the troposphere; there would be no inversion and no "layered sphere." (See Fig. 4.9.)

We are now at an elevation of 30 km (19 mi). Even though the air temperature is increasing, the air is extremely cold, −46°C (−51°F). Above polar latitudes, temperatures in this part of the stratosphere may undergo dramatic warming and cooling from one week to the next. During the onset of a warm spell, known as a *sudden warming*, the temperature may rise more than 60°C (108°F) in one week. Such a rapid warming is probably due to sinking air associated with circu-

FOCUS ON INSTRUMENTS

The Radiosonde

The vertical distribution of temperature, pressure, and humidity up to an altitude of 50 km can be obtained with an instrument called a **radiosonde**. The radiosonde is a small, lightweight box equipped with weather instruments and a radio transmitter. It is attached to a cord that has a parachute and a gas-filled balloon tied tightly at the end. As the balloon rises, the attached radiosonde measures air temperature with a small electrical resistance thermometer—a thermistor—located just outside the box. It measures humidity electrically by sending an electric current across a carbon-coated plate. Air pressure is obtained by a small aneroid barometer located inside the box. All of this information is converted to radio frequencies and transmitted to the surface. Here, a computer rapidly reconverts the various frequencies into val-

Fig. 4   The radiosonde.

ues of temperature, pressure, and moisture. Special tracking equipment at the surface may also be used to provide a vertical profile of winds. Eventually, the balloon bursts and the radiosonde returns to earth, its descent being slowed by its parachute.

At most sites, radiosondes are released twice a day, usually at the time that corresponds to midnight and noon in Greenwich, England. Releasing radiosondes is an expensive operation because many of the instruments are never retrieved, and many of those that are retrieved are often in poor working condition. For this reason, only the more affluent nations can afford the luxury of having a dense radiosonde network.

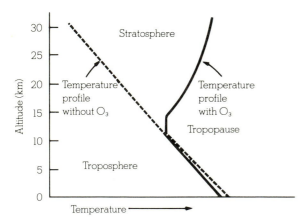

Fig. 4.9   Vertical temperature profile with and without ozone.

lation changes that often occur in late winter or early spring.

Even at this altitude of 30 km, there are clouds. These are called **nacreous**, or *mother-of-pearl*, **clouds** because they have a soft, pearly luster as viewed from the surface of the earth. Their exact composition is not known, although they appear to be composed of water in either solid or liquid (supercooled) form. They are best viewed in polar latitudes during the winter months when the sun, being just below the horizon, is able to illuminate them because of their high altitude.

As we continue to rise, the air temperature increases. At 50 km (31 mi) the temperature is −2°C (28°F). Why is the temperature so much warmer here when the maximum concentration of ozone is found near 25 km? The maximum temperature occurs in this region because most of the ultraviolet radiation responsible for the heating is absorbed here and, therefore, does not make it down to the level of ozone maximum. In addition, the air at 50 km is less dense than at 25 km. Hence, the absorption of intense solar energy at 50 km raises the temperature of fewer molecules to a much greater degree. And because of the thin atmosphere, the transfer of energy downward by molecular collisions (conduction) is very slow.

The Mesosphere Rising above 50 km (31 mi), we find the air temperature becomes isothermal and then decreases once again. As Fig. 4.10 shows, we are now out of the stratosphere and in another layer—the **mesosphere** (middle sphere). The boundary at 50 km, which separates these layers, is the **stratopause**. Like the tropopause, its height varies with latitude and season. But these variations do not appear to be as great as those experienced at the tropopause.

Just above the stratopause, we notice that the balloon is much larger now than it was when we started. The balloon expanded because the air pressure around it has decreased dramatically. We look outside at the barometer. Even though the sun is shining brightly, the sky is so dark that we can hardly read the dial. The gondola turns slightly, and the sun flashes light onto the dial, which indicates 1 mb. The air pressure is 1000 times lower than it was at the earth's surface; that is, only $\frac{1}{1000}$ of all the atmosphere's molecules are above us. Hence, 99.9 percent of the atmosphere's mass is found below us. Up here, the air is extremely thin, the sun is very bright, and the sky grows darker with every kilometer we rise.

The darkness of the sky illustrates the scarcity of molecules in the mesosphere. We know from Chapter 3 that air molecules selectively scatter shorter wavelengths of visible light, and that this light, striking the eye from all directions, makes the sky appear blue. With so few molecules, there is little scattering of light, and so the sky appears black.

(Text continues on p. 76.)

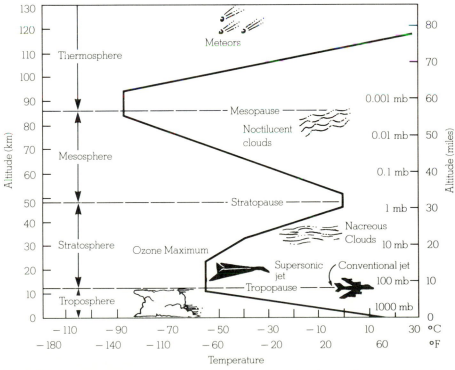

Fig. 4.10 Layers of the atmosphere as related to the vertical distribution of average temperature (heavy line).

Ozone in the Stratosphere—How it Affects Life at the Surface

Approximately 97 percent of all of the ozone in the atmosphere is found in the stratosphere. If it were brought down to the earth's surface and compressed, it would form a layer about as thick as a dime. Although thin, this layer of ozone shields the earth from harmful amounts of ultraviolet solar radiation. This protection is fortunate because ultraviolet radiation has enough energy to cause skin cancer in humans. Also, electromagnetic energy at 0.26 μm readily destroys acids on the DNA (deoxyribonucleic acid) molecule, which is responsible for transmitting the hereditary blueprint from one generation to the next. Besides preventing harmful solar energy from reaching the earth's surface, ozone strongly absorbs infrared earth radiation near 10 μm. Some of this energy is reradiated back to the surface, helping the earth maintain its heat balance. Furthermore, the absorption of ultraviolet radiation by ozone is important to the thermal balance of the stratosphere.

If the concentration of stratospheric ozone were to decrease, the following might occur:

1. An increase in the number of cases of skin cancer. Estimates are that a 1 percent reduction in the ozone layer could result in a 2 to 5 percent increase in the incidence of skin cancer.
2. An adverse impact on crops and animals due to an increase in ultraviolet radiation.
3. A change in world climate. The stratosphere would cool, while the surface of the earth

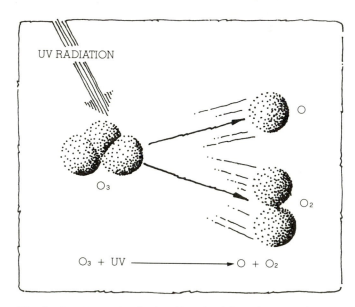

Fig. 5 An ozone molecule absorbing ultraviolet radiation can become atomic and molecular oxygen.

would slowly warm (at least initially). Because less infrared energy would be radiated from the stratosphere toward the ground, this could lead to a cooling trend. Whether warming or cooling of the atmosphere will predominate is uncertain at this time.

Ozone: Production—Destruction

Ozone (O_3) is produced naturally in the stratosphere by the combining of atomic oxygen (O) with molecular oxygen (O_2)* (in the presence of another molecule). Although it forms mainly above 25 km,

*In the troposphere, ozone is produced by lightning discharges, and it is produced in polluted atmospheres by chemical reactions that take place in sunlight.

ozone gradually drifts downward by mixing processes, producing a peak concentration near 25 km. Ozone is destroyed naturally by absorbing ultraviolet radiation with wavelengths between 0.2 and 0.3 μm:

$$O_3 + uv \rightarrow O_2 + O.$$

Ozone is also destroyed by colliding with other atoms and molecules. For example, ozone and atomic oxygen combine to form two oxygen molecules:

$$O_3 + O \rightarrow 2\,O_2.$$

Likewise, the combination of two ozone molecules destroys ozone:

$$O_3 + O_3 \rightarrow 3\,O_2.$$

These equations also represent the net result of a number of complex chemical reactions that include trace gases of nitrogen, hydrogen, and chlorine. For ex-

FOCUS ON A SPECIAL TOPIC, continued

ample, two natural destructive gases for ozone are nitric oxide (NO) and nitrogen dioxide (NO_2), which are collectively known as *oxides of nitrogen*. The origin of these gases begins at the earth's surface as soil bacteria produce N_2O (nitrous oxide), commonly known as laughing gas. This gas gradually finds its way into the stratosphere where, above about 25 km, solar energy converts it into other oxides of nitrogen. Just a small amount of nitric oxide can destroy a large amount of ozone. The following sequence of chemical reactions will show you why. In Step 1, the nitric oxide quickly combines with ozone to form nitrogen dioxide and molecular oxygen. Then, in Step 2, the nitrogen dioxide combines with atomic oxygen to form nitric oxide and molecular oxygen.

Step 1 $NO + O_3 \rightarrow NO_2 + O_2$
Step 2 $NO_2 + O \rightarrow NO + O_2$

The nitric oxide (NO) released in Step 2 is now ready to start destroying ozone again. (This is known as a catalytic reaction because the nitric oxide acts as a catalyst in that it increases the rate of chemical reaction without being consumed in the process.)

Because the ozone in the stratosphere between 25 and 35 km is maintained by a delicate natural balance between production and destruction, any increase in the concentration of nitric oxide is long lasting and a real threat to the concentration of ozone. Could this balance ever be upset?

Upsetting the Ozone Balance

The concentration of ozone may be reduced by natural events. In the upper atmosphere, both cosmic rays* and solar particles can produce secondary electrons having sufficient energy to separate molecular nitrogen (N_2) into two nitrogen atoms (N). The nitrogen atoms combine with free atomic oxygen to form nitric oxide which, in turn, rapidly destroys ozone.

In 1984, scientists, using measurements from satellites, discovered that changes in ultraviolet radiation from the sun can cause variations in the amount of stratospheric ozone, perhaps as much as 12 percent over the 11-year solar cycle of sunspot activity.

Scientists, too, are concerned that some human activities may also be altering the amount of ozone in the stratosphere. This possibility was first brought to light in the early 1970s as Congress pondered over whether or not the United States should build a supersonic jet transport. One of the gases emitted from the engines of this aircraft is nitric oxide. Although the aircraft was designed to fly in the stratosphere below the level of maximum ozone, it was feared that the nitric oxide would eventually work its way upward, where it would have an adverse effect on the ozone. This factor was one of many considered when Con-

*Cosmic rays are high-energy atomic nuclei and atomic particles that travel through space at extremely high speeds. Most cosmic rays are believed to come from supernovas, although some are produced in solar flares.

gress decided to halt the development of the United States' version of the supersonic transport.

Presently supersonic jets fly well below the region of ozone maximum. Also, these aircraft are smaller than the United States' version and, thus, emit much less nitric oxide. Because of this low emission factor, and the fact that there are relatively few of them flying, the supersonic transports (as well as military aircraft that fly in the stratosphere) do not appear to be a real threat to ozone, at least not today.

Another potentially more serious concern involves the emission of chemicals at the earth's surface. For example, nitrous oxide emitted from nitrogen fertilizers may drift into the stratosphere, where it could lead to ozone destruction. Over 36 billion kg, or 79 billion lb, of nitrogen fertilizer were used throughout the world in 1974, and, by the year 2000, this figure is expected to more than quadruple. Still another concern is the emission of chemicals known as chlorofluorocarbons, also called *fluorocarbons*. Up until 1977, when the United States imposed a ban on all nonessential use of these products, they were the most widely used propellants in spray cans. They are also used in refrigerators and car air conditioners. In the troposphere, these gases are quite safe, being nonflammable, nontoxic, and unable to chemically combine with other substances. Hence, these gases slowly diffuse

(*continued on next page*)

FOCUS ON A SPECIAL TOPIC, continued

---

upward without being destroyed. They apparently enter the stratosphere near the breaks in the tropopause, as well as in building thunderstorms that penetrate into the lower stratosphere.

Once they reach an altitude near 30 km, ultraviolet energy that is normally absorbed by ozone breaks up the fluorocarbon molecules, releasing chlorine in the process. We can see in the following sequence of reactions that chlorine, like nitric oxide, destroys ozone rapidly. In Step 1, chlorine (Cl) combines with ozone, forming a new substance (ClO) and molecular oxygen (O_2). Almost immediately, the ClO combines with free atomic oxygen (Step 2).

Step 1 $Cl + O_3 \rightarrow ClO + O_2$
Step 2 $ClO + O \rightarrow Cl + O_2$

During this process, chlorine is released to destroy more ozone.

Since the lifetime of a chlorine atom in the stratosphere is between 5 and 10 years, the destruction of ozone can take place long after the emission of fluorocarbons is halted at the surface.

Just how much ozone is being depleted by chlorine is presently being investigated by scientists. More than 5 billion kg (11 billion lb) of fluorocarbons have already been released into the troposphere and will diffuse upward during the next few decades. In a report published in 1983, the National Academy of Sciences (NAS) predicted that, due to fluorocarbons alone, stratospheric ozone could be reduced between 5 and 10 percent over the next decade, if the release of fluorocarbons continues at the 1979 level.

To further complicate the picture, theoretical calculations of ozone chemistry in the early

1980s suggested that an increase in CO_2 levels in the troposphere (resulting from fossil fuel and wood combustion) would raise the temperature in the troposphere, but lower it in the stratosphere. A cooler stratosphere would slow the process of ozone destruction. And so the question remains: How much ozone will be destroyed?

Although scientists do not yet fully understand the complex chemical nature of the stratosphere, current computer models suggest that the ozone layer may be depleted by more than 5 percent over the next several decades. Current atmospheric studies are providing more information, so that computer models will be better able to predict future levels of stratospheric ozone.

At this elevation, we dare not venture outside of the pressurized gondola. Although the composition of the atmosphere is still mainly molecular nitrogen and molecular oxygen, if we were to breathe this air, we would not live for very long. We could not survive because, even though the percentage of oxygen in the air is about the same as at sea level, the air density is very low. As a consequence, each breath contains far fewer oxygen molecules than at sea level. Without proper breathing equipment, the brain would soon become oxygen-starved—a condition known as **hypoxia**. [Pilots who fly above 3 km (10,000 ft) for too long without oxygen-breathing apparatus experience this.] With the first symptoms of hypoxia, there is usually no pain involved, just a feeling of exhaustion. Soon, visual impairment sets in, and routine tasks become difficult to perform. Some people drift into an incoherent state, neither

realizing nor caring what is happening to them. Of course, if this oxygen deficiency persists, a person will lapse into unconsciousness, and death may result. In fact, at this altitude in the mesosphere, we would suffocate in a matter of minutes.

As if the possibility of suffocating weren't bad enough, we could face another danger. Exposure to ultraviolet solar radiation at this altitude would cause severe burns on the exposed parts of the body. Due to the low air pressure, the blood in our veins would begin to boil at normal body temperatures—not a very pleasant thought.

With much less enthusiasm, we continue our journey upward into the mesosphere. The higher we go, the colder the air becomes. The decrease in temperature is due, in part, to the fact that there is little ozone in the air to absorb solar radiation. Consequently, the molecules (especially those near the top of the mesosphere) are able to emit more energy than they absorb, which results in an energy deficit and cooling. Heat from the warm stratopause is carried upward by convection to replace this energy deficit, but rising air cools. So we find vertical air motions in the mesosphere with the air becoming colder with height, up to an elevation near 85 km (53 mi). At this altitude, the temperature of the atmosphere reaches its lowest average value, −90°C (−130°F). (See Fig. 4.10.)

Near this altitude during the summer, there is sometimes enough rising air to produce clouds. These wavy, bluish-white clouds are so thin that stars shine brightly through them. Occasionally in the polar regions they can be seen from the ground near morning or evening twilight. At this time, because of their altitude, the clouds are still in sunshine, and a ground observer can see them against the background of a dark sky. For this reason, they are called **noctilucent clouds**, meaning "luminous night clouds." Although their composition is still not known, some scientists feel that these noctilucent clouds are composed of water that freezes on tiny meteoric dust particles concentrated near the top of the mesosphere.

The Thermosphere Above about 85 km, the air temperature first becomes isothermal, then it increases with height. This new layer is the **thermosphere**. The boundary that separates the lower, colder mesosphere from the warmer thermosphere is the **mesopause**.

As we climb higher into the thermosphere, the air grows warmer; we are in another inversion. In this layer, ultraviolet solar radiation below a wavelength of 0.2 μm is absorbed, mainly by molecular oxygen. This radiation supplies enough energy to break molecular oxygen into two separate oxygen atoms in a process that looks like this:

O_2 + solar radiation → 0 + 0.

The energy "left over" after the molecule separates increases the motion of the atoms. Because there are relatively few atoms and molecules up here, the absorption of a small amount of solar radiation can cause a large increase in temperature; hence, there is an inversion. Furthermore, because the amount of solar radiation affecting this region depends strongly on solar activity, temperatures in the thermosphere vary from day to day. Observe in Fig. 4.11 that, when the sun is

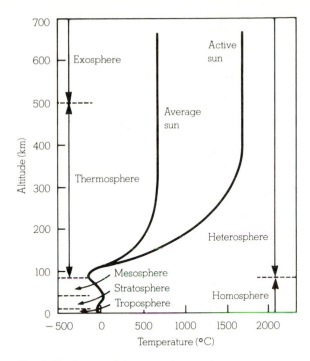

Fig. 4.11 Layers of the atmosphere and average temperatures in the thermosphere and exosphere. (The temperatures in the thermosphere are determined by measurements of the air drag on satellites.)

quiet, air temperatures near 300 km typically average about 700°C (1300°F), and that, when the sun is active, temperatures at this level are close to 1700°C (3200°F).

In the lower part of the thermosphere, we observe that the composition of air is changing. The air is becoming so thin that there are few collisions between atoms and molecules, and atomic oxygen does not rapidly recombine to form molecular oxygen. Instead of the standard concentrations of nitrogen and oxygen that existed up to about 85 km, atomic oxygen is beginning to show up in fairly large concentrations. At 110 km (68 mi), atomic oxygen is more abundant than molecular oxygen. At about 180 km (112 mi), atomic oxygen is more abundant than molecular nitrogen (see Table 4.1). With so much atomic oxygen in the thermosphere, why isn't there more ozone? For ozone to form, there must be a collision among atomic oxygen, molecular oxygen, and another molecule, which carries away the excess of energy released in the reaction. The chances of this type of a collision taking place in the low-density

Table 4.1 Abundant Gases Found in Different Regions of the Atmosphere*

ALTITUDE (km)	MOST ABUNDANT GASES		LAYER
1000	He, H, O	Light gases	
750	He, O, H		
500	O, He, N_2		Heterosphere
300	O, N_2, He		
180	O, N_2, O_2		
110	N_2, O, O_2	Heavy gases	
85	N_2, O_2, Ar		
0	N_2, O_2, Ar		Homosphere

*Notice that the composition of the atmosphere changes in the heterosphere.

region of the thermosphere is very slim, indeed. Hence, there is little ozone up here.

In the thermosphere, we observe that the thermometer reading is extremely low. But we can see from Fig. 4.11 that the temperature at this level is supposed to be many hundreds of degrees Celsius. The reason for this apparent discrepancy lies in the meaning of air temperature and how we measure it. Remember that the temperature of a gas is directly related to the average speed at which the molecules are moving—faster speeds correspond to higher temperatures. In this region, air molecules are zipping about at speeds corresponding to extremely high temperatures. However, in order to transfer enough energy to heat something up, an extremely large number of molecules must collide with the object. So, even though

Table 4.2 Mean Free Path*

ALTITUDE (km)	(mi)	REGION	MEAN FREE PATH (m)
500	310	Exosphere	10,000
250	155	Thermosphere	1000
180	111	Thermosphere	100
150	93	Thermosphere	10
100	62	Thermosphere	0.1
50	31	Stratopause	10^{-4}
0	0	Surface	10^{-7}

*The mean free path is the average distance traveled by air molecules before they collide with other molecules. Notice that the mean free paths of atoms in the exosphere are extremely long.

the air particles up here are moving extraordinarily fast, there are not enough of them bouncing against the thermometer bulb to transfer sufficient energy for it to register a high temperature. In fact, when properly shielded from the sun, the thermometer radiates away far more energy than it receives, and indicates a temperature near absolute zero. This explains how an astronaut when space walking will not only survive temperatures near 1000°C (1800°F), but will also feel a profound coldness when shielded from the sun's radiant energy. At these high altitudes, the traditional meaning of air temperature with reference to how "hot" or "cold" something feels is no longer applicable.

Although air temperatures are not measured directly in the upper thermosphere, they can, however, be determined by observing the orbital change of satellites caused by the drag of the atmosphere. Even though the air is extremely tenuous, there are enough air molecules striking a satellite to slow it down, so that it will drop into a slightly lower orbit. (This is what caused *Sky Lab* to fall to earth in July, 1979.) The amount of drag is related to the density of the air. In the thermosphere, air density varies directly with air temperature: the greater the density, the higher the temperature. Therefore, by determining air density, scientists are able to construct a vertical profile of air temperature.

The Exosphere At very high altitudes, the atmosphere becomes extremely thin. Atoms and molecules move quite a distance before they collide with one another. In Table 4.2, observe that at 250 km above the earth's surface, an atom can move an average distance, called *mean free path*, of 1000 m (3300 ft) before colliding with another atom; at 500 km, the mean free path is 10,000 m (33,000 ft). Since the chance of molecular collisions is reduced, many of the lighter, faster-moving molecules actually escape the earth's gravitational pull. The region where atoms and molecules shoot off into space is sometimes referred to as the **exosphere**. It represents the upper limit of our atmosphere. The imaginary surface of the exosphere is above the thermosphere and begins some 500 km above the surface of the earth.

Homosphere and Heterosphere   The atmospheric layers described so far have been based on the vertical profile of temperature. We can also divide the atmosphere into layers based on its gaseous composition. Below about 85 km the composition of air remains fairly uniform; it is kept homogenous by turbulent mixing. This well-mixed region is referred to as the **homosphere**. Above 85 km, the atmospheric composition varies with height. In this region, collisions between atoms and molecules are so infrequent that the air is unable to keep itself mixed. As a result, diffusion takes over as heavier atoms and molecules tend to settle to the bottom of the layer, while lighter ones float to the top. The region extending from about 85 km to the top of the atmosphere is often called the **heterosphere**.

This marks the end of our journey. On the way back to the earth's surface, let's consider a unique aspect of the upper atmosphere—the ionosphere—where ions become important enough to affect the propagation of radio waves.

The Ionosphere

The **ionosphere** is not really a layer, but rather an electrified region within the upper atmosphere—primarily the thermosphere—where fairly large concentrations of ions and free electrons exist. Ions are atoms and molecules that have lost (or gained) one or more electrons. Atoms lose electrons and become positively charged when they cannot absorb all of the energy transferred to them by solar radiation or a colliding particle. Therefore, the primary causes of ionization in the upper atmosphere are: (1) ultraviolet solar radiation; (2) high-energy cosmic rays from the sun and from supernova explosions; and (3) collisions between air molecules and energetic particles from the magnetosphere.

The rate at which atmospheric gases are ionized depends mainly on the density of the air and the energy of the solar radiation. Ions are produced most rapidly near 180 km, but at this altitude the air is dense enough that positive ions and free electrons are able to reunite, forming neutral atoms and molecules. At higher altitudes,

the air density is so low that the rapid rejoining of ions with electrons becomes nearly impossible. Consequently, the greatest number of ions and electrons occurs near 300 km, even though ion production is taking place more rapidly at lower levels.

Below an elevation of roughly 60 km (37 mi), there is insufficient ultraviolet radiation to ionize atmospheric gases to any great extent. This elevation usually represents the lower limit, or base, of the ionosphere. From here (60 km), the ionosphere extends upward to the top of the atmosphere (Fig. 4.12).

The bottom portion of the ionosphere, between 60 and 90 km, is referred to as the *D region*. The production of ions in the lower portion of the *D* region is caused mainly by cosmic rays; in the upper portion of the *D* region, ultraviolet solar radiation ionizes nitric oxide (NO). Because molecules frequently collide with each other, relatively few ions and electrons are found in this zone. Above this weakly ionized area, between about 90 to 140 km, is the *E region*. Higher yet lies the F_1 *region*, between 140 and 200 km, and the F_2 *region*, above 200 km. The *E* and *F* regions are caused by extremely shortwave ultraviolet solar energy.

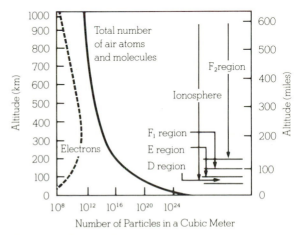

Fig. 4.12   Regions of the ionosphere along with total number of electrons and total number of air particles in a volume of 1 cubic meter at various altitudes.

Ionospheric Effects

The electrical regions of the ionosphere play a major role in radio communications. The lower parts (*D* and *E* regions) reflect standard AM radio waves back to earth. Radio waves with shorter lengths are reflected by the higher *F* region.* In fact, the shorter the wavelength, the higher the wave can penetrate into the ionosphere before being reflected (Fig. 6). Visible light, radar, television, and FM wavelengths are all too short to be reflected by the ionosphere at all. (Global communication via television is only made possible by retransmission from satellites.)

Radio waves travel in a straight-line path. The earth, however, is a sphere. Due to the earth's curvature, standard AM radio waves traveling outward from a transmitter will usually miss your radio receiver, if you are more than 160 km (100 mi) away from the station. If you are presently listening to an AM radio station whose transmitter is more than 160 km from your home, it is a good bet that the waves reaching your radio have bounced off the ionosphere. Here is something else to think about: Why can radio stations many hundreds of kilometers away be picked up clearly at night but not at all during the day?

The lower ionosphere (*D* region) is unique. It not only reflects AM radio waves, it also seriously weakens them as they are reflected. Electrons absorb

*These short radio waves are sometimes called "shortwaves." These are not to be confused with ultraviolet radiation.

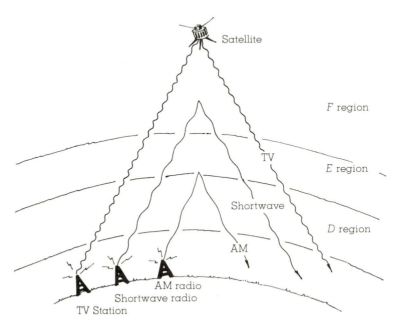

Fig. 6 Longer wavelength radio waves are reflected back to earth by the ionosphere while shorter TV waves are not.

energy from the wave as it penetrates into this zone. In the higher regions (*E* and *F*), absorption of radio waves does not take place as readily. The *D* region is strongest during the day because energy from the sun is the primary cause of ionization. At night, as electrons rapidly combine with positive ions, the *D* region gradually disappears and we are left with the *E* and *F* regions, where the recombination is much slower. Radio waves are now able to penetrate into the *F* region before being reflected back to earth. Because little absorbing of radio waves occurs in the *F* region, radio waves at night repeatedly bounce from the ionosphere to the earth's surface and back to the ionosphere again without a

significant loss of energy. In this way, standard AM radio waves are able to travel for hundreds of kilometers.

During the daylight hours the *D* region intensifies and begins to strongly absorb radio waves. As a radio wave penetrates the *D* region, part of the wave is absorbed on its way up and part on its way down. It does not take long for the radio wave to be weakened to the point where it can no longer be picked up by distant receivers.

Around sunrise and sunset, radio stations usually make "necessary technical adjustments" to compensate for the changing electrical characteristics of the *D* region. Because they can broadcast over a greater distance at night, most stations re-

FOCUS ON AN APPLICATION, continued

duce their power output near sunset. This prevents two stations, both transmitting at the same frequency, but hundreds of kilometers apart, from interfering with each other's radio programs. Imagine listening to your favorite station with Beethoven periodically butting in from 500 km away. At sunrise, the power supplied to radio transmitters is increased.

Ham operators—people who use shortwave radio transmitters and receivers—are able to communicate over thousands of kilometers during both day and night. This is made possible be-

cause these shortwaves are just that—shorter radio waves. Their wavelengths are short enough to penetrate the *D* region, but still long enough to be reflected by the higher *F* region. Occasionally, even these waves are absorbed, so that any type of radio communication is made difficult.

Radio blackouts can occur when the sun becomes quite active. An intense solar flare emits a large burst of high-energy radiation and solar wind particles that disturb the earth's magnetic field, producing a so-called **magnetic storm**. During such storms, X-rays and energetic

particles from the sun and magnetosphere intensify the *D* region by ionizing oxygen and nitrogen molecules. The increase in ions and electrons greatly enhances the *D* region's ability to absorb radio waves, often causing fadeouts and interruptions in communications that depend upon the ionosphere to reflect radio waves. Because solar flare activity is closely associated with the number of sunspots, magnetic storm activity usually increases when sunspots are numerous.

Summary

In this chapter, we examined the many ways energy from the sun affects our atmosphere. Energy travels outward from the sun in the form of radiation and as charged particles called solar wind, which have the effect of deforming the earth's magnetic field and producing auroral displays.

To better understand the sun's influence on the atmosphere, we journeyed through the atmosphere, examining the nature of each atmospheric layer. We traveled through the troposphere, where almost all weather events take place. We crossed the tropopause into the warm stratosphere, where ozone protects us from a portion of the sun's harmful ultraviolet radiation. Here, we observed that ozone is being depleted by a number of complex chemical reactions involving gases, such as chlorine and oxides of nitrogen.

Above the stratosphere we encountered the mesosphere, where little absorption of solar radiation takes place, and the air temperature drops dramatically with height. After another boundary, we found ourselves in the warmest part of the atmosphere—the thermosphere; here, the actual composition of the air begins to change as molecular oxygen splits into atomic oxygen after absorbing high-energy ultraviolet radiation. Above the thermosphere is the exosphere, where collisions between gas molecules and atoms are infrequent, and lighter, fast-moving molecules can actually escape the earth's gravitational pull and shoot off into space. Finally, we examined the ionosphere, a name given to that portion of the upper atmosphere above 60 km that contains large numbers of ions and free electrons.

Questions for Review

1. What are some characteristic features of the sun's corona, chromosphere, and photosphere?

2. What is the solar wind? How does it form?

3. Why is the magnetosphere thousands of kilometers above the earth's surface important to the plant and animal life at the surface?

4. We know from Chapter 3 that energy is transferred by conduction, convection, and radiation. Having read this chapter, name another method.

5. Explain how the aurora is produced.

6. Why does the air temperature in the troposphere normally decrease with height, while in the stratosphere it increases with height?

7. What's the average or standard lapse rate in the troposphere?

8. If you were given today's vertical temperature profile of the atmosphere up to an elevation about 20 km (12 mi), how would you be able to locate the tropopause?

9. What atmospheric layer contains all of our weather?

10. How does the height of the tropopause vary from the equator to the North Pole?

11. What types of clouds are usually found above the tropopause?

12. Even though the actual concentration of oxygen is close to 21 percent (by volume) in the upper stratosphere, explain why you would not be able to survive there.

13. If all of the ozone in the stratosphere were destroyed, what possible effects might this have on the earth's atmosphere and its inhabitants?

14. Briefly describe how the air temperature changes from the earth's surface to the lower thermosphere.

15. What natural and human-produced substances could alter the concentration of ozone in the stratosphere?

16. What is hypoxia and why can it begin at elevations as low as 3 km (10,000 ft)?

17. Why doesn't radiation with wavelengths shorter than 0.3 μm reach the earth's surface?

18. Explain how an astronaut can survive the extremely high temperatures of the upper thermosphere.

19. Explain why the lower *D* region of the ionosphere changes from day to night.

20. Why is it that you often can pick up AM radio stations hundreds and sometimes thousands of kilometers away at night, but not at all during the day?

21. Why is there little ozone above about 50 km in the atmosphere?

Questions for Thought

1. Why is it that auroral displays that can be seen in Colorado can be forecasted several days in advance?

2. Why does the aurora usually occur more frequently above Maine than above Washington State?

3. Could a liquid thermometer register a temperature of −273°C when the air temperature is actually 500°C? Where would this happen in the atmosphere, and why?

4. Why doesn't the relatively warm stratopause warm the cold mesosphere much like the earth's surface warms the air above it?

5. Explain why the upper portion of the thermosphere is much hotter than the earth's surface, yet emits much less energy.

6. Why doesn't the ionosphere extend down to the earth's surface?

7. If infrared photons carried more energy than ultraviolet photons, how do you feel this would change the structure of the ionosphere?

8. If the ionosphere extended down to the earth's surface, how would this influence AM radio communication?

9. If ozone did not exist in the stratosphere, explain why the air temperature there would decrease with height.

Problems and Exercises

1. Draw a vertical view of air temperature beginning at the earth's surface and ending in the upper thermosphere. Include in the diagram the upper and lower boundary of each atmospheric layer. Be sure to include the various regions of the iono-sphere, as well as the heterosphere and homo-sphere. Give one important characteristic of each of the layers.

2. Calculate the average lapse rate of temperature (°C/1000 m) in the mesosphere.

A coating of ice protects these grape vines from damaging low temperatures as an early spring freeze drops air temperatures well below freezing. (Photo: Steve Ferris, *Modesto Bee*)

CHAPTER 5

Seasonal and Daily Temperatures

Most of us have experienced warm summers and cold winters. Why aren't winters warm and summers cold? What causes the seasons, anyway? Before these questions can be answered, it should be noted that, as you sit quietly reading this book, you are part of a moving experience. The earth is zipping through space at thousands of kilometers per hour. As it moves, it spins. In middle latitudes, the rotational spin of the earth is hundreds of kilometers per hour. Therefore, we will begin this chapter by looking in detail at how these motions, coupled with energy from the sun, work together to produce seasonal variations in temperature. Next, we will examine the variation of temperature on a daily basis. Finally, we will conclude by examining the instruments that measure temperature.

Why the Earth Has Seasons

The earth revolves completely around the sun in an elliptical path (not quite a circle) in slightly longer than 365 days (one year). As the earth revolves around the sun, it spins on its own axis, completing one spin in 24 hours (one day). Observing the earth's North Pole from above, we can see that the earth rotates counterclockwise, or toward the east, causing the sun, moon, and stars to rise in the east and set in the west.

The average distance from the earth to the sun is 150 million km (93 million mi). Because the earth's orbit is an ellipse instead of a circle, the actual distance from the earth to the sun varies during the year. The earth comes closer to the sun in January (147 million km) than it does in July (152 million km). (See Fig. 5.1.) From this we might conclude that our warmest weather should occur in January and our coldest weather in July.

Contents

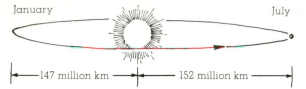

January   July

|←——147 million km ——→|←—— 152 million km ——→|

Fig. 5.1 The elliptical path (highly exaggerated) of the earth about the sun brings the earth slightly closer to the sun in January than in July.

But, in the Northern Hemisphere, we normally experience cold weather in January when we are closer to the sun and warm weather in July when we are farther away. That seems backwards, doesn't it?

The situation just described isn't really ''backwards'' because nearness to the sun is only a part of the story. Our seasons are regulated by the amount of solar energy received at the earth's surface. This amount is determined primarily by the angle at which sunlight strikes the surface, and by how long the sun shines on any latitude (daylight hours). Let's look more closely at these factors.

Radiation that strikes the earth's surface perpendicularly (directly) is much more intense than radiation that strikes the same surface at an angle. Think of shining a flashlight straight at a wall—you get a small, circular spot of light (Fig. 5.2). Now, tip the flashlight and notice how the spot of light spreads over a larger area. The same principle holds for sunlight. Radiation striking the earth at an angle spreads out and must heat a larger region than radiation impinging directly on the earth.

Everything else being equal, an area experiencing more direct solar rays will receive more heat than the same size area being struck by sunlight at an angle. As a consequence, when the sun is high in the sky, it can heat the ground to a much higher temperature than when it is low on the horizon.

The second important factor determining how warm the earth's surface becomes is the length of time the sun shines each day. Longer daylight hours, of course, mean that more energy is available from sunlight. In a given location, more solar energy reaches the earth's surface on a clear, long day than on a day that is clear but much shorter. Hence, more surface heating takes place.

From a casual observation, we know that summer days have more daylight hours than winter days. Also, the noontime summer sun is higher in the sky than is the noontime winter sun. Both of these events occur because our spinning planet is inclined on its axis (tilted) as it revolves around the sun. As Fig. 5.3 illustrates, the angle of tilt is 23½° from the perpendicular to the plane of the earth's orbit. The earth's axis points to the same direction in space all year long; thus, the Northern Hemisphere is tilted toward the sun in summer (June), and away from the sun in winter (December).

Seasons in the Northern Hemisphere

Warm Summers Note (Fig. 5.3) that, on June 22, the northern half of the world is directed toward the sun. At noon on this day, solar rays beat

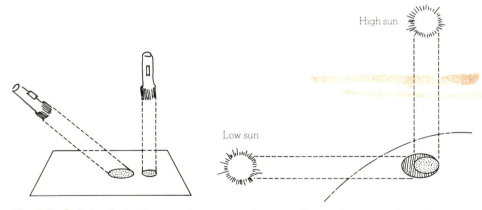

Fig. 5.2 Radiation that strikes a surface at an angle is spread over a larger area than radiation that strikes the surface directly. Oblique sun rays deliver less energy (less intense) to a surface than direct sun rays.

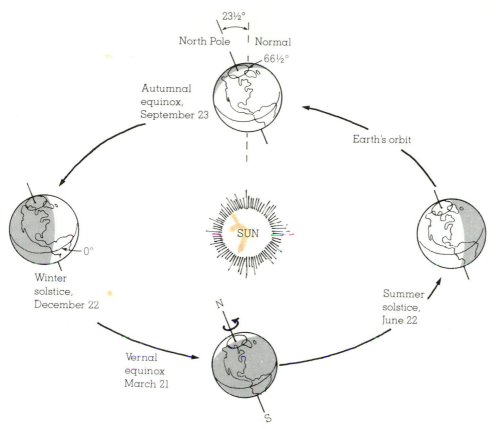

Fig. 5.3 As the earth revolves about the sun, it is tilted on its axis by an angle of 23½°. The earth's axis always points to the same area in space (as viewed from a distant star). Thus, in June, when the Northern Hemisphere is tipped toward the sun, more direct sunlight and long hours of daylight cause warmer weather than in December, when the Northern Hemisphere is tipped away from the sun. (Diagram, of course, is not to scale.)

down upon the Northern Hemisphere more directly than during any other time of year. The sun is at its highest position in the noonday sky, directly above 23½° north (N) latitude (Tropic of Cancer). If you were standing at this latitude on June 22, the sun at noon would be directly overhead. Summer* in the Northern Hemisphere officially begins on this day, called the **summer solstice**.

---

*This is the astronomical definition of the first day of summer. If summer is defined as the warmest season, and winter the coldest season, and autumn and spring as transition seasons, then the meteorological definition of the seasons in the Northern Hemisphere would be summer: June, July, August; autumn: September, October, November; winter: December, January, February; spring: March, April, May.

Study Fig. 5.3 closely and notice that, as the earth spins on its axis, the side facing the sun is in sunshine and the other side is in darkness. Thus, half of the globe is always illuminated. If the earth's axis were not tilted, the noonday sun would always be directly overhead at the equator, and there would be 12 hours of daylight and 12 hours of darkness at each latitude every day of the year. However, the earth is tilted. Since the Northern Hemisphere faces towards the sun on June 22, each latitude in the Northern Hemisphere will have more than 12 hours of daylight. The farther north we go, the longer are the daylight hours. When we reach the Arctic Circle (66½°N), daylight lasts for 24 hours. Look at Fig. 5.3 and notice how the region above 66½°N never gets into the "shadow"

Fig. 5.4   Land of the Midnight Sun. These eight exposures of the sun were taken in northern Greenland (latitude 78°N) during late July between about 11:00 P.M. and 1:00 A.M. in the morning.

zone as the earth spins. At the North Pole, the sun actually rises above the horizon on March 21 and has six months until it sets on September 23. No wonder this region is called the ''Land of the Midnight Sun''! (See Fig. 5.4.)

Do longer days near polar latitudes mean that the highest daytime summer temperatures are experienced there? Not really. Everyone knows that New York City (41°N) ''enjoys'' much hotter summer weather than Barrow, Alaska (71°N). The

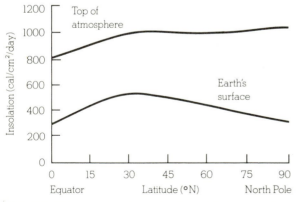

Fig. 5.5   The average amount of radiant energy received at the top of the earth's atmosphere and at the earth's surface on June 22—the summer solstice.

days in Barrow are much longer, so why isn't Barrow warmer? To figure this out, we must examine the *incoming solar radiation* (called **insolation**) on June 22. Figure 5.5 shows two curves: The upper curve represents the average amount of insolation at the top of the earth's atmosphere on June 22; the bottom curve shows the average amount of radiation that eventually reaches the earth's surface on the same day.

The upper curve increases from the equator to the pole. This increase indicates that, during the entire day of June 22, more solar radiation reaches the top of the earth's atmosphere above the poles than above the equator. True, the sun shines on these polar latitudes at a relatively large angle, but it does so for 24 hours, causing the maximum to occur there. The lower curve shows that the amount of solar radiation eventually reaching the earth's surface on June 22 is maximum near 30°N. From there, the amount of insolation reaching the ground decreases as we move poleward. Why are the two curves so different?

Above the earth's atmosphere, essentially nothing interferes with the sunlight streaming earthward. Solar energy received on a surface perpendicular to the sun's rays at the top of the atmosphere appears to stay fairly constant at nearly two calories on each square centimeter each minute (2 cal/cm²/min). This value, called the **solar constant**,* is enough energy to raise the temperature of one-fifth of a teaspoon of water about 2°C each minute.

Once sunlight enters the atmosphere, however, fine dust and air molecules scatter it, clouds reflect it, and some of it is absorbed by atmospheric gases. What remains reaches the surface. Generally, the greater the thickness of atmosphere that sunlight must penetrate, the greater are the chances that it will be either scattered, reflected, or absorbed by the atmosphere. During the summer in far northern latitudes, the sun is never very high above the horizon, so its radiant energy must pass through a thick portion of atmosphere before it reaches the earth's surface. And because of the increased cloud cover during the arctic summer, much of the sunlight is reflected before it reaches the ground.

Solar energy that eventually reaches the surface in the far north does not heat the surface effectively. A portion of the sun's energy is reflected by ice and snow, while some of it melts frozen soil. The amount actually absorbed is spread over a large area. So, even though northern cities, such as Barrow, experience 24 hours of continuous sunlight on June 22, they are not warmer than cities farther south. Overall, they receive less radiation at the surface, and what radiation they do receive does not effectively heat the surface.

In our discussion of Fig. 5.5, we saw that, on June 22, solar energy incident on the earth's surface is maximum near latitude 30°N. On this day, the sun is shining directly above latitude 23½°N. Why, then, isn't the most sunlight received here? A quick look at a world map shows that the major deserts of the world are centered near 30°N. Cloudless skies and drier air predominate near this latitude. At latitude 23½°N, the climate is more moist and cloudy, causing more sunlight to be scattered and reflected before reaching the

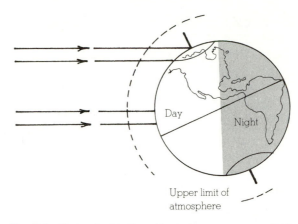

Fig. 5.6   During the Northern Hemisphere summer, sunlight that reaches the earth's surface in far northern latitudes has passed through a thicker layer of absorbing, scattering, and reflecting atmosphere than sunlight that reaches the earth's surface farther south.

surface. For this reason, more radiation falls on 30°N latitude than at the Tropic of Cancer (23½°N).

Each day past June 22, the noon sun is slightly lower in the sky. Summer days in the Northern Hemisphere begin to shorten. June eventually gives way to September, and fall begins.

Fall Weather   Look at Fig. 5.3 again and notice that, by September 23, the earth will have moved, so that the sun is directly above the equator. Except at the poles, the days and nights throughout the world are of equal length. This day is called the **autumnal** (fall) **equinox**, and it is the official beginning of fall in the Northern Hemisphere. At the North Pole, the sun appears on the horizon for 24 hours, due to the bending of light by the atmosphere. The following day (or at least within several days), the sun disappears from view, not to rise again for a long, cold six months. Throughout the northern half of the world on each successive day, there are fewer hours of daylight, and the noon sun is slightly lower in the sky. Less direct sunlight and shorter hours of daylight spell cooler weather for the Northern Hemisphere. Reduced radiation, lower air temperatures, and cooling breezes stimulate the beautiful pageantry of fall colors.

In some years around the middle of autumn, there is an unseasonably warm spell, especially in the eastern two-thirds of the United States.

*By definition, the solar constant is the rate at which radiant energy from the sun is received on a surface at the outer edge of the atmosphere perpendicular to the sun's rays when the earth is at an average distance from the sun. Recent measurements taken from a satellite suggest the solar constant is 1.97 cal/cm²/min, or 1376 W/m² in the SI system of measurement.

Does the First Frost Cause the Leaves to Change Color in Autumn?

Contrary to what many people believe, it is not the first frost that causes the leaves of deciduous trees to change color. The colors are actually in the leaves during the summer, but we do not see them because the green pigments—known as *chlorophylls*—are far more dominant.

Chlorophylls absorb and use the energy of red and violet wavelengths of sunlight to manufacture the simple sugars and starches that trees use as food. Most of the green wavelengths are not absorbed; instead, they are reflected. This, of course, causes the leaf to appear green.

The leaves remain green as long as the tree is able to replenish the chlorophylls it uses up. About several weeks before the first frost, however, a chemical change begins in the leaf, as shorter days and cooler nights cause leaf activity to diminish.

As the chlorophylls decrease, the green slowly disappears. Other pigments, such as the reddish carotenoid, which were hidden by the green of the leaf, now begin to show through.

The weather most suitable for an impressive display of fall colors is warm, sunny days followed by clear, cool nights, with temperatures dropping below 7°C (45°F).

Table 5.1 Length of Time from Sunrise to Sunset for Various Latitudes on Different Dates

	NORTHERN HEMISPHERE (READ DOWN)			
LATITUDE	**MARCH 21**	**JUNE 22**	**SEPT. 23**	**DEC. 22**
0°	12 hr	12 hr	12 hr	12 hr
10°	12 hr	12.6 hr	12 hr	11.4 hr
20°	12 hr	13.2 hr	12 hr	10.8 hr
30°	12 hr	13.9 hr	12 hr	10.1 hr
40°	12 hr	14.9 hr	12 hr	9.1 hr
50°	12 hr	16.3 hr	12 hr	7.7 hr
60°	12 hr	18.4 hr	12 hr	5.6 hr
70°	12 hr	2 months	12 hr	0 hr
80°	12 hr	4 months	12 hr	0 hr
90°	12 hr	6 months	12 hr	0 hr
LATITUDE	**SEPT. 23**	**DEC. 22**	**MARCH 21**	**JUNE 22**
	SOUTHERN HEMISPHERE (READ UP)			

This warm period, referred to as **Indian summer,*** may last from several days up to a week or more. It usually occurs when a large anticyclone (high pressure area) stalls near the southeast coast. The clockwise flow of air around this system moves warm air from the Gulf of Mexico into the central or eastern half of the nation. The warm, gentle breezes, and smoke from a variety of sources, respectively make for mild, hazy days. The warm weather ends abruptly when an outbreak of polar air reminds us that winter is not far away.

Cold Winters   Three months after the autumnal equinox, December 22, the Northern Hemisphere is tilted as far away from the sun as it will be all year (see Fig. 5.3). Nights are long and days are short. Notice in Table 5.1 that daylight decreases from 12 hours at the equator to 0 (zero) at latitudes above 66½°N. This is the shortest day of the year, called the **winter solstice**—the official beginning of winter in the northern world. On this day, the sun shines directly above latitude 23½°S (Tropic of Capricorn). In the northern half of the world, the sun is at its lowest position in the noon sky. Its rays pass through a thick section of atmosphere and spread over a large area on the surface. With shorter days and less intense sunlight, we find ourselves bundled up for the cold winter.

With so little incident sunlight, the earth's surface cools quickly. A blanket of clean snow covering the ground aids in the cooling. The snow

*The origin of the term is uncertain, as it dates back to the eighteenth century. It may have originally referred to the good weather that allowed the Indians time to harvest their crops. Normally, a period of cool autumn weather must precede the warm weather period to be called Indian summer.

reflects much of the sunlight that reaches the surface and continually radiates away infrared energy during the long nights. In northern Canada and Alaska, the arctic air rapidly becomes extremely cold as it lies poised, ready to do battle with the milder air to the south. Periodically, this cold arctic air pushes down into the northern United States, producing a rapid drop in temperature called a **cold wave**, which occasionally reaches far into the south during the winter. On January 19, 1977, cold air from northern Canada even penetrated into southern Florida, and for the first time in recorded history, it snowed in Miami!

On each winter day after December 22, the sun climbs a bit higher in the midday sky. The periods of daylight grow longer until days and nights are of equal length, and we have another equinox.

Spring—New Beginnings   The date of March 21, which marks the official arrival of spring, is called the **vernal** (spring) **equinox**. At this equinox, the noonday sun is shining directly on the equator, while, at the North Pole, the sun (after hiding for six months) peeks above the horizon. Longer days and more direct solar radiation spell warmer weather for the northern world.

Three months after the vernal equinox, it is June again. The Northern Hemisphere is tilted toward the sun, which shines high in the noonday sky. The days have grown longer and warmer, and another summer season has begun.

In summary, the seasons are controlled by solar energy striking our tilted planet, as it makes its annual voyage about its star. This tilt of the earth causes a seasonal variation in both the length of daylight and the intensity of sunlight that reaches the surface.

Seasons in the Southern Hemisphere   On June 22, the Southern Hemisphere is adjusting to an entirely different season. Because this part of the world is now tilted away from the sun, nights are long, days are short, and solar rays come in at an angle. All of these factors keep air temperatures fairly low. The June solstice marks the beginning of winter in the Southern Hemisphere. In this part of the world, summer will not officially begin until the sun is over the Tropic of Capricorn

($23\frac{1}{2}°$S)—remember that this occurs on December 22. So, when it is winter and June in the Southern Hemisphere, it is summer and June in the Northern Hemisphere. If you are tired of the hot June weather in your Northern Hemisphere city, travel to the winter half of the world and enjoy the cooler weather. The tilt of the earth as it revolves around the sun makes all this possible.

We know the earth comes nearer to the sun in January than in July. Even though this difference in distance amounts to only about 3 percent, the energy that strikes the top of the earth's atmosphere is almost 7 percent greater on January 3 than on July 4. These statistics might lead us to believe that summer should be warmer in the Southern Hemisphere than in the Northern Hemisphere, which, however, is not the case. A close examination of the Southern Hemisphere reveals that nearly 81 percent of the surface is water compared to 61 percent in the Northern Hemisphere. The added solar energy due to the closeness of the sun is absorbed by large bodies of water, becoming well mixed and circulated within them. This keeps the average summer (January) temperatures in the Southern Hemisphere cooler than summer (July) temperatures in the Northern Hemisphere. Because of water's large heat capacity, it also tends to keep winters in the Southern Hemisphere warmer than we might expect.*

Another difference between the seasons of the two hemispheres concerns their length. Because the earth describes an ellipse as it journeys around the sun, the total number of days from the vernal (March 21) to the autumnal (September 23) equinox is about 7 days longer than from the autumnal to vernal equinox (Fig. 5.7). This means that spring and summer in the Northern Hemisphere not only last about a week longer than northern fall and winter, but also about a week longer than spring and summer in the Southern Hemisphere. Hence, the shorter spring and summer of the Southern Hemisphere somewhat offset the extra insolation received due to a closer proximity to the sun.

Up to this point, we have considered the seasons on a global scale. We will now shift to more local considerations.

*For a comparison of January and July temperatures see Figs. 5.15 and 5.16.

Fig. 5.7 Because the earth travels more slowly when it is farther from the sun, it takes the earth a little more than 7 days longer to travel from March 21 to September 23 than from September 23 to March 21.

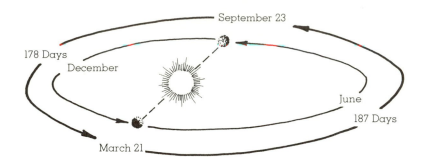

Local Seasonal Variations

Figure 5.8 shows how the sun's position changes in the middle latitudes of the Northern Hemisphere during the course of one year. Note that, during the winter, the sun rises in the southeast and sets in the southwest. During the summer, it rises in the northeast, reaches a much higher position in the sky at noon, and sets in the northwest. Clearly, objects facing south will receive more sunlight during a year than those facing north. This fact becomes strikingly apparent in hilly or mountainous country.

Hills that face south receive more sunshine, and, hence, become warmer than the partially shielded north-facing hills. Higher temperatures usually mean greater rates of evaporation and slightly drier soil conditions. Thus, south-facing

hillsides are usually warmer and drier as compared to north-facing slopes at the same elevation. In many areas of the far west, only sparse vegetation grows on south-facing slopes, while, on the same hill, dense vegetation grows on the cool, moist hills that face north. (See Fig. 5.9.) If you drive west from Denver, Colorado, into the mountains, you will find a sparse covering of heat-tolerant juniper and ponderosa pines on the south-facing side of the road. Just across the road are spruce and fir, which require cooler, moister growing conditions. In northern latitudes, hillsides that face south usually have a longer growing season. Winemakers in western New York State do not plant grapes on the north side of hills. Grapes from vines grown on the warmer south side make better wine.

Because air temperature usually decreases with height, trees found on the cooler north-facing side of mountains are often those that normally grow at higher elevations. On the other hand, the warmer south-facing side of a mountain often supports species normally found at lower elevations. Near 750 m (2500 ft) in the central Sierra Nevada of California, it is common to see digger pines (which normally grow at lower altitudes) clustered on south-facing hills. Just around the corner on the north-facing side of the same hill, incense cedar and sugar pines (both of which usually grow at higher altitudes) may be seen. (See Fig. 5.10.)

In the mountains, snow usually lingers on the ground for a longer time on north slopes than on the warmer south slopes. For this reason, ski runs are built facing north wherever possible. Also, homes and cabins built on the north side of a hill usually have a steep pitched roof, as well as a reinforced deck to withstand the added weight of snow from successive winter storms.

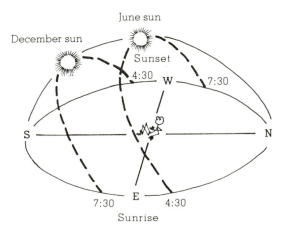

Fig. 5.8 The changing position of the sun, as observed in middle latitudes in the Northern Hemisphere.

The Tilted Earth and Its Energy Balance

In Chapter 3, we saw that the amount of incoming energy averaged over the entire earth each year is roughly equal to the amount of outgoing energy. However, given the tilt of the earth, this balance is not maintained for each latitude.

As Fig. 1 shows, high latitudes lose more radiant energy to space each year than they receive. Low latitudes, on the other hand, gain more radiant energy from the sun than they return to space through infrared energy. Only at middle latitudes near 37° does the amount of radiation received each year balance the amount lost. Thus, in one year, high latitudes experience a net loss in radiant energy, while low latitudes experience a net gain. From this situation, we might conclude that polar regions are growing colder each year, while tropical regions are becoming warmer. But this obviously does not happen.

To compensate for these gains and losses of energy, winds in the atmosphere and currents in the oceans circulate warm air and water toward the poles, and cold air and water toward the equator. In part, the poleward transfer of heat occurs when water vapor evaporates from

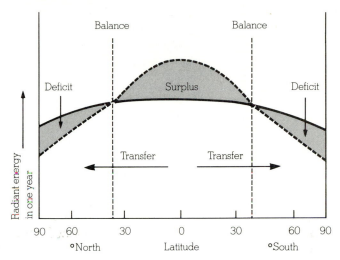

Fig. 1 The average annual incoming solar radiation (dashed line) absorbed by the earth and the atmosphere along with the average annual infrared radiation (solid line) emitted by the earth and the atmosphere.

tropical waters and is carried into high latitudes, where it condenses, releasing its latent heat. This process accounts for nearly 30 percent of the heat transported from equatorial to polar regions.

An additional 30 percent is carried northward in the form of sensible heat by winds associated with middle latitude cyclones and anticyclones, where on one side of each system warm air flows northward and

on the other side cold air moves southward. The remaining amount of heat (over 30 percent) travels poleward by ocean currents. Thus, the transfer of heat by atmospheric and oceanic circulations prevents low latitudes from steadily becoming warmer and high latitudes from steadily growing colder. These circulations are extremely important to weather and climate, and will be treated more completely in Chapter 15.

The seasonal change in the sun's position during the year can have an effect on the vegetation around the home. In winter, a large two-story home can shade its own north side, keeping it much cooler than its south side. Trees that require warm, sunny weather should be planted on the south side, where sunlight reflected from the house can even add to the warmth.

The design of a home can be important in reducing heating and cooling costs. Large windows should face south, allowing sunshine to penetrate the home in winter. To block out excess sunlight during the summer, a small eave or overhang should be built (Fig. 5.11). A kitchen with windows facing east will let in enough warm morning sunlight to help heat this area. Because the west side warms

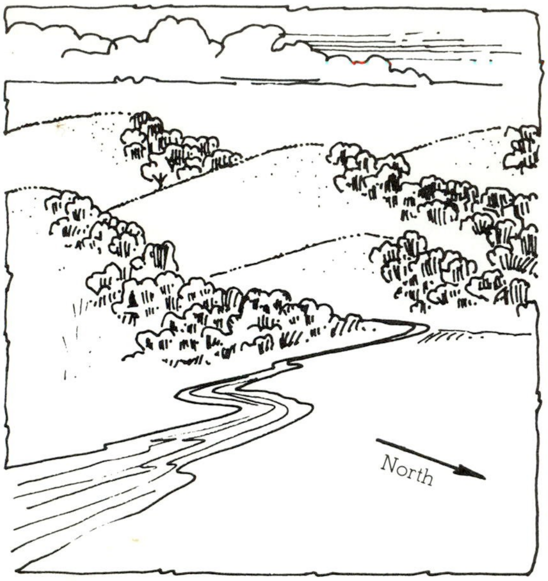

Fig. 5.9 In areas where small temperature changes can cause major changes in soil moisture, sparse vegetation on the south-facing slopes will often contrast with lush vegetation of the north-facing slopes.

rapidly in the afternoon, rooms having small windows (such as garages) should be placed here to act as a thermal buffer. Deciduous trees planted on the west side of a home provide shade in the summer. In winter, they drop their leaves, allowing the winter sunshine to warm the house. If you like the bedroom slightly cooler than the rest of the home, face it toward the north. Let nature help with the heating and air conditioning. Proper house design, orientation, and landscaping can help cut the demand for electricity, as well as for natural gas and fossil fuels, which are rapidly being depleted.

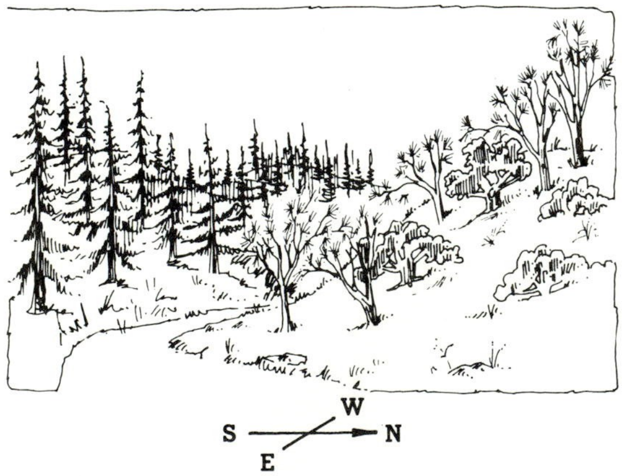

Fig. 5.10 In the mountains, the north-facing side of a hill usually has trees that typically grow at a higher (cooler) elevation. On the south-facing side of the hill, the reverse is observed.

Daily Temperature Variations

In a way, each sunny day is like a tiny season as the air goes through a daily cycle of warming and cooling. The air warms during the morning hours, as the sun gradually rises higher in the sky, spreading a blanket of heat over the ground. The sun reaches its highest point around noon, after which it begins its slow journey toward the western horizon. It is around noon when the earth's surface receives the most intense solar rays. However, somewhat surprisingly, noontime is usually not the warmest part of the day. Rather, the air continues to be heated, often reaching a maximum temperature later in the afternoon. To find out why this happens, we need to examine a shallow layer of air in contact with the ground.

Daytime Warming As the sun rises in the morning, sunlight warms the ground, and the ground warms the air in contact with it by conduction. However, air is such a poor heat conductor that this process only takes place within a few centimeters of the ground. As the sun rises higher in the sky, the air in contact with the ground be-

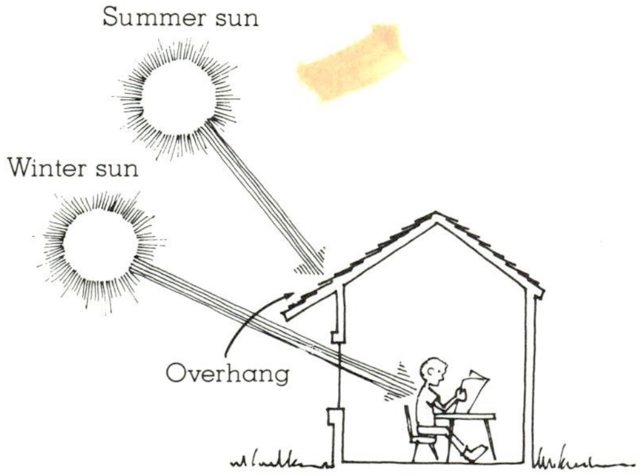

Fig. 5.11 An overhang should be long enough to block out direct summer sunlight, yet short enough to allow winter sunlight to enter through the window.

Solar Heating and the Noonday Sun

The amount of solar energy that falls on a typical American home each summer day is many times the energy needed to heat the inside for a year. Thus, many people are turning to the sun as a clean, safe, and virtually inexhaustible source of energy. If solar collectors are used to heat a home, they should be placed on south-facing roofs to take maximum advantage of the energy provided. The roof itself should be constructed as nearly perpendicular to winter sun rays as possible. To determine the proper roof angle at any latitude, we need to know how high the sun will be above the southern horizon at noon.

The noon angle of the sun can be calculated in the following manner:

1. Determine the number of degrees between your latitude and the latitude where the sun is currently directly overhead.
2. Subtract the number you calculated in step 1 from 90°. This will give you the sun's elevation above the southern horizon at noon at your latitude.

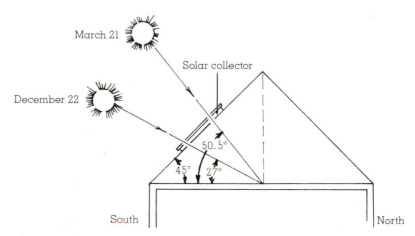

Fig. 2 The roof of a solar-heated home constructed in Denver, Colorado, at an angle of 45° absorbs the sun's energy in midwinter at nearly right angles.

For example, suppose you live in Denver, Colorado (latitude 39½°N), and the date is December 22. The difference between your latitude and where the sun is currently overhead is 63° (39½°N to 23½°S), so the sun is 27° (90°–63°) above the southern horizon at noon. On March 21 in Denver, the angle of the sun is 50½° (90°–39½°). To determine a reasonable roof angle, we must consider the average altitude of the midwinter sun (about 39° for Denver), building costs, and snow loads. Figure 2 illustrates that a roof constructed in Denver, Colorado, at an angle of 45° will be nearly perpendicular to much of the winter sun's energy. Hence, the roofs of solar-heated homes in middle latitudes are generally built at an angle between 45° and 50°.

---

comes even warmer, and there exists a thermal boundary separating the hot surface air from the slightly cooler air above. Given their random motion, some air molecules will cross this boundary: The "hot" molecules below bring greater kinetic energy to the cooler air; the "cool" molecules above bring a deficit of energy to the hot, surface air. However, on a windless day, this form of heat exchange is slow, and a substantial temperature difference usually exists just above the ground. This explains why joggers on a clear, windless, summer afternoon may experience air temperatures of over 50°C (122°F) at their feet and only 32°C (90°F) at their waist.

Near the surface, convection begins, and rising air bubbles (thermals) help to redistribute heat. In calm weather, these thermals are small and do not effectively mix the air near the surface. Thus, large vertical temperature gradients are able to exist. On windy days, however, turbulent eddies are able to mix hot, surface air with the cooler air above. This form of mechanical stirring, sometimes called *forced convection*, helps the thermals to transfer heat away from the surface more efficiently. Therefore, on windy, sunny days, the molecules near the surface are more quickly carried away than on calm, sunny days. Figure 5.12 shows a typical vertical profile of air temperature on windy days and on calm days in summer.

We can now see why the warmest part of the day is usually in the afternoon. Around noon, when the sun's rays are most intense, the ground and the air in contact with it warm rapidly. However, convection takes time to transfer this warm air upwards to a level of 1.5 m (about 5 ft), where the thermometers are. Also, even though incoming solar radiation decreases in intensity after noon, it still exceeds outgoing heat energy from the surface for a time. This yields an energy surplus for two to four hours after noon and substantially contributes to a lag between the time of maximum solar heating and the time of maximum air temperature several meters above the surface.

The exact time of the highest temperature reading varies somewhat. Where the summer sky remains cloud-free all afternoon, the maximum temperature may occur sometime between 3:00 and 5:00 P.M. Where there is afternoon cloudiness or haze, the temperature maximum occurs an hour or two earlier. In Denver, afternoon clouds, which build over the mountains, drift eastward early in the afternoon. This sometimes causes the maximum temperature to occur as early as noon. Adjacent to large bodies of water, cool air moving inland may modify the rhythm of temperature change such that the warmest part of the day occurs at noon or before. In winter, atmospheric storms circulating warm air northward can even cause the highest temperature to occur at night.

Just how warm the air becomes depends on such factors as the type of soil, its moisture content, and vegetation cover. When the soil is a poor heat conductor (as loosely packed sand is), heat energy does not readily transfer into the ground. This allows the surface layer to reach a higher temperature, availing more energy to warm the air above. On the other hand, if the soil is moist or covered with vegetation, much of the available energy evaporates water, leaving less to heat the air. As you might expect, the highest summer temperatures usually occur over desert regions, where clear skies coupled with low humidities and meager vegetation permit the surface and the air above to warm up rapidly.

Where the air is humid, haze and cloudiness lower the maximum temperature by preventing some of the sun's rays from reaching the ground. In humid Atlanta, Georgia, the average maximum temperature for July is 30.5°C (87°F). In contrast, Phoenix, Arizona—in the desert southwest at the same latitude as Atlanta—experiences an average July maximum of 40.5°C (105°F).

Nighttime Cooling As the sun lowers, its energy is spread over a larger area, which reduces the heat available to warm the ground. Sometime in late afternoon or early evening, the earth's surface and air above begin to lose more energy than they receive; hence, they start to cool.

The ground and air above both cool by radiating infrared energy. But the ground, a much better radiator than air, is able to cool more quickly. Consequently, shortly after sunset, the earth's surface is slightly cooler than the air directly above it. The surface air transfers some energy to the ground by conduction, which the ground, in turn, quickly radiates away.

As the night progresses, the ground and the air in contact with it continue to cool more rapidly than the air a few meters higher. The warmer upper air does transfer *some* heat downward, a process that is slow due to the air's poor thermal conductivity. Therefore, by late night or early morning, the coldest air is found next to the ground, with slightly warmer air above.

This measured increase in air temperature just

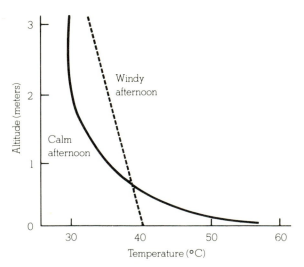

Fig. 5.12 Vertical temperature profile above an asphalt surface for a windy and a calm summer afternoon.

Radiation Inversions

A strong radiation inversion occurs when the air near the ground is much colder than the air higher up. Ideal conditions for a strong inversion exist when the air is calm, the night is long, and the air is fairly dry and cloud-free. Let's examine these ingredients one by one.

A windless night is essential for a strong radiation inversion because a stiff breeze tends to mix the colder air at the surface with the warmer air above. This mixing, along with the cooling of the warmer air as it comes in contact with the cold ground, causes a vertical temperature profile that is almost isothermal in a layer several meters thick. In the absence of wind, the cooler, more dense surface air does not readily mix with the warmer, less dense air above, and the inversion is more strongly developed. (See Fig. 3.)

A long night also contributes to a strong inversion. Generally, the longer the night (and, hence, the longer the time of surface cooling), the better are the chances that the air near the

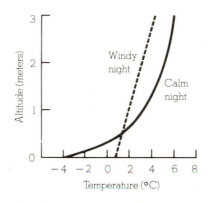

Fig. 3 Vertical temperature profile just above the ground on a windy night and a calm night. Notice that the radiation inversion develops better on the calm night.

ground will be much colder than the air above. Consequently, winter nights provide the best conditions for a strong radiation inversion, other factors being equal.

Finally, radiation inversions are more likely with a clear sky and dry air. Under such conditions, the ground and the air above are able to radiate their energy to outer space and thereby cool rapidly. However,

with cloudy weather or moist air, much of the outgoing infrared energy is absorbed and reradiated to the surface, retarding the rate of cooling. Also, on moist nights, any fog, dew, or frost will release latent heat, which warms the air.

So, radiation inversions may occur on any night. But, during long winter nights, when the air is still, cloud-free, and relatively dry, these inversions can become strong and deep. On winter nights in middle latitudes, it is common to experience below freezing temperatures near the ground, and air 5°C (9°F). warmer at your waist. In middle latitudes, the top of the inversion—the region where the air temperature stops increasing with height—is usually not more than 100 m (300 ft) above the ground. In dry, polar regions, where winter nights are measured in months, the top of the inversion is often 1000 m (about 3300 ft) above the surface. It may, however, extend to as high as 3000 m (about 10,000 ft).

above the ground is known as a **radiation inversion** because it forms mainly through radiational cooling of the surface. Because radiation inversions occur on most clear, calm nights, they are also called **nocturnal inversions**. (More information on radiation inversions is given in the focus section above.)

How cold the night air becomes depends primarily on the length of the night, the moisture content of the air, cloudiness, and the wind. Initially, wind brings cold air into a region. However, the coldest nights usually occur when the air is relatively still.

Other factors also determine how cold the air

becomes. For example, a surface that is wet or covered with vegetation can add water vapor to the air, retarding nighttime cooling. Likewise, if the soil is a good heat conductor, heat ascending toward the surface during the night adds warmth to the air, which restricts cooling. On the other hand, snow covering the ground acts as an insulating blanket that prevents heat stored in the soil from reaching the air. Snow, a good emitter of infrared energy, radiates away energy rapidly at night, which helps keep the air temperature above a snow surface quite low.

The lowest temperature on any given day is usually observed around sunrise. The cooling of

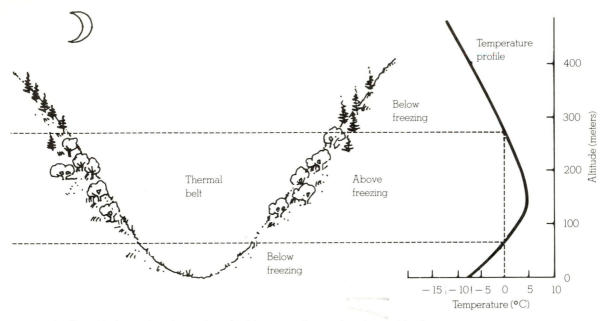

Fig. 5.13 On cold, clear nights, the settling of cold air into valleys makes them colder than surrounding hillsides.

the ground and surface air may even continue beyond sunrise for a half hour or so, as outgoing energy exceeds incoming energy. This happens because light from the early morning sun passes through a thick section of atmosphere and strikes the ground at a low angle. Consequently, the sun's energy does not effectively heat the surface. Surface heating may be reduced further when the ground is moist and available energy is used for evaporation. (Any duck hunter lying flat in a marsh knows the sudden cooling that occurs as evaporation chills the air just after sunrise.) Furthermore, once effective heating of the surface does begin, it takes time for air to transfer this heat up to where thermometers are normally placed. Hence, the lowest temperature is commonly recorded shortly after the sun has risen.

Cold, heavy surface air slowly drains downhill during the night and eventually settles in low-lying basins and valleys. Valley bottoms are thus colder than the surrounding hillsides. (See Fig. 5.13.) In middle latitudes, these warmer hillsides, called **thermal belts**, are less likely to experience freezing temperatures than the valley below. This encourages farmers to plant on hillsides those trees unable to survive the valley's low temperature.

On the valley floor, the cold, dense air is unable to rise. Smoke and other pollutants trapped in this heavy air restrict visibility. Therefore, valley bottoms are not only colder, but are also more frequently polluted than nearby hillsides. Even when the land is only gently sloped, cold air settles into lower-lying areas, such as river basins and floodplains. Because the flat floodplains are agriculturally rich areas, cold air drainage often forces farmers to seek protection for their crops. (The various methods used to protect vegetation from damaging low temperatures are given in the focus section on p. 99.)

The Use of Temperature Data

The careful recording and application of temperature data are tremendously important to us all. Without accurate information of this type, the work of farmers, power company engineers, weather analysts, and many others would be a great deal more difficult. In these next sections, we will study the ways temperature data are organized and used. We will also examine the significance of daily, monthly, and yearly temperature ranges and averages in terms of practical application to everyday living.

FOCUS ON AN APPLICATION

Protecting Crops from the Cold

On cold nights, many plants may be damaged by low temperatures. To protect small plants or shrubs, cover them with straw, cloth, or plastic sheeting. This prevents ground heat from being radiated away to the colder surroundings. If you are a household gardener concerned about outside flowers and plants during cold weather, simply wrap them in plastic or cover each with a paper cup.

Fruit trees are particularly vulnerable to cold weather in the spring when they are blossoming. The protection of such trees presents a serious problem to the farmer. Since the lowest temperatures on a clear, still night occur near the surface, the lower branches of a tree are the most susceptible to damage. Therefore, increasing the air temperature close to the ground may prevent damage. One way this can be done is to use **orchard heaters**, which warm the air around them by setting up convection currents close to the ground. Early forms of these heaters were called "*smudge pots*" because they produced large amounts of dense, black smoke that caused severe pollution. People tolerated this condition only because they believed that the smoke acted like a blanket, trapping some of the earth's heat. Studies have shown this concept to be false. Orchard heaters are now designed to produce as little smoke as possible. (See Fig. 4.)

Another way to protect trees is to mix the cold air at the ground with the warmer air above, thus raising the temperature of the air next to the ground. Such mix-

Fig. 4 Orchard heaters.

Fig. 5 Wind machine.

ing can be accomplished by using **wind machines** (Fig. 5), which are power-driven fans that resemble airplane propellers. Farmers without their own

wind machines can rent air mixers in the form of helicopters. Although helicopters are effective

(continued on next page)

FOCUS ON AN APPLICATION, continued

in mixing the air, they are expensive to operate.

If sufficient water is available, trees can be protected by irrigation. On potentially cold nights, the orchard may be flooded. Because water has a high heat capacity, it cools more slowly than dry soil. Consequently, the surface does not become as cold as it would if it were dry. Furthermore, the relatively warm water will provide some warmth by radiating heat to the branches of the trees.

So far, we have discussed protecting trees against the cold air near the ground during a radiation inversion. Farmers often face another nighttime cooling problem. For instance, when subfreezing air blows into a region, the coldest air is not found at the surface; the air actually becomes colder with height. This

condition is known as a **freeze** or *advection frost*. A single freeze in California or Florida can cause several million dollars damage to citrus crops. For example, the crippling freeze of 1983 was one of the worst of the century to hit Florida. Losses from damaged fruit and trees was estimated at more than $1 billion, as more than one-third of Florida's citrus suffered moderate to severe damage. Fortunately, freezes are not frequent; in southern California they occur about once every 10 to 15 years.

Protecting an orchard from the damaging cold air blown by the wind can be a problem. Wind machines will not help because they would only mix cold air at the surface with the colder air above. Orchard heaters and irrigation are of little value as they

would only protect the branches just above the ground. However, there is one form of protection that does work: An orchard's sprinkling system may be turned on so that it emits a fine spray of water. In the cold air, the water freezes around the branches and buds, coating them with a thin veneer of ice. As long as the spraying continues, the latent heat—given off as the water changes into ice—keeps the ice temperature at 0°C (32°F). The ice acts as a protective coating against the subfreezing air by keeping the buds (or fruit) at a temperature higher than their damaging point. Care must be taken since too much ice can cause the branches to break. The fruit may be saved from the cold air, while the tree itself may be damaged by too much protection.

Daily, Monthly, and Yearly Temperatures
As you might expect after reading the previous section, the greatest variation in daily temperature occurs right at the earth's surface. In fact, the difference between the daily maximum and minimum temperature—called the **daily** (or **diurnal**) **range of temperature**—is greatest next to the ground and becomes progressively smaller as we move away from the surface. (See Fig. 5.14.) This daily variation in temperature is also much larger on clear days than on cloudy ones.

The largest diurnal range of temperature occurs on high deserts, where the air is fairly dry and cloud-free. By day, clear summer skies allow the sun's energy to quickly warm the ground which, in turn, warms the air above to a temperature sometimes exceeding 35°C (95°F). At night, the ground cools rapidly by radiating infrared energy to space, and the minimum temperature in these

regions occasionally dips below 4°C (39°F), thus giving a daily temperature range of 31°C (56°F).

A good example of a city with a large diurnal temperature range is Reno, Nevada, which is located on a plateau at an elevation of 1350 m (4400 ft) above sea level. Here, in the dry, thin summer air, the average daily maximum temperature for July is 33°C (92°F)—short-sleeve weather, indeed. But don't lose your shirt in Reno, for you will need it at night, as the average daily minimum temperature for July is 8°C (47°F). Reno has a daily range of 25°C (45°F)!

In humid regions, the diurnal temperature range is usually small. Here, haze and clouds lower the maximum temperature by preventing some of the sun's energy from reaching the surface. At night, the moist air keeps the minimum temperature high by absorbing the earth's infrared radiation and reradiating a portion of it to the ground. An ex-

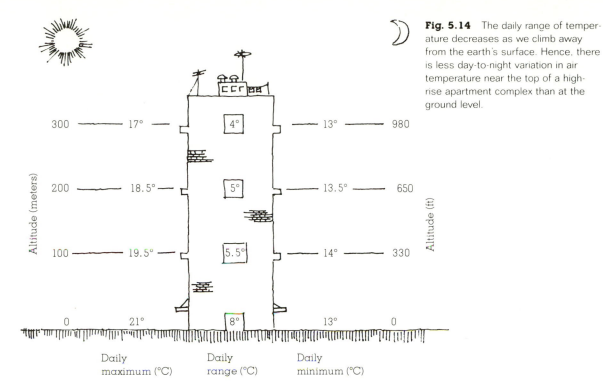

Fig. 5.14 The daily range of temperature decreases as we climb away from the earth's surface. Hence, there is less day-to-night variation in air temperature near the top of a high-rise apartment complex than at the ground level.

ample of a humid city with a small summer diurnal temperature range is Charleston, South Carolina, where the average July maximum temperature is 32°C (90°F), the average minimum is 22°C (72°F), and the diurnal range is only 10°C (18°F).

Cities near large bodies of water typically have smaller diurnal temperature ranges than cities further inland. This is caused in part by the additional water vapor in the air and by the fact that water warms and cools much more slowly than land.

The average of the highest and lowest temperature for a 24-hour period is known as the **mean daily temperature**. Most newspapers list the mean daily temperature along with the highest and lowest temperatures for the preceding day. The average of the mean daily temperatures for a particular date averaged for the past 30 years gives the average (or "*normal*") temperature for that date.

The average temperature for each month is the average of the daily mean temperature for that month. The average monthly temperatures throughout the world for January and July are given in Figs. 5.15 and 5.16. The lines on the maps are **isotherms**—lines connecting places that have the same temperature. To eliminate distor-

tions from mountain ranges and high plateaus, the isotherms are corrected to read at the same horizontal level (sea level). This is accomplished by adding to each station above sea level an amount of temperature that would correspond to the average temperature change with height. Since the energy that each latitude receives decreases from low to high latitudes, we find the highest temperatures in the tropics and subtropics and the lowest temperatures in polar regions.

Because there is a greater variation in solar radiation between low and high latitudes in winter than in summer, the isotherms in January are closer together (a tighter gradient) than they are in July. This means that if you travel from New Orleans to Detroit in January you are more likely to experience greater temperature variations than if you make the same trip in July.

At any location the difference between the average temperature of the warmest and coldest months is called the **annual range of temperature**. Usually the largest annual ranges occur over land; the smallest over water. Hence, inland cities have larger annual ranges than coastal cities. Near the equator (because daylight length varies little

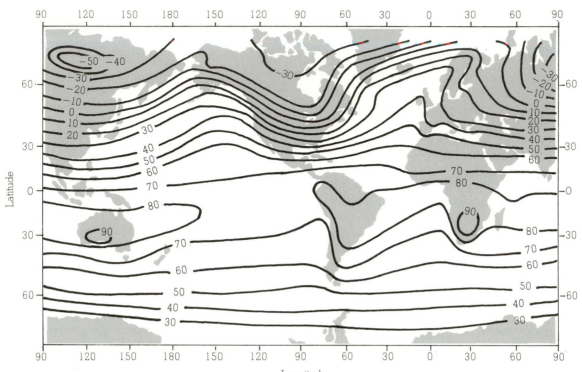

Fig. 5.15   Average air temperature near sea level in January (°F).

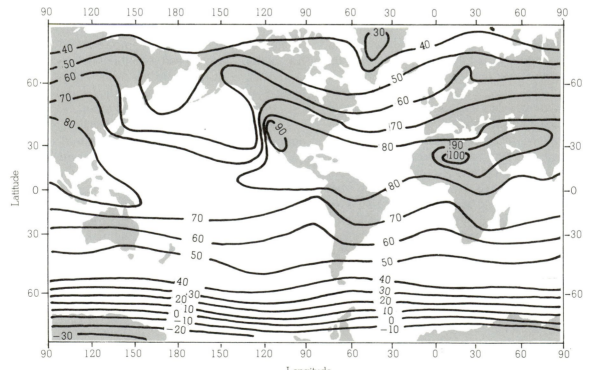

Fig. 5.16   Average air temperature near sea level in July (°F).

and the sun is always high in the noon sky), annual temperature ranges are small, usually less than 3°C (5°F). Quito, Ecuador—on the equator at an elevation of 2850 m (9350 ft)—experiences an annual range of less than 1°C. In middle and high latitudes, large seasonal variations in the amount of sunlight reaching the surface produce large temperature contrasts between winter and summer. Here, annual ranges are large, especially in the middle of a continent. Yakutsk, in northeastern Siberia near the Arctic Circle, has an extremely large annual temperature range of 62°C (144°F)!

The average temperature of any station for the entire year is the **mean annual temperature**, which represents the average of the twelve monthly average temperatures. When two cities have the same mean annual temperature, it might first seem that their temperatures throughout the year are quite similar. However, often this is not the case. For example, San Francisco, California, and Richmond, Virginia, are at the same latitude (37°N). Both have similar hours of daylight during the year; both have the same mean annual temperature— 14°C (57°F). Here, the similarities end. The temperature differences between the two cities are apparent to anyone who has traveled to San Francisco during the summer with a suitcase full of clothes suitable for summer weather in Richmond.

Tables 5.2 and 5.3 summarize the average temperatures for San Francisco and Richmond. The coldest month for both cities is January. January in Richmond averages only 8°C (14°F) colder than January in San Francisco. However, people in Richmond awaken to an average January minimum temperature of −6°C (21°F), much colder than the lowest temperature ever recorded in San Francisco, −1°C (30°F). Trees that thrive in San Francisco's weather would find it difficult surviving a winter in Richmond. So, even though San Francisco and Richmond have the same mean annual temperature, the behavior and range of their temperatures differ greatly.

Land-Water Temperature Contrasts   We have seen how both land and water can influence the daily and seasonal temperatures of a particular region. On a much larger scale, we can see in Figs. 5.15 and 5.16 that, in many places, the isotherms bend as they approach an ocean-continent boundary. On the January map, the temperatures are much lower in the middle of continents than they are at the same latitude near the oceans; on the July map, the reverse is true. The reason for these temperature variations can be attributed to the unequal heating and cooling properties of land and water. For one thing, solar energy reaching land is absorbed in a thin layer of soil; reaching water, it penetrates deeply so the heat is mixed through a much deeper layer. Also, some of the solar energy striking the water is used to evaporate water rather than heat it.

Another important reason for the temperature

Table 5.2 Temperatures for San Francisco, California (37°N)

	JAN.	FEB.	MAR.	APRIL	MAY	JUNE	JULY	AUG.	SEPT.	OCT.	NOV.	DEC.	AVERAGE YEAR
Average °C	11	12	13	14	14	15	15	15	17	16	14	11	14
Average °F	(51)	(53)	(55)	(57)	(57)	(59)	(59)	(59)	(62)	(61)	(57)	(52)	57

Record high: 38°C (101°F)   Record low: −1°C (30°F)

Table 5.3 Temperatures for Richmond, Virginia (37°N)

	JAN.	FEB.	MAR.	APRIL	MAY	JUNE	JULY	AUG.	SEPT.	OCT.	NOV.	DEC.	AVERAGE YEAR
Average °C	3	4	9	14	19	24	26	24	21	15	9	4	14
Average °F	(38)	(40)	(48)	(57)	(67)	(75)	(78)	(75)	(70)	(59)	(48)	(40)	57

Record high: 40°C (104°F)   Record low: −24°C (−12°F)

Degree-Days

The mean temperature for any day indicates the amount of fuel necessary for home or industrial heating. Estimates are that most people use their furnace when the mean daily temperature drops below 65°F (18°C). Therefore, an index used by heating engineers, called the **heating degree-day**, assumes that there will be little or no demand for fuel when the mean daily temperature is 65°F or above. Heating degree-days are determined by subtracting the mean temperature for the day from 65°F. Thus, if the mean temperature for a day is 55°F, there would be 10 (65° − 55°) heating degree-days on this day. On days when the mean temperature is above 65°F, there are no heating degree-days. Hence, the lower the average daily temperature, the more heating degree-days and the greater the predicted consumption of fuel. When the number of heating degree-days for a whole year is calculated, the heating fuel requirements for any location can be estimated. Figure 6 shows the yearly average number of heating degree-days in various locations throughout the United States.

A similar index, called the **cooling degree-day**, is used

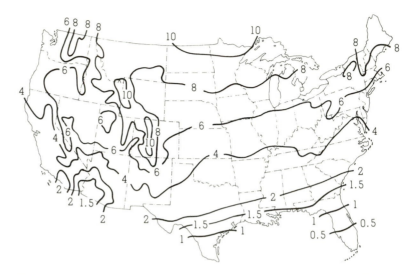

Fig. 6 Mean annual total heating degree-days in thousands of °F (base 65°F).

during warm weather to estimate the energy needed to cool indoor air to a comfortable level. The forecast of mean daily temperature is converted to cooling degree-days by subtracting 65°F from the mean. The remaining value is the number of cooling degree-days for that day. For example, a day with a mean temperature of 70°F would correspond to (70 − 65), or 5 cooling degree-days. High values indicate warm weather and high power production for cooling. (See Fig. 7.)

Knowledge of the number of

cooling degree-days in an area allows a builder to plan the size and type of equipment that should be installed to provide adequate air conditioning. Also, the forecasting of cooling degree-days during the summer gives power companies a way of predicting the energy demand during peak energy periods. A composite of heating plus cooling degree-days would give a practical indication of the energy requirements over the year.

Farmers use an index, called **growing degree-days**, as a

contrasts is that water has a high *specific heat*. As we saw in Chapter 3, it takes a great deal more heat to raise the temperature of 1 g of water 1°C than it does to raise the temperature of 1 g of soil or rock by 1°C.

Water not only heats more slowly than land, it cools more slowly as well, and so the oceans act

like huge heat reservoirs. Thus, mid-ocean surface temperatures change relatively little from summer to winter compared to the much larger annual temperature changes over the middle of continents.

Along the margin of continents, ocean currents can influence air temperatures. For example, along the eastern margins, warm ocean currents trans-

FOCUS ON AN APPLICATION, continued

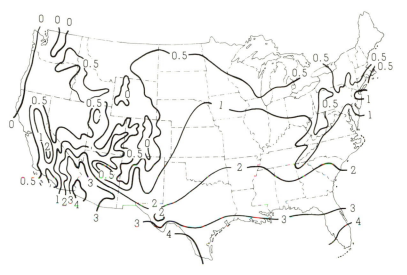

Fig. 7 Mean annual total cooling degree-days in thousands of °F (base 65°F).

Table 1 Estimated Growing Degree-Days for Certain Agricultural Crops to Reach Maturity

CROP (VARIETY, LOCATION)	BASE TEMPERATURE (°F)	GROWING DEGREE-DAYS TO MATURITY
Beans (Snap, South Carolina)	50	1200–1300
Corn (Sweet, Indiana)	50	2200–2800
Cotton (Delta Smooth Leaf, Arkansas)	60	1900–2500
Peas (Early, Indiana)	40	1100–1200
Rice (Vegold, Arkansas)	60	1700–2100
Rice (Bluebonnet, Arkansas)	60	2400–2600
Wheat (Indiana)	40	2100–2400

guide to planting and for determining the approximate dates when a crop will be ready for harvesting. A growing degree-day for a particular crop is defined as a day on which the mean daily temperature is one degree above the *base temperature*—the minimum temperature required for growth of that crop. For sweet corn, the base temperature is 50°F and, for peas, it is 40°F. (The United States Department of Agriculture uses °F.)

On a summer day in Iowa, the mean temperature might be 80°F. From Table 1, we can see that, on this day, peas would accumulate 40 growing degree-days. With the same mean temperature, a crop of sweet corn would have 30 growing degree-days. Theoretically, sweet corn can be harvested when it accumulates a total of 2200 growing degree-days. So, if sweet corn is planted in early April and each day thereafter averages about 20 growing degree-days, the corn would be ready for harvest about 110 days later, or around the middle of July. Although moisture and other conditions are not taken into account, growing degree-days nevertheless serve as a useful guide in forecasting approximate dates of crop maturity.

port warm water poleward, while, along the western margins, they transport cold water equatorward. Some coastal areas also experience upwelling, which brings even colder water to the surface. (See Chapter 15.)

Even large lakes can modify the temperature around them. In summer, the Great Lakes remain cooler than the land. As a result, refreshing breezes blow inland, bringing relief from the sometimes sweltering heat. As winter approaches, the water cools more slowly than the land. The first blast of cold air from Canada is modified as it crosses the lakes, and so the first freeze is delayed on the eastern shores of Lake Michigan.

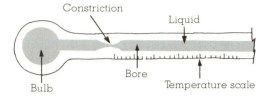

Fig. 5.17   A section of a maximum thermometer.

We have been concentrating on how air temperature changes both daily and annually. Now, we will consider how air temperatures are measured.

Measuring Air Temperature

Thermometers   Thermometers were developed to measure air temperature. Each thermometer has a definite scale, so that a thermometer reading of 0°C (32°F) in Vermont will indicate the same temperature as a thermometer with the same reading in North Dakota. If a particular reading were to represent different degrees of hot or cold, depending on location, thermometers would be useless.

Liquid-in-glass thermometers are the most commonly used instruments for measuring surface air temperature, for they are easy to read and inexpensive to construct. These thermometers have a glass bulb attached to a sealed, graduated tube about 25 cm (10 in.) long. A very small opening, or bore, extends from the bulb to the end of the tube. A liquid in the bulb (usually mercury or red-colored alcohol) is free to move from

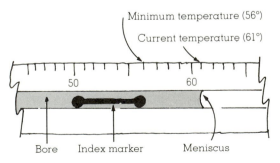

Fig. 5.18   A section of a minimum thermometer showing both the current air temperature and the minimum temperature.

the bulb up through the bore and into the tube. When the air temperature increases, the liquid in the bulb expands, and rises up the tube. When the air temperature decreases, the liquid contracts, and moves down the tube. Hence, the length of the liquid in the tube represents the air temperature. Because the bore is very narrow, a small temperature change will show up as a relatively large change in the length of the liquid column.

Maximum and minimum thermometers are liquid-in-glass thermometers used exclusively for determining daily maximum and minimum temperatures. The **maximum thermometer** looks like any other liquid-in-glass thermometer with one exception: It has a small constriction within the bore just above the bulb (Fig. 5.17). As the air temperature increases, the mercury expands and freely moves past the constriction up the tube, until the maximum temperature occurs. However, as the air temperature begins to drop, the small constriction prevents the mercury from flowing back into the bulb. Thus, the end of the stationary mercury column indicates the maximum temperature for the day. The mercury will stay at this position until either the air warms to a higher reading or the thermometer is reset by whirling it on a special holder and pivot. Usually, the whirling is sufficient to push the mercury back into the bulb past the constriction until the end of the column indicates the present air temperature.*

A **minimum thermometer** measures the lowest temperature reached during a given period. Most minimum thermometers use alcohol as a liquid, since it freezes at a temperature of −130°C compared to −39°C for mercury. The minimum thermometer is similar to other liquid-in-glass thermometers except that it contains a small dumbbell-shaped index marker in the bore (Fig. 5.18). The index marker is about 2.5 cm long and is free to slide back and forth within the liquid. It cannot move out of the liquid because the surface tension at the end of the liquid column (the meniscus) holds it in.

A minimum thermometer is mounted horizontally. As the air temperature drops, the contracting

*Thermometers that measure body temperature are maximum thermometers. It should be apparent why they are shaken both before and after you take your temperature.

liquid moves back into the bulb and brings the index marker down the bore with it. When the air temperature stops decreasing, the liquid and the index marker stop moving down the bore. As the air warms, the alcohol expands and moves freely up the tube past the stationary index marker. Because the index marker does not move as the air warms, the minimum temperature is read by observing the upper end of the marker.

To reset a minimum thermometer, simply tip it upside down. This allows the index marker to slide to the upper end of the alcohol column, which is indicating the current air temperature. The thermometer is then remounted horizontally, so that the marker will move toward the bulb as the air temperature decreases.

Highly accurate temperature measurements can be made with **electrical resistance thermometers**. A **thermistor** is a type of electrical thermometer consisting of electrical conducting wires covered with a piece of ceramic. The wires are usually platinum or nickel, occasionally copper. Thermistors do not actually measure air temperature; they measure the electrical resistance of the wire, which increases as the temperature increases. An electrical meter measures the resistance and is calibrated to represent air temperature.

Although thermistors are not widely used for everyday surface observations, they are used extensively in remote sensing of air temperature. Radiosondes use thermistors to measure the air temperature from the surface up to an altitude near 30 km (18 mi).

Another electrical thermometer is the **thermocouple**. This device operates on the principle that the temperature difference between the junction of two dissimilar metals sets up a weak electrical current. When one end of the junction is maintained at a temperature different from that of the other end, an electrical current will flow in the circuit. This current is proportional to the temperature difference between the junctions.

Air temperature may also be obtained with instruments called *infrared sensors*, or **radiometers**. Radiometers do not measure temperature directly; rather, they measure emitted radiation (usually infrared). By measuring both the intensity of radiant energy and the wavelength of maximum emission of a particular gas (either water vapor or carbon dioxide), radiometers in orbiting satellites are now able to estimate the air temperature at selected levels in the atmosphere.

A **bimetallic thermometer** consists of two different pieces of metal (usually iron and brass) welded together to form a single strip. As the temperature changes, one metal expands more than the other, causing the strip to bend. The small amount of bending is amplified through a system of levers to a pointer on a calibrated scale.

The bimetallic thermometer is usually the temperature sensing part of the **thermograph**, an instrument that measures and records temperature. On a thermograph, the pointer is a pen that sits on a circular drum (Fig. 5.19). The drum, which is covered with a piece of chart paper, is slowly rotated by a clock-drive. As the temperature of the air changes, the bimetallic thermometer bends accordingly and the pen moves up or down, marking the chart paper with the current air temperature. Since time is printed on the top of the chart

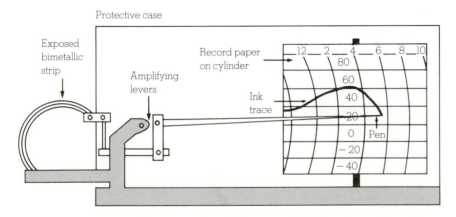

Fig. 5.19 The thermograph with a bimetallic thermometer.

Thermometers Should Be Read in the Shade

When we measure air temperature with a common liquid thermometer, an incredible number of air molecules bombard the bulb, transferring energy either to or away from it. When the air is warmer than the thermometer, the liquid gains energy, expands, and rises up the tube; the opposite will happen when the air is colder than the thermometer. The liquid stops rising (or falling) when equilibrium between incoming and outgoing energy is established. At this point, we can read the temperature by observing the height of the liquid in the tube.

It is impossible to measure air temperature accurately in direct sunlight because the thermometer absorbs radiant energy from the sun in addition to energy from the air molecules. The thermometer gains energy at a much faster rate than it can radiate it away, and the liquid keeps expanding and rising until there is equilibrium between incoming and outgoing energy. Because of the direct absorption of solar energy, the level of the liquid in the thermometer indicates a temperature much higher than the actual air temperature. Hence, a thermometer must be kept in a shady place to measure the temperature of the air accurately.

---

paper, a continuous recording of temperature is obtained. Because bimetallic thermometers are less accurate than liquid-in-glass thermometers, thermographs must be checked periodically and corrected using an accurate liquid thermometer.

Instrument Shelters Thermometers and other instruments are usually housed in an *instrument shelter* (Fig. 5.20). The shelter completely encloses the instruments, protecting them from rain, snow, and the sun's direct rays. It is painted white to reflect sunlight and has louvered sides, so that air is free to flow through it. This helps to keep the air inside the shelter at the same temperature as the air outside.

The thermometers inside a standard shelter are mounted about 1.5 m (5 ft) above the ground. As we saw in an earlier section, on a clear, calm night the air at ground level may be much colder than the air at the level of the shelter. As a result, on clear winter mornings it is possible to see ice or frost on the ground even though the minimum thermometer in the shelter did not reach the freezing point.

Because air temperatures vary considerably above different types of surfaces, shelters are usually placed over grass to insure that the air temperature is measured at the same elevation over the same type of surface. Unfortunately, some shelters are placed on asphalt, others sit on concrete, while others are located on the tops of tall buildings, making it difficult to compare air tem-

Fig. 5.20 An instrument shelter protects the instruments inside from the weather elements.

perature measurements from different locations. In fact, if either the maximum or minimum air temperature in your area seems suspiciously different from those of nearby towns, find out where the instrument shelter is situated.

Summary

The earth has seasons because the earth is tilted on its axis as it revolves around the sun. The tilt of the earth causes a seasonal variation in both the length of daylight and the intensity of sunlight that reaches the surface. When the Northern Hemisphere is tilted toward the sun, the Southern Hemisphere is tilted away from the sun. Longer hours of daylight and more intense sunlight produce summer in the Northern Hemisphere, while, in the Southern Hemisphere, shorter daylight hours and less intense sunlight produce winter. On a more local setting, the earth's inclination influences the amount of solar energy received on the north and south side of a hill, as well as around a home.

The daily variation in air temperature near the earth's surface is controlled mainly by the input of energy from the sun and the output of energy from the surface. On a clear, calm day, the surface air warms, as long as heat input (mainly sunlight) exceeds heat output (mainly convection and radiated infrared energy). The surface air cools at night, as long as heat output exceeds input. Because the ground at night cools more quickly than the air above, the coldest air is normally found at the surface where a radiation inversion usually forms. When the air temperature in agricultural areas drops to dangerously low readings, fruit trees and grape vineyards can be protected from the cold by a variety of means, from mixing the air to spraying the trees and vines with water.

The greatest daily variation in air temperature occurs at the earth's surface. Both the diurnal and annual range of temperature are greater in dry climates than in humid ones. Even though two cities may have similar average annual temperatures, the range and extreme of their temperatures can differ greatly. Temperature information influences our lives in many ways, from deciding what clothes to take on a trip to providing critical information for energy-use predictions and agricultural planning. Thermometers designed to measure air temperatures near the surface are housed in instrument shelters to protect them from direct sunlight and precipitation.

Questions for Review

1. In the Northern Hemisphere, why are summers warmer than winters, even though the earth is actually closer to the sun in winter?

2. What are the main factors that determine seasonal temperature variations?

3. During the summer, why are middle latitudes generally warmer than northern latitudes?

4. If it is winter and January in New York City, what is the season and month in Sydney, Australia?

5. Using your own words, define the solar constant.

6. Explain why Southern Hemisphere summers are not warmer than Northern Hemisphere summers.

7. Explain why the vegetation on the north-facing side of a hill is frequently different from the vegetation on the south-facing side of the same hill.

8. If you were a consultant to a large corporation constructing a new ski area, on which side of a hill would you suggest they build it? Why?

9. How is the directional orientation of a house important to heating and cooling it?

10. Draw a vertical profile of air temperature from the ground to an elevation of 3 m on a clear, windless (a) afternoon and (b) early morning just before sunrise. Explain why the temperature curves are different.

11. What are some of the factors that determine the daily fluctuation of air temperature just above the ground?

12. Explain why "thermal belts" are found along hillsides at night.

13. Why is there a lag between the time of maximum solar heating (noon) and the time of maximum air temperature several meters above the ground?

14. What weather conditions are best suited for the formation of a strong radiation inversion?

15. List some of the measures farmers use to protect their crops against the cold. Explain the physical principle behind each method.

16. Why are the lower tree branches most susceptible to damage from low temperatures?

17. Explain why large daily temperature ranges are observed (a) in dry regions and (b) over soil that is a poor thermal conductor.

18. During the winter, frost can form on the ground when the minimum thermometer indicates a low temperature above freezing. Explain.

19. Explain how the following thermometers measure air temperature:
(a) liquid-in-glass
(b) bimetallic
(c) thermistor
(d) radiometer

20. Why do the first freeze in autumn and the last freeze in spring occur in bottomlands?

Questions for Thought

1. In about 11,000 years, the Northern Hemisphere will be closer to the sun during the summer solstice. How might this situation influence the climate of the Northern Hemisphere?

2. Do you feel that the situation described in question 1 will change the length of summer and spring? What about fall and winter? Explain.

3. Explain (with the aid of a diagram) why the morning sun shines brightly through a south-facing bedroom window in December, but not in June.

4. If the tilt of the earth were decreased to only 10°, how would this change the summer and winter temperatures in your area? Explain, using a diagram.

5. If the tilt of the earth increased to 40°, explain how you feel the summer and winter temperatures would change in your area.

6. At the top of the earth's atmosphere during the early summer (Northern Hemisphere), above what latitude would you expect to receive the most solar radiation in one day? During the same time of year, where would you expect to receive the most solar radiation at the surface? Explain why the two locations are different. (If you are having difficulty with this question, refer to Fig. 5.5.)

7. If a construction company were to build a solar-heated home in middle latitudes in the Southern Hemisphere, in which direction should the solar panels on the roof be directed for maximum daytime heating?

8. Suppose you live in a middle latitude city (latitude 40°N), where the coldest part of the year is always around the middle of January. If you were to construct a solar-heated home, at what angle would you construct the roof (cost is not a factor) so that it would absorb the sun's rays at a right angle during this cold period?

9. Aside from the aesthetic appeal (or lack of such), explain why painting the outside north-facing wall of a house one color and the south-facing wall another color is not a bad idea.

10. How would the lag in daily temperature experienced over land compare to the daily temperature lag over water?

11. Where would you expect to experience the smallest variation in temperature from year to year and from month to month? Why?

12. The average temperature in San Francisco, California, for December, January, and February is 11°C (52°F). During the same three-month period the average temperature in Richmond, Virginia, is 4°C (39°F). Yet, San Francisco and Richmond have nearly the same yearly total of heating degree-days. Explain why. (Hint: See Table 5.2.)

13. On a warm summer day, one city experienced a daily range of 22°C (40°F), while another had a daily range of 10°C (18°F). One of these cities is located in New Jersey and the other in New Mexico. Which location most likely had the highest daily range, and which one had the smallest? Explain.

14. Give *two* reasons why a minimum thermometer does not use mercury as a liquid.

15. Minimum thermometers are usually read during the morning, yet they are reset in the afternoon. Explain why.

Problems and Exercises

1. Draw a graph similar to Fig. 5.5 and include the amount of solar radiation reaching the earth's surface at each latitude on the equinox and on the winter solstice.

2. Calculate the noon angle of the sun above the southern horizon at your latitude on the following dates: (a) December 22; (b) June 22; (c) September 23; and (d) March 21.

3. Each day past the winter solstice the noon sun is a little higher above the southern horizon. (a) Determine how much change takes place each day at your latitude. (b) Does the same amount of change take place at each latitude in the Northern Hemisphere? Explain.

4. On approximately what dates will the sun be overhead at noon at latitudes: (a) 10°N? (b) 15°S?

5. Design a solar-heated home that sits on the north side of an east-west running street. If the home is located at 40°N, draw a proper roof angle for maximum solar heating. Design windows, doors, overhangs, and rooms with the intent of reducing heating and cooling costs. Place trees around the home that will block out excess summer sunlight and yet let winter sunlight inside. Choose a paint color for the house that will add to the home's energy efficiency.

6. Suppose peas are planted in Indiana on May 1. If the peas need 1200 growing degree-days before they can be picked, and if the average maximum temperature for May and June is 80°F and the average minimum is 60°F, on about what date will the peas be ready to pick? (Assume a base temperature of 40°F.)

7. Use the maximum and minimum temperatures in the daily newspaper to identify those cities with large numbers of heating degree-days. Calculate the total heating degree-days for each city. With the aid of Fig. 6, p. 104, determine the percent of the yearly total heating degree-days each city received on that day. Do the same for cities with large numbers of cooling degree-days.

A quiet, humid summer afternoon is illuminated with bands of crepuscular rays. (Photo: Ross DePaola)

CHAPTER 6

Light, Color, and Atmospheric Optics

In clear weather, the sky appears blue, while the horizon appears milky white. Sunrises and sunsets can fill the sky with brilliant shades of pink, red, orange, and purple. At night, the sky is black, except for the light from the stars, planets, and the moon. The moon's size and color seem to vary during the night, and the stars twinkle. To understand what we see in the sky, we will take a closer look at sunlight, examining how it interacts with the atmosphere to produce an array of atmospheric visuals.

White and Colors

We know from Chapter 3 that nearly half of the solar radiation that reaches the atmosphere is in the form of visible light. As sunlight enters the atmosphere, it is either absorbed, reflected, scattered, or transmitted on through. How objects at the surface respond to this energy depends on their general nature (color, density, composition) and the wavelength of light that strikes them. How do we see? Why do we see various colors? What kind of visual effects do we observe because of the interaction between light and matter? In particular, what can we *see* when light interacts with our atmosphere?

We perceive light because electromagnetic waves stimulate antenna-like nerve endings in the retina of the human eye. These antennae are of two types—*rods* and *cones*. The rods respond to all wavelengths of visible light and give us the ability to distinguish light from dark. If people possessed rod-type receptors only, then only black and white vision would be possible. The cones respond to specific wavelengths of radiation between 0.4 and 0.7 μm. The cones fire an impulse through the nervous system to the brain and we

Contents

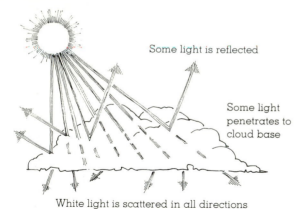

Fig. 6.1 Since tiny cloud droplets scatter visible light in all directions, light from many billions of droplets turns the underside of a cloud white.

perceive this impulse as the sensation of color. (Color blindness is caused by missing or malfunctioning cones.) Wavelengths of radiation shorter than 0.4 µm, or longer than 0.7 µm, do not stimulate color vision in humans.

White light is perceived when all visible wavelengths strike the cones of the eye with nearly equal intensity. Because the sun radiates almost half of its energy as visible light, all visible wavelengths from the midday sun reach the cones, and the sun usually appears white. A star that is cooler than our sun radiates most of its energy at slightly longer wavelengths; therefore, it appears redder. On the other hand, a star much hotter than our sun radiates more energy at shorter wavelengths and thus appears bluer. A star whose temperature is about the same as the sun's appears white.

Objects that are not hot enough to produce

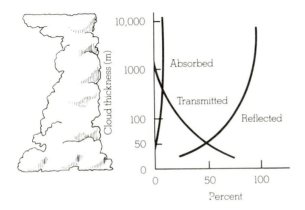

Fig. 6.2 Average percent of radiation reflected, absorbed, and transmitted by clouds of various thickness.

radiation at visible wavelengths can still have color. Everyday objects we see as red are those that absorb all visible radiation except red. The red light is reflected from the object to our eyes. Blue objects have blue light returning from them, since they absorb all visible wavelengths except blue. Some surfaces absorb all visible wavelengths and reflect no light at all. Since no radiation strikes the rods or cones, these surfaces appear black. Therefore, when we see colors, we know that light must be reaching our eyes.

White Clouds and Scattered Light

One exciting feature of the atmosphere can be experienced when we watch the underside of a puffy, growing cumulus cloud change color from white to dark gray or black. When we see this change happen, our first thought is usually, "It's going to rain." Why is the cloud initially white? Why does it change color? To answer these questions, let's investigate the concept of **scattering**.

When sunlight bounces off a surface at the same angle at which it strikes the surface, we say that the light is *reflected*, and call this phenomenon **reflection**. There are various constituents of the atmosphere, however, that tend to deflect solar radiation from its path and send it out in all directions. We know from Chapter 3 that radiation reflected in this way is said to be *scattered*. (Scattered light is also called *diffuse light*.) During the scattering process, no energy is gained or lost and, therefore, no temperature changes occur. In the atmosphere, scattering is usually caused by small objects, such as air molecules, fine particles of dust, water molecules, and some pollutants. Just as the ball in a pinball machine bounces off the pins in many directions, so solar radiation is knocked about by small particles in the atmosphere.

Cloud droplets about 20 µm or so in diameter effectively scatter visible radiation. Some of the sunlight that penetrates to the bottom of a small cloud is scattered in all directions by tiny cloud droplets. (See Fig. 6.1.) Since these droplets scatter all wavelengths of visible light just about equally, we perceive the color white when we look at the underside of many small or thin clouds. Figure 6.2 shows that, as a cloud grows larger and taller, more sunlight is reflected from the cloud and, hence, less light can penetrate all the way

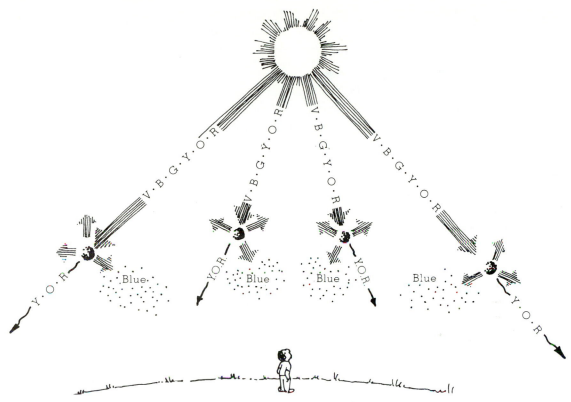

Fig. 6.3 The sky appears blue because billions of air molecules selectively scatter the shorter wavelengths of visible light more effectively than the longer ones. This causes us to see blue light coming from all directions.

through it. In fact, practically no sunlight penetrates a cloud whose thickness is 1000 m (3300 ft).

Since little sunlight reaches the underside of the cloud, little light is scattered, and the cloud base appears dark. At the same time, if droplets near the cloud base grow large, they become less effective scatterers and better absorbers. As a result, the visible light that does reach this part of the cloud is absorbed rather than scattered. Light no longer reaches the eye from the underside of the cloud, and the cloud appears black. These cloud droplets may even grow large and heavy enough to fall to the earth. From our casual observations of clouds, we know that dark, threatening ones frequently produce rain. Now, we know why they appear so dark.

Blue Skies and Hazy Days

The sky appears blue because light that stimulates the sensation of blue color is reaching the retina of the eye. How does this happen?

Individual air molecules are much smaller than cloud droplets—their diameters are small even when compared with the wavelength of visible light. Each air molecule of oxygen and nitrogen is a selective scatterer in that each scatters shorter waves of visible light much more effectively than longer waves. (This selective scattering is also known as *Rayleigh scattering*.)

As sunlight enters the atmosphere, the shorter visible wavelengths of violet, blue, and green are scattered more by atmospheric gases than are the longer wavelengths of yellow, orange, and especially red. (Violet light is scattered about 10 times more than red light.) As we view the sky, the scattered waves of violet, blue, and green strike the eye from all directions. Viewed together, these waves produce the sensation of blue light coming from all around us. (See Fig. 6.3.) Therefore, when we look at the sky it appears blue. (Earth, by the way, is not the only planet with a colorful sky. On Mars, dust in the air turns the sky red at midday and purple at sunset.)

The selective scattering of blue light by air mol-

Fig. 6.4   The scattering of sunlight by dust and haze produces these white bands of crepuscular rays.

ecules and very small particles can make distant mountains appear blue, such as the Blue Ridge Mountains of Virginia and the Blue Mountains of Australia. (See color plate 2.) In some places, a *blue haze* may cover the landscape, even in areas far removed from human contamination. Although its cause is still controversial, the blue haze appears to be the result of a particular process. Extremely tiny particles (hydrocarbons called *terpenes*) are released by vegetation to combine chemically with ozone, which filters down from the stratosphere. This reaction produces small particles (about 0.2 μm in diameter) that selectively scatter blue light.

When small particles, such as fine dust and salt, become suspended in the atmosphere, the color of the sky begins to change from blue to milky white. Although these particles are small, they are large enough to scatter all wavelengths of visible light fairly evenly. They do not show a preference for scattering a particular wavelength, and thus scatter all waves in all directions. (This type of scattering is called *Mie scattering*.) When our eyes are bombarded by all wavelengths of visible light, the sky appears milky white, and we call the day "hazy." Occasionally, water vapor will attach itself to these floating particles and further enhance their ability to scatter light. Thus, the color of the sky gives us a hint about how much material is suspended in the air: the more particles, the more scattering, and the whiter the sky becomes. On top of a high mountain, when we are above many of these haze particles, the sky usually appears a deep blue.

Haze can scatter light from the rising or setting sun, so that you see bright lightbeams, or **crepuscular rays**, radiating across the sky. A similar effect occurs when the sun shines through a break in a layer of clouds. Dust, tiny water droplets, or haze in the air beneath the clouds scatters sunlight, making that region of the sky appear bright with rays. Because these rays seem to reach downward from clouds, some people will remark that the "sun is drawing up water." In England, this same phenomenon is referred to as "Jacob's ladder." No matter what these sunbeams are called, it is the scattering of sunlight by particles in the atmosphere that makes them visible.

In summary, the scattering of light by small particles in the atmosphere causes many familiar effects: white clouds, blue skies, hazy skies, sunbeams, and colorful sunsets. In the absence of any scattering, we would simply see a white sun against a black sky—not an attractive alternative.

Twinkling and Twilight

Light that passes through a substance is said to be *transmitted*. Upon entering a denser substance, transmitted light slows in speed. If it enters the substance at an angle, the light's path also bends. This bending is called **refraction**. The amount of refraction depends primarily on two factors: the density of the material and the angle at which the light enters the material.

Refraction can be demonstrated in a darkened room by shining a flashlight into a beaker of water (Fig. 6.5). If the light is held directly above the water so that the beam strikes the surface of the water straight on, no bending occurs. But, if the light enters the water at some angle, it bends toward the normal, which is the dashed line in the diagram running perpendicular to the air-water boundary. (The normal is simply a line that intersects any surface at a right angle. We use it as a reference to see how much bending occurs as light enters and leaves various substances.) A small mirror on the bottom of the beaker reflects the light upward. This reflected light bends away from the normal as it re-enters the air. We can summarize these observations as follows: *Light that travels from a less-dense to a more-dense medium loses speed and bends toward the normal, while light that enters a less-dense medium increases in speed and bends away from the normal.*

The refraction of light within the atmosphere causes a variety of visual effects. At night, for example, the light from the stars that we see directly above us is not bent, but starlight that enters the earth's atmosphere at an angle is bent.

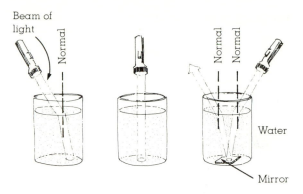

Fig. 6.5 The behavior of light as it enters and leaves a more-dense substance.

In fact, a star whose light enters the atmosphere just above the horizon has more atmosphere to penetrate and is thus refracted the most. As we can see in Fig. 6.6, the bending is toward the normal as the light enters the more-dense atmosphere. By the time this "bent" starlight reaches our eyes, the star appears to be higher than it actually is because our eyes cannot detect that the light path is bent. We see light coming from a particular direction, and interpret the star to be in that direction. So, the next time you take a midnight stroll, point to any star near the horizon and

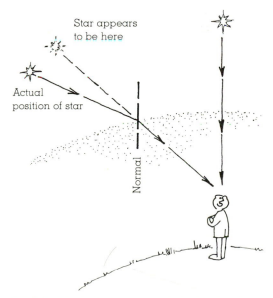

Fig. 6.6 Due to the bending of starlight by the atmosphere, stars not directly overhead appear to be higher than they really are.

Red Suns and Blue Moons

At midday, the sun seems intensely white, while at sunset it usually appears to be yellow, orange, or red. At noon, when the sun is high in the sky, light from the sun is most intense—all wavelengths of visible light are able to reach the eye with about equal intensity, and the sun appears white. (Looking directly at the sun, especially during this time of day, can cause irreparable damage to the eye. Normally, we get only glimpses or impressions of the sun out of the corner of our eye.)

Near sunrise or sunset, however, the rays coming directly from the sun strike the atmosphere at a low angle. They must pass through much more atmosphere than at any other time during the day. (When the sun is 4° above the horizon, sunlight must pass through an atmosphere more than 12 times thicker than when the sun is directly overhead.) By the time sunlight has penetrated this large amount of air, most of the shorter waves of visible light have been

scattered away by the air molecules. Just about the only waves from a setting sun that make it on through the atmosphere on a fairly direct path are the yellow, orange, and red. Upon reaching the eye, these waves produce a bright yellow-orange sunset. (See Fig. 1.)

Bright, yellow-orange sunsets only occur when the atmosphere is fairly clean, such as after a recent rain. If the atmosphere contains many fine particles whose diameters are a little larger than air molecules, slightly longer (yellow) waves also would be scattered away. Only orange and red waves would penetrate through to the eye, and the sun would appear red-orange. When the atmosphere becomes loaded with particles, only the longest red wavelengths are able to penetrate the atmosphere, and we see a red sun.

Natural events may produce red sunrises and sunsets. Over the oceans, for example, the scattering characteristics of

small, suspended salt particles and water vapor are responsible for the brilliant red suns observed from the beach. Moreover, major volcanic eruptions send vast amounts of dust and ash high into the atmosphere. These fine particles, moved by the winds aloft, circle the globe, producing beautiful sunrises and sunsets for months and even years. There were beautiful ruddy sunsets in many parts of the Northern Hemisphere after the volcano El Chichón erupted in Mexico during April 1982. (See color plate 3.)

Occasionally, the atmosphere becomes so laden with dust, smoke, and pollutants that even red waves are unable to pierce the filthy air. An eerie effect then occurs. Because no visible waves enter the eye, the sun literally disappears before it reaches the horizon! Hopefully, you have not had to witness such an atmospheric event.

The scattering of light by large quantities of atmospheric

---

remember: this is where the star appears to be. To point to the star's true position, you would have to lower your arm just a bit (about one-half a degree, according to Table 6.1).

Table 6.1   The Amount of Atmospheric Refraction (Bending) in Minutes Viewed at Sea Level under Standard Atmospheric Conditions (60 minutes equals 1°)

ELEVATION ABOVE HORIZON (degrees)	REFRACTION (minutes)
0°	35.0
5°	10.0
20°	2.6
40°	1.2
60°	0.6
90°	0.0

As starlight enters the atmosphere it often passes through regions of differing air density. Each of these regions deflects and bends the tiny beam of starlight, constantly changing the apparent position of the star. This causes the star to appear to *twinkle* or flicker, a condition known as **scintillation**. Planets, being much closer to us, appear larger, and usually do not twinkle because their size is greater than the angle at which their light deviates as it penetrates the atmosphere. Planets sometimes twinkle, however, when they are near the horizon, where the bending of their light is greatest.

The refraction of light by the atmosphere has some other interesting consequences. For example, the atmosphere gradually bends the rays

FOCUS ON A SPECIAL TOPIC, continued

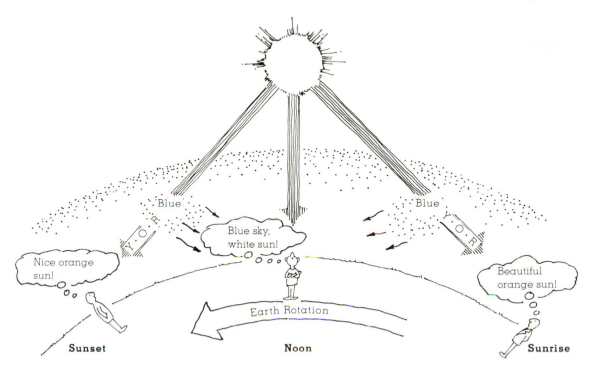

Fig. 1 Because of the selective scattering by a thick section of atmosphere, the sun at sunrise and sunset appears either yellow, orange, or red. At noon, it is usually white.

particles can cause some rather unusual sights. If the volcanic ash, dust, or smoke particles are roughly uniform in size, they can selectively scatter the sun's rays. Even at noon, various colored suns have appeared: orange suns, green suns, and even blue suns. Although rare, the same phenomenon can happen to moonlight, making the moon appear blue; thus, the expression "once in a blue moon."

from a rising or setting sun or moon. Because light rays from the lower part of the sun (or moon) are bent more than those from the upper part, the sun appears to flatten-out on the horizon, taking on an elliptical shape. Also, since light is bent most on the horizon, the sun and moon both appear to be higher than they really are (Fig. 6.7). Consequently, they both rise about two minutes earlier and set about two minutes later than they would if there were no atmosphere.

You may have noticed that on clear days the sky is often bright for some time after the sun sets. The atmosphere refracts and scatters sunlight to our eyes, even though the sun itself has disappeared from our view. **Twilight** is the name given to the time after sunset (and immediately before sunrise) when the sky remains illuminated and allows outdoor activities to continue without artificial lighting. (*Civil twilight* lasts from sunset until the sun is 6° below the horizon, while *astronomical twilight* lasts until the sky is completely dark and the astronomical observation of the faintest stars is possible.)

The length of twilight depends on season and latitude. During the summer in middle latitudes twilight adds about 30 minutes of light to each morning and evening for outdoor activities. The duration of twilight increases with increasing latitude, especially in summer. At high latitudes during the summer, morning and evening twilight may converge, producing a *white night*—a nightlong twilight.

Fig. 6.7 The bending of sunlight by the atmosphere causes the sun to rise about two minutes earlier, and set about two minutes later than it would otherwise.

In general, without the atmosphere, there would be no refraction or scattering, and the sun would rise later and set earlier than it now does. Instead of twilight, darkness would arrive immediately when the sun disappears below the horizon. Imagine the number of sandlot baseball games that would be called because of instant darkness.

The Mirage—Seeing Is Not Believing

Any good trout fisherman knows that you must squat as low as possible along the riverbank so as not to be visible to the fish. The reason for this behavior is that the light from the fisherman is bent when it enters the water in such a way that it makes the fisherman appear much taller to the fish than he actually is and, therefore, more visible. Conversely, to the fisherman, the fish appears to be much closer to the surface than it actually is (Fig. 6.8). The fisherman and the fish see each other displaced from their true positions. Light,

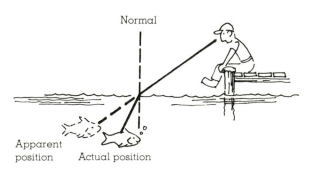

Fig. 6.8 The refraction of light as it passes from the water into the less-dense air causes a fish to appear closer to the surface than it actually is.

being refracted as it passes between air and water, creates this effect. In the atmosphere, when an object appears to be displaced from its true position, we call this phenomenon a **mirage**. A mirage is not a figment of the imagination—our minds are not playing tricks on us, but the atmosphere is.

Atmospheric mirages are created by light passing through and being refracted by air layers of different densities. Such changes in air density are usually caused by sharp changes in air temperature. The greater the rate of temperature change, the greater the light rays are bent. For example, on a warm, sunny day, black road surfaces absorb a great deal of solar energy and become very hot. Air in contact with these hot surfaces warms by conduction and, because air is a poor thermal conductor, we find much cooler air only a few meters higher. On hot days, these road surfaces often appear wet. (See Fig. 6.9.) Such "puddles" disappear as we approach them, and advancing cars seem to swim in them. Yet, we know the road is dry. The apparent wet pavement above a road is the result of blue skylight refracting up into our eyes as it travels through air of different densities. A similar type of mirage occurs in deserts during the hot summer. Many thirsty travelers have been disappointed to find that what appeared to be a water hole was in actuality hot desert sand.

Sometimes, these "watery" surfaces appear to shimmer. The shimmering results as rising and sinking air near the ground constantly change the air density. As light moves through these regions, its path also changes, causing the shimmering effect.

Fig. 6.9 Your author is not standing on water, nor are his shoes really that large. The wet-surface effect is the result of blue skylight bending up into the camera as the light passes through air of different densities. The shoes appear distorted because the light from them is passing through the region of greatest density (temperature) change.

When the air near the ground is much warmer than the air above, objects may not only appear to be lower than they really are, but also (often) inverted. These mirages are called **inferior** (lower) **mirages**. The tree in Fig. 6.10 certainly doesn't grow upside down. So why does it look that way? It appears to be inverted because light reflected from the top of the tree moves outward in all directions. Rays that enter the hot, less-dense air above the sand are refracted upward, entering the eye from below. The brain is fooled into thinking that these rays came from below the ground, which makes the tree appear upside down. Some light from the top of the tree travels directly toward the eye through air of nearly constant density and, therefore, bends very little. These rays reach the eye "straight-on," and the tree appears upright. Hence, off in the distance, we see a tree and its upsidedown image beneath it. (Some of the poles in Fig. 6.9 show this effect.)

The atmosphere can play optical jokes on us in extremely cold areas, too. In polar regions, air next to a snow surface can be much colder than the air many meters above. Because the air in this

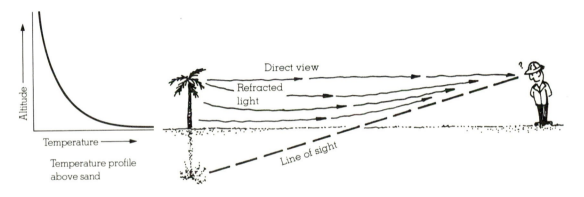

Fig. 6.10 Inferior mirage over hot desert sand.

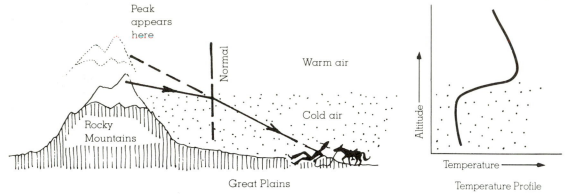

Fig. 6.11 The formation of a superior mirage. When cold air lies close to the surface with warm air aloft, light from distant mountains is refracted away from the normal as it enters the cold air. This causes an observer on the ground to see mountains higher and closer than they really are.

cold layer is very dense, light from distant objects entering it bends toward the normal in such a way that the objects can appear to be shifted upward. This phenomenon is called a **superior** (upward) **mirage**. Figure 6.11 shows the conditions favorable for a superior mirage.

A special type of superior mirage is the **Fata Morgana**, a mirage that transforms a fairly uniform horizon into one of vertical walls and columns with spires. According to legend, *Fata Morgana* (Italian for fairy Morgan) was the half-sister of King Arthur. Morgan, who was said to live in a crystal palace beneath the water, had magical powers that could build fantastic castles out of thin air. Looking across the Straits of Messina (between Italy and Sicily), residents of Reggio, Italy, on occasion would see buildings, castles, and sometimes whole cities appear, only to vanish again in minutes. The *Fata Morgana* is observed where the air temperature increases with height above the surface, slowly at first, then more rapidly, then slowly again. Consequently, mirages like the *Fata Morgana* are frequently seen where warm air rests above a cold surface, such as in polar regions. (See color plate 5.)

Halos, Sundogs, and Sun Pillars

A ring of light encircling and extending outward from the sun or moon is called a **halo** (Fig. 6.12). Such a display is produced when sunlight or moonlight is refracted as it passes through ice

crystals. Hence, the presence of a halo indicates that *cirriform clouds* are present.

The most common type of halo is the 22° halo—a ring of light 22° from the sun or moon.* Such a halo forms when tiny suspended ice crystals with diameters less than 20 μm become randomly oriented as air molecules constantly bump against them. The refraction of light rays through these crystals forms a halo like the one shown in Fig. 6.12. Less common is the 46° halo, which forms in a similar fashion to the 22° halo. (See Fig. 6.13.) With the 46° halo, however, the light is refracted through hexagonal column-type ice crystals that have the shape of tiny pencils, with diameters in a narrow range between about 15 and 25 μm.

Occasionally, a bright arc of light may be seen at the top of a 22° halo. (See color plate 6.) Since the arc is tangent to the halo, it is called a **tangent arc**. Apparently, the arc forms as large hexagonal pencil-shaped ice crystals fall with their long axes horizontal to the ground. Refraction of sunlight through the ice crystals produces the bright arc of light. When the sun is on the horizon, the arc that forms at the top of the halo is called an *upper tangent arc*. When the sun is above the horizon, a *lower tangent arc* may form on the lower part of the halo beneath the sun. The shape of the arcs change greatly with the position of the sun.

A halo is usually seen as a bright, white ring,

*For an outstretched arm, an angle of 22° is about the distance from the end of the thumb to the little finger.

Fig. 6.12 A 22° halo around the sun.

but there are refraction effects that can cause it to have color. To understand this, we must first examine refraction more closely.

When white light passes through a glass prism, it is refracted and split into a spectrum of visible colors (Fig. 6.14). Each wavelength of light is slowed by the glass, but each is slowed a little differently. Because longer wavelengths (red) slow the least and shorter wavelengths (violet) slow the most, red light bends the least, and violet light bends the most. The breaking up of white light by "selective" refraction is called **dispersion**. As light passes through ice crystals, dispersion causes red light to be on the inside of the halo and blue light on the outside.

When hexagonal platelike ice crystals with diameters larger than about 30 μm are present in the air, they tend to fall slowly and orient themselves horizontally (Fig. 6.15). (The horizontal orientation of these ice crystals prevents a ring halo.) In this position, the ice crystals act as small prisms, refracting and dispersing sunlight that passes through them. If the sun is near the horizon in

such a configuration that itself, ice crystals, and observer are all in the same horizontal plane, the observer will see a pair of brightly colored spots, one on either side of the sun. These colored spots are called **sundogs**, *mock suns*, or *parhelia*

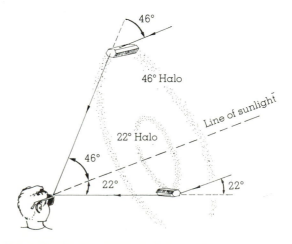

Fig. 6.13 The formation of a 22° and a 46° halo.

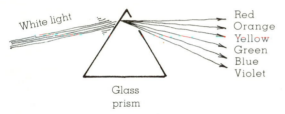

White light

Red
Orange
Yellow
Green
Blue
Violet

Glass
prism

Fig. 6.14 Refraction and dispersion of light through a glass prism.

(meaning "with the sun"). The colors usually grade from red (bent least) on the inside closest to the sun to blue (bent more) on the outside.

While sun dogs, tangent arcs, and halos are caused by *refraction* of sunlight *through* ice crystals, **sun pillars** are caused by *reflection* of sunlight *off* ice crystals. Sun pillars appear most often at sunrise or sunset as a vertical shaft of light extending upward or downward from the sun. (See color plate 8.) Pillars may form as hexagonal plate-like ice crystals fall with their flat bases oriented horizontally. As the tiny crystals fall in still air, they tilt from side to side like a falling leaf. This allows sunlight to reflect off the tipped surfaces of the

crystals, producing a relatively bright area in the sky above the sun. Pillars may also form as sunlight reflects off hexagonal pencil-shaped ice crystals that fall with their long axes oriented horizontally. As these crystals fall, they are able to rotate about their horizontal axes, producing many orientations that reflect sunlight. So, look for sun pillars when the sun is low on the horizon and cirriform clouds are present.

Rainbows

Now we come to one of the most spectacular light shows observed on the earth—the rainbow. Rainbows occur when rain is falling in one part of the sky, and the sun is shining in another. (Rainbows also may form by the sprays from waterfalls and water sprinklers.) To see the rainbow, we must face the falling rain with the sun at our backs. Look at Fig. 6.16 closely and note that, when we see a rainbow in the morning, we are facing a rainshower in the west; behind us—in the east—it is clear. Because clouds tend to move from west

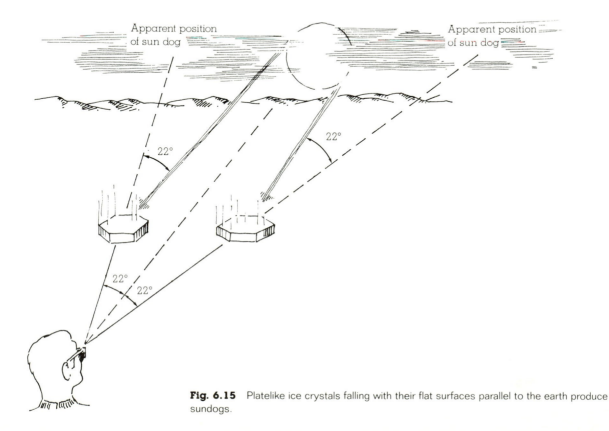

Apparent position of sun dog

Apparent position of sun dog

22°

22°

22°
22°

Fig. 6.15 Platelike ice crystals falling with their flat surfaces parallel to the earth produce sundogs.

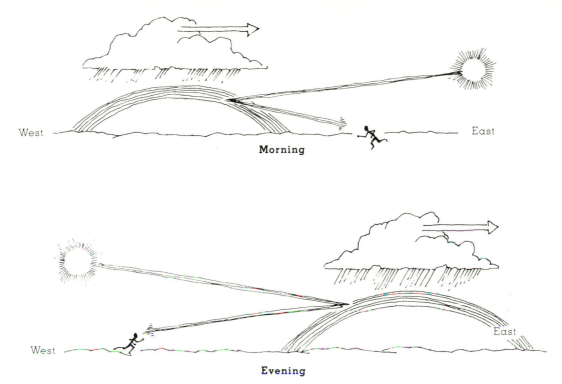

Fig. 6.16 When you observe a rainbow, the sun is always to your back. In middle latitudes, a rainbow in the morning suggests rain is on the way, while a rainbow at night (in the evening) indicates that clearing weather is ahead.

to east in middle latitudes, it is a good bet that they will move toward us and that it will rain soon. However, when we see a rainbow in the evening, we are facing east toward the rainshower. Clear skies in the west (behind us) suggest that the showers will give way to clearing weather. These observations explain why the following weather rhyme became popular:

> Rainbow in morning, sailors take warning
> Rainbow at night, a sailor's delight.*

When we look at a rainbow we are looking at sunlight that has entered the falling drops, and, in effect, has been redirected back toward our eyes. Exactly how this process happens requires some discussion.

*This rhyme is often used with the words "red sky" in the place of rainbow. The red sky makes sense when we consider that it is the result of red light from a rising or setting sun being reflected from the underside of clouds above us. In the morning, a red sky indicates that it is clear to the east and cloudy to the west. A red sky in the evening suggests the opposite.

As sunlight enters a raindrop, it slows and bends, with violet light refracting the most and red light the least (Fig. 6.17). Although most of this light passes right on through the drop and is not seen by us, some of it strikes the backside of the drop at such an angle that it is reflected within the drop. The angle at which this occurs is called the *critical angle*. For water, this angle is 48°. Light that strikes the back of a raindrop at an angle exceeding the critical angle, bounces off the back of the drop and is internally reflected toward our eyes. Because each light ray bends differently from the rest, each ray emerges from the drop at a slightly different angle. The light leaving the drop is, therefore, dispersed into a spectrum of colors from violet to red. Since we see only a single color from each drop, it takes myriads of raindrops (each refracting and reflecting light back to our eyes at slightly different angles) to produce the brilliant colors of a *primary rainbow*.

Figure 6.17 might lead us erroneously to believe that red light should be at the bottom of the bow and violet at the top. A more careful observation

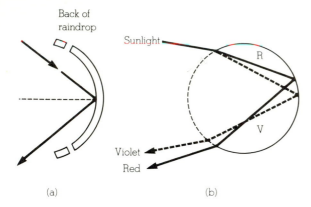

Fig. 6.17 Sunlight internally reflected and dispersed by a raindrop. (a) The light ray is internally reflected only when it strikes the backside of the drop at an angle greater than the critical angle for water. (b) Refraction of the light as it enters the drop causes the point of reflection (on the back of the drop) to be different for each color. Hence, the colors are separated from each other when the light emerges from the raindrop.

secondary bow is much fainter than the primary one. The *secondary bow* is caused when sunlight enters the raindrops at an angle that allows the light to make two internal reflections in each drop. Each reflection weakens the light intensity and makes the bow dimmer. Figure 6.19 shows that the color reversals—with red now at the bottom and violet on top—are due to the way the light emerges from each drop after going through two internal reflections.

As you look at a rainbow, keep in mind that only one ray of light is able to enter your eye from each drop. Everytime you move, whether it be up, down, or sideways, the rainbow moves with you. The reason why this happens is that, with every movement, light from different raindrops enters your eye. The bow you see is not exactly the same rainbow that the person standing next to you sees. In effect, each of us has a personal rainbow to ponder and enjoy!

of the behavior of light leaving two drops (Fig. 6.18) shows us why the reverse is true. Violet light from the upper drop and red light from the lower drop reach the eye in such a way that they cause the colors of a primary rainbow to change from violet on the inside (bottom) to red on the outside (top).

Frequently, a larger second (secondary) rainbow with its colors reversed can be seen above the primary bow. (See color plate 9.) Usually this

Coronas, Glories, and Heiligenschein

When the moon is seen through a thin veil of clouds composed of tiny spherical water droplets, a bright ring of light, called a **corona** (meaning crown), may appear to rest on the moon. The same effect can occur with the sun, but, due to the sun's brightness, it is usually difficult to see. (See color plates 10 and 11.)

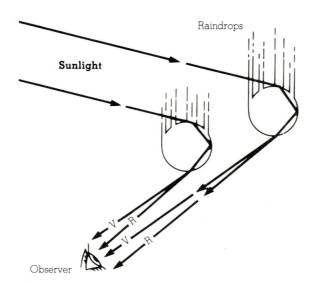

Fig. 6.18 The formation of a primary rainbow.

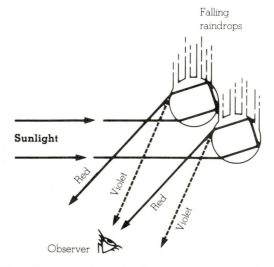

Fig. 6.19 Two internal reflections are responsible for the weaker secondary rainbow.

The Green Flash

Occasionally, a flash of green light may be seen near the upper rim of a rising or setting sun. Remember from our earlier discussion that, when the sun is near the horizon, its light must penetrate a thick section of atmosphere. This thick atmosphere refracts sunlight, with purple and blue light bending the most, and red light the least. Because of this bending, more blue light should appear along the top of the sun. But because the atmosphere selectively scatters blue

light, very little reaches us, and we see green light instead.

Usually, the green light is too faint to see with the human eye. However, under certain atmospheric conditions, such as when the surface air is very hot or when an upper-level inversion exists, the green light is magnified by the atmosphere. When this happens, a momentary flash of green light appears, often just before the sun disappears from view.

The flash usually lasts about a

second, although in polar regions it can last longer. Here, the sun slowly changes in elevation and the flash may exist for many minutes. Members of Admiral Byrd's expedition in the south polar region reported seeing the green flash for 35 minutes in September as the sun slowly rose above the horizon, marking the end of the long winter.

The corona is due to **diffraction**—the bending of light as it passes around objects. To understand the corona, imagine water waves moving around a small stone in a pond. As the waves spread around the stone, the trough of one wave may meet the crest of another wave. This situation results in the waves canceling each other, thus producing calm water. (This is known as *destructive interference*.) Where two crests come together (*constructive interference*), they produce a much larger wave. The same thing happens when light passes around tiny cloud droplets. Where light waves constructively interfere, we see bright light; where destructive interference occurs, we see darkness. Sometimes, the corona appears white, with alternating bands of light and dark. On other occasions, the rings have color.

The colors appear when the cloud droplets are of uniform size. Because the amount of bending due to diffraction depends upon the wavelength of light, the shorter wavelength blue light appears on the inside of a ring, while the longer wavelength red light appears on the outside. These colors may repeat over and over, becoming fainter as each ring is farther from the moon or sun. Also, the smaller the cloud droplets, the larger the ring diameter. Therefore, clouds that have recently formed (such as thin altostratus and altocumulus) are the best corona producers.

When different size droplets exist within a cloud, the corona becomes distorted and irregular. Sometimes the cloud exhibits patches of color, often pastel shades of pink, blue, or green. These bright areas produced by diffraction are called **iridescence**. Iridescence is most often observed within 20° of the sun.

Like the corona, the **glory** is also a diffraction phenomenon. When an aircraft flies above a cloud layer composed of water droplets less than 50 μm in diameter, a set of colored rings, called the *glory*, may appear around the shadow of the aircraft. The same effect can happen when you stand with your back to the sun and look into a cloud or fog bank, as a bright ring of light may be seen around the shadow of your head. In this case, the glory is called the **brocken bow**, after the Brocken Mountains in Germany, where it is particularly common.

For these optical phenomena to occur, the sun must be to your back, so that sunlight can be returned to your eye from the water droplets. Sunlight that enters the small water droplet along its edge is refracted, then reflected off the backside of the droplet. The light then exits at the other side of the droplet, being refracted once again (Fig. 6.20). However, in order for the light to be returned to your eyes, the light actually clings ever so slightly to the edge of the droplet—

Fig. 6.20 Light that produces the glory follows this path in a water droplet.

the light actually skims along the surface of the droplet as a *surface wave* for a short distance. Diffraction of light coming from the edges of the droplets produces the ring of light we see as the glory and the brocken bow. The colorful rings may be due to the various angles at which different colors leave the droplet.

On a clear morning with dew on the grass, stand facing the dew with your back to the sun and observe that, around the shadow of your head, is a bright area—the **Heiligenschein** (German for halo). The *Heiligenschein* forms when sunlight, which falls on nearly spherical dew drops, is focused and reflected back toward the sun along nearly the same path that it took originally. (Light reflected in this manner is said to be *retroreflected*.) The light, however, does not travel along the exact path; it actually spreads out just enough to be seen as bright white light about your head on a dew-covered lawn. (See color plate 12.)

Summary

Sunlight in the atmosphere can produce a variety of atmospheric visuals from hazy days and blue skies to crepuscular rays and blue moons. Refraction of light by the atmosphere causes stars near the horizon to appear higher than they really are. It also causes the sun and moon to rise earlier and set later than they otherwise would. Mirages form when refraction of light displaces objects from their true positions. Inferior mirages cause objects to appear lower than they really are, while superior mirages displace objects upward. The *Fata Morgana* transforms a fairly flat landscape into one of castles and columns.

Halos and sundogs form from the *refraction* of light through ice crystals. Sun pillars are the result of sunlight *reflecting* off gently falling ice crystals. The refraction, reflection, and dispersion of light in raindrops create a rainbow. To see a rainbow, the sun must be to your back, and rain must be falling in front of you.

Diffraction of light produces coronas, glories, and cloud iridescence. We can see the *Heiligenschein* when morning sunlight falls on nearly spherical dew drops. Under certain atmospheric conditions, the amplification of green light near the upper rim of a rising or setting sun produces the illusive green flash.

Questions for Review

1. Why does the underside of cumulus clouds in the process of building frequently change color from white to dark gray or even black?

2. The sky of the earth is blue; the sky of Mars is red. What accounts for the difference?

3. What can make a setting sun appear red?

4. Why do stars "twinkle"?

5. How does light bend as it enters a more-dense substance at an angle? How does it bend upon leaving the more-dense substance? Make a sketch to illustrate your answer.

6. Since twilight occurs without the sun being visible, how does it tend to lengthen the day?

7. On a clear, dry, warm day, why do dark road surfaces frequently appear wet?

8. What atmospheric conditions are necessary for an inferior mirage? A superior mirage?

9. How is a halo different from a sundog?

10. Explain how sun pillars form.

11. Explain the rhyme below:

Rainbow in morning, joggers take warning
Rainbow at night (evening), jogger's delight.

12. Why are secondary rainbows higher and much dimmer than primary rainbows? Explain your answer with the aid of diagram.

13. Explain how light is able to reach your eyes when you see (a) a corona, and (b) a glory.

14. How does a green flash form?

1 Red sky in morning, sailors take warning. The rising sun is casting ominous red and purple light onto clouds moving in from the west. (Photo by Pat Badalamente Zawadzki.)

2 Blue Ridge Mountains in Virginia. (Photo by author.)

3 Red twilight sky over California produced by the sulfur-rich particles from the volcano El Chichón during October, 1982. (Photo by author.)

4 Blue skies and white clouds. The selective scattering of blue light by air molecules (Rayleigh scattering) produces the blue sky; while the scattering of all wavelengths of visible light by liquid cloud droplets (Mie scattering) produces the white clouds. (Photo by author.)

5 The Fata Morgana mirage over the Arctic ice. (From *Rainbows, Halos, and Glories* by Robert Greenler, Cambridge University Press, 1980; photo by the author.)

6 Halo with a tangent arc. (Photo by author.)

7 A sun dog photographed at Bartlett Cove, Glacier Bay, Alaska. (Photo by J. L. Medeiros.)

8 Sun Pillar. (From *Rainbows, Halos, and Glories* by Robert Greenler, Cambridge University Press, 1980; photo by the author.)

9 A primary and a secondary rainbow. (Photo by author.)

10 Corona around the sun. This type of corona, called *Bishop's Ring*, is the result of diffraction of sunlight by tiny volcanic particles emitted from the volcano El Chichón in 1982. (Photo by Elizabeth Beaver Burnett.)

11 Corona around the moon, resulting from the diffraction of light by tiny altocumulus cloud droplets of uniform size. (Photo by author.)

12 The Heiligenschein. (Photo by author.)

13 The Aurora Borealis in Alaska. (Photo by Robert A. Bumpas, NCAR.)

14 Noctilucent clouds. (Ben Fogle, NCAR.)

5 The illumination of clouds produced by the scattering of twilight and moonlight. (The moon is the white dot in the upper corner of the photo.) (Photo by Ross DePaola.)

Questions for Thought

1. Explain why on a cloudless day the sky will usually appear milky-white before it rains and a deep blue after it rains.

2. How long does twilight last on the moon?

3. Why is it often difficult to see the road while driving on a foggy night with your high beam lights on?

4. What would be the color of the sky if air molecules scattered the longest wavelengths of visible light and passed the shorter wavelengths straight through? (Use a diagram to help explain your answer.)

5. Explain why the colors of the planets are not related to the temperatures of the planets, while the colors of the stars are related to the temperatures of the stars.

6. If there were no atmosphere surrounding the earth, what color would the sky appear at sunrise? At sunset? What color would the sun appear at noon? At sunrise? At sunset?

7. Why are we able to see the sun during the day, but not the other stars in the sky?

8. Why are rainbows seldom observed at noon?

9. On a cool, clear summer day, a blue haze often appears over the Great Smoky Mountains of Tennessee. Explain why the blue haze usually changes to a white haze as the humidity of the air increases.

10. During a lunar eclipse, the earth, sun, and moon are aligned as shown in the diagram below. The earth blocks sunlight from directly reaching the moon's surface, yet the surface of the moon will often appear a pale red color during a lunar eclipse. How can you account for this phenomenon?

11. Explain why smoke rising from a cigarette often appears blue, yet appears white when blown from the mouth.

12. Explain why it is easier to get sunburned on a high mountain than in the valley below. (The answer is not that you are closer to the sun on top of the mountain.)

13. Why are stars more visible on a clear night when there is no moon than on a clear night with a full moon?

Problems and Exercises

1. At least 5 times a day go outside and gaze at the sky. In a notebook record what you see. You will be amazed at what is there—crepuscular rays, halos, coronas, cloud iridescence, red sunsets, white horizons, and more.

2. Make your own rainbow. On a sunny afternoon or morning, turn on the water sprinkler to create a spray of water drops. Stand as close to the spray as possible (without getting soaked) and observe that, as you move up, down, and sideways, the bow moves with you. (a) Explain why this happens. (b) Also, explain with the use of a diagram why the sun must be at your back in order to see the bow.

3. Take a large beaker or bottle and fill it with water. Add a small amount of nonfat powdered milk and stir until the water turns a faint milky white. Shine white light into the beaker, and, on the opposite side, hold a white piece of paper. (a) Explain why the milk has a blue cast to it and why the light shining on the paper appears ruddy. (b) What do you know about the size of the milk particles? (c) Is this a form of Rayleigh scattering or Mie scattering? Explain. (d) How does this demonstration relate to the color of the sky and the color of the sun, at sunrise and sunset?

Moon

Earth

Sun

Drought-resistant vegetation struggles to survive another hot day on the arid plateau of central Nevada. Yet, on any summer day, there is actually more water vapor in the air of this desert than there is in the air of a wet New England snowstorm. (Photo by author)

CHAPTER 7

Humidity

Humidity refers to any one of a number of ways of specifying the amount of water vapor in the air. To most of us, a moist day suggests high humidity. However, there is usually more water vapor in the hot, "dry" air of the Sahara Desert than in the cold, "damp" polar air in New England. Does this mean that desert air has a higher humidity? The answer is both yes and no, depending on the type of humidity we mean. There are several ways to express atmospheric water vapor content; hence, there are several meanings for the concept of atmospheric humidity.

Absolute Humidity

Suppose we measure the water vapor in a parcel of air about the size of a large balloon. With a chemical drying agent, we can extract the vapor from the air, weigh it, and obtain its mass. If we then compare the vapor's mass with the volume of air in the parcel, we would have determined the **absolute humidity** of the air—that is, the mass of water vapor in a given volume of air:

$$\text{Absolute humidity} = \frac{\text{mass of water vapor}}{\text{volume of air}}.$$

Absolute humidity represents the density of water vapor in the parcel and, normally, is expressed as grams of water vapor in a cubic meter of air. For example, if the water vapor in 1 cubic meter of air weighs 25 grams, the absolute humidity of the air is 25 grams per cubic meter (25 g/m^3).

We learned in Chapter 2 that a rising or descending parcel of air will experience a change in its volume because of the changes in surrounding air pressure. Consequently, when a volume of air fluctuates, the absolute humidity changes—even

Contents

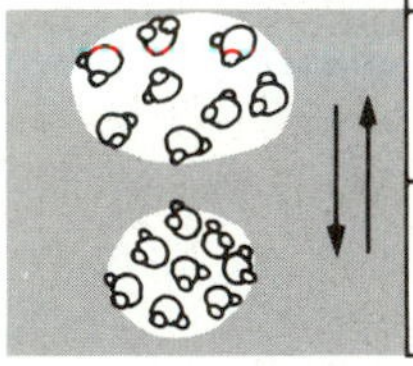

Parcel size	H₂O vapor content	Absolute humidity
2m³	10g	5g/m³
1 m³	10g	10 g/m³

Fig. 7.1 With the same amount of water vapor in a parcel of air, an increase in volume decreases absolute humidity, while a decrease in volume increases absolute humidity.

though the air's vapor content has remained constant (Fig. 7.1). For this reason, the absolute humidity is not commonly used in atmospheric studies.

Specific Humidity and Mixing Ratio

Humidity, however, can be expressed in ways that are not influenced by changes in air volume. When the mass of the water vapor in an air parcel is compared with the mass of all the air in the parcel (including vapor), the result is called the **specific humidity**:

$$\text{Specific humidity} = \frac{\text{mass of water vapor}}{\text{total mass of air}}.$$

Another convenient way to express humidity is to compare the mass of the water vapor in the parcel to the mass of the remaining dry air. This is called the **mixing ratio**:

$$\text{Mixing ratio} = \frac{\text{mass of water vapor}}{\text{mass of dry air}}.$$

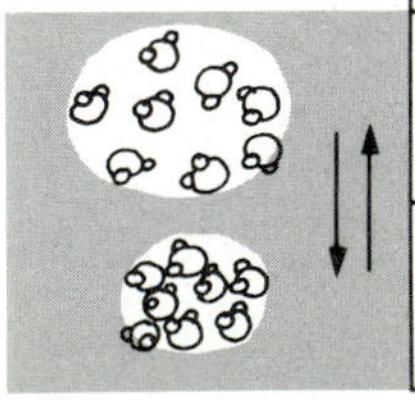

Weight of parcel	Weight of H₂O vapor	Specific humidity
1 kg	1 g	1 g/kg
1 kg	1 g	1 g/kg

Fig. 7.2 The specific humidity does not change as air rises and descends.

Both specific humidity and mixing ratio are expressed as grams of water vapor per kilogram of air (g/kg).

The specific humidity and mixing ratio of an air parcel remain constant *as long as the moisture content of the parcel does not change*. This happens because the total number of molecules (and, hence, the mass of the parcel) remains constant, even as the parcel expands or contracts (Fig. 7.2). Since changes in parcel size do not affect specific humidity and mixing ratio, these two concepts are used extensively in the study of the atmosphere.

In Chapter 2, we learned an important fact: *Warm air has a greater capacity for water vapor than does cold air.* The reason for this phenomenon is that the temperature of the air is a measure of the average kinetic energy (average speed) of its molecules. Higher temperatures correspond to higher average speeds. Therefore, at high air temperatures, condensation—that is, the joining of many billions of water vapor molecules to make droplets of liquid water—is less likely because most of the molecules have sufficient speed (sufficient energy) to remain as a vapor. As the air temperature lowers, the average speed of the molecules decreases and, hence, fewer molecules have sufficient energy to remain in the air as a vapor. However, even when the air temperature drops to very low readings, there will always be a few molecules moving fast enough to remain as water vapor. Notice in Fig. 7.3 how rapidly the air's capacity for water vapor increases as the air temperature goes up. A cubic meter of air at 35°C (95°F) can hold about 4 times as much water vapor as air at 10°C (50°F).

Figure 7.4 shows how specific humidity varies with latitude. The average specific humidity is highest in the warm, muggy tropics. As we move away from the tropics, it decreases, reaching its lowest average value in the polar latitudes. Although the major deserts of the world are located near latitude 30°, Fig. 7.4 shows that, at this latitude, the average air contains nearly twice the water vapor as does the air at latitude 50°N. Hence, the air of a desert is certainly not "dry," nor is the water vapor content very low. Since the hot, desert air of the Sahara often contains more moisture than the cold, polar air farther north, we can say that summertime Sahara air has a higher specific humidity. (We will see later in what sense we consider desert air to be "dry.")

Vapor Pressure

The air's moisture content may also be described by measuring the pressure exerted by the water vapor in the air. Suppose we have a parcel of air with some water vapor. We know that the total pressure inside the parcel is due to the collision of all the molecules against the inside surface of the parcel. In other words, the total pressure inside the parcel is equal to the sum of the pressures of the individual gases.* If the total pressure inside the parcel is 1000 mb and the gases inside include nitrogen (78 percent), oxygen (21 percent), and water vapor (1 percent), then the partial pressure exerted by nitrogen would be 780 mb and that exerted by oxygen, 210 mb. The partial pressure of water vapor, called the **actual vapor pressure**, would only be 10 mb. Therefore, because the number of water vapor molecules in any volume of air is small compared with the total number of air molecules in the volume, the actual vapor pressure is normally a small fraction of the total air pressure.

Everything else being equal, the more air molecules in a parcel, the greater the total air pressure. When you blow up a balloon, you increase its pressure by putting in more air. Similarly, an increase in the number of water vapor molecules will increase the total vapor pressure. Hence, the actual vapor pressure is a fairly good measure of the total amount of water vapor in the air: *High actual vapor pressure indicates large numbers of vapor molecules, while low actual vapor pressure indicates comparatively small numbers of vapor molecules.*†

Figure 7.5 shows the vapor pressures (expressed in millibars) over the United States for an average January. The amount of vapor in the air is greatest in the Gulf Coast states and lowest in

*This is known as *Dalton's law of partial pressure.*

†Remember that actual vapor pressure is only an approximation of the total vapor content. A change in total air pressure will affect the actual vapor pressure even though the total amount of water vapor in the air remains the same.

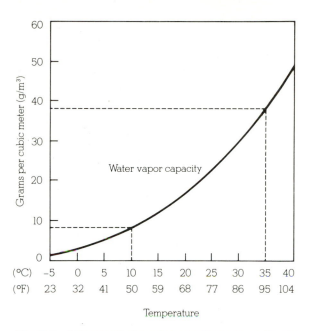

Fig. 7.3 Air's capacity for water vapor increases as the air temperature increases. The heavy line shows the saturation absolute humidity. Note that air with a temperature of 35°C can hold about 4 times more water vapor than can air at 10°C.

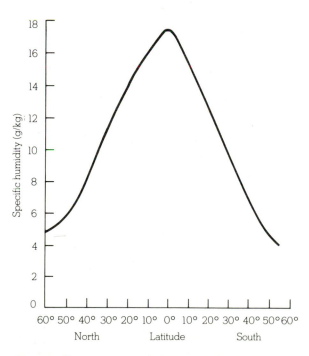

Fig. 7.4 The average specific humidity for each latitude.

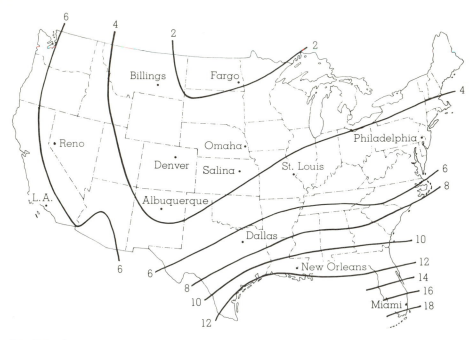

Fig. 7.5 Average water vapor pressure near the earth's surface in millibars for the month of January.

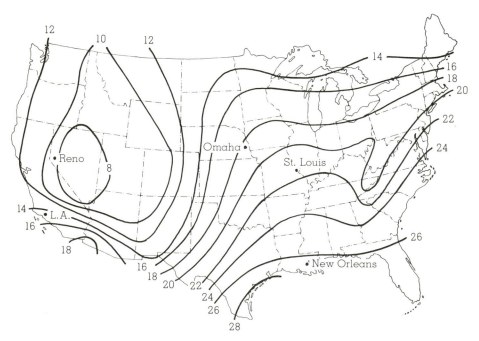

Fig. 7.6 Average water vapor pressure near the earth's surface in millibars for the month of July.

the interior. Compare New Orleans with Omaha. Cold, dry winds from northern Canada flow relentlessly into the Central Plains during the winter, keeping this area dry. But warm, moist air from the Gulf of Mexico helps maintain a higher vapor pressure in the South.

Figure 7.6 is a similar diagram showing the average vapor pressure for July. Again, the greatest vapor pressures are observed along the Gulf Coast. Note, too, that air over the eastern and central portion of the country in July contains between 3 and 6 times more moisture than it does in January. Compare St. Louis in July and January. Since cold air no longer moves in from Canada, the warm July air has a much greater capacity for moisture, and constantly receives water vapor from the moist breezes that move northward from the Gulf of Mexico. The Great Basin region averages the lowest values with a minimum occurring over Nevada, an area surrounded by mountains that effectively shield it from significant amounts of moisture moving in from the southwest and northwest.

Actual vapor pressure indicates the air's total vapor content. **Saturation vapor pressure** describes how much water vapor the air could hold at any temperature. Put another way, saturation vapor pressure is the maximum pressure that water vapor molecules would exert if the air were saturated with vapor at a given temperature. The saturation vapor pressure, then, depends primarily on air temperature. If we examine Fig. 7.7 we see that, as the air temperature increases, more vapor is required to saturate the air. The increased vapor exerts more pressure, so the saturation vapor pressure goes up. In cooler air, fewer vapor molecules are required to saturate air, and the saturation vapor pressure goes down. From the graph in Fig. 7.7, determine the saturation vapor pressure for air at a temperature of 10°C (50°F).*

The insert in Fig. 7.7 shows that, when both water and ice exist at the same temperature below freezing, *the saturation vapor pressure of the air just above the water is greater than the saturation vapor pressure of the air over the ice.* In other words, at any temperature below freezing, it takes more vapor molecules to saturate air di-

*Answer: Approximately 14 mb.

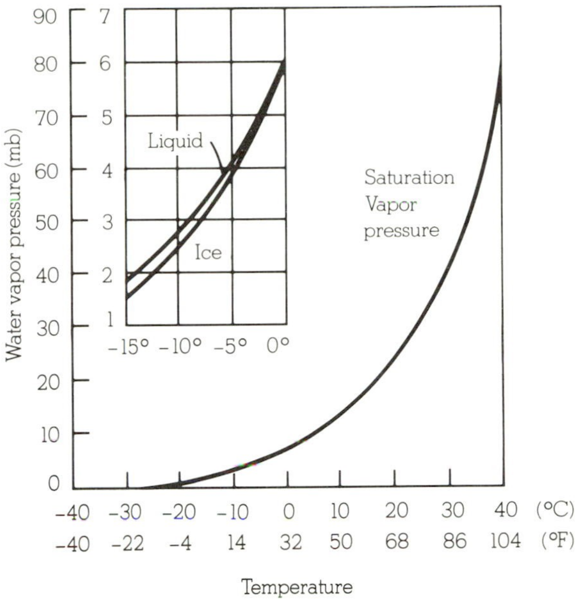

Fig. 7.7 Saturation vapor pressure increases with rising temperature. Observe that, at below freezing temperatures, the saturation vapor pressure is greater over water than over ice.

rectly above water than it does to saturate air directly above ice. This fact is extremely important, and—as we will see in the chapter on precipitation—one that plays a major role in the process of rain formation.

So far, we've described the amount of moisture actually in the air. If we want to report the moisture content of the air around us, we have several options:

1. Absolute humidity tells us the *mass* of water vapor in a fixed volume of air, or the *water vapor density.*
2. Specific humidity measures the *mass* of water vapor in a fixed *total mass* of air, and the mixing ratio describes the mass of water vapor in a fixed mass of the remaining dry air.
3. The actual vapor pressure of air expresses the amount of water vapor in terms of the amount of *pressure* that the water vapor molecules exert.

Each of these measures has its uses but, as we will see, the concepts of vapor pressure and saturation vapor pressure are critical to an under-

FOCUS ON A SPECIAL TOPIC

Why Is the Gulf Coast So Humid?

Why does the air over the Gulf states in July contain more than twice as much water vapor as the air along the southern California coast, even though both locations are adjacent to large bodies of water? A partial answer is that the air is warmer over the Gulf states. The average July maximum temperature in New Orleans is 4°C (7°F) warmer than in Los Angeles. But this is not sufficient to account for the fact that New Orleans' air contains about 55 percent more water vapor. What, then causes the difference?

Figure 1 shows a summertime situation where air from the Pacific Ocean is moving into southern California and air from the Gulf of Mexico is moving into the southeastern states. Notice that the Pacific water is much cooler than the Gulf water. Westerly winds, blowing across the Pacific, cool to just about the same temperature as the water. Likewise, air over the warmer Gulf reaches a temperature near that of the water below it.

The air at both locations is nearly saturated with water vapor. Since warm air can hold more water vapor than can cold air, the Gulf air contains more vapor molecules and, hence, has a higher vapor pressure. As the air moves inland, away from the source of moisture, the air temperature in both cases increases, while the amount of moisture in the air (vapor pressure) hardly changes. Therefore, air that moves inland from the Gulf of Mexico contains a great deal more moisture because the Gulf of Mexico is much warmer than the Pacific Ocean.

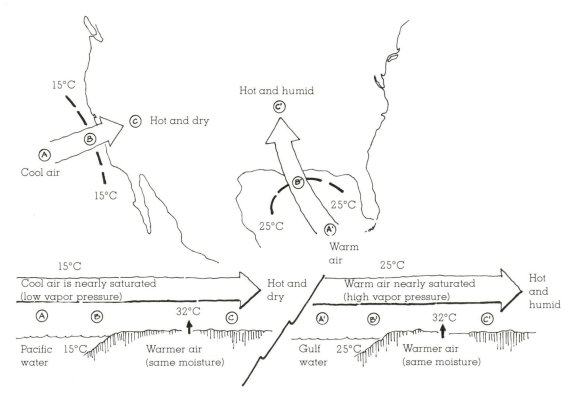

Fig. 1 Air from the Pacific Ocean produces hot and dry conditions, while air from the Gulf of Mexico produces hot and muggy conditions.

standing of the sections that follow. Now, let's look at the most commonly used moisture variable—relative humidity.

Relative Humidity

While relative humidity is the most commonly used way of describing atmospheric moisture, it is also, unfortunately, the most misunderstood. The concept of relative humidity may at first seem confusing because it does not indicate the actual amount of water vapor in the air. Instead, the **relative humidity** is the ratio of the amount of water vapor actually in the air compared to the maximum amount of water vapor the air can hold at that particular temperature (and pressure). It is the *ratio* of the air's water vapor *content* to its *capacity*:

$$\text{Relative humidity} = \frac{\text{amount of water vapor in the air}}{\text{amount of water vapor the air can hold}}.$$

Relative humidity is usually given as a percent. Air with a 50 percent relative humidity actually contains one-half the amount of water vapor it could hold. Air with a 100 percent relative humidity is said to be *saturated* because it is filled to capacity with vapor. Since relative humidity is used so much in the everyday world, let's examine it more closely.

If we increase or decrease the amount of water vapor in the air, the relative humidity will change. Suppose an air parcel has a maximum capacity of 10 g of water vapor, and further suppose that we keep the parcel's temperature constant, so that its capacity for water vapor does not change. Assume the air parcel is initially completely dry and, thus, has a relative humidity of 0 percent. Now, we add a single gram of water vapor to the air. The air now holds one-tenth of its capacity for water vapor, and so the relative humidity is 10 percent. If 4 g more of vapor are added, the air now contains 5 g of water vapor and the relative humidity increases to 50 percent. Adding 5 g more would make a total of 10 g of water vapor. The air parcel would be saturated, and its relative humid-

ity would be 100 percent. So, adding water vapor into air of constant temperature will increase the air's relative humidity. Likewise, removing water vapor from the same air will decrease the relative humidity.

It is also possible to change the relative humidity without changing the air's water vapor content. For example, a change in air temperature can bring about a change in relative humidity. This happens because a change in air temperature alters the air's capacity for holding water vapor. We can see from the left-hand column of Fig. 7.8 that, at 10°C, the water vapor capacity of the air parcel is 9 g, producing a relative humidity of 9/9, or 100 percent. Suppose the air warms to 20°C (middle column, Fig. 7.8). We can see that, at 20°C, the air's capacity for holding water vapor increases to 17 g. If the actual water vapor content remains at 9 g, then the air is holding less of what it could hold, and the relative humidity decreases to 9/17, or 53 percent. Warming the air to 30°C (right-hand column, Fig. 7.8) raises the air's capacity for water vapor to 30 g, which lowers the relative humidity to 9/30, or 30 percent. The opposite effect occurs as air cools: As the air temperature drops (with no change in water vapor content), the relative humidity increases because the air is approaching saturation.

In many places, the air's total vapor content varies only slightly during an entire day, and so it is the changing air temperature that primarily regulates the daily variation in relative humidity. As the air cools during the night, the relative humidity increases. Normally, the highest relative humidity occurs in the early morning, during the coolest part of the day. As the air warms during the day, the relative humidity decreases, with the lowest values usually occurring during the warmest part of the afternoon.

These changes in relative humidity are important in determining the amount of evaporation from vegetation and wet surfaces. Warm air with a low relative humidity has a large capacity for additional water vapor. If you water your lawn on a hot afternoon, when the relative humidity is low, much of the water will evaporate quickly from the lawn, instead of soaking into the ground. Watering the same lawn in the evening, when the relative humidity is higher, will cut down the evaporation and increase the effectiveness of the watering.

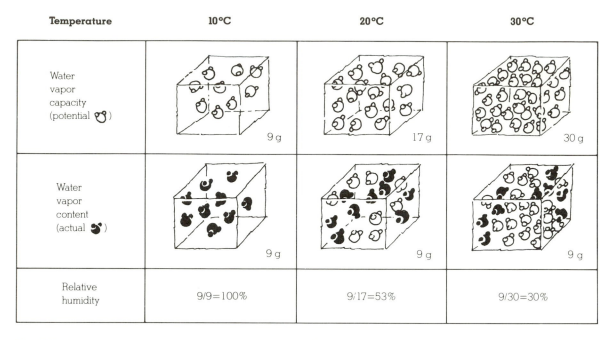

Temperature	10°C	20°C	30°C
Water vapor capacity (potential 🌧)	9 g	17 g	30 g
Water vapor content (actual ●)	9 g	9 g	9 g
Relative humidity	9/9=100%	9/17=53%	9/30=30%

Fig. 7.8 A change in air temperature (with no change in the air's water vapor content) will change the relative humidity.

Relative Humidity in the Home Question: How does the relative humidity of the winter air in your home compare with that in the Sahara Desert? Some homes actually have a lower relative humidity than the desert, and the inhabitants are usually unaware of it. Remember that cold arctic air contains only a little water vapor. Even when saturated, air with a temperature of −25°C (−13°F) only holds 0.5 g of water vapor in each kilogram of air. When this air is brought indoors and heated to 20°C (68°F), its water vapor capacity increases to 14.7 g/kg—over 29 times what it was outside. The relative humidity of the heated air inside the house drops to 3 percent.* This relative humidity is much lower than you would normally experience in a desert during the hottest time of the day!

Very low relative humidities in a house can have an adverse effect on things living inside. For example, house plants have a difficult time surviving because the moisture from their leaves and the soil evaporates rapidly. Hence, house plants usually need watering more frequently in winter than in summer. People suffer, too, when the relative humidity is quite low. The rapid evaporation of moisture from exposed flesh causes skin to crack, dry, flake, or itch. These low humidities also irritate the mucous membranes in the nose and throat, producing an "itchy" throat. Similarly, dry nasal passages permit inhaled bacteria to incubate, causing persistent infections. The remedy for most of these problems is simply to increase the relative humidity. But how?

The relative humidity in a home can be increased just by heating water and allowing it to

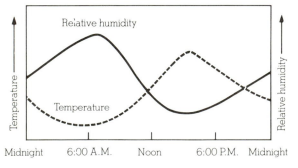

Fig. 7.9 When the air is cool (morning), the relative humidity is high. When the air is warm (afternoon), the relative humidity is low.

$$\frac{*0.5 \text{ g} = \text{actual vapor present}}{14.7 \text{ g} = \text{capacity}} = .03, \text{ or 3 percent.}$$

FOCUS ON A SPECIAL TOPIC

Vapor Pressure and Boiling—The Higher You Go, the Longer It Takes

If you camp in the mountains, you may have noticed that, the higher you camp, the longer it takes vegetables to cook in boiling water. To understand this, we need to examine the relationship between vapor pressure and boiling. As water boils, bubbles of water vapor rise to the top of the liquid and escape. For this to occur, the saturation vapor pressure exerted by the bubbles must equal the pressure of the atmosphere; otherwise, the bubbles would collapse. Boiling, therefore, occurs when the saturation vapor pressure of the escaping bubbles is equal to the total atmospheric pressure.

Because the saturation vapor pressure is directly related to the temperature of the liquid, higher water temperatures produce higher vapor pressures. Hence, any change in atmospheric pressure will change the temperature at which water boils: An increase in air pressure raises the boiling point, while a decrease in air pressure lowers it. Notice in Fig. 2 that, to make pure water boil at sea level, the water must be heated to a temperature of 100°C (212°F). At Denver, Colorado, which is situated about 1500 m (5000 ft) above sea level, the air pressure is near 850 mb, and water boils at 95°C (203°F).

Once water starts to boil, its temperature remains constant, even if you continue to heat it. This happens because energy supplied to the water is used to convert the liquid to a gas (steam). Now we can see why vegetables take longer to cook in the mountains. To be thor-

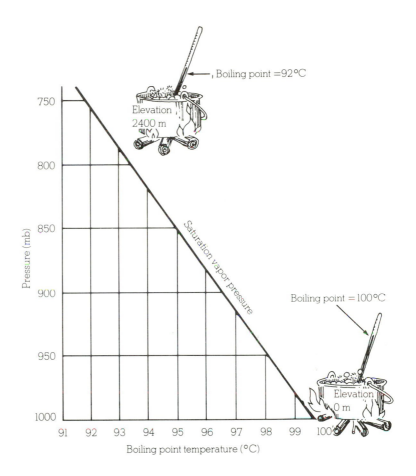

Fig. 2 The lower the air pressure, the lower the saturation vapor pressure and, hence, the lower the boiling point temperature.

oughly cooked, they must boil for a longer time because the boiling water is cooler than at lower levels. In New York City, which is near sea level, it takes about five minutes to hard boil an egg. An egg boiled for five minutes in the "mile high city" of Denver, Colorado, turns out to be runny.

evaporate into the air, which will raise the relative humidity to a more comfortable level. In modern homes, a humidifier, installed near the furnace, adds moisture to the air at a rate of about one gallon per room per day. The air, with its added moisture, is circulated throughout the home by a forced air heating system. In this way, all rooms get their fair share of moisture—not just the room where the vapor is added.

The high relative humidity of summer can pose a different problem. With a high relative humidity, the air is nearly saturated with water vapor. In this weather, body moisture does not readily evaporate into the air; instead, it collects on the skin as beads of perspiration. The reduced evaporational cooling in hot, moist weather makes most people feel hotter than it actually is and, for this reason, people often remark that it's not the heat, it's the humidity. To lower the air's moisture content, as well as the air temperature, many homes are air conditioned. Outside air cools as it passes through a system of cold coils located in the air conditioning unit. The cooling increases the air's relative humidity, and the air reaches saturation. The water vapor condenses into liquid water, which is carried away. The cooler, dehumidified air is now forced into the home.

In hot regions, where the relative humidity is low, evaporative cooling systems can be used to cool the air. These systems operate by having a fan blow hot, dry outside air across a large container of water. Evaporation of the water extracts heat from the fan-blown air above the water surface. This cooler air is forced into the home, bringing some relief from the hot weather.

Evaporative coolers are also known as "swamp" coolers. They work best when the relative humidity is low and the air is warm. They do not work well in hot, muggy weather because a high relative humidity greatly reduces the rate of evaporation. Besides, swamp coolers add water vapor to the air—something that is not needed when the air is already uncomfortably humid. That is why swamp coolers may be found on homes in Arizona, but not on homes in Alabama.

Dew Point Consider a volume of air whose temperature is 10°C (50°F) and relative humidity is 100 percent. Suppose the air warms to 20°C

(68°F), with no change in vapor content or air pressure. The relative humidity drops and the air is no longer saturated. To what temperature must the 20°C air be cooled so that it is once again saturated? Of course, the answer is 10°C. For this amount of moisture, 10°C is called the *dew-point temperature*, or simply, the **dew point**. It represents the temperature to which air would have to be cooled (with no change in air pressure or moisture content) for saturation to occur. (When the dew-point temperature is below freezing, it is also referred to as the **frost point**.)

The dew point is an important measurement used to predict the formation of dew, frost, fog, and even the minimum temperature. When used with an empirical formula (see Chapter 10), the dew point can help determine the height of the base of a cumulus cloud. Since atmospheric pressure varies only slightly at the earth's surface, the *dew point is a good indicator of the air's actual water vapor content*. High dew points indicate high water vapor content; low dew points, low water vapor content. Addition of water vapor to the air increases the dew point; removing water vapor lowers it.

The difference between air temperature and dew point can indicate whether the relative humidity is low or high. When the air temperature and dew point are far apart, the relative humidity is low; when they are close to the same value, the relative humidity is high. When the air temperature and dew point are equal, the relative humidity is 100 percent.

Comparing Humidities Figure 7.10 shows how the average relative humidity varies from the equator to the poles. High relative humidities are normally found in the tropics, where there is little separation between air temperature and dew point. The average relative humidity is low near latitude 30° and high near the poles.

Since polar air is often described as being "dry," why does it usually have a high relative humidity? In the polar region, the dew point and the air temperature are usually close together. A low dew point means the air contains little water vapor, and the low temperatures tell us that the polar air's capacity for water vapor is low as well. With low dew point and small capacity for water vapor, polar air is often close to saturation. This produces a high relative humidity—the average is 80 percent—in air that contains very little water vapor.

Remember that we raised the question as to which air has the higher humidity—polar air or desert air? Take a minute now and compare Fig. 7.10 with Fig. 7.4. In Fig. 7.10, we see that polar air has a higher *relative humidity* than does desert air at 30° latitude, where the average spread between air temperature and dew point is large. However, in Fig. 7.4, we see that, on the average, the specific humidity is higher at 30° than at the poles. This means, of course, that a given amount of desert air normally contains more water vapor than the same amount of polar air. So, does desert air have a higher humidity than polar air? For relative humidity, the answer is no; for specific humidity, the answer is yes.

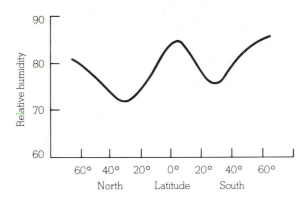

Fig. 7.10 Relative humidity averaged for latitudes north and south of the equator.

Measuring Humidity

The common instrument used to obtain dew point and relative humidity is a **psychrometer**, which consists of two liquid-in-glass thermometers mounted side by side and attached to a piece of metal that has either a handle or chain at one end (Fig. 7.12). The thermometers are exactly alike except that one has a piece of cloth (wick) cov-

ering the bulb. The wick-covered thermometer—called the *wet bulb*—is dipped in clean water, while the other thermometer is kept dry. Both thermometers are ventilated for a few minutes, either by whirling the instrument (*sling psychrometer*), or by drawing air past it with an electric fan (*aspiration psychrometer*). Water evaporates from the wick and that thermometer cools. The drier the air, the greater the amount of evaporation and cooling. After a few minutes, the wick-covered thermometer will cool to the lowest value possible. This is called the **wet-bulb temperature**—the coolest temperature that can be obtained by evaporating water into the air.

Air temperature: −9°C
Dew point: −9°C

Air temperature: 35°C
Dew point: 7°C

Fig. 7.11 Which situation—the snowstorm or the desert air—has the highest relative humidity? Which actually contains the most water vapor? (Answer: the snowstorm has the highest relative humidity and the desert air contains the most water vapor.)

FOCUS ON A SPECIAL TOPIC

Computing Relative Humidity and Dew Point

Suppose we want to compute the air's relative humidity. We already know that the relative humidity is the ratio of the air's actual water vapor content compared to its capacity. The actual vapor pressure (e) is a measure of the air's actual water vapor content, and the saturation vapor pressure (e_s) is a measure of the air's total capacity for water vapor. Therefore, the actual and saturation vapor pressures can be used to determine the relative humidity of the air. Relative humidity may be expressed as:

$$Relative\ humidity = \frac{e}{e_s} \times 100\%.^*$$

Table 1   Saturation Vapor Pressure Over Water for Various Air Temperatures

AIR TEMPERATURE (°C)	(°F)	SATURATION VAPOR PRESSURE (mb)	AIR TEMPERATURE (°C)	(°F)	SATURATION VAPOR PRESSURE (mb)
−18	(0)	1.5	18	(65)	21.0
−15	(5)	1.9	21	(70)	25.0
−12	(10)	2.4	24	(75)	29.6
−9	(15)	3.0	27	(80)	35.0
−7	(20)	3.7	29	(85)	41.0
−4	(25)	4.6	32	(90)	48.1
−1	(30)	5.6	35	(95)	56.2
2	(35)	6.9	38	(100)	65.6
4	(40)	8.4	41	(105)	76.2
7	(45)	10.2	43	(110)	87.8
10	(50)	12.3	46	(115)	101.4
13	(55)	14.8	49	(120)	116.8
16	(60)	17.7	52	(125)	134.2

Let's look at a practical example of using vapor pressures to measure relative humidity. Suppose the air temperature in a room is 27°C (80°F). Suddenly the room is cooled with no change in moisture content. At successively lower temperatures the air's capacity to hold water vapor decreases. The lowering saturation vapor pressure (e_s) approaches the actual vapor pressure (e) and the relative hu-

*Relative humidity may also be expressed as RH = w/w_s × 100%, where w is the actual mixing ratio and w_s is the saturation mixing ratio. Relative humidity computations using mixing ratio and adiabatic charts are given in Appendix F.

midity increases. With an actual vapor pressure of 25 mb, 100 percent relative humidity will be reached at a temperature of 21°C (70°F). This temperature (21°C) must then be the dew-point temperature of the air. If, then, we know the actual vapor pressure in a room, we can determine the dew point by using Table 1 to locate the temperature at which air will be saturated with that amount of vapor. Similarly, if we are told that the dew point in the room has some value, we can look up that temperature in Table 1 and find the actual vapor pressure.

In essence, Table 1 can be used to obtain the saturation vapor pressure (e_s) and the actual vapor pressure (e) if the air temperature and dew point of the air are known. With this information we can calculate relative humidity. For example, what is the relative humidity of air with a temperature of 29°C and a dew point of 18°C? Answer: At 29°C, Table 1 shows e_s = 41 mb. For a dew point of 18°C the actual vapor pressure (e) is 21 mb; therefore, the relative humidity = 21/41 × 100 percent = 51 percent.

The dry thermometer (commonly called the *dry bulb*) gives the current air temperature. The temperature difference between the dry bulb and the wet bulb is known as the **wet-bulb depression**. A large depression indicates that a great deal of water vapor can evaporate into the air and that the relative humidity is low. A small depression indicates that little evaporation of water vapor is possible, so the air is close to saturation and the relative humidity is high. If there is no depression, the dry bulb and wet bulb are the same; the air is saturated and the relative humidity is 100 percent.

(Tables used to compute relative humidity and dew point are given in Appendix D.)

The length of human hair changes by about 2.5 percent between a relative humidity of 0 and 100 percent. Hair becomes shorter with low humidities and longer with high humidities. (In moist air, people with naturally curly hair experience the "frizzies" as their hair increases in length. Under the same conditions, people with long, straight hair find it going "limp.")

The fact that the length of hair changes with the relative humidity is used in an instrument called the **hair hygrometer**. This instrument uses human (or horse) hair to measure relative humidity. A number of strands of hair (with oils removed) are attached to a system of levers. A small change in hair length is magnified by a linkage system and transmitted to a dial (Fig. 7.13) calibrated to show relative humidity, which can then be read directly or recorded on a chart. (Often, the chart is attached to a clock-driven rotating drum that gives a continuous record of relative humidity.) Because the hair hygrometer is not as accurate as the psychrometer (especially at very high and very low relative humidities), it requires frequent calibration, especially in areas that experience large daily variations in relative humidity.

The **electrical hygrometer** is another instrument used to measure humidity. It consists of a flat plate coated with a film of carbon. An electric current is sent across the plate. As the moisture content of the air changes, the electrical resist-

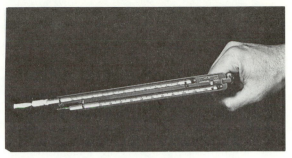

Fig. 7.12   The sling psychrometer.

ance of the carbon coating changes. These changes are translated into relative humidity. This instrument is commonly used in the radiosonde, which gathers atmospheric data at various levels above the earth. Finally, still another instrument—the **infrared hygrometer**—measures atmospheric humidity by measuring the amount of infrared energy absorbed by water vapor in a sample of air.

Summary

In this chapter, we have examined the many ways of describing humidity. We saw that the absolute humidity represents the density of water vapor in a given volume of air. Specific humidity measures the mass of water vapor in a fixed mass of air, while the mixing ratio expresses humidity as the

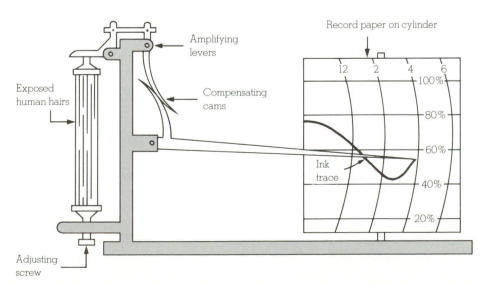

Fig. 7.13   The hair hygrometer measures relative humidity by amplifying and measuring changes in the length of human (or horse) hair.

mass of water vapor in the fixed mass of remaining dry air. The actual vapor pressure indicates the air's total water vapor content by expressing the amount of water vapor in terms of the amount of pressure that the water vapor molecules exert. The saturation vapor pressure describes how much water vapor the air could hold at any given temperature in terms of how much pressure the water vapor molecules would exert if the air were saturated at that temperature. A good indicator of the air's actual water vapor content is the dew point—the temperature to which air would have to be cooled for saturation to occur.

The relative humidity is the ratio of the air's water vapor content to the air's water vapor capacity. Air with a high relative humidity is not necessarily holding a great deal of water vapor, it is simply holding close to its capacity for water vapor. With a constant water vapor content, cooling the air causes the relative humidity to increase, while warming the air causes the relative humidity to decrease. When the air temperature and dew point are close together, the relative humidity is high, and, when they are far apart, the relative humidity is low. High relative humidity in hot weather makes us feel hotter than it really is by retarding the evaporation of perspiration. Although relative humidity can be confusing (because it can change with either air temperature or moisture content), it is nevertheless the most widely used way of describing the air's moisture content.

Questions for Review

1. Why are specific humidity and mixing ratio more commonly used in representing atmospheric moisture than absolute humidity?

2. Describe the relationship between air temperature and the air's capacity to hold water vapor. (See Fig. 7.3.)

3. During the summer, in terms of actual amount of water vapor in the air, where is the driest part of the United States?

4. Why does the saturation vapor pressure increase as the air temperature increases?

5. Why is the specific humidity of Gulf air much higher than the specific humidity of Pacific air?

6. Explain why boiling vegetables take longer to cook in the mountains.

7. What does the relative humidity represent?

8. When the relative humidity is given, why is it also important to know the air temperature?

9. Explain two ways the relative humidity may be changed.

10. Explain why, during a summer day, the relative humidity will change as shown in Fig. 7.9.

11. How is the difference between dew point and air temperature related to the relative humidity?

12. Why is cold polar air described as ''dry'' when the relative humidity of that air is very high?

13. How can a region have a high specific humidity and a low relative humidity? Give an example.

14. Why does the relative humidity of cold air decrease when brought into a warm home?

15. Why are evaporative coolers used in Arizona, Nevada, and California but not in Florida, Georgia, or Indiana?

16. How are the dew-point temperature and wet-bulb temperature different?

17. Name three instruments that may be used to measure atmospheric moisture content. Briefly explain how each one works.

Questions for Thought

1. Explain how and why each of the following will change as a parcel of air with an unchanging amount of water vapor rises, expands, and cools:
   (a) absolute humidity
   (b) relative humidity
   (c) actual vapor pressure
   (d) saturation vapor pressure

2. Where in the United States would you go to experience the *least* variation in dew point (actual moisture content) from January to July?

3. On a clear, calm morning, water condenses on the ground in a thick layer of dew. As the water slowly evaporates into the air, you measure a slow increase in dew point. Explain why.

4. Two cities have exactly the same amount of water vapor in the air. The 6:00 A.M. relative humidity in one city is 93 percent, while the 3:00

P.M. relative humidity in the other city is 28 percent. Explain how this can come about.

5. Suppose the dew point of cold outside air is the same as the dew point of warm air indoors. If the door is opened, and cold air replaces some of the warm inside air, would the new relative humidity indoors be (a) lower than before, (b) higher than before, or (c) the same as before? Explain your answer.

6. On a warm, muggy day the air is described as "close." What are several plausible explanations for this expression?

7. Outside, on a very warm day, you swing a sling psychrometer for about a minute and read a dry-bulb temperature of 38°C and a wet-bulb temperature of 24°C. After swinging the instrument again, the dry bulb is still 38°C, but the wet bulb is now 26°C. Explain how this could happen.

8. In Yellowstone National Park, there are numerous ponds of boiling water. If Yellowstone is about 2200 m (7200 ft) above sea level (where the air pressure is normally about 775 mb), what is the normal boiling point of water in Yellowstone?

Problems and Exercises

1. During July, how much more water vapor (in percent) is there normally in the air over New York City than in the air over Seattle, Washington? (Hint: See Fig. 7.6.)

2. If the air temperature in a room is 18°C and the dew point is 7°C, what is the relative humidity in the room?

3. On a bitter cold, snowy morning, the air temperature and dew point of the outside air are −7°C. If this air is brought indoors and warmed to 21°C, with no change in vapor content, what is the relative humidity of the air inside the home?

4. (a) With the aid of Fig. 7.6 and Fig. 7.7, determine the average summer dew points in St. Louis, Mo.; New Orleans, La.; and Los Angeles, Ca. (b) If the high temperature on a particular summer day in all three cities is 32°C (90°F), then calculate the afternoon relative humidity at each of the three cities. (Hint: Either Fig. 7.7 or Table 1, p. 142, will be helpful.)

5. Suppose with the aid of a sling psychrometer you obtain an air temperature of 30°C and a wet-bulb temperature of 25°C. What is (a) the wet-bulb depression, (b) the dew point, and (c) the relative humidity of the air? (Use the tables in Appendix D.)

6. If the air temperature is 35°C and the dew point is 21°C, determine the relative humidity using (a) Table 1, p. 142, (b) Fig. 7.7, and (c) Tables D-1 and D-2 in Appendix D.

7. In Fig. 7.6, the average vapor pressure in Nevada is about 8 mb. (a) Use Table 1, p. 142, to determine the average dew point of this air. (b) Much of the state is above an elevation of 1500 m (5000 ft). At 1500 m, the normal pressure is about 12.5 percent less than at sea level. If the air over Nevada were brought down to sea level, without any change in vapor content, what would be the new vapor pressure of the air?

8. In Fig. 7.11, determine the actual vapor pressure, saturation vapor pressure, and relative humidity of the air (a) in the snowstorm and (b) in the desert.

As evening arrives in Yosemite Valley, cool air settles close to the ground, producing a thin blanket of radiation fog. (Photo by author)

CHAPTER 8

Dew, Frost, and Fog

Have you walked barefoot across a lawn on a summer morning and felt the wet grass under your feet? Did you ever wonder how those glistening droplets of dew could form on a clear summer night? Or why they formed on grass but not on bushes several meters above the ground? In this chapter, we will investigate the formation of dew, as well as other forms of condensation that occur close to the earth's surface.

How Does Dew Form on Clear Nights?

On clear, calm nights, objects near the earth's surface cool rapidly by radiation. The ground and objects on it often become much colder than the surrounding air. Air that comes in contact with these cold surfaces cools by conduction, and its ability to hold water vapor decreases. Eventually, the air cools to the dew point—the temperature at which saturation occurs. As the air cools slightly below this temperature, water vapor condenses onto the nearest available surface (twigs, leaves, and blades of grass), forming tiny visible specks of water called **dew**. Because the coolest air is usually at ground level, dew is more likely to form on blades of grass than on objects several meters above the surface. This thin coating of dew not only dampens bare feet, but is also a valuable source of moisture for many plants during periods of low rainfall. Averaged for an entire year in middle latitudes, dew yields a blanket of water between 12 and 50 mm (0.5 and 2 in.) thick.

Dew is more likely to form on nights that are clear and calm than on nights that are cloudy and windy. Clear nights allow objects near the ground to cool rapidly by emitting infrared radiation, and calm winds mean that the coldest air will be lo-

Contents

cated at ground level. These atmospheric conditions are usually associated with large fair-weather, high-pressure systems. On the other hand, the cloudy, windy weather that inhibits rapid cooling near the ground and the forming of dew often signifies the approach of a rain-producing storm system. These observations inspired the following folk-rhyme:

When the dew is on the grass,
rain will never come to pass.
When grass is dry at morning light,
look for rain before the night!

Frozen Dew and Frost

Beads of frozen water observed on objects after a clear, cold, windless night are **frozen dew**. In

Fig. 8.1   These are the delicate ice-crystal patterns that frost exhibits on a window during a cold winter morning.

order for it to form, the air temperature near the ground must reach the dew point while the dew point is above freezing. Dew forms and then freezes, becoming tiny beads of ice as the temperature drops below the freezing point.

Visible white frost, like frozen dew, forms on cold, clear, calm mornings. But unlike frozen dew, the dew-point temperature (now called the *frost point*) is at or below freezing. When the air temperature cools to the frost point and further cooling occurs, water vapor can change directly to ice without becoming a liquid first—a process called *deposition* (it is also called *sublimation*). The delicate, white crystals of ice that form in this manner are called *hoarfrost*, *white frost*, or simply **frost**. Frost has a treelike branching pattern that easily distinguishes it from the nearly spherical beads of frozen dew.

On cold winter mornings, frost may form on the inside of a windowpane in much the same way as it does outside, except that the cold glass chills the indoor air adjacent to it. When the temperature of the inside of the window drops below freezing, water vapor in the room forms a light, feathery deposit of frost (Fig. 8.1).

In very dry weather, the air temperature may become quite cold and drop below freezing without ever reaching the frost point, and no visible frost forms. *Freeze* and *black frost* are words denoting this situation. These conditions can severely damage crops.

So, dew, frozen dew, and frost form in the rather shallow layer of air near the ground on clear, calm nights. But what happens to air as a deeper layer adjacent to the ground is cooled? We have learned that if air cools without any change in water vapor content, the relative humidity increases. When air cools to the dew point, the relative humidity becomes 100 percent and the air is saturated. Continued cooling condenses some of the vapor into tiny cloud droplets.

Condensation Nuclei

Actually, the condensation process that produces clouds is not quite so simple. Just as dew and frost need a surface to form on, there must be air-borne particles on which water vapor can condense to produce cloud droplets.

Although the air may look clean, it never really is. On an ordinary day, a volume of air about the size of your index finger contains between 1000 and 150,000 particles. Since many of these serve as surfaces on which water vapor can condense, they are called **condensation nuclei**. Without them, relative humidities of several hundred percent would be required before condensation could begin.

Some condensation nuclei are quite small and have a radius less than 0.2 μm; these are referred to as **Aitken nuclei**, after the British physicist who discovered that water vapor condenses on nuclei. Particles ranging in size from 0.2 to 1 μm are called **large nuclei**, while others, called **giant nuclei**, are much larger and have radii exceeding 1 μm. (See Table 8.1.) The condensation nuclei most favorable for producing clouds have a radius of 0.1 μm or more. Usually, between 100 and 1000 nuclei of this size exist in a cubic centimeter of air. These particles enter the atmosphere in a variety of ways: dust, volcanoes, factory smoke, forest fires, or salt from ocean spray. Because most are released into the atmosphere near the ground, the largest concentrations of nuclei are observed in the lower atmosphere near the earth's surface.

Condensation nuclei are extremely light (many have a mass less than one-trillionth of a gram), so they can remain suspended in the air for many days. They are most abundant over industrial cities, where highly polluted air may contain nearly 1 million particles per cubic centimeter. They decrease in cleaner "country" air and over the oceans, where concentrations may dwindle to only a few nuclei per cubic centimeter.

Some particles are **hygroscopic** ("water-seeking"), and water vapor condenses upon these surfaces when the relative humidity is considerably lower than 100 percent. Ocean salt is hygroscopic, as is common table salt. In humid weather, it is difficult to pour salt from a shaker because water vapor condenses onto the salt crystals, sticking them together. Other hygroscopic nuclei include sulfuric and nitric acid particles. Not all particles serve as good condensation nuclei. Some are **hydrophobic** ("water-repelling") and resist condensation even when the relative humidity is above 100 percent. As we can see, condensation may begin on some particles when the relative humidity is well below 100 percent and on others only when the relative humidity is much higher

Table 8.1 Characteristic Sizes and Concentrations of Condensation Nuclei and Cloud Droplets

TYPE OF PARTICLE	APPROXIMATE RADIUS (μm)	NO. OF PARTICLES (per cm³)	
		Range	Typical
Small (Aitken) condensation nuclei	<0.2	1000 to 10,000	1000
Large condensation nuclei	0.2 to 1.0	1 to 1000	100
Giant condensation nuclei	>1.0	<1 to 10	1
Fog and cloud droplets	>10	10 to 1000	300

than 100 percent. However, at any given time there are usually many nuclei present, so that haze, fog, and clouds will form at relative humidities near or below 100 percent.

Haze

Suppose you visit an area that contains a large concentration of suspended dust or salt particles. There, you may notice that distant objects are usually more visible in the afternoon than in the morning, even when the concentration of particles in the air has not changed. Why? During the warm afternoon, the relative humidity of the air is often below the point where water vapor begins to condense, even on active hygroscopic nuclei. There-

Hygroscopic nuclei   Hydrophobic nuclei

Fig. 8.2 Hygroscopic nuclei are "water-loving," and water vapor rapidly condenses on their surfaces. Hydrophobic nuclei are "water-repelling" and resist condensation.

Fig. 8.3   The high relative humidity of the cold air above the lake is causing a layer of haze to form on a still spring morning.

fore, the floating particles remain small—usually no larger than about one-tenth of a micrometer. These tiny, dry, haze particles selectively scatter some rays of sunlight, while allowing others to penetrate the air. The scattering effect of **dry haze** produces a bluish color when viewed against a dark background and a yellowish tint when viewed against a light-colored background.

As the air cools during the night, the relative humidity increases. When the relative humidity reaches 75 percent, condensation may begin on the most active hygroscopic nuclei, producing a **wet haze**. As water collects on the nuclei, their size increases and the particles, although still small, become large enough to scatter light much more efficiently. (Recall from Chapter 6 that, when sunlight strikes wet haze, nearly all visible waves are scattered away evenly, causing the haze layer to appear white.) In fact, as the relative humidity increases from about 60 percent to 80 percent, the scattering effect increases by a factor of nearly 3. Since relative humidities are normally high during cool mornings, much of the light from distant objects is scattered away by the wet haze parti-

cles before reaching you; hence, it is difficult to see these distant objects.

Not only does wet haze restrict visibility more than dry haze, it also appears dull gray or white. Near seashores and in clean air over the open ocean, large salt particles suspended in air with a high relative humidity often produce a thin white veil across the horizon.

Fog

By now, it should be apparent that condensation is a continuous process beginning when water vapor condenses onto hygroscopic nuclei at relative humidities as low as 75 percent. As the relative humidity of the air increases, the visibility decreases, and the landscape becomes masked with a grayish tint. As the relative humidity gradually approaches 100 percent, the haze particles grow larger, and condensation begins on the less-active nuclei. Now a large fraction of the available nuclei have water condensing onto them, causing the droplets to grow even bigger, until eventually they become visible to the naked eye. The increasing size and concentration of droplets further restrict visibility. When the visibility lowers to less than 1 km (0.62 mi), and the air is wet with countless millions of tiny floating water droplets, the wet haze becomes a cloud resting near the ground, which we call **fog**.

With the same water content, fog that forms in dirty city air often is thicker than fog that forms over the ocean. Normally, the smaller number of condensation nuclei over the middle of the ocean produce fewer, but larger, fog droplets. City air with its abundant nuclei produces many tiny fog droplets, which greatly increase the thickness of the fog and reduce visibility. A dramatic example of a thick fog forming in air with abundant nuclei occurred in London, England, during the early 1950s. The fog became so thick, and the air so laden with smoke particles, that sunlight could not penetrate the smoggy air, requiring that street lights be left on at midday.

Fog that forms in polluted air can turn acidic as the tiny liquid droplets combine with gaseous impurities, such as oxides of sulfur and nitrogen. **Acid fog** poses a threat to human health, especially to people with pre-existing respiratory problems. It also adversely affects plants and structures subject to corrosion. In the Los Angeles basin during December, 1982, acid fog reached a level of acidity comparable to that of bathroom cleanser. In Chapter 22, we will examine this problem in more detail.

As tiny fog droplets grow larger, they become heavier and tend to fall toward the earth. A fog droplet with a diameter of 25 μm settles toward the ground at about 5 cm (2 in.) each second. At this rate, most of the droplets in a fog layer 180 m (about 600 ft) thick would reach the ground in less than one hour. Therefore, two questions arise: How does fog form? How is fog maintained once it does form?

Fog, like any cloud, usually forms in one of two ways: (1) by cooling—air is cooled to its saturation point (dew point); and (2) by evaporation—water vapor is added to the air by evaporation, thus raising the dew point until the air becomes saturated. Once fog forms it is maintained by new fog droplets, which constantly form on available nuclei. In other words, the air must maintain its

degree of saturation either by continual cooling or evaporation of vapor into the air. Let's examine both processes.

Radiation Fog How can the air cool so that a cloud will form near the surface? Radiation and conduction are the primary means for cooling nighttime air near the ground. Fog produced by the earth's radiational cooling is called **radiation fog**, or **ground fog**. It forms best on clear nights when a shallow layer of moist air near the ground is overlain by drier air. Under these conditions, the ground cools rapidly since the shallow, moist layer does not absorb much of the earth's outgoing infrared radiation. As the ground cools, so does the air directly above it, and a surface inversion forms. The moist, lower layer (chilled rapidly by the cold ground) quickly becomes saturated, and fog forms. The longer the night, the longer the time of cooling and the greater the likelihood of fog. Hence, radiation fogs are most common over land in late fall and winter.

Another factor promoting the formation of radiation fog is a light breeze of less than 5 knots. The slight air movement brings more of the moist air in direct contact with the cold ground and the transfer of heat occurs more rapidly. A strong breeze would prevent a radiation fog from forming by mixing the air near the surface with the drier air above. The ingredients of clear skies and light winds are associated with large high-pressure areas (anticyclones). Consequently, during the winter, when a high becomes stagnant over an area, radiation fog may form on many consecutive days.

Because cold, heavy air drains downhill and collects in valley bottoms, we normally see radiation fog forming in low-lying areas. Hence, radiation fog is frequently called **valley fog**. The cold air and high moisture content in river valleys make them susceptible to radiation fog. Since radiation fog normally forms in lowlands, hills may be clear all day long, while adjacent valleys are fogged in.

Radiation fogs form upward from the ground as the night progresses and are usually deepest around sunrise. Often a shallow fog layer will dissipate or "burn off" by the afternoon. Of course, the fog does not "burn"; rather, sunlight penetrates the fog and warms the ground, causing the air temperature in contact with the ground to increase. The warm air rises and mixes with the

Fig. 8.4 This shallow layer of fog formed after the sun rose. In the still night air, radiational cooling brought the air temperature almost to the dew point. However, the air adjacent to the ground became saturated and a thick blanket of dew covered the grass. At daybreak, the sun's rays evaporated the dew, adding water vapor to the air. A light breeze stirred the moist air sufficiently to cause saturation and, hence, fog to form in a shallow layer near the ground.

Fig. 8.5 Radiation fog nestled in a valley.

foggy air above, which increases the temperature and water vapor capacity of the foggy air. In the slightly warmer air, some of the fog droplets evaporate, allowing more sunlight to reach the ground, which produces more heating, and soon the fog completely disappears.

Satellite photographs show that a blanket of radiation fog tends to "burn off," first around its periphery, where the fog is usually thinnest. Sunlight rapidly warms this region, causing the fog to dissipate as the warmer air mixes in towards the denser foggy area.

If the fog is thick, with little sunlight penetrating it, and there is little mixing along the outside edges, the fog may not dissipate. This is often the case in the Central Valley area of California during the late fall and winter. A fog layer over 500 m (1700 ft) thick settles between two mountain ranges, while a strong inversion normally keeps the warmest air above the top of the fog. During the day, much of the light from the low winter sun reflects off the top of the fog, allowing only a small amount of sunlight to penetrate the fog and warm the ground. As the air heats from below, the fog dissipates upward from the surface in a rather shallow layer less than 150 m (500 ft), creating the illusion that the fog is lifting. Since the fog no longer touches the ground, and a strong inversion exists above it, the fog is called a **high inversion fog**. (The low cloud above the ground is also called *stratus*, or, simply, *high fog*.) As soon as the sun sets, radiational cooling lowers the air temperature, and the fog once again forms on the ground. This daily lifting and lowering of the fog without the sun ever breaking through it may last for many days or even weeks during winter in California's Central Valley.

Advection Fog   Cooling surface air to its saturation point may be accomplished by warm, moist air moving over a cold surface. The surface must be sufficiently cooler than the air above, so that the transfer of heat from air to surface will cool the air to its dew point and produce fog. Fog that forms in this manner is called **advection fog**.

A good example of advection fog may be ob-

Fig. 8.6   The leading edge of advection fog (called a *fog bank*) lies poised off shore, ready to move inland.

Why Are Headlands Usually Foggier than Beaches?

If you drive along a highway that parallels an irregular coastline, you may have observed that advection fog is more likely to form in certain regions. For example, headlands that protrude seaward usually experience more fog than do beaches that are nestled in the mouth of bays. Why?

As air moves on shore it crosses the coastline at nearly a right angle. This causes the air to flow together or converge in the vicinity of the headlands (Fig. 1). This area of weak convergence causes the surface air to rise and cool just a little. If the rising air is close to being saturated, it will cool to its dew point and fog will form.

Meanwhile, near the beach area, the surface air spreads apart or diverges as it crosses the coastline. This area of weak

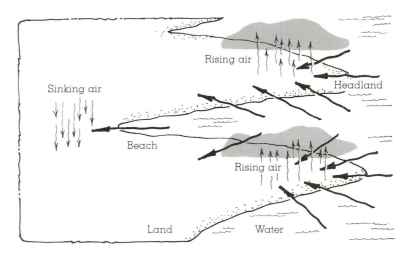

Fig. 1 Along an irregular coastline, advection fog is more likely to form at the headland where moist surface air converges and rises than at the beach where air diverges and sinks.

divergence creates sinking and slightly warmer air. Because the sinking of air increases the separation between air temperature and dew point, fog is less

likely to form in this region. Hence, the headlands can be shrouded in fog, while beaches are basking in sunshine.

served along the Pacific Coast during summer (Fig. 8.7). The main reason fog forms in this region is that the surface water near the coast is much colder than the surface water farther offshore. Warm, moist air from the Pacific Ocean is advected by westerly winds over the cold, coastal waters. Chilled from below, the air temperature drops to the dew point, and fog is produced. Advection fog, unlike radiation fog, always involves the movement of air, so when there is a stiff summer breeze in San Francisco, it's common to watch advection fog roll in past the Golden Gate Bridge.

As summer winds carry the fog inland over the warmer land, the fog near the ground dissipates, leaving a sheet of low-lying gray clouds that block out the sun. Further inland, the air is sufficiently warm, so that even these low clouds evaporate and disappear. Since the fog is more likely to burn off during the warmer part of the day, a typical

summertime weather forecast for coastal areas would read, "Fog and low cloudiness along the coast extending locally inland both night and mornings with sunny afternoons."

Because they provide moisture to the coastal redwood trees, advection fogs are important to the scenic beauty of the Pacific Coast. Fog moisture is collected by the needles and branches of the redwoods. Some is directly absorbed by the needles, while much of the remaining water drips to the ground, where it is utilized by the tree's shallow root system. Without the summer fog, coastal redwood trees would have trouble surviving the dry California summers. Hence, we find them nestled in the fog belt along the coast.

Advection fogs also prevail where two ocean currents with different temperatures flow next to one another. Such is the case in the Atlantic Ocean off the coast of Newfoundland, where the cold

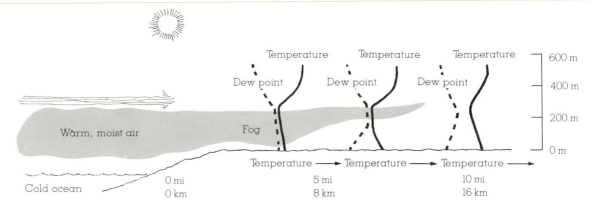

Fig. 8.7 Advection fog forms as warm, moist air cools to its dew point while moving over a cold surface. As fog moves inland along the Pacific Coast in summer, the air warms and the fog lifts above the surface. Eventually, the air becomes warm enough to totally evaporate the fog.

southward-flowing Labrador Current lies almost parallel to the warm northward-flowing Gulf Stream. Warm southerly air moving over the cold water produces fog in that region—so frequently that fog occurs on about two out of three days during summer.

Advection fog also forms over land. In winter, warm, moist air from the Gulf of Mexico moves northward over progressively colder land. As the air cools to its saturation point, a fog forms in the southern or central United States. Because the cold ground is often the result of radiation cooling, fog that forms in this manner is sometimes called **advection–radiation fog**. During this same time of year, air moving across the warm Gulf Stream encounters the colder land of the British Isles and produces the thick fogs of England. Similarly, fog forms as marine air moves over an ice or snow surface. In extremely cold arctic air, ice crystals form instead of water droplets, producing an **ice fog**.

Upslope Fog Fog that forms as moist air flows up along an elevated plain, hill, or mountain is called **upslope fog**. Typically, upslope fog forms during the winter and spring on the eastern side of the Rockies, where the eastward sloping plains are nearly a kilometer higher than the land further east. Occasionally, cold air moves from the lower eastern plains westward. The air gradually rises, expands, becomes cooler, and—if sufficiently moist—a fog forms. Upslope fogs that form over an extensive area may last for many days.

Up to now, we have seen how the cooling of air produces fog. But remember that fog may also form if enough water vapor is added to the air by evaporation. Fogs that form in this way are called **evaporation fogs**.

Evaporation Fog On a cold day, you may have unknowingly produced evaporation fog. When moist air from your mouth or nose meets the cold air and saturates it with vapor, a tiny cloud forms with each exhaled breath.

A common form of evaporation fog is the **steam fog**, which forms when cold air moves over warm water. This type of fog forms above a heated outside swimming pool in winter. As long as the

Fig. 8.8 Tiny drops, each one made from many fog droplets, drip from the needles of this tree and provide a valuable source of moisture during the otherwise dry summer along the coast of California.

water is warmer than the unsaturated air above, water will evaporate from the pool into the air. The increase in water vapor raises the dew point, and, if mixing is sufficient, the air becomes saturated. With continued evaporation from the warm water, the excess vapor condenses onto nuclei and produces a fog. The colder air directly above the water is heated from below and becomes warmer than the air directly above it. This warmer air rises and, from a distance, the rising condensing vapor appears as "steam."

It is common to see steam fog forming over lakes on autumn mornings, as cold air settles over water still warm from the long summer. On occasion, over the Great Lakes, columns of condensed vapor rise from the fog layer, forming whirling "*steam devils*," which appear similar to the dust devils on land. If you travel to Yellowstone National Park, you will see steam fog forming above thermal ponds all year long (Fig. 8.9). Over the ocean in polar regions, steam fog is referred to as **arctic sea smoke**.

Steam fog may form above a wet surface on a sunny day. This is commonly observed after a rain shower as sunlight shines on a wet road, heats the asphalt, and quickly evaporates the water. This vapor saturates the already moist air, producing steam fog. Fog that forms in this manner is short-lived and disappears as the road surface dries.

A warm rain falling through a layer of cold, moist air can produce fog. Remember that the saturation vapor pressure depends on temperature: Higher temperatures correspond to higher saturation vapor pressures. When a warm raindrop falls into a cold layer of air, the saturation vapor pressure over the raindrop is greater than that of the air. This vapor pressure difference causes water to evaporate from the raindrop into the air, and, when the cold surface air becomes sufficiently moist, fog forms. Fog of this type is often associated with warm air riding up and over a mass of colder surface air. The fog usually develops in the shallow layer of cold air just ahead of an approaching warm front or behind a cold front, which is why this type of evaporation fog is also known as **frontal fog**. Snow covering the ground is an especially favorable condition for frontal fog to form. The melting snow extracts heat from the environment, thereby cooling the already rain-saturated air.

Fig. 8.9 Even in summer, warm air rising above thermal pools in Yellowstone National Park condenses into a type of steam fog.

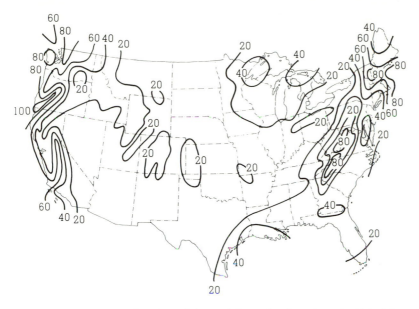

Fig. 8.10 Average annual number of days with heavy fog throughout the United States.

Foggy Weather

Notice in Fig. 8.10 that heavy fog is more prevalent in coastal margins—especially those regions lapped by cold currents—than in the center of the continent. In fact, three regions stand out as having the most days with heavy fog: (1) the Pacific Coast states; (2) the Appalachian highland region; and (3) New England. Areas experiencing more than 20 days a year with heavy fog include the Gulf and Atlantic coastal states, portions of the Great Plains and Rocky Mountains, and the regions around the Great Lakes. The least number of foggy days occur in the Southwest.

The foggiest spot near sea level in the United States is Cape Disappointment, Washington. Located at the mouth of the Columbia River, it averages 2556 hours of heavy fog each year. Anyone who travels to this spot hoping to enjoy the sun during August and September would find its name appropriate indeed. Along the Atlantic Ocean just off the coast of Maine, heavy fog shrouds Mistake Island for an average of 1580 hours each year. And Mt. Washington, New Hampshire, nestled in the clouds at 1908 m (6262 ft) above sea level experiences fog on more than 300 days each year!

Although fog is basically a nuisance, it has many positive aspects. For example, the California Central Valley fog that many people scorn is extremely important to the economy of that area. Fruit and nut trees that have finished growing during the summer and fall require **winter chilling**—a large number of hours with the air temperature below 7°C (45°F) before trees will begin to grow again. The winter fog blocks out the sun and helps keep daytime temperatures quite cool, while keeping nighttime temperatures above freezing: The more continuous the fog, the more effective the chilling. Consequently, the agricultural economy of the region depends heavily on the fog, for without it and the winter chill it stimulates, many of the fruit and nut trees would not grow well. During the spring, when trees are in bloom, fog prevents nighttime air temperatures from dipping to dangerously low readings by trapping infrared energy radiated by the earth and releasing latent heat to the air as fog droplets form.

Unfortunately, fog also has many negative aspects. Along a gently sloping highway, the elevated sections may have excellent visibility, while in lower regions—only a few kilometers away—fog may cause poor visibility. Driving from the clear area into the fog on a major freeway can be extremely dangerous. In fact, every winter many

Fog Dispersal

In any airport fog-clearing operation the problem is to improve visibility, so that aircraft can take off and land. Experts have tried various methods, which can be grouped into four categories: (1) increase the size of the fog droplets, so that they become heavy and settle to the ground as a light drizzle; (2) seed cold fog with dry ice (solid carbon dioxide), so that fog droplets are converted into ice crystals; (3) heat the air, so that the fog evaporates; and (4) mix the cooler saturated air near the surface with the warmer unsaturated air above.

To date, only one of these methods has been reasonably successful—the seeding of cold fog. **Cold fog** forms when the air temperature is below freezing, and most of the fog droplets remain as liquid water. (Liquid fog in below-freezing air is also called **supercooled fog**.) Cities such as Denver, Salt Lake City, and Spokane experience cold fog each winter. The fog can be cleared by injecting several hundred pounds of dry ice into it. As the tiny pieces of cold ($-78°C$) dry ice descend, they freeze some of the supercooled fog droplets in their path, producing larger ice crystals. These fall to the ground, leaving a "hole" in the fog for aircraft takeoffs and landings. Unfortunately, most of the fogs that close airports in the United States are **warm fogs** that form when the air temperature is above freezing. These afflict airports in cities such as Seattle, Los Angeles, New Orleans, New York City, and Philadelphia. Since dry ice seeding does not work in warm fog, other techniques must be tried.

One method involves injecting hygroscopic particles into the fog. Large salt particles and other chemicals absorb the tiny fog droplets and form into larger drops. More large drops and fewer small drops improve the visibility; plus, the larger drops are more likely to fall as a light drizzle. Since the chemicals are expensive and the fog clears for only a short time, this method of fog dispersal is not economically feasible.

Another technique for fog dispersal is to warm the air enough, so that the fog droplets evaporate and visibility improves. An early account of this method comes from England. There, during World War II, fuel oil was ignited along runways to warm the air enough to dissipate the fog and allow returning bombers to land. Tested at Los Angeles International Airport in the early 1950s, this technique was abandoned because it was smoky, expensive, and not very effective. In fact, the burning of hundreds of dollars worth of fuel only cleared the runway for a short time. And the smoke particles, released during the burning of the fuel, provided abundant nuclei for the fog to recondense upon.

At large airports in France, a sophisticated fog-clearing tech-

people are involved in fog-related auto accidents. These usually occur when a car enters the fog and, because of the reduced visibility, the driver puts on the brakes to slow down. The car behind then slams into the slowed vehicle, causing a chain-reaction accident with many cars involved. In fact, one such accident occurred near Sacramento, California, during February, 1984, when 70 cars and trucks smashed into each other on a foggy highway, covering it with debris for 3 kilometers. Although 15 people were injured, there were fortunately no fatalities.

Extremely limited visibility exists while driving at night in heavy fog with the high-beam lights on. The light scattered back to the driver's eyes from the fog droplets makes it difficult to see very far down the road. However, even in thick fog, there is usually a drier and therefore clearer region extending about 35 cm (14 in.) above the road surface. People who drive a great deal in foggy weather take advantage of this by installing extra head lamps—called *fog lamps*—just above the front bumper. These lights are directed downward into the clear space where they provide improved visibility.

Fog-related problems are not confined to land. Even with today's sophisticated electronic equipment, dense fog in the open sea hampers navi-

FOCUS ON A SPECIAL TOPIC, continued

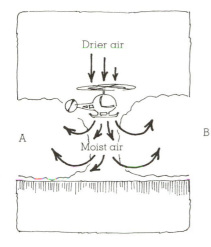

Fig. 2 Helicopters hovering above an area of shallow fog (diagram A) can produce a clear area (photograph B).

nique called **turboclair** is being used. In this system, a series of jet engines installed underground along the edge of the runway pump hot exhaust gases into the air to warm and stir it, causing the fog droplets to evaporate. But the turbulence created by the exhaust makes takeoffs and landings unsafe for small aircraft.

A final method of warm fog dispersal uses helicopters to mix the air. The chopper flies across the fog layer, and the turbulent downwash created by the rotor blades brings drier air above the fog into contact with the moist fog layer (Fig. 2). The aim, of course, is to evaporate the fog. Experiments show that this method works well, as long as

the fog is a shallow radiation fog with a relatively low liquid water content. But many fogs are thick, have a high liquid water content, and form by other means. An inexpensive and practical method of dispersing warm fog has yet to be discovered.

gation. A Swedish liner rammed the luxury liner *Andrea Doria* in thick fog off Nantucket Island on July 25, 1956, causing 52 casualties. On a fog-covered runway in the Canary Islands, two *747* jet airliners collided, taking the lives of over 570 people in March, 1977.

Airports suspend flight operations when fog causes visibility to drop below a prescribed minimum. The resulting delays and cancellations become costly to the airline industry and irritate passengers. With fog-caused problems such as these, it is no wonder that scientists have been seeking ways to disperse, or at least "thin," fog. (For more information on fog-thinning techniques, read the focus section entitled "Fog Dispersal.")

Summary

In this chapter, we have looked at the different forms of condensation that occur on or near the earth's surface. We saw that dew forms when the air temperature cools to the dew point in a shallow layer of air near the surface. If the dew should freeze, it produces tiny beads of ice called frozen dew. Frost forms when the air cools to a dew point that is at 0°C or below.

As the air cools in a deeper layer near the surface, the relative humidity increases and water vapor begins to condense on hygroscopic condensation nuclei, forming wet haze. As the relative humidity approaches 100 percent, condensation occurs on most nuclei, and the air becomes filled with tiny liquid droplets (or ice crystals) called fog.

Fog forms in two primary ways: cooling of air and evaporating water into the air. Radiation fog, advection fog, and upslope fog form by the cooling of air, while steam fog and frontal fog are two forms of evaporation fog. Although fog has some beneficial effects—providing winter chilling for fruit trees and water for thirsty redwoods—in many places it is a nuisance, for it disrupts air traffic and it is the primary cause of a number of auto accidents.

Questions for Review

1. Why is dew more likely to form on a clear, still night than on a cloudy, windy night?

2. Explain how dew, frozen dew, and visible frost each form.

3. What is the difference between hygroscopic and hydrophobic nuclei?

4. Distinguish among dry haze, wet haze, and fog.

5. Why is fog that forms in industrial areas normally thick?

6. How can fog form when the air's relative humidity is less than 100 percent?

7. List three primary ways in which fog forms. Describe each one.

8. What conditions are necessary for the formation of radiation fog?

9. Why is radiation fog often observed at the bottom of valleys?

10. Why do ground fogs usually "burn off" by early afternoon?

11. List as many positive consequences of fog as you can.

12. What atmospheric conditions are necessary for the development of advection fog?

13. How can evaporation produce fog?

14. What are some of the negative consequences of fog?

15. List and discuss several methods used in the dispersal of warm fog.

16. What method is used to disperse cold fog?

Questions for Thought

1. Explain the reasoning behind the wintertime expression, "Clear moon, frost soon."

2. Explain why icebergs are frequently surrounded by fog.

3. During a summer visit to New Orleans, you stay in an air-conditioned motel. One afternoon, you put on your sunglasses, step outside, and within no time your glasses are "fogged up." Explain what has apparently caused this.

4. While driving from cold air (well below freezing) into much warmer air (well above freezing), frost forms on the windshield of the car. Does the frost form on the inside or outside of the windshield? How can the frost form when the air is so warm?

5. What are the economic and social benefits of being able to dissipate fog?

6. Why are really clean atmospheres and really dirty atmospheres undesirable?

7. Why do relative humidities seldom reach 100 percent in polluted air?

8. Why are advection fogs rare over tropical water?

9. A January snowfall covers central Arkansas with 5 in. of snow. The following day, a south wind brings heavy fog to this region. Explain what has apparently happened.

10. If all fog droplets gradually settle earthward, explain then how fog can last (without disappearing) for many days at a time.

11. On a clear morning, you can see a layer of wet haze forming above a field. What do you know about the following meteorological elements in that haze layer: relative humidity, air temperature, dew point, wind speed, and visibility?

12. Near the shore of an extremely large lake, explain why steam fog is more likely to form during the autumn and advection fog in early spring.

13. On a bitter cold morning, two homes each have ice on their inside windows. One house has

A wide variety of cumuliform clouds are forming over the foothills of eastern Colorado. The sprouting clouds with pronounced vertical growth are cumulus congestus. Cumulus clouds with much smaller vertical extent are cumulus humilis. The very small scattered ragged-looking clouds are cumulus fractus. (Photo by author)

frozen beads of water, the other one frost. Which of the two homes probably has the lowest relative humidity?

14. The air temperature during the night cools to the dew point in a deep layer, producing fog. Before the fog formed, the air temperature cooled each hour about 2°C. After the fog formed, the air temperature cooled by only 0.5°C each hour. Give two reasons why the air cooled more slowly after the fog formed.

15. On a winter night, the air temperature cooled to the dew point and fog formed. Before the formation of fog, the dew point remained almost constant. After the fog formed, the dew point began to decrease. Explain why.

16. Why can you see your breath on a cold morning? Does the air temperature have to be below freezing for this to occur?

Problems and Exercises

1. The data in the chart below represent the dew point and expected minimum temperature near the ground for various clear winter mornings in a southeastern city. Assume that the dew point remains constant throughout the night. Answer the following questions about the data.

(a) On which morning would there be the greatest likelihood of observing visible frost? Explain why.

(b) On which morning would frozen dew most likely form? Explain why.

(c) On which morning would there be black frost with no sign of visible frost, dew, or frozen dew? Explain.

(d) On which morning would you probably only observe dew on the ground? Explain why.

	MORNING 1	MORNING 2	MORNING 3	MORNING 4	MORNING 5
Dew point	2°C (35°F)	−7°C (20°F)	1°C (34°F)	−4°C (25°F)	3°C (38°F)
Expected minimum temperature	4°C (40°F)	−3°C (27°F)	0°C (32°F)	−4.5°C (24°F)	2°C (35°F)

CHAPTER 9

Clouds: Identification and Observation

Clouds are aesthetically appealing and add excitement to the atmosphere. Without them, there would be no rain or snow, thunder or lightning, rainbows or halos. How monotonous if one had only a clear blue sky to look at. A cloud is a visible aggregate of tiny water droplets or ice crystals suspended in the air. Some are found only at high elevations, while others nearly touch the ground. Clouds can be thick or thin, big or little—they exist in a seemingly endless variety of forms. To impose order on this variety, we divide clouds into 10 basic types. With a careful and practiced eye, you can become reasonably proficient in correctly identifying them. First, we will see how clouds are named, then we will see how cloud formations are identified.

Classification of Clouds

Although ancient astronomers named the major stellar constellations about 2000 years ago, clouds were not formally identified and classified until the early nineteenth century. The French naturalist Lamarck (1744–1829) proposed the first system for classifying clouds in 1802; however, his work did not receive wide acclaim. One year later, Luke Howard, an English naturalist, developed a cloud classification system that found general acceptance. In essence, Howard's innovative system employed Latin words to describe clouds as they appear to a ground observer. He named a sheet-like cloud *stratus* (Latin for "layer"); a puffy cloud *cumulus* ("heap"); a wispy cloud *cirrus* ("curl of hair") and a rain cloud *nimbus* ("violent rain"). In Howard's system, these were the four basic cloud forms. Other clouds could be described by combining the basic types. For example, nimbostratus

Contents

is a rain cloud that shows layering, whereas cumulonimbus is a rain cloud having pronounced vertical development.

In 1887, Abercromby and Hildebrandsson expanded Howard's original system and published a classification system that, with only slight modification, is still used today. Ten principal cloud forms are divided into four primary cloud groups. Each group is identified by the height of the cloud's base above the surface: high clouds, middle clouds, and low clouds. The fourth group contains clouds showing more vertical than horizontal development. Within each group, cloud types are identified by their appearance. Table 9.1 lists these four groups and their cloud types.

The approximate base height of each cloud group is given in Table 9.2. Note that the altitude separating the high and middle cloud groups overlaps and varies with latitude. Large temperature changes cause most of this latitudinal variation. For example, high cirriform clouds are composed almost entirely of ice crystals. In tropical regions, air temperatures low enough to freeze all liquid water usually occur only above 6000 m (about 20,000 ft). In polar regions, however, these same temperatures may be found at altitudes as low as 3000 m (about 10,000 ft). Hence, while you may observe cirrus clouds at 3600 m (about 12,000 ft) over northern Alaska, you will not see them at that elevation above southern Florida.

Clouds cannot be accurately identified strictly on the basis of elevation. Other visual clues are necessary. Some of these are explained in the following section.

Cloud Identification

High Clouds High clouds in middle and low latitudes generally form above 6000 m (20,000 ft). Because the air at these elevations is quite cold and "dry," high clouds are composed almost exclusively of ice crystals and are also rather thin.*

---

*Studies conducted above Boulder, Colorado, during the early 1980s discovered small quantities of liquid water in cirrus clouds at temperatures as low as $-36°C$ ($-33°F$).

High clouds usually appear white, except near sunrise and sunset, when the unscattered (red, orange, and yellow) components of sunlight are reflected from the underside of the clouds.

The most common high clouds are the **cirrus**, which are thin, wispy clouds blown by high winds into long streamers called *mares' tails*. Notice in Fig. 9.1 that they can look like a white, feathery patch with a faint wisp of a tail at one end. Cirrus clouds usually move across the sky from west to east, indicating the prevailing winds at their elevation, and they generally point to fair, pleasant weather.

Cirrocumulus clouds, seen less frequently than cirrus, appear as small, rounded, white puffs that may occur individually, or in long rows. (See color plate 16.) When in rows, the cirrocumulus cloud has a rippling appearance that distinguishes it from the silky look of the cirrus and the sheetlike cirrostratus. Cirrocumulus seldom cover more than a small portion of the sky. The dappled cloud elements that reflect the red or yellow light of a setting sun make this one of the most beautiful of all clouds. The small ripples in the cirrocumulus strongly resemble the scales of a fish; hence, the expression "mackerel sky" commonly describes a sky full of cirrocumulus clouds.

The thin, sheetlike, high clouds that often cover the entire sky are **cirrostratus** (color plate 17), which are so thin that the sun and moon can be clearly seen through them. The ice crystals in these clouds refract the light passing through them, and will often produce a halo. In fact, the veil of cirrostratus may be so thin that a halo is the only clue to its presence. Thick cirrostratus clouds give the sky a glary white appearance and frequently form ahead of an advancing storm; hence, they can be used to predict rain or snow within 12 to

Table 9.1 The Four Major Cloud Groups and Their Types

1. High clouds
    Cirrus (Ci)
    Cirrostratus (Cs)
    Cirrocumulus (Cc)
	3. Low clouds
	    Stratus (St)
	    Stratocumulus (Sc)
	    Nimbostratus (Ns)

2. Middle clouds
    Altostratus (As)
    Altocumulus (Ac)
	4. Clouds with vertical
	    development
	    Cumulus (Cu)
	    Cumulonimbus (Cb)

Table 9.2 Approximate Height of Cloud Bases above the Surface for Various Locations

CLOUD GROUP	TROPICAL REGION	TEMPERATE REGION	POLAR REGION
High	6000 to 18,000 m (20,000 to 60,000 ft)	5000 to 13,000 m (16,000 to 43,000 ft)	3000 to 8000 m (10,000 to 26,000 ft)
Middle	2000 to 8000 m (6500 to 26,000 ft)	2000 to 7000 m (6500 to 23,000 ft)	2000 to 4000 m (6500 to 13,000 ft)
Low	surface to 2000 m (0 to 6500 ft)	surface to 2000 m (0 to 6500 ft)	surface to 2000 m (0 to 6500 ft)

24 hours, especially if they are followed by middle type clouds.

Middle Clouds   The middle clouds have bases between 2000 and 7000 m (6500 to 23,000 ft) in the temperate zone. These clouds are composed of water droplets and—when the temperature becomes low enough—some ice crystals.

Altocumulus clouds are middle clouds that appear as gray, puffy masses, sometimes rolled out in parallel waves or bands. (See color plate 18.) Usually, one part of the cloud is darker than another, which helps to separate it from the higher cirrocumulus. Also, the individual puffs of the al-

tocumulus are generally larger than those of the cirrocumulus. A layer of altocumulus may sometimes be confused with altostratus; in case of doubt, clouds are called altocumulus if there are rounded masses or rolls present. Altocumulus clouds that look like ''little castles'' (*castellanus*) in the sky indicate the presence of rising air at cloud level. The appearance of these clouds on a warm, humid summer morning often portends thunderstorms by late afternoon.

The **altostratus** is a gray or blue-gray (never white) cloud that often covers the entire sky over an area that extends over many hundreds of square kilometers. In the thinner section of the cloud, the

Fig. 9.1 Cirrus clouds.

Fig. 9.2 Altostratus cloud. The appearance of a dimly visible sun through a deck of gray clouds is usually a good indication that the clouds are altostratus.

sun (or moon) may be dimly visible as a round disk, which is sometimes referred to as a "watery sun" (Fig. 9.2). Thick cirrostratus clouds are occasionally confused with thin altostratus clouds. The gray color, height, and dimness of the sun are good clues to identifying an altostratus. The fact that halos only occur with cirriform clouds also helps one distinguish them. Another way to separate the two is to look at the ground for shadows. If there are none, it is a good bet that the cloud is altostratus because cirrostratus are usually transparent enough to produce them. Altostratus clouds often form ahead of storms having widespread and relatively continuous precipitation.

Low Clouds Low clouds, their bases lying below 2000 m (6500 ft), are almost always composed of water droplets; however, in cold weather, they may contain ice particles and snow.

The **nimbostratus** is a dark gray, "wet"-looking cloud layer associated with more or less continuously falling rain or snow. (See color plate 20.) The intensity of this precipitation is usually light or moderate—it is never of the heavy, showery

variety. The base of the nimbostratus cloud is normally impossible to identify clearly, and is easily confused with the altostratus. Thin nimbostratus is usually darker gray than thick altostratus, and you cannot see the sun or moon through a layer of nimbostratus. Visibility below a nimbostratus cloud deck is usually quite poor because rain will evaporate and saturate the air in this region, producing a lower layer of cloud or fog beneath the original base. Since these lower clouds drift rapidly with the wind, they form irregular shreds with a ragged appearance and are called *stratus fractus*, or *scud*.

A low, lumpy cloud layer is the **stratocumulus**. It appears in rows, in patches, or as rounded masses with blue sky visible between the individual cloud elements. The color of stratocumulus ranges from light to dark gray. It differs from altocumulus in that it has a lower base and larger individual cloud elements. (Compare color plate 18 with color plate 21.) To distinguish between the two, hold your hand at arm's length and point toward the cloud. Altocumulus cloud elements will generally be about the size of your thumbnail;

stratocumulus cloud elements will usually be about the size of your fist. Rain or snow rarely fall from stratocumulus.

Stratus is a uniform grayish cloud that often covers the entire sky. It resembles a fog that does not reach the ground (Fig. 9.3). Actually, when a thick fog ''lifts,'' the resulting cloud is a deck of low stratus. Normally, no precipitation falls from the stratus, but sometimes it is accompanied by a light mist or drizzle. This cloud commonly occurs over Pacific and Atlantic coastal waters in summer. A thick layer of stratus might be confused with nimbostratus, but the distinction between them can be made by observing the base of the cloud. Often, stratus has a more uniform base than does nimbostratus. Also, a deck of stratus may be confused with a layer of altostratus. However, if you remember that stratus are lower and darker gray, the distinction can be made.

Clouds with Vertical Development Familiar to almost everyone, the puffy **cumulus** cloud takes on a variety of shapes, but most often it looks like a piece of floating cotton with sharp outlines and a flat base (Fig. 9.4). The base appears white to light gray, and, on a humid day, may be only 1000 m above the ground and a kilometer or so wide. The top of the cloud—often in the form of rounded towers—denotes the limit of rising air and is usually not very high. These clouds can be distinguished from stratocumulus by the fact that cumulus clouds are detached (usually a great deal of blue sky between each cloud) while stratocumulus usually occur in groups or patches. Also, the cumulus has a dome- or tower-shaped top as opposed to the generally flat tops of the stratocumulus. Cumulus clouds that show only slight vertical growth are called *cumulus humilis* and are associated with fair weather; therefore, we call these clouds ''fair weather cumulus.'' Ragged-edge cumulus clouds that are smaller than cumulus humilis and scattered across the sky are called *cumulus fractus*.

Harmless-looking cumulus often develop on warm, summer mornings, and, by afternoon, become much larger and more vertically developed.

Fig. 9.3 A layer of low-lying stratus clouds.

Fig. 9.4 Fair weather cumulus clouds, known technically as *cumulus humilis*.

Fig. 9.5 Cumulus congestus. This line of cumulus congestus clouds is building along Maryland's eastern shore.

When the growing cumulus resembles a head of cauliflower, it becomes a *cumulus congestus*, or *towering cumulus*. Most often, it is a single large cloud, but, occasionally, several grow into each other, forming a line of towering clouds, as shown in Fig. 9.5. Precipitation that falls from a cumulus congestus is always showery.

If a cumulus congestus continues to grow vertically, it develops into a giant **cumulonimbus**—a thunderstorm cloud (Fig. 9.6). While its dark base may be no more than 300 m (1000 ft) above the earth's surface, its top may extend upward to the tropopause, over 15,000 m (50,000 ft) higher. A cumulonimbus can occur as an isolated cloud, or as part of a line or "wall" of clouds.

Tremendous amounts of energy are released by the condensation of water vapor within a cumulonimbus and result in the development of violent up- and down-drafts, which may exceed 50 knots. The lower (warmer) part of the cloud is usually composed of only water droplets. Higher up in the cloud, water droplets and ice crystals both abound, while, toward the cold top, there are only ice crystals. Swift winds at these higher altitudes can reshape the top of the cloud into a huge flattened anvil (*cumulonimbus incus*). These great thunderheads may contain all forms of precipitation—large raindrops, snowflakes, snow pellets, and sometimes hailstones—all of which can fall to earth in the form of heavy showers. Lightning, thunder, and even violent tornadoes are associated with the cumulonimbus.

Cumulus congestus and cumulonimbus frequently look alike, making it difficult to distinguish between them. However, you can usually distinguish them by looking at the top of the cloud. If the sprouting upper part of the cloud is sharply defined and not fibrous, it is usually a cumulus congestus; conversely, if the top of the cloud loses its sharpness and becomes fibrous in texture, it is usually a cumulonimbus. Compare Fig. 9.5 with Fig. 9.6. The weather associated with

Fig. 9.6 A cumulonimbus cloud. Strong upper-level winds blowing from right to left produce a well-defined anvil. Sunlight scattered by falling ice crystals produces the white (bright) area beneath the anvil. Notice the precipitation falling from the base of the cloud.

these clouds also differs: lightning, thunder, and large hail only occur with cumulonimbus.

So far, we have discussed the 10 primary cloud forms, summarized pictorially in Fig. 9.7. This figure, along with the cloud photographs and descriptions, should help you identify the more common cloud forms. Don't worry if you find it hard to estimate cloud heights. This is a difficult procedure, requiring much practice. You can use local objects (hills, mountains, tall buildings) of known height as references on which to base your height estimates.

To better describe a cloud's shape and form, a number of descriptive words may be used in conjunction with its name. We mentioned a few in the previous section; for example, a stratus cloud with a ragged appearance is a stratus fractus, and a cumulus cloud with marked vertical growth is a cumulus congestus. Table 9.3 lists some of the more common terms that are used in cloud identification.

Cloud Observations

Determining Sky Conditions Often, a daily weather forecast will include a phrase such as, "overcast skies with clouds becoming scattered by evening." To the average person, this means that the cloudiness will diminish; to the meteorologist, the terms "overcast" and "scattered" have a more specific meaning. Descriptions of sky conditions are defined by the fraction of sky covered by clouds. A *clear* sky, for example, is less than one-tenth covered by clouds. When cloudiness increases to between one-tenth and five-tenths, the sky is *scattered* with clouds. Partly cloudy (or partly sunny) also describes this sky condition. Clouds covering between six-tenths and nine-tenths produce a sky with *broken* clouds, and *overcast* conditions exist when more than nine-tenths of the sky is covered with clouds.

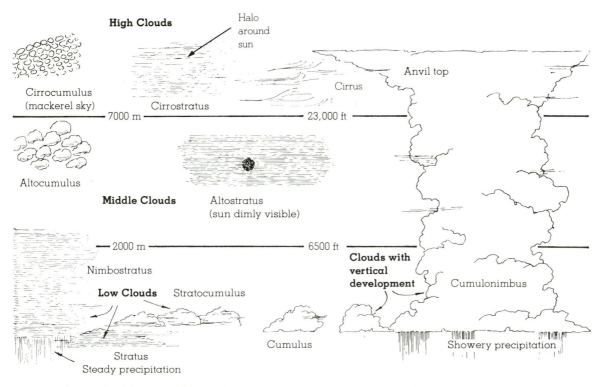

Fig. 9.7 A generalized illustration of basic cloud types based on height above the surface and vertical development.

Table 9.3 Common Terms Used in Identifying Clouds

TERM	LATIN ROOT AND MEANING	DESCRIPTION
Lenticularis	(*lens, lenticula*, lentil)	Clouds having the shape of a lens; often elongated and usually with well-defined outlines. This term applies mainly to cirrocumulus, altocumulus, and stratocumulus
Fractus	(*frangere*, to break or fracture)	Clouds that have a ragged or torn appearance; applies only to stratus and cumulus
Humilis	(*humilis*, of small size)	Cumulus clouds with generally flattened bases and slight vertical growth
Congestus	(*congerere*, to bring together; to pile up)	Cumulus clouds of great vertical extent that from a distance may resemble a head of cauliflower
Calvus	(*calvus*, bald)	Cumulonimbus in which at least some of the upper part is beginning to lose its cumuliform outline
Capillatus	(*capillus*, hair; having hair)	Cumulonimbus characterized by the presence in the upper part of cirriform clouds with fibrous or striated structure
Undulatus	(*unda*, wave; having waves)	Clouds in patches, sheets, or layers showing undulations
Translucidus	(*translucere*, to shine through; transparent)	Clouds that cover a large part of the sky and are sufficiently translucent to reveal the position of the sun or moon
Incus	(*incus*, anvil)	The smooth cirriform mass of cloud in the upper part of a cumulonimbus that is anvil-shaped
Mammatus	(*mamma*, mammary)	Baglike clouds that hang like a cow's udder on the underside of a cloud; may occur with cirrus, altocumulus, altostratus, stratocumulus, and cumulonimbus
Pileus	(*pileus*, cap)	A cloud in the form of a cap or hood above or attached to the upper part of a cumuliform cloud, particularly during its developing stage
Castellanus	(*castellum*, a castle)	Clouds that show vertical development and produce towerlike extensions, often in the shape of small castles

Table 9.4 presents a summary of sky cover conditions.

Observing sky conditions far away can sometimes fool even the trained observer. A broken cloud deck near the horizon usually appears as overcast because the open spaces between the clouds are less visible at a distance (Fig. 9.8). Therefore, cloudiness is usually over-estimated when clouds are near the horizon. Viewed from afar, clouds not normally associated with precipitation may appear darker and thicker than they actually are. The reason for this is that light from a distant cloud travels through more atmosphere and is more attenuated than the light from the same type of cloud closer to the observer.

Up to this point, we have seen how clouds look from the ground. We will now look at how ground-based sky and ceiling observations are made and at clouds from a different vantage point—the satellite view.

Table 9.4   Description of Sky Conditions

DESCRIPTION	MEANING
Clear (CLR)	less than one-tenth sky covered by clouds
Scattered (SCT) or partly cloudy	from one-tenth to five-tenths sky covered by clouds
Broken (BKN) or cloudy	from six-tenths to nine-tenths sky covered by clouds
Overcast (OVC)	more than nine-tenths sky covered by clouds
Sky obscured*	all of sky is hidden by surface-based phenomena

*This condition is due to fog, blowing snow, smoke, and so forth, rather than cloud cover.

Satellite Observations   The weather satellite is a cloud-observing platform in earth orbit. It provides extremely valuable cloud photographs of areas where there are no ground-based observations. Because water covers over 70 percent of the earth's surface, there are vast regions where few (if any) surface cloud observations are made. Before weather satellites were used, severe storms, such as hurricanes and typhoons, went undetected until they moved dangerously near inhabited areas. Residents of the regions affected had little advance warning. Today, satellites spot these storms while they are still far out in the ocean and track them accurately.

Continuously improved detection devices make our satellites more versatile than ever. Early satellites, such as *TIROS I*, launched on April 1, 1960, used television cameras to photograph clouds. Contemporary satellites use radiometers, which can observe clouds during both day and night.

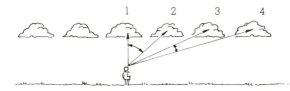

Fig. 9.8   Clouds on the horizon appear closer together than clouds overhead. Note that the amount of clear space between each cloud is the same. To the observer, however, there appears to be more space between clouds 1 and 2 than between clouds 3 and 4.

Fig. 9.9   Artist's view of the *SMS/GOES* meteorological satellite.

Radiometers detect the radiation coming from an object at specific wavelength bands. For example, the sophisticated visible and infrared spin scan radiometer (VISSR) used in the *Synchronous Meteorological Satellite (SMS)/Geostationary Operational Environmental Satellite (GOES)* series picks up visible images between 0.55 and 0.90 μm and infrared images from 10.5 to 12.6 μm. The infrared wavelength band is used because it corresponds to the atmospheric window described in Chapter 3.

Information on cloud thickness and height can be deduced from satellite photographs. Visible photographs show the sunlight reflected from a cloud's upper surface. Because thick clouds have a higher albedo (reflectivity) than thin clouds, they appear brighter on a visible satellite photograph. However, high, middle, and low clouds have just about the same albedo, so it is difficult to distinguish among them simply by using visible light photographs. To make this distinction, infrared cloud pictures are used. They produce a better image of the actual radiating surface because they do not show the strong visible reflected light. Since warm objects radiate more energy than cold ob-

FOCUS ON INSTRUMENTS

Measuring Cloud Ceilings

In addition to knowing about sky conditions, it is usually important to have a good estimate of the height of cloud bases. Aircraft could not operate safely without accurate cloud height information, particularly at lower elevations.

The term **ceiling** is defined as the height of the lowest layer of clouds above the surface that are either broken or overcast, but not thin. Direct information on cloud height can be obtained from pilots who report the altitude at which they encounter the ceiling. Less directly, **ceiling balloons** are used to measure the height of clouds. A small balloon filled with a known amount of hydrogen or helium rises at a fairly constant and known rate. The ceiling is determined by measuring the time required for it to enter the lowest cloud layer. For example, if the balloon rises 125 m (about 400 ft) each minute, and it takes three minutes to enter a broken layer of stratocumulus, the ceiling would be 375 m (about 1200 ft). Ceiling balloon observations can be made at night simply by attaching a small battery-operated light to the balloon.

If you drive by a large airport on a cloudy night you may see an intense light shining upward into the clouds (or rapidly sweeping across the cloud base). Such lights are an integral part of an instrument called a **ceilometer**, which is used to measure the ceiling from the ground. There are two types of ceilometers: one with a rotating light beam and the other with a fixed beam. The rotating-beam

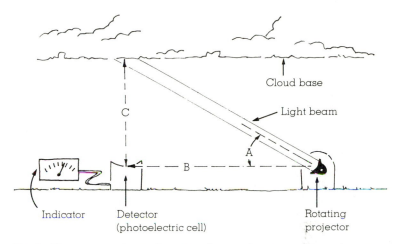

Fig. 1   The main ingredients of a rotating-beam ceilometer: projector, detector, and indicator; if the angle A and baseline B are both known, then the ceiling C can be determined.

ceilometer consists of a projector, detector, and indicator (Fig. 1). The projector rotates vertically from horizon to horizon and sends out a powerful light beam that moves along the base of the cloud. The light-sensitive detector, usually a photoelectric cell, is a known distance from the projector and points straight up. When the light reflected from the cloud base shines directly into the detector it excites the photoelectric cell, and the angle of the projector beam above the horizon is then recorded. This information is transmitted to an indicator located either inside the weather office or in the control tower. By knowing the projector angle and its distance from the detector, the cloud height is determined mathematically. Automatic operation and continuous information make this instrument useful.

The older fixed-beam ceilometer operates in a similar way except that the projector lamp points straight up at the cloud base, while the detector continuously sweeps out an angle from horizon to horizon. This instrument scans much more slowly than the rotating-beam ceilometer and is, therefore, being phased out.

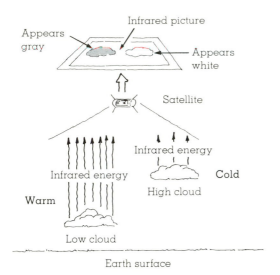

Fig. 9.10 Generally, the lower the cloud, the warmer its top. Warm objects emit more infrared energy than cold objects. Thus, an infrared satellite picture can distinguish warm, low (gray) clouds from cold, high (white) clouds.

jects, high temperature regions can be artificially made to appear darker on an infrared photograph. Because the tops of low clouds are warmer than those of high clouds, cloud observations made in the infrared can distinguish between warm, low clouds (dark) and cold, high clouds (light).

Compare the visible satellite photograph of clouds in the eastern Pacific on April 23, 1978 (Fig. 9.11a), with the infrared photograph taken on the same day (Fig. 9.11b). Notice that all of the clouds in the visible photograph appear white, while, in the infrared, they appear as many shades of gray. In the visible photograph, the clouds covering part of Oregon and northern California appear relatively thin compared to the thicker, brighter clouds to the west. Furthermore, these thin clouds must be high because they are bright in the infrared picture. The elongated band of clouds off the coast marks the position of an approaching weather front. Here, the clouds appear white and

(a)   (b)

Fig. 9.11 A visible picture (a) and an infrared picture (b) of the eastern Pacific taken on the same day (April 23, 1978) at just about the same time.

bright in both pictures, indicating a zone of thick, heavy clouds. Behind the front, the lumpy clouds are probably cumulus because they appear gray in the infrared photo, indicating that their tops are low and relatively warm.

When temperature differences are small, it is difficult to directly identify significant cloud and surface features on an infrared picture. Some way must be found to increase the contrast between features and their backgrounds. This can be done by a process called *computer enhancement.* Certain temperature ranges in the infrared photograph are assigned specific shades of gray—grading from black to white. These shades of gray are then color-contoured to make specific features, such as deep cloud layers and the freezing level, more obvious. Figure 9.12 is an infrared-enhanced picture for the same date and area as shown in Fig. 9.11. Note the dark and light contouring in the picture. Clouds with cold tops, and those with tops near freezing, are assigned the darkest color. Hence, the dark gray areas embedded along the front represent the region where the coldest and, therefore, highest and thickest clouds are found. It is here where the stormiest weather is probably occurring. Also notice that, near the southern tip of the picture, the dark gray blotches surrounded by areas of white are thunderstorms that have developed over warm tropical waters. They show up clearly as white, thick clouds in both the visible and infrared photographs.

Early *TIROS* satellites took pictures that only covered about one-fifth of the earth each day. Contemporary satellites can photograph a much larger section of the earth. In fact, there are satellites positioned over the equator at an altitude of nearly 36,000 km (22,300 mi) that can observe almost one-third of the earth's surface in a single photograph. These satellites, which orbit the equator at the same rate the earth spins, are called **geostationary satellites** (or *synchronous satellites*) because they remain above a fixed spot on the earth's surface. This allows continuous monitoring of a specific region.

Geostationary satellites are also important because they use a "real time" data system, meaning that the satellites transmit photographs to the receiving system on the ground as soon as the camera takes the picture. Successive cloud photographs from these satellites can be put into a

Fig. 9.12   An enhanced infrared picture of the eastern Pacific taken on April 23, 1978.

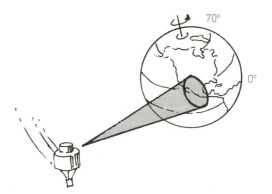

Fig. 9.13   The geostationary satellite moves through space at the same rate that the earth rotates, so it remains above a fixed spot on the equator and monitors one area constantly.

Satellites Do More than Observe Clouds

The use of satellites to monitor weather is not restricted to observing clouds. For example, satellites measure radiation from the earth's surface and atmosphere, giving us information about the earth-atmosphere energy budget and the vertical distribution of temperature and moisture. Radiation intensities from the ocean surface are translated into temperature readings. This information is valuable to the fishing industry, as well as to the meteorologist. Satellites also monitor the amount of snow cover in winter, the extent of ice fields in the Arctic and Antarctic, and the movement of large icebergs that drift into shipping lanes. One polar orbiting satellite actually carries equipment that can detect faint distress signals anywhere on the globe and relay them to rescue forces on the ground.

Infrared sensors on polar orbiting satellites are able to assess conditions of crops, areas of deforestation, and regions of extensive drought. Some satellites are equipped with a water vapor sensing channel near 6 μm that can profile the distribution of water vapor in the atmosphere. This has become a valuable tool for determining the location of middle tropospheric swirling wind patterns and jet streams. Satellites are also able to detect volcanic eruptions and follow the movement of ash clouds, such as those that came from Mount St. Helens in 1980, El Chichón in 1982, and the Indonesian volcanoes Galunggung and Una Una during the summers of 1982 and 1983, respectively. During the winter, GOES satellites are able to monitor the southward progress of freezing air in Florida and Texas, allowing forecasters to warn growers of impending low temperatures, so that they can take necessary measures to protect sensitive crops.

Geostationary satellites, such as GOES, are equipped with systems that receive environmental information from remote data collection platforms on the surface. These platforms include instrumented buoys, river gauges, automatic weather stations, seismic and tsunami ("tidal" wave) stations, and ships. This information is transmitted to the satellite, which relays it to a central receiving station at Wallop's Island, Virginia.

Some examples of data collected include information on earthquakes, wind speed and direction, humidity and rainfall, tides, ocean currents, water levels, water and air temperatures, and tsunamis. Satellite transmission of this data to one center allows for faster data analysis and a more comprehensive picture of the physical state of our world.

Today, a network of five geostationary satellites positioned over the equator give nearly a complete global coverage from

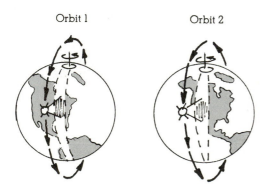

Orbit 1   Orbit 2

Fig. 9.14 Polar orbiting satellites scan from north to south, and on each successive orbit the satellite scans an area further to the west.

time-lapse movie sequence to show the cloud movement, dissipation, or development associated with weather fronts and storms. This is a great help in forecasting the progress of large weather systems. Wind directions and speeds at various levels may also be approximated by monitoring cloud movement with the geostationary satellite.

To complement the geostationary satellites, there are **polar orbiting** weather **satellites**, which closely parallel the earth's meridian lines. They pass over the north and south polar regions on each revolution. As the earth rotates to the east beneath the satellite, each pass monitors an area to the about latitude 55°N to 55°S. Japan's *Geostationary Meteorology Satellite* monitors eastern Asia and the western Pacific. The United States' *GOES West*, positioned above 135°W longitude, observes the weather over western North America and much of the Pacific, while *GOES East* (another United States satellite), positioned above 85°W longitude, monitors North and South America and most of the Atlantic Ocean.* The European Space Agency's *METEOSAT* observes Europe, Africa, the Mediterranean, and the eastern Atlantic, while India's *INSAT* monitors the Indian Ocean and Asia.

In July, 1972, NASA launched the first satellite specifically designed to monitor the natural resources of the earth. Originally

*During July, 1984, *GOES East* lost all imaging capabilities. As a result, *GOES West* was moved to a new location: 98°W during the hurricane season and 108°W during the remainder of the year. By February, 1986, a new *GOES East* should be in orbit.

called *ERTS-I* (*Earth Resources Technology Satellite*), it (along with its sister satellites launched in 1976, 1978, 1982, and 1984, respectively) now goes by the name *LandSat*—short for "land satellite." *LandSat* circles the earth every 103 minutes—14 times a day—in a near-polar circular orbit about 918 km (570 mi) above sea level. Each pass photographs a strip 185 km (115 mi) wide. These photographs, taken in several wavelength bands, provide valuable information about this planet's geology, hydrology, oceanography, and ecology. *LandSat* also collects data transmitted from remote ground stations in North America. These stations monitor a variety of environmental data, with water quality, rainfall amount, and snow depth of particular interest to the meteorologist and hydrologist.

Satellite information is not confined to the lower atmosphere. The *Solar Mesosphere Explorer* satellite, launched in 1981, uses a radiometer to determine air

temperature, water vapor content, and ozone distribution in the stratosphere. Launched the same year, the *Dynamics Explorer Satellite* flies into the ionosphere and magnetosphere sampling ionized gases with a mass spectrometer (basically an instrument that determines the masses of atoms and molecules). And both geostationary and polar orbiting satellites carry instruments that monitor solar activity.

By the late 1980s, new and improved satellites will provide additional information. For example, the National Oceanic and Atmospheric Administration plans to launch a polar orbiting satellite called *Windsat* that will carry a laser instrument designed to obtain global wind information. In addition, geosynchronous satellites may be equipped with a *lightning mapper sensor*—an instrument designed to continuously monitor lightning activity over broad areas of the earth.

west of the previous pass (Fig. 9.14). Eventually, the satellite covers the entire earth. The *Nimbus* and subsequent *National Oceanic and Atmospheric Administration (NOAA)* satellite series are all polar orbiters.

Polar orbiting satellites have the advantage of photographing clouds directly beneath them. Thus, they provide sharp pictures in polar regions, where photographs from a geostationary satellite are distorted because of the low angle at which the satellite "sees" this region. Polar orbiters also circle the earth at a much lower altitude (about 850 km or 530 mi) than geostationary satellites and provide detailed photographic information about

objects, such as violent storms and cloud systems, whose size is as small as 1 km (about 0.5 mi). For sharp detailed photographs from a geostationary satellite, the diameter of the object must be at least 4 km (about 2.5 mi).

Summary

In this chapter, we have seen that, when clouds are classified according to their height and physical appearance, they are divided into four main groups: high, middle, low, and clouds with vertical development. Since each cloud has physical characteristics that distinguish it from all the others, careful cloud observations normally lead to correct identification.

We also saw that satellites enable scientists to obtain a bird's-eye view of clouds on a global scale. Polar orbiting satellites obtain data covering the earth from pole to pole, while geostationary satellites located above the equator continuously monitor a desired portion of the earth. Both types of satellites use radiometers that detect emitted radiation. As a consequence, clouds can be observed both day and night.

Visible satellite pictures, which show sunlight reflected from a cloud's upper surface, can distinguish thick clouds from thin clouds. Infrared pictures show an image of the cloud's radiating top and can distinguish low clouds from high clouds. To increase the contrast between cloud features, infrared photographs are enhanced.

Satellites do a great deal more than simply photograph clouds. They provide us with a wealth of physical information about the earth and the atmosphere.

Questions for Review

1. Clouds are most generally classified by height. What are the major height categories and the cloud type associated with each?

2. List at least two distinguishable characteristics of each of the 10 basic clouds.

3. Why are high clouds normally thin? Why are they composed almost entirely of ice crystals?

4. How can you distinguish altostratus from cirrostratus?

5. Which clouds are associated with each of the following characteristics:
  (a) lightning
  (b) heavy rain showers
  (c) mackerel sky
  (d) mares' tails
  (e) halos
  (f) light continuous snow
  (g) hailstones

6. Why does a broken layer of clouds near the horizon often appear as overcast?

7. Draw a diagram to explain how a rotating-beam ceilometer works.

8. If clouds are described as broken, what percent of the sky is covered by clouds?

9. How is a ceiling balloon used to determine the height of the base of a cloud?

10. List and explain the various types of environmental information obtained from satellites.

11. How do geostationary satellites differ from polar orbiting satellites?

12. Explain why visible and infrared images can be used to distinguish: (a) high clouds from low clouds; (b) thick clouds from thin clouds.

13. Why are infrared images enhanced?

Questions for Thought

1. Explain why altocumulus clouds might be observed at 6400 m (21,000 ft) above the surface in Mexico City, Mexico, but never at that altitude above Fairbanks, Alaska.

2. How are satellites able to locate the boundaries of ocean currents?

3. The sky is overcast and it is raining. Explain how you could tell if the cloud above you is a nimbostratus or a cumulonimbus.

4. Suppose it is raining lightly from a deck of nimbostratus clouds. Beneath the clouds are small, ragged, puffy clouds that are moving rapidly with the wind. What would you call these clouds? How did they probably form?

5. What is the name of the meteorological satellite that photographs the clouds shown on your local TV news program?

Problems and Exercises

1. Sketch how each of the clouds listed below would appear if you were to observe them from the ground. (Refer to Table 9.3 for assistance.)

(a) cumulus humilis
(b) cumulus congestus
(c) altocumulus castellanus
(d) altocumulus lenticularis
(e) stratus fractus
(f) stratocumulus undulatus
(g) altostratus translucidus
(h) cumulonimbus incus
(i) cumulus mammatus

2. If a ceiling balloon rises at 120 m (about 400 ft) each minute, what is the ceiling of an overcast deck of stratus clouds 1500 m (about 5000 ft) thick if the balloon disappears into the clouds in 5 minutes?

3. Compare the visible satellite picture (Fig. 9.11a) with the infrared picture (Fig. 9.11b). With the aid of the infrared photograph, label on the visible picture the regions of middle, high, and low clouds. On the enhanced infrared photograph (Fig. 9.12), label where the highest and thickest clouds appear to be located.

A stable mass of moist air, gliding up and over the Sierra Nevada, condenses into lenticular clouds near Verdi, Nevada. (Photo: Dick Hilton)

CHAPTER 10

Clouds: Stability and Development

Clouds, spectacular features in the sky, add beauty and color to the natural landscape. Yet, clouds are important for nonaesthetic reasons, too. As they form, vast quantities of heat are released into the atmosphere. Clouds help regulate the earth's energy balance by reflecting and scattering solar radiation, and by absorbing the earth's infrared energy. And, of course, without clouds there would be no precipitation. But clouds are also significant because they visually indicate the physical processes taking place in the atmosphere; to a trained observer, they are signposts in the sky. This chapter examines the atmospheric processes these signposts point to, the first of which is atmospheric stability.

Atmospheric Stability

We know from Chapter 2 that most clouds form as air rises and cools. Why does air rise on some occasions and not on others? And why do the size and shape of clouds vary so much when the air does rise? Let's see how knowing about the air's stability will help us to answer these questions.

When we speak of atmospheric stability, we are referring to a condition of equilibrium. For example, rock A resting in the depression in Fig. 10.1 is in *stable* equilibrium. If the rock is pushed up along either side of the hill and then let go, it will quickly return to its original position. On the other hand, rock B, resting on the top of the hill, is in a state of *unstable* equilibrium, as a slight push will set it moving away from its original position. Applying these concepts to the atmosphere, we can see that air is in stable equilibrium when, after being lifted or lowered, it tends to return to its original position—it resists upward and downward

Contents

Fig. 10.1 When rock A is disturbed, it will return to its original position; rock B, however, will accelerate away from its original position.

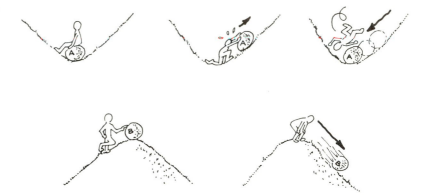

air motions. Air that is in unstable equilibrium will, when given a little push, move farther away from its original position—it favors vertical air currents.

In order to explore the behavior of rising and sinking air, we must first review some concepts from Chapter 2. Recall that a rising air parcel expands and cools because it moves into a region of lower pressure. A sinking parcel of air, on the other hand, moves into a region of greater air pressure, where it is compressed into a smaller volume and the air inside warms. If a parcel of air expands and cools, or compresses and warms, with no interchange of heat with its surroundings, this situation is called an **adiabatic process**. As long as the air in the parcel is unsaturated (the relative humidity is less than 100 percent), the rate of adiabatic cooling or warming remains constant. This rate of heating or cooling is about 10°C for every 1000 m of change in elevation (5.5°F per 1000 ft), and applies only to unsaturated air. For this reason, it is called the **dry adiabatic rate**. For example, suppose a parcel of unsaturated surface air with a temperature of 30°C (86°F) rises and cools dry adiabatically to an elevation of 3000 m (about 10,000 ft) above the ground. Since it cools at a rate of 10°C for each 1000 m, the total change in temperature will be 30°C. So the parcel's final temperature will be a rather cool 0°C (32°F). If the same parcel were brought back down to sea level, it would have the same temperature as it did before it was lifted.

As the rising air cools, its capacity to hold water vapor decreases, so that the relative humidity of the rising air increases. If the air cools to its dew point, the relative humidity becomes 100 percent. Further lifting results in condensation, a cloud forms, and latent heat is released into the rising air. Be-

cause the heat added during condensation offsets some of the cooling due to expansion, the air no longer cools at the dry adiabatic rate but at a lesser rate called the **moist adiabatic rate**.* (Because latent heat is added to the rising saturated air, the process is not really adiabatic.) If a saturated parcel containing water droplets were to sink, it would compress and warm at the moist adiabatic rate because evaporation of the liquid droplets would offset the rate of compressional warming. Hence, the rate at which rising or sinking saturated air changes temperature—the moist adiabatic rate—is less than the dry adiabatic rate.

Unlike the dry adiabatic rate, the moist adiabatic rate is not constant, but varies greatly with temperature, and, hence, with moisture too, as warm saturated air holds more water vapor than cold saturated air. A rising parcel of warm saturated air will liberate more heat than will a parcel of cold saturated air rising the same distance. As a result, the moist adiabatic rate is much less than the dry adiabatic rate for warm air; however, the two rates are nearly the same for very cold air. (See Table 10.1.) Although the moist adiabatic rate does vary, we will use an average of 6°C per 1000 m (3.3°F per 1000 ft) in most of our examples and calculations.

Determining Stability

We determine the stability of the air by comparing the temperature of a rising parcel to that of its surroundings. If the rising air is colder than its

*If condensed water or ice is removed from the rising saturated air, the cooling process is called *pseudoadiabatic*.

environment, it will be more dense (heavier) and tend to sink back to its original level. In this case, the air is *stable* because it resists upward displacement. If the rising air is warmer and, therefore, less dense (lighter) than the surrounding air, it will continue to rise until it reaches the same temperature as its environment. This is an example of *unstable* air. To figure out the air's stability, we need to measure the temperature both of the rising air and of its environment at various levels above the earth.

Stable Air Suppose we release a radiosonde and it sends back temperature data as shown in Fig. 10.2. (Such a vertical profile of temperature is called a *sounding*.) We measure the air temperature in the vertical and find that it decreases by 4°C for every 1000 m (2°F per 1000 ft). Remember from Chapter 2 that the rate at which the air temperature changes with elevation is called

Table 10.1 The Moist Adiabatic Rate for Different Temperatures and Pressures in °C/1000 m and °F/1000 ft

PRESSURE (mb)	TEMPERATURE (°C)					TEMPERATURE (°F)				
	-40	-20	0	20	40	-40	-5	30	65	100
1000	9.5	8.6	6.4	4.3	3.0	5.2	4.7	3.5	2.4	1.6
800	9.4	8.3	6.0	3.9	2.8	5.2	4.6	3.3	2.2	1.5
600	9.3	7.9	5.4	3.5	2.6	5.1	4.4	3.0	1.9	1.4
400	9.1	7.3	4.6	3.0	2.4	5.0	4.0	2.5	1.6	1.3
200	8.6	6.0	3.4	2.5	2.0	4.7	3.3	1.9	1.4	1.1

the lapse rate. Because this is the rate at which the air temperature surrounding us would be changing if we were to climb upward into the atmosphere, we will refer to it as the **environmental lapse rate**. Now suppose in Fig. 10.2a that a parcel of unsaturated air with a temperature of 30°C is lifted from the surface. As it rises, it

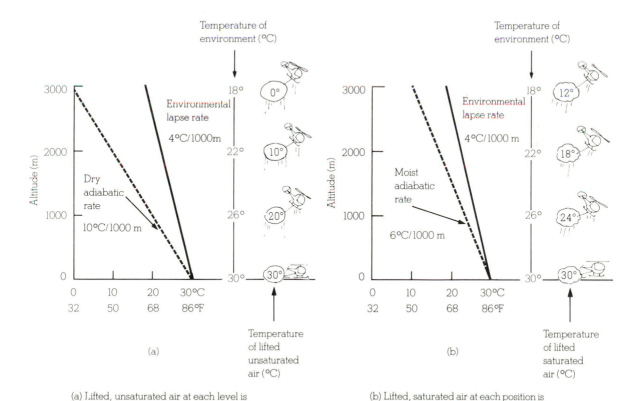

(a) Lifted, unsaturated air at each level is colder and heavier than the air around it. If given the chance, the parcel would return to its original position.

(b) Lifted, saturated air at each position is colder and heavier than the air surrounding it. If released, the parcel would return to its original position.

Fig. 10.2 An absolutely stable atmosphere occurs when the environmental lapse rate is less than the moist adiabatic rate.

Sinking Air and Subsidence Inversions

Another way the atmosphere becomes more stable is if an entire layer of air sinks. If a layer of unsaturated air over 1000 m thick and covering a large area subsides, the entire layer will warm by adiabatic compression. As the layer subsides, it becomes compressed by the weight of the atmosphere and shrinks vertically. The upper part of the layer sinks farther, and, hence, heats more than the bottom part. This is illustrated in Fig. 1. After subsiding, the top of the layer is actually warmer than the bottom, and an inversion is formed. Inversions that form by slowly sinking air over a large area are called **subsidence inversions**. They sometimes occur at the surface, but more frequently, they are observed aloft and are often associated with large high-pressure areas because of the sinking air motions associated with these systems. An inversion represents an atmosphere that is absolutely stable. Why? Within the inversion, warm air overlies

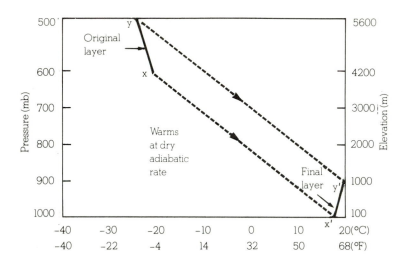

Fig. 1 The layer x–y is initially 1400 m thick. If the entire layer slowly subsides, it shrinks in the more dense air near the surface. As a result of the shrinking, the top of the layer warms more than the bottom, and the entire layer (x'–y') becomes more stable.

cold air, and, if air rises into the inversion, it is becoming colder, while the air around it is getting warmer. Obviously, the colder air would tend to sink. Inversions, therefore, act as lids on vertical

air motion. When an inversion exists near the ground, stratus, fog, haze, and pollutants are all kept close to the surface. In fact, most air pollution alerts occur with subsidence inversions.

cools at the dry adiabatic rate (10°C/1000 m), and the temperature inside the parcel at 1000 m would be 20°C, or 6°C lower than the air surrounding it. Look at Fig. 10.2a closely and notice that, as the air parcel rises higher, the temperature difference between it and the surrounding air becomes ever greater. Even if the parcel is initially saturated (Fig. 10.2b), it would cool at the moist rate—6°C per 1000 m—and would be colder than its environment at all levels. In both cases, the rising air is colder and heavier than the air surrounding it. In this example, the atmosphere is **absolutely stable**. *The air is always absolutely stable when the environmental lapse rate is less than the moist adiabatic rate.*

Since absolutely stable air strongly resists upward vertical motion, it will, *if forced to rise*, tend to spread out horizontally. If clouds form in this rising air they, too, will spread horizontally in relatively thin layers and usually have flat tops and bases. We might expect to see clouds—such as cirrostratus, altostratus, nimbostratus, or stratus—forming in stable air.

A Stable Atmosphere What conditions are necessary to bring about a stable atmosphere? As we have just seen, air is stable when the environmental lapse rate is small. This can occur when the air aloft warms, or the surface air cools. If the air aloft is being replaced by warmer air (warm advection), and the surface air is not changing appreciably, the environmental lapse rate decreases (weakens) and the air becomes more stable. Similarly, the environmental lapse rate decreases and air becomes more stable when the lower layer cools. This can be caused by an influx of cold air (cold advection), radiational cooling, or air moving over a cold surface. These conditions can produce a persistent fog where the stable air keeps the fog from rising and dissipating.

Before we turn our attention to unstable air, let's first examine a condition known as **neutral stability**. If the lapse rate is exactly equal to the dry adiabatic rate, rising or sinking unsaturated air will cool or warm at the same rate as the air around it. At each level, it would have the same temperature and density as the surrounding air. Because this air tends neither to continue rising nor sinking,

the atmosphere is said to be neutrally stable. For saturated air, *neutral stability* exists when the environmental lapse rate is equal to the moist adiabatic rate.

Unstable Air Suppose a radiosonde sends back the temperatures above the earth as plotted in Fig. 10.3. Once again, we determine the air's stability by comparing the environmental lapse rate to the moist and dry adiabatic rates. In this case, the environmental lapse rate is 11°C per 1000 m (6°F per 1000 ft). A rising parcel of unsaturated surface air will cool at the dry adiabatic rate. Because the dry adiabatic rate is less than the environmental lapse rate, the parcel will be warmer than the surrounding air and will continue to rise, constantly moving upward, away from its original position. The atmosphere is unstable. Of course,

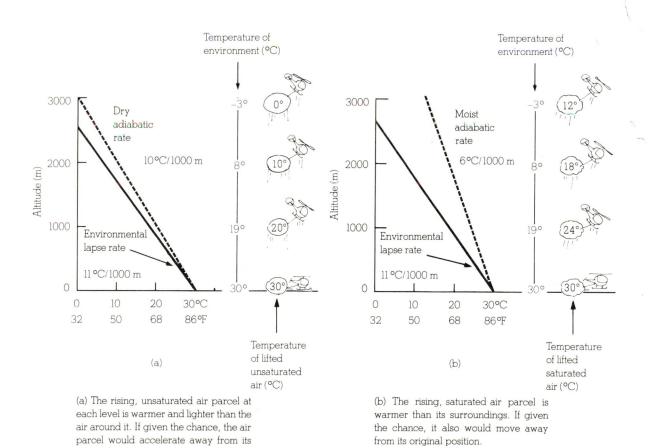

(a) The rising, unsaturated air parcel at each level is warmer and lighter than the air around it. If given the chance, the air parcel would accelerate away from its original position.

(b) The rising, saturated air parcel is warmer than its surroundings. If given the chance, it also would move away from its original position.

Fig. 10.3 An absolutely unstable atmosphere occurs when the environmental lapse rate is greater than the dry adiabatic rate.

a parcel of saturated air cooling at the lower moist adiabatic rate will be even warmer than the air around it (Fig. 10.3b). In both cases, the air parcels, once they start upward, will continue to rise on their own because they are warmer and less dense than the air around them. The atmosphere in this example is said to be **absolutely unstable**. *Absolute instability results when the environmental lapse rate is greater than the dry adiabatic rate.* Usually, absolute instability is limited to a shallow layer of surface air on hot, sunny days.

So far, we have seen that the atmosphere is absolutely stable when the environmental lapse rate is less than the moist adiabatic rate and absolutely unstable when the environmental lapse rate is greater than the dry adiabatic rate. How-

ever, a typical type of instability exists when the lapse rate lies between the moist and dry adiabatic rates.

The environmental lapse rate in Fig. 10.4 is 7°C per 1000 m (4°F per 1000 ft). When a parcel of unsaturated air rises, it cools dry adiabatically and is colder at each level than the air around it (Fig. 10.4a). It will, therefore, tend to sink back to its original level because it is in a stable atmosphere. Now, suppose the rising parcel is saturated. As we can see in Fig. 10.4b, the rising air is warmer than its environment at each level. Once the parcel is given a push upward, it will tend to move in that direction; the atmosphere is unstable for the saturated parcel. In this example, the atmosphere is said to be **conditionally unstable**. This type of stability depends upon whether or not the rising

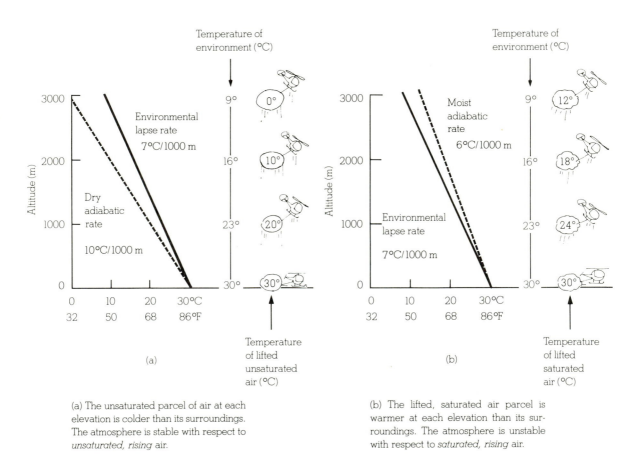

(a) The unsaturated parcel of air at each elevation is colder than its surroundings. The atmosphere is stable with respect to *unsaturated, rising* air.

(b) The lifted, saturated air parcel is warmer at each elevation than its surroundings. The atmosphere is unstable with respect to *saturated, rising* air.

Fig. 10.4 Conditionally unstable air. The atmosphere is stable if the rising air is unsaturated (a), but unstable if the rising air is saturated (b).

air is saturated. When the rising parcel of air is unsaturated, the atmosphere is stable; when the parcel of air is saturated, the atmosphere is unstable. Conditional instability means that, if unsaturated air could be lifted to a level where it becomes saturated, instability would result.

Conditional instability occurs whenever the environmental lapse rate is between the moist adiabatic rate and the dry adiabatic rate. Recall from Chapter 4 that the average lapse rate in the troposphere is about 6.5°C per 1000 m (3.6°F per 1000 ft). Since this value lies between the dry adiabatic rate and the average moist rate, the atmosphere is ordinarily in a state of conditional instability.

As the atmosphere becomes more unstable, vertical motions increase, and thicker cumuliform clouds begin to form. Two questions arise: (1) What mechanisms cause the atmosphere to become more unstable? (2) What causes air to rise so that clouds are able to form? (Unstable air normally needs a "trigger" to start it moving upward.)

Causes of Instability   The atmosphere becomes more unstable as the environmental lapse rate steepens; that is, as the air temperature drops rapidly with increasing height. This may be caused by an influx of cold air aloft, or by the warming of air near the surface. The surface air may warm due to daytime heating, warm advection, or by moving over a surface warmer than itself. The combination of cold air aloft and warm surface air can produce a steep lapse rate and an unstable atmosphere.

A layer of air may also be made more unstable by either mixing or lifting. Let's look at mixing first. In Fig. 10.5, the environmental lapse rate before mixing is less than the moist rate, and the layer is stable. Now, suppose the air in the layer is mixed either by convection or wind-induced turbulent eddies. Air is cooled adiabatically as it is brought up from below and heated adiabatically as it is mixed downward. The up and down motion in the layer redistributes the air in such a way that the temperature at the top of the layer decreases, while, at the base, it increases. This steepens the environmental lapse rate and makes the layer more unstable. If this mixing continues for some time, and the air remains unsaturated, the vertical tem-

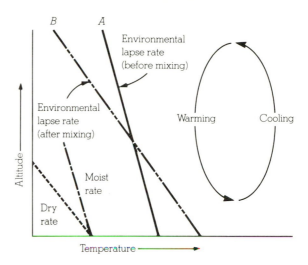

Fig. 10.5   Mixing tends to steepen the lapse rate. Rising, cooling air lowers the temperature toward the top of the layer, while sinking, warming air increases the temperature near the bottom.

perature distribution would eventually be equal to the dry adiabatic rate.

Just as lowering an entire layer of air makes it more stable, the lifting of a layer makes it more unstable. In Fig. 10.6, the air lying between the 1000 mb and 900 mb is initially absolutely stable since the environmental lapse rate of layer x–y is less than the moist adiabatic rate. The layer is

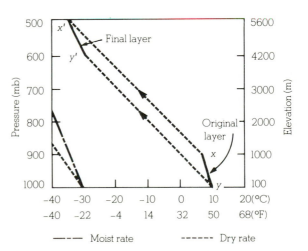

Fig. 10.6   The lifting of an entire layer of air (x–y) tends to increase the instability of the layer.

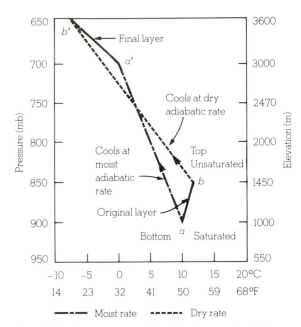

Fig. 10.7 Convective instability. The layer *a–b* is initially absolutely stable. The lower part of the layer is saturated, and the upper part is "dry." After lifting, the entire layer (*a'–b'*) becomes absolutely unstable.

lifted, and, as it rises, the rapid decrease in air density aloft causes the layer to stretch out vertically. If the layer remains unsaturated, the entire layer cools at the dry adiabatic rate. Due to the stretching effect, however, the top of the layer cools more than the bottom. This steepens the environmental lapse rate. Note that the absolutely stable layer *x–y*, after rising, has become conditionally unstable between 500 and 600 mb (layer *x'–y'*).

A very stable air layer may be converted into an absolutely unstable layer when the lower portion of a layer is moist and the upper portion is quite dry. In Fig. 10.7, the inversion layer between 900 and 850 mb is absolutely stable. Suppose the bottom of the layer is saturated, while the air at the top is unsaturated. If the layer is forced to rise, even a little, the upper portion of the layer cools at the dry adiabatic rate and grows cold quite rapidly, while the air near the bottom cools more slowly at the moist adiabatic rate. It does not take much lifting before the upper part of the layer is much colder than the bottom part; the

environmental lapse rate steepens and the entire layer becomes absolutely unstable (layer *a'–b'*). The potential instability, brought about by the lifting of a stable layer, whose surface is humid and whose top is "dry," is called **convective instability**. Convective instability is associated with the development of severe storms, such as thunderstorms and tornadoes.

Up to now, we have looked briefly at stability as it relates to cloud development; that is, layered clouds tend to form in stable air, while cumuliform clouds tend to form in unstable air. The following section describes how atmospheric stability influences the physical mechanisms responsible for the development of individual cloud types.

Cloud Development

Most clouds form as air rises, expands, and cools. Basically, the following mechanisms are responsible for the development of the majority of clouds we observe: (a) convection; (b) topography; (c) widespread ascent due to convergence of surface air; and (d) uplift along weather fronts.

Convection and Clouds Some areas of the earth's surface are better absorbers of solar radiation than others and, therefore, heat up more quickly. The air in contact with these "hot spots" becomes warmer than its surroundings. A hot "bubble" of air—a *thermal*—breaks away from the warm surface and rises, expanding and cooling as it ascends. As the thermal rises, it mixes with the cooler, drier air around it and gradually loses its identity. Its upward movement now slows. Frequently, before it is completely diluted, subsequent rising thermals penetrate it and help the air rise a little higher. If the rising air cools to its saturation point, the moisture will condense, and the thermal becomes visible to us as a cumulus cloud.

Observe in Fig. 10.9 that the air motions are downward on the outside of the cumulus cloud. The downward motions are caused in part by evaporation around the outer edge of the cloud, which cools the air, making it heavy. Another rea-

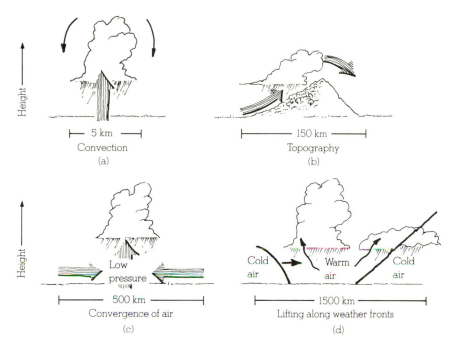

Fig. 10.8 The primary ways clouds form.

son for the downward motion is the completion of the convection current started by the thermal. Cool air slowly descends to replace the rising warm air. Therefore, we have rising air in the cloud and sinking air around it. Since subsiding air greatly inhibits the growth of thermals beneath it, small cumulus clouds usually have a great deal of blue sky between them.

As the cumulus clouds grow, they shade the ground from the sun. This, of course, cuts off surface heating and upward convection. Without the continual supply of rising air, the cloud begins to erode as its droplets evaporate. Unlike the sharp outline of a growing cumulus, the cloud now has indistinct edges, with cloud fragments extending from its sides. As the cloud dissipates (or moves along with the wind), surface heating begins again and regenerates another thermal, which becomes a new cumulus. This is why you often see cumulus clouds form, gradually disappear, then reform in the same spot.

Suppose that it is a warm, humid summer afternoon and the sky is full of cumulus clouds. The cloud bases are all at nearly the same level above the ground and the cloud tops extend only about a thousand meters higher. The development of these clouds depends primarily upon the air's stability and moisture content. To illustrate how these factors influence the formation of a convective

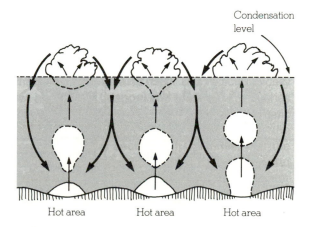

Fig. 10.9 Cumulus clouds form as hot, invisible air bubbles detach themselves from the surface, then rise and cool to the condensation level. Below and within the cumulus clouds the air is rising. Around the cloud, the air is sinking.

FOCUS ON AN APPLICATION

Clouds of Smoke and Stability

During the course of a day, the stability of the atmosphere, especially near the ground, changes. Occasionally, this can be visualized by observing the behavior of smoke leaving a stack. Figure 2 illustrates the type of smoke plumes that may develop with different temperature profiles and, hence, different conditions of stability. In diagram A, it is early morning and the atmosphere is stable with a radiation inversion extending from the surface to well above the height of the stack. In this stable environment, there is little vertical mixing and so the smoke spreads horizontally rather than vertically. When viewed from above, the smoke plume resembles the shape of a fan, and it is referred to as a **fanning plume**.

Later in the morning, the air near the surface warms quickly and becomes unstable (diagram B). Surface heating has not quite destroyed the inversion, however. Consequently, air is stable above the stack and unstable below. Because of the stable layer, vertical motions are confined to the region near the surface. Hence, the smoke mixes downward, increasing the concentration of pollution near the surface, sometimes to dangerously high levels. This effect is called **fumigation**.

As daytime heating continues, the atmosphere may become unstable. Rising and sinking air cause the smoke to move up and down in a looping pattern (diagram C). The continued rising of warm air and sinking of cool air can cause the temperature profile to equal that of the

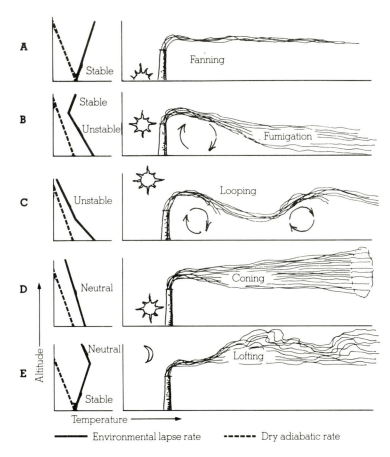

Fig. 2 Smoke-plume patterns that are related to various types of atmospheric stability.

dry adiabatic rate (diagram D). In this neutral atmosphere, the smoke from the stack tends to take on the shape of a cone because vertical and horizontal motions are about equal. After sunset, the ground cools rapidly and the radiation inversion reappears. When the top of the inversion extends upward to slightly above the stack, there is stable air near the ground with neutral (or slightly unstable) air

above (diagram E). Because the stable air in the inversion prevents the smoke from mixing downward, the smoke is carried upward, producing a **lofting smoke plume**.

As you watch the behavior of smoke plumes, keep in mind that the atmospheric conditions they signify are often a clue as to the type of cloud formations that may occur.

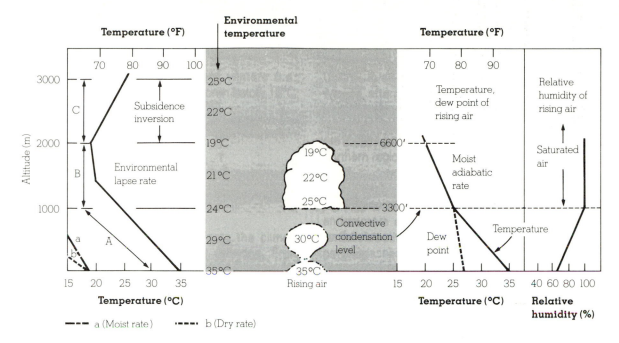

Fig. 10.10 The development of a cumulus cloud.

cloud, we will examine the temperature and moisture characteristics within a rising bubble of air. Since the actual air motions that go into forming a cloud are rather complex, we will simplify matters by making these assumptions:

1. No mixing takes place between the rising air and its surroundings.
2. Only a single thermal produces the cumulus cloud.
3. The cloud forms when the relative humidity is 100 percent.
4. The rising air in the cloud remains saturated.

The environmental lapse rate on this particular day is plotted in Fig. 10.10. The changing environmental lapse rate indicates changes in the air's stability. The environmental lapse rate in layer A is greater than the dry adiabatic rate, so the layer is absolutely unstable. The air layers above it—layer B and layer C—are both absolutely stable since the environmental lapse rate in each layer is less than the moist adiabatic rate. However, the overall environmental lapse rate from the surface up to the base of the inversion (2000 m) is 7.5°C

per 1000 m (4.1°F per 1000 ft), which indicates conditional instability.

Now, suppose that a warm bubble of air with an air temperature and dew point of 35°C and 27°C (95°F and 80.5°F), respectively, breaks away from the surface and begins to rise. Notice that, a short distance above the ground, the air inside the bubble is warmer than the air around it, so it is buoyant and rises freely. The rising air cools at the dry adiabatic rate and the dew point falls, but not as rapidly.* The rate at which the dew point drops varies with the moisture content of the rising air, but an approximation of 2°C per 1000 m (1°F per 1000 ft) is commonly used. So, as unsaturated rising air cools, the air temperature and dew point approach each other at the rate of 8°C per 1000 m (4.5°F per 1000 ft). This process causes an increase in the air's relative humidity.

*The decrease in dew point is caused by the rapid decrease in air pressure within the rising air. Since the dew point is directly related to the actual vapor pressure of the rising air, a decrease in total air pressure causes a corresponding decrease in vapor pressure and, hence, a lowering of the dew point.

Determining Convective Cloud Bases

Once a convective cloud forms, stability, humidity, and entrainment all play a major part in its vertical development. The level at which the cloud initially forms, however, is determined primarily by the temperature and moisture content of the original thermals. On a cool, moist day cumulus clouds will form closer to the surface than when the air is hot and relatively dry.

The bases of cumulus clouds that form by convection can be estimated quite easily when the surface air temperature and dew point are known. If the air is not too windy, we can assume that entrainment of air will not change the characteristics of a rising thermal. Since the rising air cools at the dry adiabatic rate of about 10°C per 1000 m, and the dew point drops at about 2°C per 1000 m, the air temperature and dew point ap-

proach each other at the rate of 8°C for every 1000 m of rise. Rising surface air with an air temperature and dew point spread of 8°C would produce saturation and a cloud at an elevation of 1000 m. Another way to look at it is that a 1°C difference between the surface air temperature and the dew point produces a cloud base at 125 m. Therefore, by finding the difference between surface air temperature (T) and dew point (T_d), and multiplying this value by 125, we can estimate the base of the convective cloud forming overhead:

$$H_{meter} = 125 (T - T_d), \quad (1)$$

where H is the cloud base in meters above the surface, with both T and T_d measured in degrees Celsius. If T and T_d are in °F, H can be calculated with the formula:

$$H_{feet} = 222 (T - T_d). \quad (2)$$

To illustrate the use of Formula 1, let's determine the base of the cumulus cloud in Fig. 10.10. Recall that the surface air temperature and dew point were 35°C and 27°C, respectively. The difference, $T - T_d$, is 8°C. This value multiplied by 125 gives us a cumulus cloud with a base at 1000 m above the ground. This agrees with the convective condensation level we originally calculated.

Along the East Coast in summer, the air is warm and muggy, and the separation between air temperature and dew point may be smaller than 9°C (16°F). The bases of afternoon cumulus clouds over cities, such as Philadelphia and Baltimore, are typically about 1000 m (3300 ft) above the ground. Further west, in the Central Plains, where the

At an elevation of 1000 m (3300 ft), the air has cooled to the dew point, the relative humidity is 100 percent, condensation begins, and a cloud forms. The elevation where the cloud forms is called the **convective condensation level** (CCL). Above the CCL the rising air is saturated and cools at the moist adiabatic rate. Condensation continues to occur, and the dew point within the cloud drops more rapidly with height than before. The air remains saturated as both the air temperature and dew point decrease at the moist adiabatic rate.*

*A more sudden decrease in the dew point occurs above the CCL because the vapor content of the air lowers as liquid cloud droplets form.

The rising air remains warmer than the environment and continues its spontaneous rise upward through layer B (Fig. 10.10). The top of the bulging cloud at 2000 m (about 6600 ft) represents the top of the rising air, which has now cooled to a temperature equal to its surroundings. The air would have a difficult time rising much above this level because of the stable subsidence inversion directly above it. The subsidence inversion, associated with the downward air motions of a high-pressure system, prevents the clouds from building very high above their bases. Hence, an afternoon sky full of flat-base cumuli with little vertical growth indicates fair weather. (Recall from Chapter 9 that the proper name of these fair-weather cumulus clouds is cumulus humilis.)

16 Cirrocumulus clouds. (Photo by author.)

17 Cirrostratus with a halo. (Photo by author.)

18 Altocumulus clouds. (Photo by author.)

19 An unusual display of altocumulus billow clouds. (Photo by author.)

20 Nimbostratus with scud. (Photo by author.)

21 Stratocumulus clouds. (Photo by author.)

22 Stratocumulus with crepuscular rays. (Photo by author.)

23 Lenticular clouds forming one on top of the other on the eastern side of the Sierra Nevada Mountains. (Photo by J. L. Medeiros.)

24 Mammatus clouds photographed after sunset. (Photo by author.)

26 The cloud forming over and downwind of Mt. Rainier is called a *banner cloud*. (Photo by author.)

27 Strong downward air currents, possibly from a mountain wave, are shearing off a portion of this sprouting cumulus cloud and causing its ice particles to fall toward the earth. (Photo by Dick Hilton.)

28 Cumulus clouds developing into thunderstorms over the Great Plains on a humid summer afternoon. (Photo by author.)

29 A time exposure of lightning with an intense thunderstorm. (Photo by D. Baumhefner, NCAR.)

30 A dramatic display of virga falling from an instense thunderstorm at sunset. (Photo by Elizabeth Beaver Burnett.)

31 Cirrus clouds associated with strong upper level winds (perhaps a jet stream) over Egypt and the Red Sea. (NASA photograph.)

32 Cumulus clouds and thunderstorms building over Florida during a summer afternoon. Observe that the clouds are notably absent over water. (NASA photograph.)

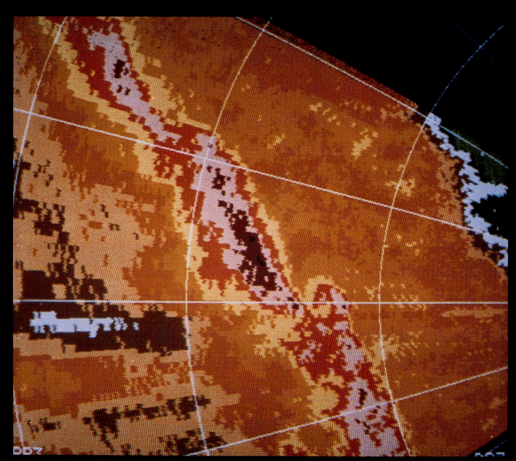

33 Doppler radar display showing precipitation in a severe thunderstorm. The "hook echo" just above the middle horizontal coordinate shows where a tornado has just formed. (NCAR photograph.)

FOCUS ON OBSERVATION, continued

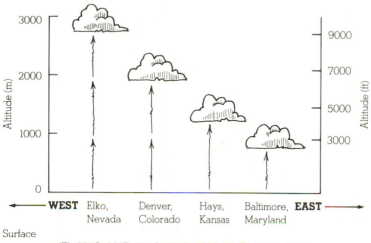

Surface
temperature **(T)** 32°C (90°F) 32°C (90°F) 32°C (90°F) 32°C (90°F)

Dew point **(Td)** 10°C (50°F) 15°C (59°F) 20°C (68°F) 24°C (75°F)

Fig. 3 During the summer, cumulus cloud bases typically increase in elevation above the ground as one moves westward into the drier air of the Central Plains.

air is drier and the spread between surface air temperature and dew point is greater, the cloud bases are higher. For example, west of Salina, Kansas, the cumulus cloud bases are generally greater than 1500 m (about 5000 ft) above the surface. On a summer afternoon in central Nevada, it is not uncommon to observe cumulus forming at 2400 m (about 8000 ft). In the Central Valley of California, where the summer afternoon spread between air temperature and dew point usually exceeds 22°C (40°F), the air must rise to almost 2700 m (about 9000 ft) before a cloud forms. Due to sinking air aloft, thermals in this area are unable to rise to that elevation, and afternoon cumulus clouds are seldom observed forming overhead.

As we can see, the stability of the air above the condensation level plays a major role in determining the vertical growth of a cumulus cloud. Notice in Fig. 10.11 that, when a deep stable layer begins a short distance above the cloud base, only cumulus humilis are able to form. If a deep unstable layer exists above the cloud base, cumulus congestus are likely to grow, with billowing cauliflowerlike tops. When the unstable layer is extremely deep—usually greater than 4 km (2.5 mi)—the cumulus congestus may even develop into a cumulonimbus.

Seldom do cumulonimbus clouds extend very far above the tropopause. The stratosphere is quite stable, so once a cloud penetrates the tropopause, it usually stops growing vertically and spreads horizontally. The low temperature at this altitude produces ice crystals in the upper section of the cloud. High winds near the tropopause blow the ice crystals horizontally, producing the flat anvil-shaped top so characteristic of cumulonimbus clouds.

The vertical development of a convective cloud also depends upon the mixing that takes place around its periphery. The rising, churning cloud mixes cooler air into it. Such mixing is called **entrainment**. If the environment around the cloud is very dry, the cloud droplets quickly evaporate. The effect of entrainment, then, is to increase the rate at which the rising air cools by the injection of cooler air into the cloud and the subsequent evaporation of the cloud droplets. If the rate of cooling

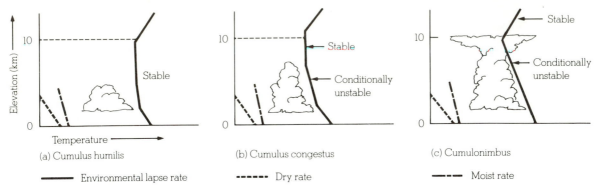

Fig. 10.11 The air's stability greatly influences the growth of cumulus clouds.

approaches the dry adiabatic rate, the air stops rising and the cloud no longer builds, even though the lapse rate may indicate a conditionally unstable atmosphere.

Up to now, we have looked at convection over land. Convection and the development of cumulus clouds also occur over large bodies of water. As cool air flows over a body of relatively warm water, the lowest layer of the atmosphere becomes warm and moist. This induces instability—convection begins and cumulus clouds form. If the air moves over progressively warmer water, as is sometimes the case in the open ocean, more active convection occurs and a cumulus cloud can build into cumulus congestus and finally into cumulonimbus. This sequence of cloud development is observed from satellites as cold northerly winds move southward over the northern portions of the Atlantic and Pacific oceans.

Topography and Clouds Horizontally moving air obviously cannot go through a large obstacle, such as a mountain, so the air must go over it. Forced lifting along a topographic barrier is called **orographic uplift**. Often, large masses of air rise when they approach a long chain of mountains like the Sierra Nevada or Rockies. This lifting produces cooling, and, if the air is humid, clouds form. Clouds produced in this manner are called **orographic clouds**. The type of cloud that forms will depend on the air's stability and moisture con-

tent. On the other side of the mountain, as the air moves downhill, it warms. This sinking air is now drier, since much of its moisture was removed in the form of clouds and precipitation on the windward side. It is, therefore, more common to see clouds forming on the windward side of a mountain than on the leeward side. However, under certain atmospheric conditions, clouds may form on the leeward side as well. Stable air flowing over a mountain often moves in a series of waves that may extend for several hundred kilometers on the leeward side. These waves resemble the waves that form in a river downstream from a large boulder. Waves in the atmosphere can produce wave clouds on the leeward side of a mountain range.

As air rises on the upwind side of the wave, it cools and may condense, producing a cloud. On the downwind side of the wave, the air sinks and warms; the cloud evaporates. Clouds that form in the wave directly over the mountain are called *mountain wave clouds*, and those that form downwind of the mountain are called *lee wave clouds*. Since wave clouds often have a characteristic lens shape, they are known as **lenticular clouds**. Viewed from the ground, the clouds appear motionless as the air rushes through them and are often referred to as *standing wave clouds*. Since they most frequently form at elevations where middle clouds form, they are called *altocumulus standing lenticulars*.

Lenticular clouds frequently form one above the other like a stack of pancakes. They do this when the air between the cloud-forming layer is too dry to produce clouds. When a strong wind blows almost perpendicular to a high mountain range, mountain waves may extend into the stratosphere, producing a spectacular display. At a distance, these clouds resemble a fleet of hovering spacecraft. (See color plate 23.) No wonder a large number of UFO sightings take place when lenticular clouds form! In Fig. 10.12 you can see that, beneath the lee wave, a large swirling eddy forms. The rising part of the eddy may cool enough to produce **rotor clouds**. The air in the rotor is extremely turbulent and presents a major hazard to aircraft in the vicinity. Dangerous flying conditions also exist near the leeside of the mountain, where strong downward motions are present. (These types of winds will be covered in more detail in Chapter 13.)

Widespread Ascent and Clouds Just as mountains force air to rise, the convergence—flowing together—of air in the lower troposphere will initiate lifting and possibly the formation of clouds. (See Fig. 10.8.) A cyclonic storm system is the most common cause of the widespread ascent of air. Associated with this kind of system are weather fronts that separate air with contrasting temperatures and densities. For example, along a warm front, warm air gently rides up and over colder surface air, producing clouds that may cover an area of many hundred, and even thousands, of square kilometers. The clouds that form are usually layered, cirrus, cirrostratus, altostratus, and nimbostratus. Along a cold front, where usually warm moist air is forced upward, lines of cumuliform clouds often form. Even in the absence of fronts, clouds may form in regions of surface low pressure. Recall that surface air flows in toward the center of a low. This air, converging from all directions, must go somewhere. Since it cannot go into the ground, it rises, resulting in clouds.

Thus far, we have examined the primary ways air may be forced to rise so that clouds are able to form. We will now look at some of the other ways in which clouds develop.

Changing Cloud Forms Under certain conditions, a layer of altostratus may change into altocumulus. This happens if the top of the original cloud deck cools while the bottom warms. Because clouds are such good absorbers and emitters of infrared radiation, the top of the cloud will often cool as it radiates infrared energy to space more rapidly than it absorbs solar energy. Meanwhile, the bottom of the cloud will warm as it absorbs infrared energy from below more quickly than it radiates this energy away. This makes the cloud layer unstable to the point that small convection cells begin within the cloud itself. The up and down motions in a layered cloud produce globular elements that give the cloud a lumpy appearance. The cloud forms in the rising part of a cell and clear spaces appear where descending currents occur.

Cirrocumulus and stratocumulus may form in a similar way. When the wind is fairly uniform throughout a cloud layer, these new cloud elements appear evenly distributed across the sky. However, if the wind speed or direction changes with height, the horizontal axes of the convection cells align with the average direction of the wind. The new cloud elements then become arranged in rows and are given the name **cloud streets**. When the changes in wind speed and direction reach a critical value, and an inversion caps the cloud-forming layers, wavelike clouds called **billows** may form along the top of the cloud layer (Fig. 10.13).

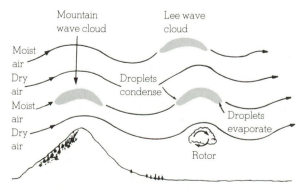

Fig. 10.12 The formation of lenticular clouds.

Fig. 10.13   Billow clouds forming in a region of rapidly changing wind speed.

Fig. 10.14   An example of altocumulus castellanus.

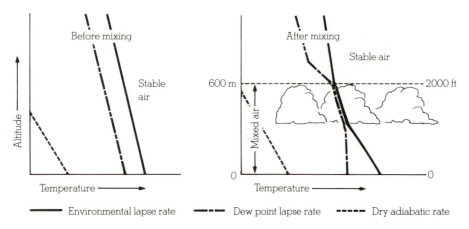

Environmental lapse rate —— Dew point lapse rate ----- Dry adiabatic rate

Fig. 10.15 The mixing of a moist layer of air near the surface can produce a deck of stratocumulus clouds.

Occasionally, altocumulus show vertical development and produce towerlike extensions. The clouds often resemble floating castles and, for this reason, they are called *altocumulus castellanus* (Fig. 10.14). They form when rising currents within the cloud extend into conditionally unstable air above the cloud. Apparently, the buoyancy for the rising air comes from the latent heat released during condensation within the cloud. This process can occur in cirrocumulus clouds, producing *cirrocumulus castellanus*.

Mixing and Stratocumulus Occasionally, the stirring of a moist layer of stable air will produce a deck of stratocumulus clouds. In Fig. 10.15, the air is stable and close to saturation. Suppose a strong wind mixes the layer from the surface up to an elevation of 600 m (2000 ft). As we saw earlier, the lapse rate will steepen as the upper part of the layer cools and the lower part warms. At the same time, mixing will make the moisture distribution in the layer more uniform. The warmer temperature and decreased moisture content causes the lower part of the layer to dry out. On the other hand, the decrease in temperature and increase in moisture content saturates the top of the mixed layer, producing a layer of stratocumulus clouds. Figure 10.15 indicates that the air above the region of mixing is still stable and inhibits

further mixing. However, if the surface warms substantially, rising thermals may penetrate the stable region and the stratocumulus clouds may change into more widely separated clouds, such as cumulus or cumulus congestus. A stratocumulus layer changing to a sky dotted with growing cumulus clouds often occurs as surface heating increases on a warm, humid, summer day.

Unusual Clouds Jet aircraft flying at high altitudes often produce a cirruslike trail of condensed vapor called a **contrail** (Fig. 10.16). The condensation may come directly from the water vapor added to the air from engine exhaust. In this case, there must be sufficient mixing of the hot exhaust gases with the cold air to produce saturation. The release of particles in the exhaust may even provide nuclei on which ice crystals form. Contrails evaporate rapidly when the relative humidity of the surrounding air is low. If the relative humidity is high, however, contrails may persist for many hours. Contrails may also form by a cooling process. The reduced pressure produced by air flowing over the wing causes the air to cool. This cooling may supersaturate the air, producing an *aerodynamic contrail*. This type of trail usually disappears quickly in the turbulent wake of the aircraft.

A towering cumulus cloud is occasionally seen

Fig. 10.16   A contrail forming behind a jet aircraft.

with a thin cloud resembling a silken scarf capping its top (Fig. 10.17). This cloud is known as a *cap cloud*, or **pileus**. Pileus clouds form when a moist stable layer exists above a developing cumulus congestus or cumulonimbus. The building cumulus will deflect horizontal air up and over its top, causing a wave in the stable layer. If the air flowing over the top of the cloud condenses, a pileus forms. This, of course, is similar to the formation of the mountain wave clouds we investigated earlier.

Most clouds form in rising air, yet there are clouds that form in descending air. A typical example is the **mammatus**. This cloud derives its name from its appearance—baglike sacks that hang beneath a cloud resembling a cow's udder. (See color plate 24.) Mammatus most frequently form on the underside of cumulonimbus; however, they may develop beneath cirrus, cirrocumulus, altostratus, altocumulus, and stratocumulus. For mammatus to form, the sinking air must be cooler

than the air around it and have a high liquid water or ice content. As saturated air sinks, it warms, but the warming is retarded because of the heat taken from the air to evaporate the liquid or melt ice particles. Hence, the sinking air warms at the moist adiabatic rate. If the lapse rate of the surrounding air is unstable with respect to saturated air, the descending air will be cooler than its environment and continue to sink. When sinking blobs of air remain saturated below the base of the cloud, we see them as rounded masses and call them mammatus.

Summary

In this chapter, we have tied together the concepts of stability and the formation of clouds. We learned that rising unsaturated air cools at the dry adiabatic rate and, due to the release of latent heat, rising saturated air cools at the moist adiabatic rate. In a stable atmosphere, a lifted parcel of air will be colder (heavier) than the air surrounding it at each new level, and it will sink back to its original position. Because stable air tends to resist upward vertical motions, clouds forming in a stable atmosphere often spread horizontally and have a

Fig. 10.17   A pileus cloud forming above a developing cumulus cloud.

stratified appearance, such as cirrostratus and altostratus. Stable air may be caused by either cooling the surface air, warming the air aloft, or by the sinking (subsidence) of an entire layer of air, in which case a very stable subsidence inversion usually forms.

In an unstable atmosphere, a lifted parcel of air will be warmer (lighter) than the air surrounding it at each new level, and it will continue to rise upward away from its original position. In an unstable atmosphere, rising air tends to form clouds that develop vertically, such as cumulus congestus and cumulonimbus. Instability may be caused by warming the surface air, cooling the air aloft, or by the lifting of an entire layer of air.

On warm, humid days the instability generated by surface heating can produce cumulus clouds at a height determined by the temperature and moisture content of the surface air. Instability may cause changes in existing clouds as convection changes an altostratus into an altocumulus. Also, mixing can change a clear day into a cloudy one.

In a conditionally unstable atmosphere, an unsaturated parcel of air can be lifted to a level where condensation begins, latent heat is released, and instability results, as the temperature inside the rising parcel becomes warmer than the air surrounding it.

Questions for Review

1. What is an adiabatic process?

2. Why are the moist and dry adiabatic rates of cooling different?

3. Under what conditions would the moist adiabatic rate of cooling be almost equal to the dry adiabatic rate?

4. Explain the difference between environmental lapse rate and dry adiabatic rate.

5. What is a stable atmosphere and how can it form?

6. Describe the general characteristics of clouds associated with stable and unstable air.

7. List and explain several processes by which stable air can be made unstable.

8. If the atmosphere is conditionally unstable, what condition is necessary to bring on instability?

9. Explain why cumulus clouds are conspicuously absent over a cool water surface.

10. Why are cumulus clouds more frequently observed during the afternoon than at night?

11. Explain why an inversion represents an absolutely stable atmosphere.

12. How and why does lifting or lowering a layer of air change its stability?

13. Why do cumulonimbus clouds often have flat tops?

14. Describe the conditions necessary to produce stratocumulus clouds by mixing.

15. Why are there usually large spaces of blue sky between cumulus clouds?

16. List four primary ways clouds form, and describe the formation of one cloud type by each method.

17. Why are lenticular clouds also called standing wave clouds?

18. Why, in summer, are cumulus humilis more likely to form than cumulus congestus in the center of a large high-pressure area?

19. How can a layer of altostratus change into one of altocumulus?

20. Briefly describe how each of the following clouds form:
  (a) lenticular
  (b) rotor
  (c) billow
  (d) castellanus
  (e) contrail
  (f) pileus
  (g) mammatus

Questions for Thought

1. How is it possible for a layer of air to be convectively unstable and absolutely stable at the same time?

2. Are the bases of convective clouds generally higher during the day or the night? Explain.

3. Explain how a mountain wave cloud may form without the formation of lee wave clouds at the same level.

4. Use Fig. 7.6 (Chapter 7) to help you explain why the bases of cumulus clouds, which form from rising thermals during the summer, increase in height above the surface as you move due west of a line that runs north-south through central Kansas.

5. Cumulus mammatus are said to form by upside-down convection. Explain what that means.

6. Suppose two saturated air parcels are lifted upward from the surface. Parcel A has a temperature and dew point of −10°C, while parcel B has a temperature and dew point of 20°C. The decrease in air temperature inside which parcel comes closest to the dry adiabatic rate? Explain.

7. For least polluted conditions, what would be the best time of day for a farmer to burn agricultural debris?

Problems and Exercises

1. Under which set of conditions would a cumulus cloud base be observed at the highest level above the surface? Surface air temperatures and dew points are as follows: (a) 35°C, 14°C; (b) 30°C, 19°C; (c) 34°C, 9°C; (d) 29°C, 7°C; (e) 32°C, 6°C.

2. If the height of a cumulus cloud is 1000 m above the surface, and the dew point at the earth's surface beneath the cloud is 20°C, determine the air temperature at the earth's surface beneath the cloud.

3. The convective condensation level over New Orleans, Louisiana, on a warm, muggy afternoon is 2000 ft. If the dew point of the rising air at this level is 73°F, what is the approximate dew point and air temperature at the surface? Determine the surface relative humidity. (Hint: See Chapter 7.)

4. Suppose the air pressure outside a conventional jet airliner flying at an altitude of 10 km (about 33,000 ft) is 250 mb. Further, suppose the air inside the aircraft is pressurized to 1000 mb. If the outside air temperature is −50°C (−58°F), what would be the temperature of this air if brought inside the aircraft and compressed at the dry adi-abatic rate to a pressure of 1000 mb? (Assume that a pressure of 1000 mb is equivalent to an altitude of 0 m.)

5. In Fig. 10.18 a radiosonde is released and sends back temperature data as shown in the diagram.
(a) Calculate the environmental lapse rate from the surface up to 3000 m.
(b) What type of atmospheric stability does the sounding indicate?
Suppose the wind is blowing from the west and a parcel of surface air with a temperature of 20°C and a dew point of 12°C begins to rise upward along the western (windward) side of the mountain.
(c) What is the relative humidity of the air parcel at 0 m (pressure 1013 mb) before rising? (Hint: See Chapter 7.)
(d) As the air parcel rises, at approximately what elevation would condensation begin and a cloud start to form?
(e) What is the air temperature and dew point of the rising air at the base of the cloud?
(f) What is the air temperature and dew point of the rising air inside the cloud at an elevation of 3000 m? (Use moist rate of 6°C per 1000 m.)
(g) At an altitude of 3000 m, how does the air temperature inside the cloud compare with the air temperature outside the cloud, as measured by the radiosonde? What type of atmospheric stability (stable or unstable) does this suggest? Explain.
(h) At an elevation of 3000 m, would you expect the cloud to continue to develop vertically? Explain.
(i) What would be the name of the cloud that is forming?
Suppose that a parcel of air inside the cloud descends from the top of the mountain at 3000 m (pressure 700 mb) down the eastern (leeward) side of the mountain to an elevation of 0 m (pressure 1013 mb).
(j) If the descending air warms at the dry adiabatic rate from the top of the mountain all the way down to 0 m, what is the sinking air's temperature and dew point when it reaches 0 m?
(k) What would be the relative humidity of the sinking air at 0 m?

(l) What accounts for the sinking air being warmer at the base of the mountain on the eastern side?

(m) Explain why the sinking air is drier (its dew point is lower) on the eastern side at 0 m.

6. Answer the same questions in problem 5 except, this time, use the adiabatic chart provided in Appendix F. (Chart F.5)

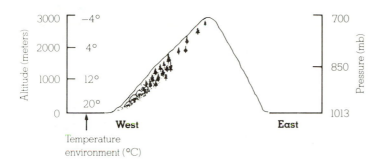

Fig. 10.18   Data from a radiosonde. (The information in this illustration is to be used in answering questions 5 and 6 of the Problems and Exercises section.)

Heavy, wet snowflakes falling through air with a temperature just below freezing stick together, producing larger flakes that cover everything with a blanket of white. (Photo by author)

CHAPTER 11

Precipitation

Is it ever "too cold to snow"? Although many believe in this expression, the fact remains that it is *never* too cold to snow. True, colder air cannot hold as much water vapor as warmer air; but, no matter how cold the air becomes, it always contains some water vapor that could produce snow. At Fort Yellowstone, Wyoming, for example, three inches of snow fell on February 2, 1899, when the maximum temperature reached only $-28°C$ ($-18°F$). In fact, tiny ice crystals have been observed falling at temperatures as low as $-47°C$ ($-53°F$). We usually associate extremely cold air with "no snow" because the coldest winter weather occurs on clear, calm nights—conditions that normally prevail with strong high-pressure areas that have few if any clouds.

This chapter raises a number of interesting questions to consider regarding precipitation: Why, for example, does the largest form of precipitation—hail—fall during the warmest time of the year? Why does it sometimes rain on one side of the street but not on the other? What is "sleet" and how does it differ from hail? First, we will examine the processes that produce rain and snow; then, we will look closely at the other forms of precipitation. Our discussion will conclude with a section on how precipitation is measured.

Precipitation Processes

As we all know, cloudy weather does not necessarily mean that it will rain or snow. In fact, clouds may form, linger for many days, and never produce precipitation. In Eureka, California, the August daytime sky is overcast more than 50 percent of the time, yet the average precipitation there for August is merely one-tenth of an inch. We know

Contents

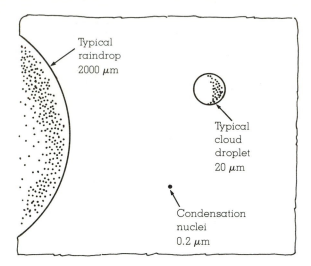

Fig. 11.1 Relative sizes of raindrops, cloud droplets, and condensation nuclei.

that clouds form by condensation, yet apparently condensation alone is not sufficient to produce rain. Why not? To answer this question we need to closely examine the tiny world of cloud droplets.

How Do Cloud Droplets Grow Larger? An ordinary cloud droplet is extremely small, having an average diameter of 20 μm (0.002 cm). Notice in Fig. 11.1 that a typical cloud droplet is 100 times smaller than a typical raindrop. If a cloud droplet is in equilibrium with its surroundings, the size of the droplet does not change because the water molecules condensing onto the droplet will be exactly balanced by those evaporating from it. If, however, it is not in equilibrium, the droplet will

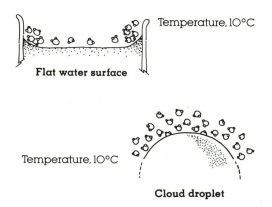

Fig. 11.2 At equilibrium, the vapor pressure over a curved droplet of water is greater than that over a flat surface.

either increase or decrease, depending on whether condensation or evaporation predominates.

Consider a cloud droplet in equilibrium with its environment. The total number of vapor molecules around the droplet remains fairly constant and defines the droplet's saturation vapor pressure. Since the droplet is in equilibrium, the saturation vapor pressure is also called the **equilibrium vapor pressure**. Figure 11.2 shows a cloud droplet and a flat water surface, both of which are in equilibrium. Because more vapor molecules surround the droplet, it has a greater equilibrium vapor pressure. The reason for this is that water molecules are less strongly attached to a curved water surface; hence, they evaporate more readily.

To keep the droplet in equilibrium, more vapor molecules are needed around it to replace those molecules that are constantly evaporating from its surface. Smaller cloud droplets exhibit a greater curvature, which causes a more rapid rate of evaporation. As a result of this process (called the **curvature effect**), smaller droplets require an even greater equilibrium vapor pressure to keep them from completely evaporating away. Therefore, when air is saturated with respect to a flat surface, it is unsaturated with respect to a curved droplet of pure water, and the droplet evaporates. So, to keep cloud droplets in equilibrium with the surrounding air, the air must be **supersaturated**; that is, the relative humidity must be greater than 100 percent. The smaller the droplet, the higher the supersaturation needed to keep it in equilibrium.

Figure 11.3 shows the curvature effect for pure water. The dark, heavy line represents the relative humidity needed to keep a droplet with a given diameter in equilibrium with its environment. Note that when the droplet's size is less than 2 μm, the relative humidity must be well above 100.1 percent for the droplet to survive. As droplets become larger, the effect of curvature lessens; for a droplet whose diameter is greater than 10 μm, the curvature effect is so small that the droplet behaves as if it were a flat surface.

Just as relative humidities less than that required for equilibrium permit a water droplet to evaporate and shrink, those greater than the equilibrium value allow the droplet to grow by condensation. From Fig. 11.3, we can see that a droplet whose diameter is 1 μm will grow larger as the relative humidity approaches 101 percent. But relative humidities, even in clouds, rarely become greater than 101 percent. How, then, do tiny cloud droplets of less than 1 μm grow to the size of an average cloud droplet?

Recall from Chapter 8 (fog formation discussion) that condensation begins on tiny particles called condensation nuclei. Because many of these nuclei are hygroscopic (water-attracting), condensation may begin on such particles when the relative humidity is well below 100 percent. As condensation begins, the salt dissolves, forming a solution. The dissolving of hygroscopic particles reduces the relative humidity necessary for the onset of condensation. This is called the **solute effect**. As a result of the solute effect, a droplet containing salt (a hygroscopic nucleus) can be in equilibrium with its environment when the atmospheric relative humidity is much lower than 100 percent.

Imagine that we place condensation nuclei of varying sizes into moist but unsaturated air. As the air cools, the relative humidity increases and, because of the curvature effect, condensation first begins on the larger nuclei. If these larger nuclei are salt particles, they will absorb water until equilibrium is established with the surroundings. As the air cools further, the relative humidity increases, and the droplets will grow larger until a new equilibrium radius is reached. As the relative humidity of the environment approaches 100 percent, the droplets grow still larger. When each radius reaches a few micrometers, the droplets are so diluted that they behave as if they were pure water.

If the air remains supersaturated with respect to the droplets, condensation would continue but more slowly. The growth rate decreases because the hygroscopic material in the droplet has become diluted and less attractive to water vapor. The available water vapor would begin to condense onto smaller, more attractive nuclei. Therefore, in supersaturated air, large cloud droplets grow slowly, while many smaller cloud droplets are just beginning to form.*

*Over land masses where large concentrations of nuclei exist, there may be many hundreds of droplets per cubic centimeter, all competing for the available supply of water vapor. Over the oceans where the concentration of nuclei is less, there are normally fewer (typically less than 100 per cm³) but larger cloud droplets.

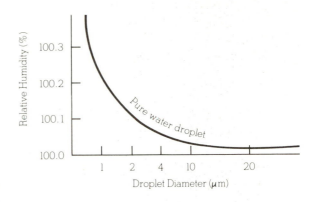

Fig. 11.3 The curved line represents the relative humidity needed to keep a droplet in equilibrium with its environment. For a given droplet size, the droplet will evaporate and shrink when the relative humidity is less than that given by the curve. The droplet will grow by condensation when the relative humidity is greater than the value on the curve.

We now have a cloud composed of many small droplets—too light to fall as rain. These minute droplets require only slight upward air currents to keep them suspended. Those droplets that do fall descend slowly, and evaporate in the drier air beneath the cloud. It is evident, then, that most clouds cannot produce precipitation. The condensation process by itself is entirely too slow to produce rain. Even under ideal conditions, it would take several days for this process alone to create a raindrop. However, observations show that clouds can develop and begin to produce rain in less than an hour. Since it takes about 1 million average sized (20 μm) cloud droplets to make an average size (2000 μm) raindrop, there must be some other process by which cloud droplets grow large and heavy enough to fall as precipitation. Even though all of the intricacies of how rain is produced are not yet fully understood, two important processes stand out: (1) the collision-coalescence process and (2) the ice-crystal process.

Collision and Coalescence Process Clouds that form at temperatures above freezing are called *warm clouds*. In such clouds, collisions between droplets play a significant part in producing precipitation. To produce the many collisions necessary to form a raindrop, some cloud droplets must be larger than others. Larger drops could have

Table 11.1 Terminal Velocity of Different Size Particles Involved in Condensation and Precipitation Processes

DIAMETER (μm)	TERMINAL VELOCITY		TYPE OF PARTICLE
	m/sec	ft/sec	
0.2	0.0000001	0.0000003	condensation nuclei
20	0.01	0.03	typical cloud droplet
100	0.27	0.9	large cloud droplet
200	0.70	2.3	large cloud droplet or drizzle
1000	4.0	13.1	small raindrop
2000	6.5	21.4	typical raindrop
5000	9.0	29.5	large raindrop

formed on large hygroscopic nuclei, such as salt particles. Random bumping and touching of droplets could also account for size variation.

Air retards the falling drops. The amount of air resistance depends on the size of the drop and on its rate of fall: The greater its speed, the more air molecules the drop encounters each second. The speed of the falling drop increases until the air resistance equals the pull of gravity. At this point, the drop continues to fall, but at a steady speed, which is called its **terminal velocity**. Because larger drops have a smaller surface-area-to-weight ratio, they fall farther and faster before reaching their terminal velocity. The result of this is that larger drops have a greater terminal velocity than smaller drops. (See Table 11.1.) Note that, in calm air, a typical raindrop falls over 600 times faster than a typical cloud droplet!

Large droplets overtake and collide with smaller drops in their path. This merging of cloud droplets

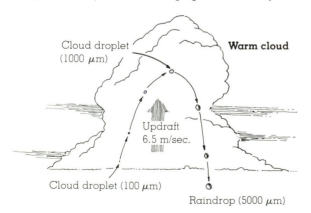

Fig. 11.4 A cloud droplet rising then falling through a warm cumulus cloud can grow by collision and coalescence and emerge from the cloud as a large raindrop.

by collision is called **coalescence**. Laboratory studies show that collision does not always guarantee coalescence; sometimes, the droplets actually bounce apart during collision. Coalescence appears to be enhanced if colliding droplets have opposite (and, hence, attractive) electrical charges. Therefore, atmospheric electricity seems to play a role in the growth of cloud droplets and in the production of rain. Another important factor that determines cloud droplet growth by the collision process is the amount of time the droplet spends in the cloud. A very large cloud droplet of 200 μm falling in still air will take about 12 minutes to travel through a cloud 500 m (1640 ft) thick and over an hour if the cloud thickness is 2500 m (8200 ft). Rising air currents in a forming cloud slow the rate at which droplets fall. A thick cloud with strong updrafts will maximize the time cloud droplets spend in the cloud and, hence, the size to which they can grow.

A warm stratus cloud is typically less than 500 m thick and has slow, upward air movement (generally less than 0.1 m/sec). Under these conditions, a large droplet would be in the cloud for a relatively short time and grow by coalescence to only about 200 μm. If the air beneath the cloud is moist, the droplets may reach the ground as drizzle, the lightest form of rain. If, however, the stratus cloud base is fairly high above the ground, the drops will evaporate before reaching the surface, even when the relative humidity is 90 percent.

In tropical regions, where warm cumulus clouds build to great heights, convective updrafts of at least 1 m/sec (and some exceeding many tens of meters per second) occur. Look at Fig. 11.4. Suppose a cloud droplet of 100 μm is caught in an updraft whose velocity is 6.5 m/sec. As it rises, it collides with and captures smaller drops in its path, and grows until it reaches a size of about 1000 μm. At this point, the updraft in the cloud is just able to balance the pull of gravity on the drop. Here, the drop remains suspended until it grows just a little bigger. Once the weight of the drop is greater than the force of the updraft, the drop slowly descends, growing larger as it falls. By the time it reaches the bottom of the cloud, it will be a large raindrop with a diameter of over 5000 μm (5 mm). Raindrops of this size typically occur at the beginning of a rain shower originating in these warm, convective cumulus clouds.

So far, we have examined the way cloud droplets grow large enough to fall as raindrops by the collision-coalescence process in warm clouds. In a cloud with sufficient water content, four significant factors in raindrop production are:

1. the relative droplet size
2. the electric charge of the droplets and the electric field in the cloud
3. the cloud thickness
4. the updrafts of the cloud.

Relatively thin stratus clouds with slow upward air currents are, at best, only able to produce drizzle, while the towering cumulus clouds associated with rapidly rising air can cause heavy showers. Now, let's turn our attention to the ice-crystal process of rain formation.

Ice-Crystal Process   The ice-crystal process of rain formation is extremely important in middle and high latitudes, where clouds extend upward into regions where the air temperature is below freezing. Such clouds are called *cold clouds*. Figure 11.5 illustrates a typical cold cloud that has formed over the Great Plains, where the "cold" part is above the 0°C isotherm.

Suppose we take an imaginary balloon flight up through the cumulonimbus cloud in Fig. 11.5. Entering the cloud, we observe cloud droplets growing larger by processes described in the previous section. As expected, only water droplets exist here, for the base of the cloud is warmer than 0°C. Surprisingly, in the cold air just above the 0°C isotherm, almost all of the cloud droplets are still composed of liquid water. Water droplets ex-

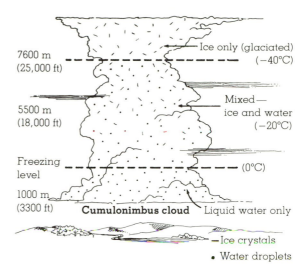

Fig. 11.5   The distribution of ice and water in a cumulonimbus cloud.

isting at temperatures below freezing are referred to as **supercooled**. Even at higher levels, where the air temperature is −10°C (14°F), there is only one ice crystal for every million liquid droplets. Near 5500 m (18,000 ft), where the temperature becomes −20°C (−4°F), ice crystals become more numerous, but are still outnumbered by water droplets. Clouds composed of both ice and water are called *mixed clouds*.

Not until we reach an elevation of 7600 m (25,000 ft), where temperatures drop below −40°C (also −40°F), do we find *only* ice crystals. The region of a cloud where only ice particles exist is called *glaciated*. Why are there so few ice crystals in the middle of the cloud, even though temperatures there are well below freezing?*

Pure ice melts when the temperature is raised to 0°C, but water does not necessarily freeze until temperatures fall well below that value. Over large bodies of fresh water, ice ordinarily forms when the temperature drops just slightly below 0°C. Yet, laboratory studies show that, the smaller the amount of pure water, the lower the temperature

*In continental clouds, such as this one, where there are many small cloud droplets less than 20 μm in diameter, the onset of ice-crystal formation begins at temperatures between −9°C and −15°C. In clouds where larger but fewer cloud droplets are present, ice crystals begin to form at temperatures between −4°C and −8°C. In some of these clouds, glaciation can occur at −8°C, which may be only 2500 m (8200 ft) above the surface.

FOCUS ON AN ISSUE

Does Cloud Seeding Enhance Precipitation?

The primary goal in many cloud-seeding experiments is to inject (or seed) a cloud with small particles that will act as nuclei, so that the cloud particles will grow large enough to fall to the surface as precipitation. The first ingredient in any seeding project is, of course, the presence of clouds. (Seeding does not generate clouds.) However, not just any cloud will do. For optimum results, the cloud must be cold; that is, at least a portion of it (preferably the upper part) must be supercooled because cloud seeding uses the ice-crystal process to cause the cloud particles to grow.

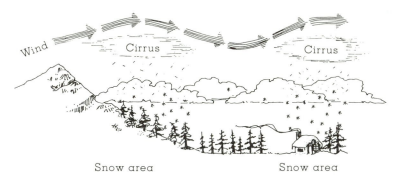

Fig. 1 Natural seeding by cirrus clouds may form bands of precipitation downwind of a mountain chain.

Some of the first experiments in cloud seeding were conducted by Vincent Schaefer and Irving Langmuir during the late 1940s. To seed a cloud, they dropped crushed pellets of dry ice (solid carbon dioxide) from a plane. Because dry ice has a temperature of −78°C (−108°F), it acts as a cooling agent. Small pellets dropped into the cloud cool the air to the point where new liquid droplets are able to form. These new droplets (and the original ones) are now able to freeze in an instant at temperatures below −40°C. The newly formed ice crystals then grow larger by deposition at the expense of the nearby liquid droplets and, upon reaching a sufficiently large size, fall as precipitation.

In 1947, Bernard Vonnegut demonstrated that silver iodide (AgI) could be used as a cloud-seeding agent. Because silver iodide has a crystalline structure similar to an ice crystal, it acts as an effective ice nucleus at temperatures of −4°C (25°F) and lower. Silver iodide causes ice crystals to form in two ways:

1. Ice crystals form when silver iodide crystals come in contact with supercooled droplets.
2. Ice crystals grow by deposition directly on the silver iodide crystal.

Silver iodide is much easier to handle than dry ice, since it can be supplied to the cloud from burners located either on the ground or on the wing of a small aircraft. Although other substances, such as lead iodide and cupric sulfide, are also effective ice nuclei, silver iodide still remains the most commonly used substance in cloud-seeding projects.

at which it freezes. A cloud droplet about 25 μm in diameter freezes spontaneously at −36°C, while a droplet whose diameter is only a few micrometers does not freeze until the temperature is near −40°C.

The freezing of pure water is called **spontaneous nucleation**. For spontaneous nucleation to occur, enough molecules within the water droplet must join together in a rigid pattern to form a tiny ice crystal, or *embryo*. Once the embryo grows to a critical size, it acts as a nucleus. Other molecules in the droplet then attach themselves to this nucleus, and the whole water droplet freezes. The chance of an ice crystal growing to a critical size decreases as the volume of water decreases. Tiny ice crystals form at water temperatures just below freezing, but, for small volumes of water, molecular motions remain large enough to weaken their structure. The ice crystals simply form and break apart. So, for these smaller droplets, lower temperatures are required if the ice crystal is to grow to critical size before being broken apart by

FOCUS ON AN ISSUE, continued

Just how effective is artificial seeding with silver iodide in increasing precipitation? This is a much-debated question among meteorologists. First of all, it is difficult to evaluate the results of a cloud-seeding experiment. When a seeded cloud produces precipitation, the question always remains as to how much precipitation would have fallen had the cloud not been seeded.

Other factors must be considered when evaluating cloud-seeding experiments: the type of cloud, its temperature, moisture content, and droplet size distribution.

Although some experiments suggest that cloud seeding does not increase precipitation, others seem to indicate that seeding *under the right conditions* may enhance precipitation between 10 and 20 percent. And so the controversy continues.

Some cumulus clouds show an "explosive" growth after being seeded. The latent heat given off when the droplets freeze functions to warm the cloud, causing it to become more buoyant. It grows rapidly, and becomes a longer-lasting cloud, which may produce more precipitation.

The business of cloud seeding can be a bit tricky, since overseeding can produce too many ice crystals. When this occurs, the cloud becomes glaciated and the ice particles, being very small, do not fall as precipitation. Since few liquid droplets exist, the ice crystals cannot grow by the ice-crystal process; rather, they evaporate, leaving a clear area in a thin stratified cloud. Because dry ice can produce the most ice crystals in a supercooled cloud, it is the substance most suitable for deliberate overseeding. Hence, it is the substance most commonly used to dissipate cold fog at airports. (See Chapter 8.)

Warm clouds have also been seeded in an attempt to produce rain. Tiny water drops and particles of hygroscopic salt are injected into the base of the cloud. These particles when carried into the cloud by updrafts create large cloud droplets, which grow even larger by the collision-coalescence process. To date, the results obtained using this method are inconclusive.

Under certain conditions, clouds may be seeded naturally. For example, when cirriform clouds lie directly above a lower cloud deck, ice crystals may de-

scend from the higher cloud and seed the cloud below. As the ice crystals mix into the lower cloud, supercooled droplets are converted to ice crystals, and the precipitation process is enhanced. When the cirrus clouds form waves downwind from a mountain chain (Fig. 1), bands of precipitation often form.

Cloud seeding may be inadvertent. Some industries emit large concentrations of condensation nuclei and ice nuclei into the air. Studies have shown that these particles are at least partly responsible for increasing precipitation in, and downwind of, cities. On the other hand, studies have also indicated that the burning of certain types of agricultural waste may produce smoke containing many condensation nuclei. These produce clouds that yield less precipitation because they contain numerous, but very small, droplets.

In summary, cloud seeding in certain instances leads to more precipitation; in others, to less precipitation, and, in still others, to no change in precipitation amounts. Many of the questions about cloud seeding have yet to be resolved.

thermal agitation. Thus, large cloud droplets will freeze by spontaneous nucleation at higher temperatures. At temperatures below $-40°C$, it is almost certain that an ice crystal will form and grow to critical size in even the smallest cloud droplet. Therefore, any cloud that forms in extremely cold air (below $-40°C$) will almost certainly be composed of ice, since the cloud droplets will freeze spontaneously.

Ice crystals may also form at temperatures below 0°C, if there are foreign particles present called **ice nuclei**. Certain ice nuclei allow water vapor to deposit as ice directly onto them in cold, saturated air. These are called *deposition nuclei* (or *sublimation nuclei*) because water vapor changes directly into ice without going through the liquid phase. Although cloud condensation nuclei are abundant in the atmosphere, deposition nuclei are not. Even in cold, supersaturated air, the formation of ice crystals by direct deposition is rare. However, if the geometry of an ice nucleus resembles that of an ice crystal, rapid freezing of super-

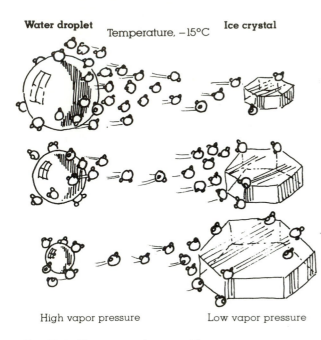

Water droplet Temperature, −15°C **Ice crystal**

High vapor pressure   Low vapor pressure

Fig. 11.6 The ice-crystal process. The greater vapor pressure around the liquid droplets forces water vapor molecules toward the ice crystals. The ice crystals absorb the water vapor and grow larger, while the water droplets grow smaller.

cooled liquid droplets may occur. Such particles are called *freezing nuclei*. Water condenses onto the surface of the nucleus and freezes immediately. The number of freezing nuclei available in the atmosphere is small, especially at temperatures above −10°C (14°F). As the temperature decreases, more particles become active freezing nuclei. Although some uncertainty exists regarding the principal source of freezing nuclei, it is known that clay minerals, such as kaolinite and montmorillonite, become effective freezing nuclei at temperatures about −15°C (5°F). Meteoric dust that settles after meteorites evaporate in the upper atmosphere and ice crystals themselves are also excellent freezing nuclei.

Supercooled droplets may also freeze if they collide with freezing nuclei, a process called **contact nucleation**. The particles are called *contact nuclei*. Recent studies suggest that contact nuclei can be just about any substance, and that contact nucleation may be the dominant force in the production of ice crystals in clouds.

We can now understand why there are so few ice crystals in the cold mixed region of some clouds. Cloud droplets may freeze sponta-

neously, but only at the very low temperatures usually found at high altitudes. Ice nuclei may initiate the growth of ice crystals, but they do not abound in nature. Therefore, we are left with a cold cloud that contains many more liquid droplets than ice particles, even at temperatures as low as −10°C. Neither the tiny liquid nor solid particles are large enough to fall as precipitation. How, then, does the ice-crystal process produce rain and snow?

In a normal mixed cloud many supercooled droplets will surround each ice crystal. Suppose that the ice particle and liquid droplet in Fig. 11.6 are part of a cloud whose temperature is −15°C. Observe that more water molecules surround each supercooled water droplet than each ice particle. This reflects the important fact discussed briefly in Chapter 7: *The saturation vapor pressure over a water surface is greater than that over an ice surface at the same subfreezing temperature*

This difference in vapor pressure establishes a force that causes water vapor molecules to move from the liquid drops toward the ice crystals. The removal of vapor molecules reduces the vapor pressure above the droplet. Since the droplet is now out of equilibrium with its surroundings, it evaporates to replenish the diminished supply of water vapor above it. This provides a continuous source of moisture for the ice crystals, which absorb the water vapor and grow rapidly by deposition. Hence, the ice crystals grow at the expense of the surrounding water droplets.

The constant supply of moisture to the ice crystal allows it to enlarge rapidly. At some point, the ice crystal will become heavy enough to overcome updrafts in the cloud and, then, it will begin to fall. But a single falling ice crystal does not make a snowstorm; consequently, other ice crystals must quickly form. The falling ice crystals collide with supercooled droplets, which freeze on contact and stick together—a process called **accretion**, or **riming**. The delicate ice crystals may also *splinter* into smaller ice particles, or tiny seeds, which serve as contact nuclei for hundreds of supercooled droplets. A chain reaction soon develops, producing many ice crystals (Fig. 11.7). As they fall, they may collide and stick to one another, forming the aggregate of ice crystals called a *snowflake*. If the snowflake melts before reaching the ground, it continues its fall as a raindrop.

Therefore, much of the rain falling in middle and northern latitudes—even in summer—begins as snow.

The first person to formally propose the theory of ice-crystal growth due to differences in the vapor pressure between ice and supercooled water was Alfred Wegener (1880–1930), the German climatologist who also proposed the geological theory of continental drift. In the early 1930s, important additions to this theory were made by the Swedish meteorologist Tor Bergeron, who proposed that essentially all raindrops begin as ice crystals. Several years later the German meteorologist Walter Findeisen made additional contributions to Bergeron's theory; hence, the ice-crystal theory of rain formation has come to be known as the *Wegener-Bergeron-Findeisen process*, or, simply, the *Bergeron process*.

Precipitation in Clouds   In cold, strongly convective clouds, precipitation may begin only minutes after the cloud forms, and may be initiated by either the collision-coalescence or the ice-crystal process. Once either process begins, most pre-

cipitation growth is by accretion. Although precipitation is commonly absent in warm-layered clouds, it is often associated with such cold-layered clouds as nimbostratus and altostratus. This precipitation is thought to form principally by the ice-crystal process because the liquid water content of these clouds is generally lower than that in convective clouds, thus making the collision-coalescence process much less effective. Nimbostratus clouds are normally thick enough to extend to levels where air temperatures are quite low, and they usually last long enough for the ice-crystal process to initiate precipitation.

Precipitation Types

Up to now, we have seen how cloud droplets are able to grow large enough to fall to the ground as rain or snow. While falling, raindrops and snowflakes may be altered by atmospheric conditions encountered beneath the cloud and transformed into other forms of precipitation that can profoundly influence our environment.

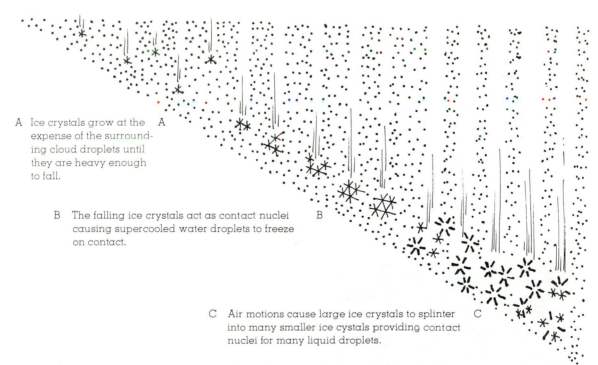

A Ice crystals grow at the expense of the surrounding cloud droplets until they are heavy enough to fall.

B The falling ice crystals act as contact nuclei causing supercooled water droplets to freeze on contact.

C Air motions cause large ice crystals to splinter into many smaller ice cystals providing contact nuclei for many liquid droplets.

Fig. 11.7 A chain reaction develops as ice crystals grow by deposition, increase in size by freezing supercooled droplets in their path, and break apart, thus producing many ice crystals that will grow into larger crystals called snowflakes.

Rain Most people consider rain to be any falling drop of liquid water. To the meteorologist, however, that falling drop must have a diameter equal to, or greater than, 500 μm (0.5 mm) to be considered rain. Fine uniform drops of water whose diameters are smaller than 0.5 mm are called **drizzle**. Most drizzle falls from stratus clouds; however, small raindrops may fall through air that is unsaturated, partially evaporate, and reach the ground as drizzle. Occasionally, the rain falling from a cloud never reaches the surface because the low humidity causes rapid evaporation. As the drops become smaller, their rate of fall decreases, and they appear to hang in the air as a rain streamer. These evaporating streaks of water are called **virga** (Fig. 11.8).

Raindrops may also fall from a cloud and not reach the ground, if they encounter rapidly rising air. Large raindrops have a terminal velocity of about 9 m/sec (20 mi/hr), and, if they encounter rising air whose speed is greater than 9 m/sec, they will not reach the surface. If the updraft weakens or changes direction and becomes a down-

draft, the suspended drops will fall to the ground as a sudden rain **shower**. The showers falling from cumuliform clouds are usually brief and sporadic, as the cloud moves overhead and then drifts on by. If the shower is excessively heavy—producing rainfall at the rate of 100 mm (about 4 in.) per hour—it is termed a *cloudburst*. Beneath a cumulonimbus cloud, which normally contains large convection currents, it is entirely possible that one side of a street may be dry (updraft side), while a heavy shower is occurring across the street (downdraft side). Continuous rain, on the other hand, usually falls from a layered cloud that covers a large area and has smaller vertical air currents. These are the conditions normally associated with nimbostratus clouds.

Normal raindrops are rarely larger than 6000 μm (6 mm). Once the drops reach this size they become unstable and break apart into smaller drops. In a cloud with strong updrafts, large drops break into smaller drops, which then rise with the vertical currents. These, in turn, grow to an unstable size, break apart, and produce more small drops, which are again carried upward. This chain reaction can take place rapidly, producing a vast number of drops in a short time. If the updrafts weaken, a heavy shower falls from the cloud.

Rain is usually transparent, but, occasionally, it takes on various colors. For example, brown rain is fairly common, especially during a drought when strong winds carry dirt and silt into a cloud. In extreme cases, where many tons of dust or volcanic ash are carried aloft, tiny mud or ash balls have actually fallen from clouds.

After a rainstorm, visibility usually improves primarily because precipitation removes (scavenges) many of the suspended particles. When rain combines with gaseous pollutants, such as oxides of sulfur and nitrogen, it becomes acidic. *Acid rain*, which has an adverse effect on plants and water resources, is becoming a major problem in many industrialized regions of the world. (See Chapter 22.)

It is important to know the interval of time over which rain falls. Did it fall over several days, gradually soaking into the soil? Or did it come all at once in a cloudburst, rapidly eroding the land, clogging city gutters, and causing floods along creeks and rivers unable to handle the sudden increased flow? The *intensity* of rain is the amount

Fig. 11.8 In the dry air beneath the cloud, the falling rain evaporates, producing streaks of water called virga.

Are Raindrops Tear-Shaped?

As rain falls, the drops take on a characteristic shape. Choose the shape in Fig. 2 that you feel most accurately describes that of a falling raindrop. Did you pick number 1? The tear-shaped drop has been depicted by artists for many years. Unfortunately, raindrops are not tear-shaped. Actually, the shape depends on the drop size. Raindrops less than 2 mm in diameter are nearly spherical, and look like raindrop number 2. The attraction among the molecules of the liquid (surface tension) tends to squeeze the drop

Fig. 2 Which of the three drops drawn here represents the real shape of a falling raindrop?

into a shape that has the smallest surface area for its total volume—a sphere.

Large raindrops, with diameters exceeding 2 mm, take on a different shape as they fall. Believe it or not, they look like

number 3, slightly elongated, flattened on the bottom, and rounded on top. As the larger drop falls, the air pressure against the drop is greatest on the bottom and least on the sides. The pressure of the air on the bottom flattens the drop, while the lower pressure on its sides allows it to expand a little. This shape has been described as everything from a falling parachute to a loaf of bread, or even a hamburger bun. You may call it what you wish, but remember: it is not tear-shaped.

that falls in a given period; intensity is always based on the accumulation during a certain interval of time. For example, *heavy* rain means that water is accumulating at a rate greater than 0.76 cm (0.30 in.) each hour. One can see that heavy rainfall reported for many consecutive hours can lead to substantial precipitation totals.

Snow We have learned that much of the precipitation reaching the ground actually begins as snow. In summer, the freezing level is usually above 3600 m (12,000 ft), and the snowflakes falling from a cloud melt before reaching the ground. However, in winter, the freezing level is much lower, and falling snowflakes have a better chance of survival. Snowflakes can generally fall about 300 m (1000 ft) below the freezing level before completely melting. Occasionally, you can spot the melting level when you look in the direction of the sun, if it is near the horizon. Because snow scatters incoming sunlight better than rain, the darker region beneath the cloud contains falling snow, while the lighter region is falling rain. The melting zone, then, is the transition between the light and dark areas (Fig. 11.9).

The sky will look different, however, if you are

looking directly up at the precipitation. Because snowflakes are such effective scatterers of light, they redirect the light beneath the cloud in all directions—some of it eventually reaching your eyes, making the region beneath the cloud look a lighter shade of gray. Falling raindrops, on the other hand, scatter very little light toward you, and the underside of the cloud appears dark. It is this change in shading that enables some observers to predict with uncanny accuracy whether precipitation will be in the form of rain or snow.

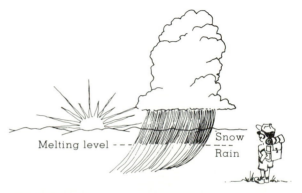

Fig. 11.9 Because snow scatters sunlight more effectively than rain, the sky appears darker above the melting level than below it.

Fig. 11.10 The dangling white streamers of ice crystals beneath these cirrus clouds are known as *fall streaks*. The bending of the streaks is due to the changing wind speed with height.

Falling ice crystals and snowflakes from high cirrus clouds—called **fall streaks**—behave in much the same way as virga. As the ice particles fall into drier air, they usually sublimate. Because the winds at higher levels move the cloud and ice particles horizontally more quickly than do the slower winds at lower levels, fall streaks appear as dangling white streamers (Fig. 11.10).

Snowflakes and Snowfall Snowflakes that fall through moist air that is slightly above freezing slowly melt as they descend. A thin film of water forms on the edge of the flakes, which acts like glue when other snowflakes come in contact with it. In this way, several flakes join to produce giant snowflakes often measuring several centimeters or more in diameter. These large, soggy snowflakes are associated with moist air and temperatures near freezing. However, when snowflakes fall through extremely cold air with a low moisture content, small, powdery flakes of "dry" snow accumulate on the ground.

If you catch falling snowflakes on a dark object and examine them closely, you will see that the most common snowflake form is a fernlike branching star shape called a *dendrite* (Fig. 11.11). Since many types of ice crystals grow, why is the dendrite crystal the most common shape for snowflakes? The type of crystal formed, as well as its growth rate, depends on the air temperature and relative humidity. Table 11.2 summarizes the crystal forms that develop when supercooled water and ice coexist in a saturated environment. Note that dendrites are common at temperatures between $-12°C$ to $-16°C$. The maximum growth rate of ice crystals depends on the difference in saturation vapor pressure between water and ice, and this difference reaches a maximum in the

Plate

Column

Dendrite

Fig. 11.11 Common ice-crystal forms.

temperature range where dendrite crystals are most likely to grow. Therefore, this type of crystal grows more rapidly than the other crystal forms. As ice crystals fall through a cloud, they are constantly exposed to changing temperature and moisture conditions. Since many ice crystals can join together to form a much larger snowflake, ice crystals may assume many complex patterns (Fig. 11.12).

Snow falling from developing cumulus clouds is often in the form of **flurries**. These are usually light showers that fall intermittently for short durations and produce only light accumulations. A more intense snow shower is called a **snow squall**. These brief but heavy falls of snow are comparable to summer rain showers and, like snow flurries, usually fall from cumuliform clouds. A more continuous snowfall (sometimes steadily, for sev-

Table 11.2   Ice-Crystal Shapes that Form at Various Temperatures

| ENVIRONMENTAL TEMPERATURE | | |
°C	°F	CRYSTAL FORM
0 to −4	32 to 25	thin plates
−4 to −10	25 to 14	columns
−10 to −12	14 to 10	plates
−12 to −16	10 to 3	dendrites
−16 to −22	3 to −8	plates
−22 to −50	−8 to −58	hollow columns

eral hours) accompanies nimbostratus and altostratus clouds.

The intensity of snow is based on its reduction of horizontal visibility at the time of observation. Heavy snow means the visibility due to falling snow is less than 0.5 km (⁵/₁₆ mi). As a general

Fig. 11.12   The many patterns of dendrite snow crystals.

When Is It Too Warm to Snow?

Just when is it too warm to snow? A person who has never been in a snowstorm might answer, "When the air temperature rises above freezing." However, anyone who lives in a climate that experiences cold winters knows that snow may fall when the air temperature is considerably above freezing. In fact, in some areas, snowstorms often begin with a surface air temperature near 2°C (36°F). Why doesn't the falling snow melt in this air? Actually, it does melt, at least to some degree. Let's examine this in more detail.

In order for falling snowflakes to survive in air with temperatures much above freezing, the air must be unsaturated (relative humidity is less than 100 percent), and the wet-bulb temperature must be at freezing or below. You may recall from our discussion on humidity (Chapter 7) that the wet-bulb temperature is the lowest temperature that can be obtained by evaporating water into the air. Consequently, it is a measure of the amount of cooling that can occur in the atmosphere as water evaporates into the air. When rain falls into a layer of dry air with a low wet-bulb temperature, rapid evaporation and cooling occur. This is why the

air temperature often decreases when it begins to rain. During the winter, as raindrops evaporate in this dry air, rapid cooling may actually change a rainy day into a snowy one. This same type of cooling allows snowflakes to survive above freezing (melting) temperatures.

Suppose it is winter and the sky is overcast. At the surface, the air temperature is 2°C (36°F), the dew point is −6°C (21°F), and the wet-bulb temperature is 0°C (32°F).* The air temperature drops sharply with height, from the surface up to the cloud deck. Soon, flakes of snow begin to fall from the clouds into the unsaturated layer below. In the air where temperatures are above freezing, the snowflakes begin to partially melt. The air, however, is dry, so the water quickly evaporates, cooling the air. In addition, evaporation cools the falling snowflake and retards its rate of melting. As snow continues to fall, evaporative cooling causes the air temperature to continue to drop. The addition of water vapor to the air increases the

*The wet-bulb temperature is always higher than the dew point, except when the air is saturated. At that point, the air temperature, wet-bulb temperature, and dew point are all the same.

dew point, while the wet-bulb temperature remains essentially unchanged. Eventually, the entire layer of air cools to the wet-bulb temperature and becomes saturated at 0°C. As long as the wind doesn't bring in warmer air, the precipitation remains as snow.

We can see that when snow falls into warmer air (say at 8°C or 46°F), the air must be extremely dry in order to have a wet-bulb temperature at freezing or below. In fact, with an air temperature of 8°C (46°F) and a wet-bulb temperature of 0°C (32°F), the dew point would be −23°C (−9°F) and the relative humidity 11 percent. Conditions such as these are extremely unlikely at the surface before the onset of precipitation. Actually, the highest air temperature possible with a below freezing wet-bulb temperature is about 10°C (50°F). Hence, snowflakes will melt rapidly in air with a temperature above this value. However, it is still possible to see flakes of snow at temperatures greater than 10°C (50°F), especially if the snowflakes are swept rapidly earthward by the cold, relatively dry downdraft of a thunderstorm.

rule, heavy snow means a total accumulation of 10 or more centimeters (4 in.) during a 12-hr period, or a fall of 15 cm (6 in.) or more during a 24-hr period. There are, however, some variations to this general rule. For example, where average 10-cm snowfalls are common, a heavy snowfall is typically 15 cm or more. Where snow is infre-

quent, a heavy snow may be an accumulation of 5 or 6 cm (about 2 or 3 inches).

When a strong wind is blowing at the surface, snow can be picked up and deposited into huge drifts. Drifting snow is usually accompanied by *blowing snow*; that is, snow lifted from the surface by the wind and blown about in such quantities that horizontal visibility is greatly restricted. The combination of drifting and blowing snow, after falling snow has ended, is called a *ground blizzard*. A true **blizzard** is a weather condition characterized by low temperatures and strong winds (greater than 28 knots) bearing large amounts of fine, dry, powdery particles of snow, which can reduce visibility to only a few meters.

A Blanket of Snow A mantle of snow covering the landscape is much more than a beautiful setting—it is a valuable resource provided by nature. A blanket of snow is a good insulator (poor heat conductor). In fact, the more air spaces there are between the individual snowflake crystals, the better insulator they become. A light, fluffy covering of snow protects sensitive plants and their root systems from damaging low temperatures by retarding the loss of ground heat.

On winter nights, ground that is covered with dry snow maintains a warmer temperature than ground that is exposed to the cold air (Fig. 11.13). In this way, snow can prevent the ground from freezing downward to great depths. In cold climates that receive little snow, it is often difficult to grow certain crops because the frozen soil makes spring cultivation almost impossible. Frozen ground also prevents early spring rains from percolating downward into the soil, leading to rapid water runoff and flooding. If subsequent rains do not fall, the soil could even become moisture-deficient. If you become lost in a cold and windy snowstorm, build a snow cave and climb inside. It not only will protect you from the wind, but it also will protect you from the extreme cold by slowing the escape of heat your body generates.

The accumulation of snow in mountains provides winter recreation areas, and the melting snow in spring and summer is of great economic value in that it supplies streams and reservoirs with much-needed water.

Winter snows may be beautiful, but they are not without hardships and potential hazards. As spring approaches, rapid melting of the snowpack may flood low-lying areas. Too much snow on the side of a steep hill or mountain may become an avalanche as the spring thaw approaches. The added weight of snow on the roof of a building may cause it to collapse, leading to costly repairs and even loss of life. Each winter, heavy snows clog streets and disrupt transportation. To keep traffic moving, streets must be plowed and sanded, or salted to lower the temperature at which the snow freezes. This can be expensive, especially if the snow is heavy and wet. Cities unaccustomed to snow are usually harder hit by a moderate snowstorm than cities that frequently experience snow. A January snowfall of several centimeters in New Orleans, Louisiana, can bring traffic to a standstill, while a snowfall of several centimeters in Buffalo, New York, would go practically unnoticed.

Sleet and Freezing Rain Consider the falling snowflake in Fig. 11.14. As it falls into warmer air,

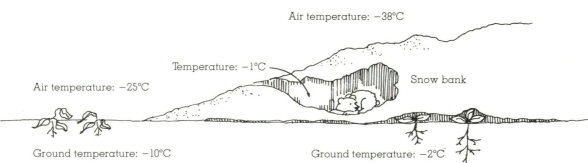

Air temperature: −38°C

Temperature: −1°C

Air temperature: −25°C

Snow bank

Ground temperature: −10°C

Ground temperature: −2°C

Fig. 11.13 A blanket of snow prevents ground heat from escaping into the cold air. As a consequence, the ground remains warmer beneath a covering of snow.

Sounds of Silence

Once snow settles on the ground, it can affect the way sound propagates. You may have noticed that, after a snowfall, it seems quieter than usual: Freshly fallen snow can absorb sound—just like acoustic tiles. As the snow gets deeper, this absorption increases. Anyone who has walked through the woods on a snowy evening knows the quiet created by a thick blanket of snow. As snow becomes older and more densely packed, its ability to absorb sound is reduced.

New snow covering a pavement will sometimes squeak as you walk in it. The sound produced is related to the snow's temperature. When the air and snow are only slightly below freezing, the pressure from the heel of a boot partially melts the snow. The snow can then flow under the weight of the boot and no sound is made. However, on cold days when the snow temperature drops below −10°C (14°F), the heel of the boot will not melt the snow, and the ice crystals are crushed. The crunching of the crystals produces the creaking sound.

it begins to melt. When it falls through the deep subfreezing surface layer of air, it turns back into ice, not as a snowflake, but as a tiny ice pellet called **sleet**. Generally, these ice pellets are transparent (or translucent), with diameters of 5 mm (0.2 in.) or less. They bounce when striking the ground and produce a tapping sound when they hit a window or piece of metal.

The cold surface layer beneath a cloud may be too shallow to freeze raindrops as they fall. In this case, they reach the surface as supercooled liquid drops. Upon striking a cold object, the drops spread out and almost immediately freeze, forming a thin veneer of ice. This form of precipitation is called

freezing rain, *glaze*, or *silver thaw*. If the drops are small (less than 0.5 mm in diameter), the precipitation is called **freezing drizzle**.

Freezing rain can create a beautiful winter wonderland by coating everything with silvery, glistening ice. At the same time, highways turn into skating rinks for automobiles, and the destructive weight of the ice—which can be many tons on a single tree—breaks tree branches, power lines, and telephone cables. (See Fig. 11.15.) A case in point is the ice storm of January, 1983, which left more than 250,000 people from Mississippi to New England without power, and caused an estimated 25 deaths. As much as 4 cm (1.6 in.) of ice covered sections of Oregon and Washington in January, 1970, causing $6 million in damage to fruit trees in the Hood River Valley. The area most frequently hit by these storms extends over a broad region from Texas into Minnesota and eastward into the middle Atlantic states and New England. Such storms are extremely rare in southern California and Florida.

In summary, Fig. 11.16 shows various winter temperature profiles and the type of precipitation associated with each. In profile (a), the air temperature is below freezing at all levels, and snowflakes (*) reach the surface. In (b), a zone of above-freezing air causes snowflakes to partially melt; then, in the deep, subfreezing air at the surface, the liquid freezes into sleet (△). In the

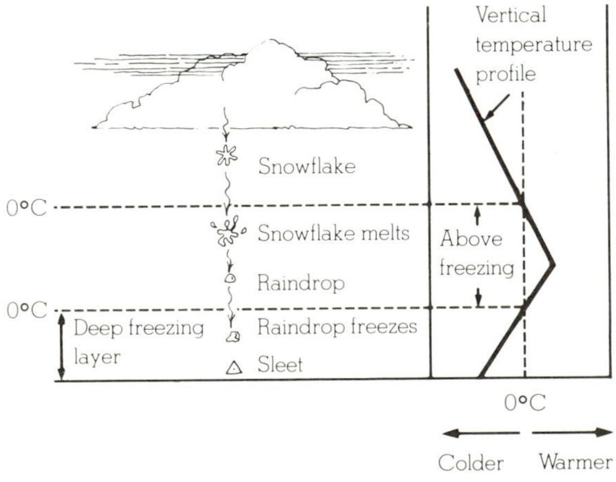

Fig. 11.14 The formation of sleet.

Fig. 11.15 A heavy coating of freezing rain during this ice storm caused tree limbs to break and power lines to sag.

Vertical temperature profiles

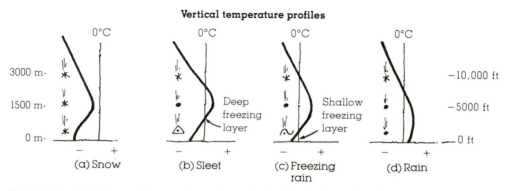

Fig. 11.16 Vertical temperature profiles associated with different forms of precipitation.

FOCUS ON AN APPLICATION

Aircraft Icing

Consider an aircraft flying through an area of freezing rain. As the large, supercooled drops strike the leading edge of the wing, they break apart and form a film of water, which quickly freezes into a solid sheet of ice. This smooth, transparent ice—called **clear ice**—is similar to the glaze that coats trees during ice storms. Clear ice can build up quickly; it is heavy and difficult to remove, even with modern deicers.

When an aircraft flies through a cloud composed of tiny, supercooled liquid droplets, **rime ice** may form. Rime ice forms when some of the cloud droplets strike the wing and freeze before they have time to spread, thus leaving a rough and brittle coating of ice on the wing. Because the small, frozen droplets trap air between them, rime ice usually appears white. Even though rime ice redistributes the flow of air over the wing more than clear ice does, it is lighter in weight and is more easily removed with deicers.

Because the raindrops and cloud droplets in most clouds vary in size, a mixture of clear and rime ice usually forms on aircraft. Also, because concentrations of liquid water tend to be greatest in warm air, icing is usually heaviest and most severe when the air temperature is between 0°C and −10°C (32°F to 14°F).

A major hazard to aviation, icing reduces aircraft efficiency by increasing weight. Icing has other adverse effects, depending on where it forms. On a wing or fuselage, ice can disrupt the air flow and decrease the plane's flying capability. When ice forms in the air intake of the engine, it robs the engine of air, causing a reduction in power. Icing may also affect the operation of brakes, landing gear, and instruments.

---

shallow subfreezing surface air in (c), the melted snowflakes freeze on contact, producing freezing rain (∼). In (d), the air temperature is above freezing in a sufficiently deep layer so that precipitation reaches the surface as rain (·). (Note: Other weather symbols are presented in Appendix B.)

Snow Grains and Snow Pellets Snow grains are small, opaque grains of ice, the solid equivalent of drizzle. They are fairly flat or elongated, with diameters generally less than 1 mm (0.04 in.). They fall in small quantities from stratus clouds, and never in the form of a shower. Upon striking a hard surface, they neither bounce nor shatter. **Snow pellets**, on the other hand, are white, opaque grains of ice, with diameters between 2 and 5 mm (0.1 and 0.2 in.). They are sometimes confused with snow grains. The distinction is easily made, however, by remembering that, unlike snow grains, snow pellets are brittle, crunchy, and bounce (or break apart) upon hitting a hard surface. They usually fall as showers, especially from cumulus congestus clouds.

Consider the cumulus congestus cloud in Fig.

11.17. The freezing level is near the surface and, since the atmosphere is unstable, the air temperature drops quickly with height. An ice crystal in the cold (−23°C) middle region of the cloud would be surrounded by many supercooled cloud droplets and ice crystals. In the very cold air, the crystals tend to rebound after colliding rather than sticking to one another. However, when the ice crystals collide with the supercooled water droplets, they immediately freeze the droplets, producing a spherical aggregate of icy matter (rime) containing many tiny air spaces. These small air bubbles have two effects on the growing ice particle: (1) They keep its density low; and (2) they scatter light, making the particle opaque. By the time the ice particle reaches the lower half of the cloud, it has grown in size, and its original shape is gone. When the ice particle accumulates so much rime that it can no longer be recognized as an ice crystal (or snowflake), it is called **graupel**. Since the freezing level is at a low elevation, the graupel reaches the surface as a light, round clump of snowlike ice—a snow pellet. In summer, when the freezing level is well above the cloud base, the graupel melts before reaching the earth. In vigorously convective clouds, however, the graupel may develop into a full-fledged hailstone.

Hail **Hailstones** are pieces of ice either transparent or partially opaque, ranging in size from that of small peas to that of golf balls or larger. Some are round, others take on irregular shapes. The largest authenticated hailstone fell on Coffeyville, Kansas, in September, 1970. (See Fig. 11.18.) This giant weighed 757 grams (1.67 lb) and had a measured diameter of over 14 cm (5.5 in.). Needless to say, large hailstones are quite destructive. They can break windows, dent cars, batter roofs of homes, and cause extensive damage to livestock and crops. In fact, a single hailstorm can destroy a farmer's crop in a matter of minutes.

Estimates are that, in the United States alone, hail accounts for some $700 million in damage annually. Although hailstones are potentially lethal, only two fatalities due to hail have been documented in the United States during this century. Near Lubbock, Texas, in May, 1930, a farmer died of injuries incurred after having been caught in the middle of a violent hailstorm. More recently, a three-month-old boy died from a fractured skull

Fig. 11.18 The Coffeyville hailstone. This giant hailstone—the largest ever reported—fell on the community of Coffeyville, Kansas, on September 3, 1970.

caused by large hailstones, which pelted Fort Collins, Colorado, on July 30, 1979. In April of 1888, in northern India, 246 people died in a hailstorm, and, in southeast China, some 200 people lost their lives and thousands were injured in a devastating hailstorm in June, 1932.

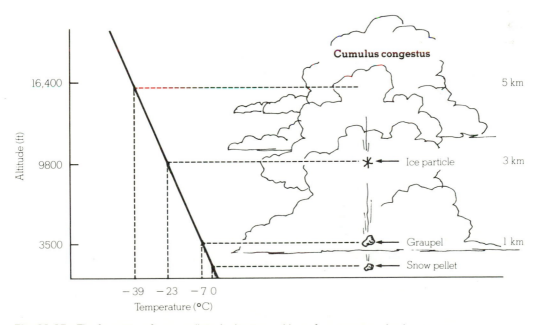

Fig. 11.17 The formation of snow pellets. In the very cold air of a convective cloud, ice particles grow into graupel, which remains frozen and reach the surface as snow pellets.

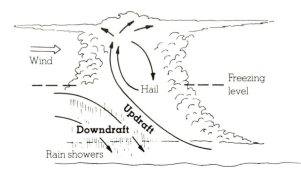

Fig. 11.19 The formation of hailstones. The violent up-drafts in this cumulonimbus cloud keep hailstones suspended in the cloud for several up and down cycles. As the hailstone moves up and down inside the cloud, it grows larger by accretion.

Hail is produced in a cumulonimbus cloud, as graupel, large frozen raindrops, or just about any particles (even insects) act as embryos that grow by accretion. It takes a million cloud droplets to form a single raindrop, but it takes about 10 billion cloud droplets to form a golfball-size hailstone. For a hailstone to grow to this size, it must remain in the cloud for a substantial period of time. Violent, upsurging air currents within the cloud carry small embryos high above the freezing level, where they drop, only to be lifted again by other updrafts (Fig. 11.19). During each up and down cycle, a coating of ice grows around the embryos. If this cycle is repeated several times (as is the case in violent thunderstorms), the hailstones may grow to an appreciable size. As soon as they become too large and heavy to be supported by the rising air, they escape and fall into the warmer air below the cloud. Small hailstones often melt before

reaching the ground, but, in the violent thunderstorms of summer, hailstones may grow large enough to reach the ground before completely melting. Strangely, then, we find the largest form of frozen precipitation occurring during the warmest time of the year.

Summertime accumulation of hail in some areas can be substantial. In June, 1959, a hailstorm lasting 85 minutes covered the town of Selden, Kansas, with hail 46 cm (18 in.) deep. Maryville, Missouri, received 30 cm (about 12 in.) of hail from a storm in September, 1898. Occasionally, in the Great Plains states, snow plows have to be called on to clear highways clogged by summer hailstorms.

Figure 11.20 shows a large hailstone that has been cut in half. Notice that it has concentric layers of milky-white and clear ice. We have already seen that, during a hailstone's repeated up and down trip within the cloud, it grows larger by accretion. If the air is very cold, supercooled droplets freeze immediately on the stone, producing a coating of opaque rime ice containing many air bubbles. As the water freezes, heat is released, which keeps the surface of the hailstone warmer than its environment. As the stone falls into the wet lower half of the cloud, droplets may collect so rapidly that the hailstone's surface temperature remains above 0°C. Now, the supercooled droplets no longer freeze on impact; instead, they spread a coating of water around the hailstone. As the hailstone re-enters a colder section of the cloud, the surface water cools and slowly freezes. Trapped air escapes, and the ice that forms is clear. Therefore, as a hailstone passes through a cloud of changing temperature and liquid water content, alternating layers of opaque and clear ice form. Observing these layers can help us understand much about the hailstone's environment during its formation.

Fig. 11.20 A hailstone cut open. Photographed under polarized light (*left*) and regular light (*right*), its layered structure is revealed.

Measuring Precipitation

Instruments A *standard* 20 cm (8 in.) *rain gauge* is commonly used to measure rainfall. This instrument consists of a funnel-shaped collector 20 cm in diameter attached to a 50 cm (20 in.)

FOCUS ON AN ISSUE

Preventing a Hailstorm

Because large hailstones are so damaging to crops, various methods have been tried to prevent them from forming in thunderstorms. One method employs the seeding of clouds with large quantities of silver iodide. These nuclei freeze supercooled water droplets and convert them into ice crystals. The ice crystals grow larger as they come in contact with additional supercooled cloud droplets. In time, the ice crystals grow large enough to be called graupel, which then becomes a hailstone embryo. Large numbers of embryos are produced by seeding in hopes that competition for the remaining supercooled droplets may be so great that none of the embryos would be able to grow into large and destructive hailstones. Soviet scientists claim great success in suppressing hail using silver iodide. However, their experiments are performed in such a way that statistical evaluation is not possible. In the United States, the results of most hail-suppression experiments are still inconclusive.

---

long measuring tube (Fig. 11.21). The cross-sectional area of the collector is exactly 10 times that of the tube. Hence, rain falling into the collector is amplified tenfold in the tube, permitting measurements of great precision. A wooden scale, calibrated to allow for the vertical exaggeration, is inserted into the tube and withdrawn. The wet portion of the scale indicates the depth of water. So, 25 cm (10 in.) of water in the tube would be measured as 2.5 cm (1 in.) of rainfall. Because of this amplification, rainfall measurements can be made when the amount is as small as 0.025 cm (0.01 in.). An amount less than this is called a **trace**.

As we can see, the measuring tube can only collect 5 cm (2 in.) of rain. Rainfall of more than 5 cm causes an overflow into an outer cylinder. Here, the excess rainfall is stored and protected from appreciable evaporation. When the gauge is emptied, the overflow is carefully poured into the tube and measured.

Another instrument that measures rainfall is the *tipping bucket rain gauge.* In Fig. 11.22, notice that this gauge has a receiving funnel leading to two small metal collectors (buckets). The bucket beneath the funnel collects the rain water. When it accumulates the equivalent of 0.025 cm (0.01 in.) of rain, the weight of the water causes it to tip and empty itself. The second bucket immediately moves under the funnel to catch the water. When it fills, it also tips and empties itself, while the original bucket moves back beneath the funnel. Each time a bucket tips, an electric contact is made, causing a pen to register a mark on a remote recording chart. Adding up the total number of marks gives the rainfall for a certain time period.

Remote recording of precipitation can also be made with a *weighing-type rain gauge.* With this gauge, precipitation is caught in a cylinder and accumulates in a bucket. The bucket sits on a

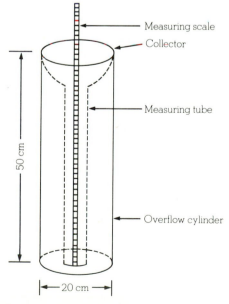

Fig. 11.21   Components of the standard rain gauge.

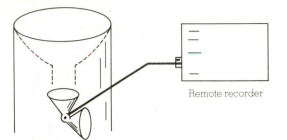

Fig. 11.22 The tipping bucket rain gauge. Each time the bucket fills with 0.01 inch of rain, it tips, sending an electric signal to the remote recorder.

Remote recorder

sensitive weighing platform. Special gears translate the accumulated weight of rain or snow into millimeters or inches of precipitation. The precipitation totals are recorded by a pen on chart paper, which covers a clock-driven drum. By using special electronic equipment, this information can be transmitted from rain gauges located in remote areas to satellites or land-based stations, thus providing precipitation totals from previously inaccessible regions.

The depth of snow in a region is determined by measuring its depth at three or more representative areas. The amount of snowfall is defined as the average of these measurements. Since snow often blows around and accumulates into drifts, finding a representative area can be a problem. Determining the actual depth of snow can include considerable educated guesswork. Snow depth may also be measured by removing the collector and inner cylinder of a standard rain gauge and allowing snow to accumulate in the outer tube. Turbulent air around the edge of the tube often blows flakes away from the gauge. This makes the amount of snow collected less than the actual snowfall. To remedy this, slatted windshields are placed around the cylinder to block the wind and ensure a more correct catch.

Precipitation is a highly variable weather element. A huge thunderstorm may drench one section of a town while leaving another section completely dry. Given this variability, it should be apparent that a single rain gauge on top of a building cannot represent the total precipitation for any particular region.

Water Equivalent   The depth of water that would result from the melting of a snow sample is called the **water equivalent**. In a typical fresh snow pack, about 10 cm of snow will melt down to about 1 cm of water, giving a water equivalent ratio of 10:1. This ratio, however, will vary greatly, depending on the texture and packing of the snow. Very wet snow falling in air near freezing may have a water equivalent of 6:1. On the other hand, in dry, powdery snow, the ratio may be as high as 30:1. Toward the end of the winter, large compacted drifts representing the accumulation of many storms may have a water equivalent of less than 2:1.

Determining the water equivalent of snow is a fairly straightforward process: The snow accumulated in a rain gauge is melted and its depth is measured. Another method uses a long tube pushed into the snow to a desired depth. This snow sample is then melted and poured into a rain gauge for measuring its depth.

Summary

In this chapter, we have seen that cloud droplets are too small and light to reach the ground as rain. Cloud droplets do grow larger by condensation, but this process by itself is much too slow to produce substantial precipitation. Because larger cloud droplets fall faster and farther than smaller ones, they grow larger as they fall by coalescing with drops in their path. If the temperature in a cloud drops below freezing, then ice crystals play an important role in producing precipitation. Some crystals may form on ice nuclei, or they may form when they come in contact with a foreign particle. At temperatures below −40°C, liquid droplets spontaneously freeze. Because of differences in vapor pressures between water and ice, an ice crystal surrounded by water droplets grows larger at the expense of the droplets. As it begins to fall, it grows even larger by colliding with liquid droplets, which freeze on contact. In an attempt to coax more precipitation from clouds, some are seeded.

Precipitation can reach the surface in a variety of forms. In winter, raindrops may freeze on impact, producing freezing rain that can disrupt electrical service by downing power lines. Raindrops may freeze into tiny pellets of ice above the ground

FOCUS ON INSTRUMENTS

Weather Radar and Precipitation

After it showers, ask someone what time it began to rain. More than likely the answer will be, "When the first raindrops hit the ground." Did the rain really begin at this time? Actually, raindrops begin to fall within clouds as soon as they are large enough to overcome any updrafts. Raindrops falling from the middle of a cloud 6100 m (about 20,000 ft) thick take over 9 minutes to reach the surface. And what about updrafts? Falling drops can become caught in rising currents and remain suspended for a long time. So, when *does* rain begin to fall? Until atmospheric scientists began using weather radar to study precipitation, the answer to this question was indeed difficult to obtain.

Radar (*radio detection and ranging*) has become an essential tool of the atmospheric scien-tist, for it gathers information about storms and precipitation in previously inaccessible regions. Atmospheric scientists use weather radar to examine the inside of a cloud much like physicians use X-rays to examine the inside of a human body. Essentially, the radar unit consists of a transmitter that sends out short, powerful microwave pulses having wavelengths between 1 and 20 cm. When this energy encounters a foreign object—called a *target*—a fraction of the energy is scattered back toward the transmitter and is detected by a receiver. The returning signal is amplified and displayed on a screen, producing an image or "echo" from the target. The elapsed time between transmission and reception indicates the target's distance.

Smaller targets require detec-tion by shorter wavelengths. Cloud droplets are detected by radar using wavelengths of 1 cm, whereas longer wavelengths (between 3 and 10 cm) penetrate tiny cloud droplets, but are scattered by larger precipitation particles. The brightness of the echo is directly related to the amount of rain falling in the cloud. So, the radar screen shows not only where precipitation is occurring, but also how intense it is. In recent years, the radar image has been displayed using various colors to denote the intensity of precipitation within the range of the radar unit. Some television weather shows use this observation technique and call it "color radar."

---

and reach the surface as sleet. Depending on conditions, snow may fall as pellets, grains, or flakes, all of which can influence how far we see and hear. Strong updrafts in a cumulonimbus cloud may carry ice particles high above the freezing level, where they acquire a further coating of ice and form destructive hailstones. Although the rain gauge is still the most commonly used method of measuring precipitation, radar has recently become an important instrument for determining precipitation intensity.

Questions for Review

1. What is the main difference between a cloud droplet and a raindrop?

2. Why do typical cloud droplets seldom reach the ground as rain?

3. What are warm clouds? How is precipitation believed to form in them?

4. List and describe three ways in which ice crystals can form in a cloud.

5. Briefly explain the collision and coalescence process of producing precipitation.

6. Explain in your own words how the ice-crystal process of producing rain works. Illustrate with a diagram.

7. Would the collision-and-coalescence process work better in producing rain in (a) a warm, thick nimbostratus cloud, or (b) a warm, towering cumulus congestus cloud? Explain.

8. Why is it foolish to seed a clear sky with silver iodide?

9. How would you be able to distinguish between virga and fall streaks?

10. Explain why heavy showers usually accompany cumuliform clouds, while steady precipitation normally falls from stratiform clouds.

11. Why are large snowflakes usually observed when the air temperature near the ground is just below freezing?

12. Dendrite ice crystals are the most common form of ice crystals. Why?

13. What is the difference between snow flurries and a snow squall?

14. How do the atmospheric conditions that produce sleet differ from those that produce hail?

15. Why is hail more common in summer than in winter?

16. How is radar used to study precipitation?

17. List and then explain how several common precipitation gauges measure rain and snow.

Questions for Thought

1. Ice crystals that form by accretion are fairly large. Explain why they fall slowly.

2. Why is a warm, tropical cumulus cloud more likely to produce precipitation than a cold, stratus cloud?

3. Explain why very small cloud droplets of pure water evaporate even when the relative humidity is 100 percent.

4. Suppose a thick nimbostratus cloud contains ice crystals and cloud droplets all about the same size. Which precipitation process will be most important in producing rain from this cloud? Why?

5. Clouds that form over water are usually more efficient in producing precipitation than clouds that form over land. Why?

6. Everyday in summer a blizzard occurs over the Great Plains. Explain where and why.

7. During a recent snowstorm, Denver, Colorado, received 7 cm (3 in.) of snow. Sixty kilo-meters east of Denver, a city received no measurable snowfall, while 150 kilometers east of Denver another city received 10 cm (4 in.) of snow. If Denver is located to the east of the Rockies, and the upper-level winds were westerly during the snowstorm, give an explanation as to what *could* account for this snowfall pattern.

8. Explain how seeding a cloud with silver iodide might actually decrease precipitation.

9. Lead iodide is an effective freezing nuclei. Why do you think it has not been used for that purpose?

10. When cirrus clouds are above a deck of altocumulus clouds, occasionally a clear area, or "hole," will appear in the altocumulus cloud layer. What do you suppose could cause this to happen?

11. On a cold, clear day, ice crystals may actually fall from the sky. Can you explain what could account for this? (Because the crystals sparkle in the sunlight, they are called *diamond dust*.)

Problems and Exercises

1. In the daily newspaper, a city is reported as receiving 1.32 cm (0.52 in.) of precipitation over a 24-hour period. If all the precipitation fell as snow, and if we assume a normal water equivalent ratio of 10:1, how much snow did this city receive?

2. How many times faster does a large raindrop (diameter 5000 μm) fall than a cloud droplet (diameter 20 μm), if both are falling at their terminal velocity in still air?

3. (a) How many minutes would it take drizzle with a diameter of 200 μm to reach the surface if it falls at its terminal velocity from the base of a cloud 1000 m (about 3300 ft) above the ground? (Assume the air is saturated beneath the cloud, the drizzle does not evaporate, and the air is still.)
(b) Suppose the drizzle in problem 3a evaporates on its way to the ground. If the drop size is 200 μm for the first 450 m of descent, 100 μm for the next 450 m, and 20 μm for the final 100 m, how long will it take the drizzle to reach the ground if it falls in still air?

4. Suppose a large raindrop (diameter 5000 μm) falls at its terminal velocity from the base of a cloud 1500 m (about 5000 ft) above the ground.

(a) If we assume the raindrop does not evaporate, how long would it take the drop to reach the surface?

(b) What will be the shape of the falling raindrop just before it reaches the ground?

(c) What type of cloud would you have expected this raindrop to fall from? Explain.

5. Describe a plausible life history for the large hailstone in the accompanying figure. Explain how the areas of clear ice and rime ice probably formed, and the number of up and down trips the stone made in the cumulonimbus cloud.

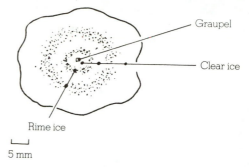

Graupel

Clear ice

Rime ice

5 mm

A persistent strong wind lifts clouds over the east face of Alaska's Mt. McKinley. (Photo: Bradford Washburn, Boston Museum of Science)

CHAPTER 12

The Atmosphere in Motion— Charts, Forces, and Winds

December 19, 1980, was a cool day in Lynn, Massachusetts, but not cool enough to dampen the spirits of more than 2000 people who gathered in Central Square—all hoping to catch at least one of the 1500 dollar bills that would be dropped from a small airplane at noon. Right on schedule, the aircraft circled the city and dumped the money onto the people below. However, to the dismay of the onlookers, a westerly wind caught the currency before it reached the ground and carried it out over the cold Atlantic Ocean. Had the pilot or the sponsoring leather manufacturer examined the weather charts beforehand, they might have been able to predict that the wind would ruin their advertising scheme.

This little scenario raises two questions: (1) Why does the wind blow; and (2) how can one tell its direction by looking at weather charts? Chapter 1 has already answered the first question: Air moves in response to horizontal differences in pressure. This happens when we open a vacuum-packed can—air rushes from the higher pressure region outside the can toward the region of lower pressure inside. In the atmosphere, the wind blows in an attempt to equalize imbalances in air pressure. Does this mean that the wind always blows directly from high to low pressure? Not really, because the movement of air is controlled not only by pressure differences but by other forces as well. In this chapter, we will consider the forces that influence atmospheric motions at the surface and aloft. Through studying these forces, we will be able to tell how the wind should blow in a particular region by examining surface and upper-air charts.

Contents

Surface and Upper-Level Charts

In order to understand how pressure influences winds, we must first examine charts that show variations in pressure. We need to know how such charts are constructed and what they represent.

Station Pressure Chapter 2 developed a number of concepts about atmospheric pressure. Two significant ones were (1) that atmospheric pressure is a measure of the weight of the air above the point of observation; and (2) that the height of the mercury column in a mercury barometer is a measure of air pressure. Keep these important concepts in mind as you read this section.

The seemingly simple task of reading the height of the mercury column to obtain the air pressure is actually not all that simple. Being a fluid, mercury is sensitive to changes in temperature; it will expand when heated and contract when cooled. Consequently, to obtain accurate pressure readings without the influence of temperature, all mercury barometers are corrected as if they were read at the same temperature. Because the earth is not a perfect sphere, the force of gravity is not a constant. Although small, gravity differences do influence the height of the mercury column and must be considered when reading the barometer. Finally, each barometer has its own "built-in" error, called *instrument error*, which is caused, in part, by the surface tension of the mercury against the glass tube. After being corrected for temperature, gravity, and instrument error, the barometer reading at a particular location is termed **station pressure**.

Figure 12.1a gives the station pressure measured at four locations only a few hundred kilometers apart. We might ask why these pressure readings vary so much, since the cities are so close together. Horizontal pressure variations such as these would produce winds of hundreds of kilometers per hour.

We might guess from Fig. 12.1a that the differences in station pressure of the four cities are due primarily to their variation in altitude. This becomes even clearer when we realize that atmospheric pressure changes much more quickly when we move upward than it does when we move sideways. As an example, the vertical change in air pressure from the base to the top of the Empire State Building—a distance of a little more than ½ km—is typically much greater than the horizontal difference in air pressure from New York City to Miami, Florida—a distance of over 1600 km. Therefore, we can see that a small vertical difference between two observation sites can yield a large difference in station pressure. Thus, to properly monitor *horizontal* changes in pressure, barometer readings must be corrected for altitude.

Sea Level Pressure Altitude corrections are made so that a barometer reading taken at one elevation can be compared with a barometer reading taken at another. Pressure observations are normally adjusted to an altitude of 0 m, which is referred to as *mean sea level*—the level representing the average surface of the ocean. The size of the correction depends primarily on how high the station is above sea level.

Near the earth's surface, atmospheric pressure decreases by about 10 mb for every 100 m increase in elevation (about 1 in. of mercury for each 1000-ft rise) in an atmosphere where the air temperature decreases at the standard lapse rate of 6.5°C/1000 m. Under these conditions, suppose a barometer located at sea level and reading a station pressure of 1010 mb, were lifted into the atmosphere. At 100 m, it would read 1000 mb and, at 200 m, the reading would be 990 mb. Therefore, at 200-m altitude, we simply add 20 mb to the station pressure to obtain the sea level pressure. In general, we add 10 mb for every 100 m altitude to obtain sea level pressure.

Because a standard atmosphere seldom exists, this value (10 mb per 100 m) only approximates actual conditions. For example, in Chapter 2 we saw that atmospheric pressure decreases more rapidly with altitude in cold air than it does in warm air. Hence, in cold air, the vertical rate of pressure change is typically greater than 10 mb per 100 m, while in warm air, it is less. Moreover, since the vertical lapse rate changes slightly from day to day, temperature and moisture corrections made for "average" conditions may introduce additional errors. However, if we simplify matters and assume a standard lapse rate, we can correct each of the station pressures in Fig. 12.1a to sea level simply by adding 10 mb per 100 m to each station pressure. When we plot these corrected values

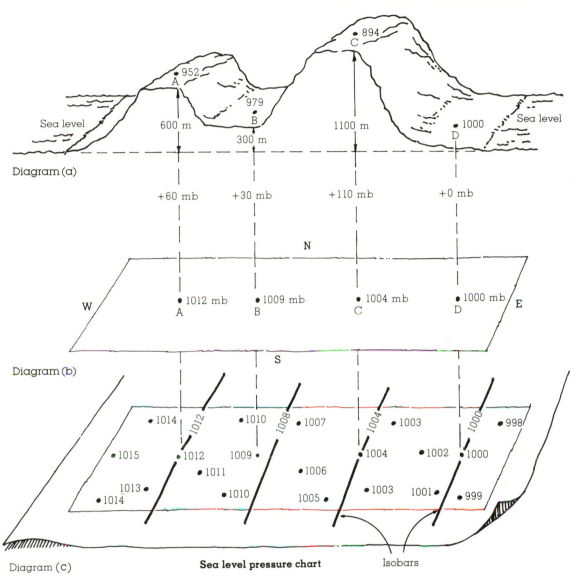

Fig. 12.1 The top diagram (a) shows four stations (A, B, C, and D) at varying elevations above sea level, all with different station pressures. The middle diagram (b) represents sea level pressures of the four stations plotted on a sea level chart. The bottom diagram (c) shows isobars drawn on the chart (dark lines) at intervals of 4 mb.

on the sea level pressure chart (Fig. 12.1b), we are able to see the horizontal variation in sea level pressure. Notice that the pressure varies from higher pressure on the left side of the map to lower pressure on the right—something impossible to see from the uncorrected station pressures we initially examined in Fig. 12.1a.

When more pressure data are added (Fig. 12.1c),

the chart can be analyzed and the pressure pattern visualized. **Isobars** (lines connecting points of equal pressure) are drawn at intervals of 4 mb,* with 1000 mb being the base value. Note that the isobars do not pass through each point, but, rather,

*An interval of 2 mb would put the lines too close together, and an 8-mb interval would spread them too far apart.

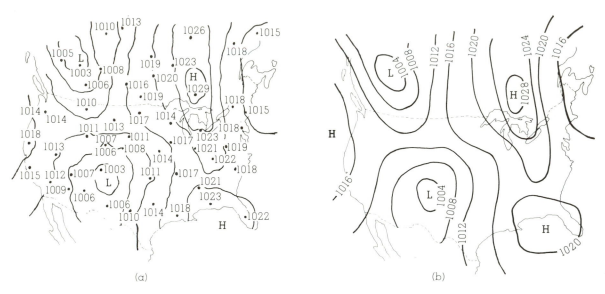

Fig. 12.2   (a) Sea level isobars drawn so that each observation is taken into account. Not all observations are plotted. (b) Sea level isobars after smoothing.

between many of them, with the exact values being interpolated from the data given on the chart. For example, follow the 1008-mb line from the top of the chart southward and observe that there is no plotted pressure of 1008 mb. The 1008-mb isobar, however, comes closer to the station with a sea level pressure of 1007 mb than it does to the station with a pressure of 1010 mb.

The isobars in Fig. 12.2a are drawn precisely, with each individual observation taken into account. Notice that many of the lines are irregular, especially in mountainous regions over the Rockies. The reason for the wiggle is due, in part, to small-scale local variations in pressure and to errors introduced by correcting observations that were taken at high-altitude stations. An extreme case of this type of error occurs at Leadville, Colorado (elevation 3096 m), the highest city in the United States. Here, the station pressure is typically near 700 mb. This means that nearly 300 mb must be added to obtain a sea level pressure reading! A mere 1 percent error in estimating the exact correction would result in a 3-mb error in sea level pressure. For this reason, isobars are smoothed through readings from high-altitude stations and from stations that might have small observational errors. Figure 12.2b shows how the isobars appear on the surface map after they are smoothed.

Constant Height and Constant Pressure Charts   The sea level pressure chart described so far is called a **constant height chart** because it represents the atmospheric pressure at a constant level—in this case, sea level. The same type of chart could be drawn to show the horizontal variations in pressure at any level in the atmosphere; for example, at 3000 m (Fig. 12.3).

Another type of chart is commonly used in studying the weather; namely, the **constant pressure chart**. Instead of showing pressure variations at a constant altitude, these charts are constructed to show height variations along an equal pressure (*isobaric*) surface. Constant pressure charts are convenient to use because the height variables they show are easier to deal with in meteorological equations than the variables of pressure. Since isobaric charts are in common use, let's examine them in detail.

The dots in Fig. 12.4 represent air molecules from sea level up to the tropopause. To simplify matters, we assume that the air density is constant throughout the entire air layer and that all of the air molecules are squeezed into this layer. If we climb halfway up the air column and stop, then draw a sheetlike surface representing this level, we will have made a constant height surface. This altitude (5600 m) is where we would, under standard conditions, measure a pressure of 500 mb. Observe that everywhere along this surface there are an equal number of molecules above it. This means that the level of constant height also represents a level of constant pressure. At every point on this **isobaric surface**, the height is 5600 m above sea level and the pressure is 500 mb. Within the air column, we could draw any number of horizontal slices, and each slice would represent both an isobaric and constant height surface. A map of any one of these surfaces would be blank, since there are no horizontal variations in either pressure or altitude.

If the air temperature should change in any portion of the column, the air density and pressure would change along with it (Fig. 12.5). Note that we have colder air to the north and warmer air to the south. To simplify this situation, we will keep the surface pressure from varying as the density changes. To do this, the total number of molecules in the column above each region must remain constant. Therefore, in the cold, more-dense air the column contracts, while in the warm, less-dense air the column expands.

Now, in Fig. 12.5 look at the surface of constant 500-mb pressure extending from south to north. Observe that in the warm air at the extreme southern end of the diagram a pressure of 500 mb is observed at an altitude of 5800 m, considerably above the average height of this pressure level (5600 m). In this region, note that the pressure at 5600 m will be greater than 500 mb. From this, we can conclude that when there is warm air aloft, constant pressure surfaces are typically found at higher elevations than normal, and the pressure at any one altitude is considered higher than normal. Stated another way: *High elevations on an isobaric chart correspond to higher-than-normal pressures at any given altitude.*

Use the same reasoning to examine the colder air in the north end of Fig. 12.5. Here, the air column has shrunk and the height of the 500-mb surface is at 5500 m, lower than the "normal" altitude of 5600 m. If we move upward from the 500-mb level to an altitude of 5600 m, we would find the air pressure to be lower than 500 mb. From this, we can conclude that when there is cold air aloft, constant pressure surfaces are typically found at lower altitudes than normal, and the pressure at each level is lower than normal. In other words: *Low elevations on an isobaric chart*

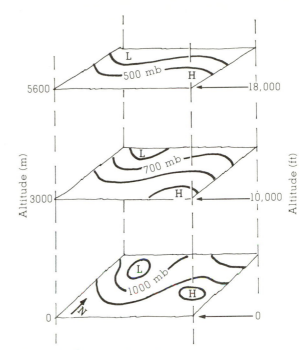

Fig. 12.3   Each map shows isobars on a constant height chart. The isobars represent variations in horizontal pressure at that altitude. An average isobar at sea level would be about 1000 mb; at 3000 m, about 700 mb; and at 5600 m, about 500 mb.

correspond to lower-than-normal pressures at any given altitude.

The variations in height of the constant pressure surface in Fig. 12.5 are shown in Fig. 12.6. Note that where the constant altitude lines intersect the 500-mb pressure surface, **contour lines**

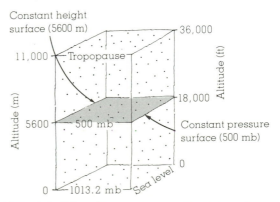

Fig. 12.4   When there are no horizontal variations in pressure, constant pressure surfaces are parallel to constant height surfaces. In the diagram, a measured pressure of 500 mb is 5600 m above sea level everywhere.

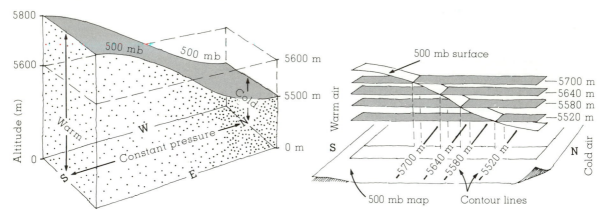

Fig. 12.5 Because of the changes in air density, a surface of constant pressure (shaded area) rises in warm air and lowers in cold air. Where the horizontal temperature changes most quickly, the constant pressure surface changes elevation most rapidly.

Fig. 12.6 Changes in elevation of a constant pressure surface (500 mb) show up as contour lines on a constant pressure (500 mb) map. Where the surface dips most rapidly, the lines are closer together.

(lines connecting points of equal elevation) are drawn on the 500-mb map below. Each contour line, of course, tells us the altitude above sea level at which one obtains a pressure reading of 500 mb. In the warmer air to the south, the elevations are high, while in the cold air to the north, the elevations are low. The contour lines are crowded together in the middle of the chart, where the pressure surface dips rapidly due to the changing air temperature. Where there is little horizontal temperature change, there are also few contour lines.

On upper-air charts, contour lines and isobars usually decrease in value from south to north be-

cause the average air temperature aloft typically decreases in that direction. The lines, however, are not straight, they bend and turn, indicating *ridges* where the air is warm and indicating depressions, or *troughs*, where the air is cold. In Fig. 12.7, we can see how the wavy contours on the map relate to the changes in altitude of the pressure surface.

A 500-mb map averaged for the month of January is shown in Fig. 12.8. The heights of the contour lines are given in meters. The difference in elevation between each contour line (called the *contour interval*) is 60 m. As we would expect, the average height of the 500-mb level decreases as we move north. Superimposed on this map are dashed lines, which represent lines of equal temperature (isotherms). Observe how the contour lines tend to parallel the isotherms.

Although we have examined only the 500-mb chart, other isobaric charts are commonly used. Table 12.1 lists these charts and their approximate heights above sea level.

Upper-level charts are a valuable tool. As we will see, they show wind-flow patterns that are extremely important in forecasting the weather. They can also be used to determine the movement of weather systems and to predict the behavior of surface pressure areas. To the pilot of a small aircraft, a constant pressure chart can help determine whether the plane is flying at an altitude either higher or lower than its altimeter indicates.

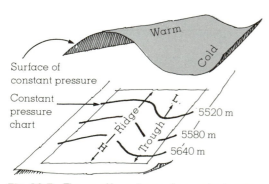

Fig. 12.7 The wavelike patterns of a constant pressure surface reflect the changes in air temperature. An elongated region of warm air aloft shows up on an isobaric map as higher heights and a ridge; the colder air shows as lower heights and a trough.

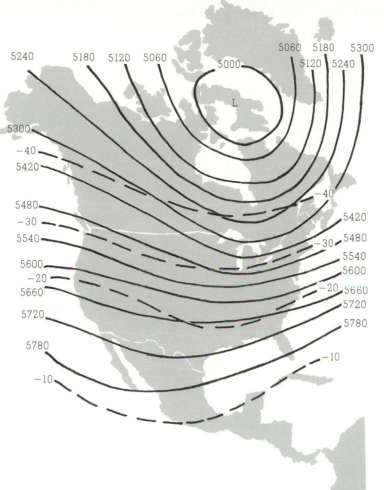

Fig. 12.8 Average 500-mb chart for January. Solid lines are contour lines in meters. Dashed lines are isotherms in °C.

(For more information on this topic, read the focus section ''Flying on a Constant Pressure Surface—High to Low, Look Out Below,'' p. 236.)

Now that we have examined surface and upper-air charts, we will be able to use them in our study of the forces that produce and affect winds.

Newton's Laws of Motion

Our understanding of why the wind blows stretches back through several centuries, with many scientists contributing to our knowledge. When we think of the movement of air, however, one great scholar stands out—Isaac Newton (1642–1727), who formulated several fundamental laws of motion.

Newton's first law of motion states that *an ob-*

ject at rest will remain at rest and an object in motion will remain in motion (and travel at a constant velocity along a straight line) as long as no force is exerted on the object. For exam-

Table 12.1 Common Isobaric Charts and Their Approximate Elevation above Sea Level

ISOBARIC SURFACE (mb)	APPROXIMATE ELEVATION	
	(m)	(ft)
1000	120	400
850	1,460	4,800
700	3,000	9,800
500	5,600	18,400
300	9,180	30,100
200	11,800	38,700
100	16,200	53,200
50	20,600	67,600

FOCUS ON AN APPLICATION

Flying on a Constant Pressure Surface—High to Low, Look Out Below

Aircraft that use pressure altimeters typically fly along a constant pressure surface rather than a constant altitude surface. They do this because the *altimeter* is simply an aneroid barometer calibrated to convert atmospheric pressure to an approximate elevation. The altimeter elevation indicated by an altimeter assumes the air temperature decreases at the standard lapse rate of 6.5°C/1000 m (3.6°F/1000 ft). Since the air temperature seldom, if ever, decreases at this rate, altimeters generally indicate an altitude different from their true elevation.

Figure 1 shows a standard column of air bounded on each side by air with a different temperature and density. On the left side, the air is warm; on the right, it is cold. The heavy dashed line represents a constant pressure surface of 700 mb as seen from the side. In the standard air, the 700-mb surface is located at 10,000 ft above sea level.

In the warm air, the 700-mb

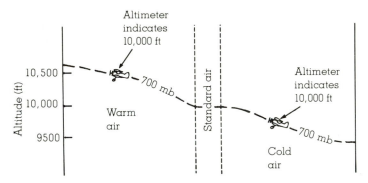

Fig. 1 An aircraft flying along a surface of constant pressure may change altitude as the air temperature changes. Without being corrected for the temperature change, a pressure altimeter will continue to read the same elevation.

surface rises; in the cold air, it descends. An aircraft flying along the 700-mb surface would be at an altitude less than 10,000 ft in the cold air, equal to 10,000 ft in the standard air, and greater than 10,000 ft in the warmer air. With no corrections for temperature, the altimeter would indicate the same altitude at all three positions because the air pressure does not change. We can see that, if no temperature corrections are made, an aircraft flying into warm air will increase in alti-

tude and fly higher than its altimeter indicates. Put another way: The altimeter inside the plane will read an altitude lower than the plane's true elevation.

Flying from standard air into cold air represents a potentially dangerous situation. As an aircraft flies into cold air, it flies along a lowering pressure surface. If no correction for temperature is made, the altimeter shows no change in elevation even though the aircraft is losing altitude; hence, the plane will

---

ple, a baseball in a pitcher's hand will remain there until a force (a push) acts upon the ball. Once the ball is pushed (thrown), it would continue to move in that direction forever if it were not for the force of air friction (which slows it down), the force of gravity (which pulls it toward the ground), and the catcher's mitt (which exerts an equal but opposite force to bring it to a halt). Similarly, to start air moving, to speed it up, to slow it down, or even to change its direction requires the action of an external force. This brings us to Newton's second law.

Newton's second law states that *the force exerted on an object equals its mass times the*

acceleration produced. In symbolic form, this law is written as

$$F = ma.$$

From this relationship we can see that, when the mass of an object is constant, the force acting on the object is directly related to the acceleration that is produced. A force in its simplest form is a push or a pull. Acceleration is the speeding up, the slowing down, or the changing of direction of an object. (More precisely, acceleration is the change of velocity* over a period of time.)

*Velocity specifies both the speed of an object and its direction of motion.

FOCUS ON AN APPLICATION, continued

be flying lower than the altimeter indicates. This can be a serious problem, especially for planes flying above mountainous terrain with poor visibility and where high winds and turbulence can reduce the air pressure drastically. To ensure adequate clearance under these conditions, pilots fly their aircraft higher than they normally would, consider air temperature, and compute a more realistic altitude by resetting their altimeters to reflect these conditions.

Even without sharp temperature changes, pressure surfaces may dip suddenly. This is especially true close to the ground (see Fig. 2.) An aircraft flying into an area of decreasing pressure will lose altitude unless corrections are made. For example, suppose a pilot has set the altimeter for sea level pressure above station A. At this location, the plane is flying along an isobaric surface at a true altitude of 500 ft. As the plane flies toward station B, the pressure surface (and the plane) dips but the altimeter continues to read 500 ft, which is too high. To cor-

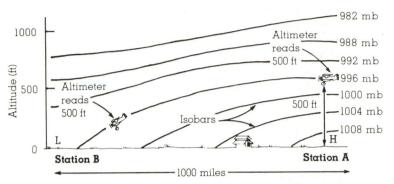

Fig. 2 In the absence of horizontal temperature changes, pressure surfaces can dip toward the surface. An aircraft flying along the pressure surface will either lose or gain altitude depending on the direction of flight.

rect for such changes in pressure, a pilot can radio the ground station for a current altimeter setting. With this additional information, the altimeter reading will more closely match the aircraft's actual altitude. Pilots, therefore, periodically radio for new altimeter settings while in flight and especially when landing.

Because of the inaccuracies inherent in the pressure altimeter, many high performance and commercial aircraft are now equipped with a radio altimeter. This device is like a small radar unit that measures

the altitude of the aircraft by sending out radio waves, which bounce off the terrain. The time it takes these waves to reach the surface and return is a measure of the aircraft's altitude. If used in conjunction with a pressure altimeter, a pilot can determine the variations in a constant pressure surface simply by flying along that surface and observing how the true elevation measured by the radio altimeter changes.

Because more than one force may act upon an object, Newton's second law always refers to the *net*, or total, force that results. An object will always accelerate in the direction of the total force acting on it. Therefore, to determine in which direction the wind will blow, we must identify and examine all of the forces that affect the horizontal movement of air. These forces include:

1. pressure gradient force
2. Coriolis force
3. centripetal force
4. friction

We will first study the forces that influence the flow of air aloft. Then we will see which forces modify winds near the ground.

Forces that Influence the Winds Aloft

We already know that horizontal differences in atmospheric pressure cause air to move and, hence, the wind to blow. Since air is an invisible gas, it may be easier to see how pressure differences cause motion if we examine a visible fluid, such as water.

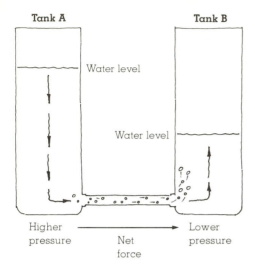

Fig. 12.9 The higher fluid pressure at the bottom of tank A creates a net force directed toward the lower fluid pressure at the bottom of tank B. This net force causes water to move from higher pressure toward lower pressure.

In Fig. 12.9, the two large tanks are connected by a pipe. Tank A is two-thirds full and tank B is only one-half full. Since the water pressure at the bottom of each tank is proportional to the weight of water above, the pressure at the bottom of tank A is greater than the pressure at the bottom of tank B. Moreover, since fluid pressure is exerted equally in all directions, there is a greater pressure in the pipe directed from tank A toward tank B than from B toward A.

Since pressure is force per unit area, there must also be a net force directed from tank A toward tank B. This force causes the water to

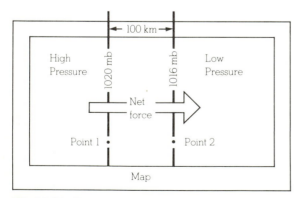

Fig. 12.10 The pressure gradient between point 1 and point 2 is 4 mb per 100 km. The net force directed from higher toward lower pressure is the pressure gradient force.

flow from left to right, from higher pressure toward lower pressure. The greater the pressure difference, the stronger the force, and the faster the water moves. In a similar way, horizontal differences in atmospheric pressure cause air to move.

Pressure Gradient Force Figure 12.10 shows a region of higher pressure on the map's left side, lower pressure on the right. The isobars show how the horizontal pressure is changing. If we compute the amount of pressure change that occurs over a given distance, we have the **pressure gradient**:

$$\text{Pressure gradient} = \frac{\text{difference in pressure}}{\text{distance}}.$$

If we let the symbol delta (Δ) mean "a change in," we can simplify the expression and write the pressure gradient as

$$PG = \frac{\Delta p}{d},$$

where Δp is the pressure difference between two places some horizontal distance (*d*) apart. In Fig. 12.10 the pressure gradient between points 1 and 2 is 4 mb per 100 km.

Suppose the pressure goes up in the region of high pressure, and the isobar at point 1 becomes 1024 mb. This would increase the pressure gradient to 8 mb per 100 km. If the increment of pressure between each isobar remains at 4 mb, then there would be 3 isobars over the space of 100 km. The crowding of isobars over a relatively small distance is termed a *steep* (or *strong*) *pressure gradient*. If the pressure in Fig. 12.10 changes such that the isobars are spread more widely apart, then the difference in pressure would be small over a relatively large distance. This condition is called a *gentle* (or *weak*) *pressure gradient*.

Notice in Fig. 12.10 that when differences in horizontal air pressure exist there is a net force acting on the air. This force, called the **pressure gradient force** (*PGF*), is directed from higher toward lower pressure at right angles to the isobars. The magnitude of the force is directly related to the pressure gradient. Steep pressure gradients correspond to strong pressure gradient forces and vice versa. Figure 12.11 shows the relationship between pressure gradient and pressure gradient force.

The *PGF* causes the wind to blow. Because of

this, closely spaced isobars on a weather chart indicate steep pressure gradients, strong forces, and high winds. On the other hand, widely spaced isobars indicate gentle pressure gradients, weak forces, and light winds.

If the *PGF* were the only force acting upon air, we would always find winds blowing directly from higher toward lower pressure. However, the moment air starts to move, it is deflected in its path by the Coriolis force.

Coriolis Force The Coriolis force describes an apparent force that is due to the rotation of the earth. To understand how it works, consider two people playing catch as they sit opposite one another on the rim of a merry-go-round (Fig. 12.12, platform A). If the merry-go-round is not moving, each time the ball is thrown, it moves in a straight line to the other person.

Suppose the merry-go-round starts turning counterclockwise—the same direction the earth spins as viewed from above the North Pole. If we watch the game of catch from above, we see that the ball moves in a straight-line path just as before. However, to the people playing catch on the merry-go-round, the ball seems to veer to its right each time it is thrown, always landing to the right of the point intended by the thrower (Fig. 12.12, platform B). This is due to the fact that, while the ball moves in a straight-line path, the merry-go-round rotates beneath it; by the time the ball reaches the opposite side, the catcher has moved. To anyone on the merry-go-round, it seems as if there is some force causing the ball to deflect to the right. This apparent force is called the **Coriolis force** after Gaspard Coriolis, a nineteenth-century French scientist who first worked it out mathematically. (Because it is an *apparent* force due to the rotation of the earth, it is also called the *Coriolis effect*.) This effect occurs on the rotating earth, too. All free-moving objects, such as ocean currents, aircraft, artillery projectiles, and air molecules seem to deflect from a straight-line path because the earth rotates under them.

The Coriolis force *causes the wind to deflect to the right of its path in the Northern Hemisphere and to the left in the Southern Hemisphere.** To illustrate this, consider a satellite in

*The Coriolis force only changes the *direction* of the wind, not the *speed* of the wind.

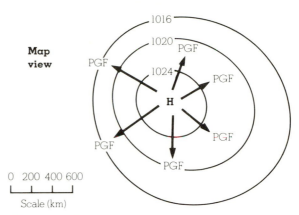

Fig. 12.11 The closer the spacing of the isobars, the greater the pressure gradient. The greater the pressure gradient, the stronger the pressure gradient force (*PGF*). The arrows represent the relative magnitude of the force, which is always directed from higher toward lower pressure.

polar circular orbit. If the earth were not rotating, the path of the satellite would be observed to move directly north-south, parallel to the earth's meridian lines. However, the earth *does* rotate, carrying us and meridians eastward with it. Because of this, in the Northern Hemisphere we see the satellite moving southwest instead of due south; it seems to veer off its path and move toward *its right*. In the Southern Hemisphere, the earth's direction of rotation is clockwise as viewed from above the South Pole. Consequently, a satellite moving northward from the South Pole would appear to move northwest and, hence, would veer to the *left* of its path.

The magnitude of the Coriolis force varies with the speed of the moving object and the latitude. Figure 12.13 shows this variation for various wind

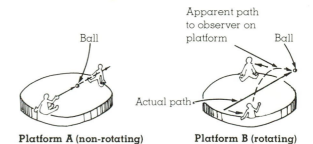

Platform A (non-rotating)   **Platform B (rotating)**

Fig. 12.12 On nonrotating platform A, the thrown ball moves in a straight line. On platform B, which rotates counterclockwise, the ball continues to move in a straight line. However, platform B is rotating while the ball is in flight; thus, to anyone on platform B, the ball appears to deflect to the right of its intended path.

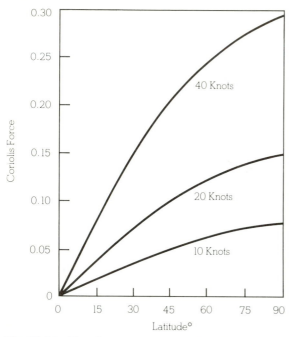

Fig. 12.13 The relative variation of the Coriolis force at different latitudes with different wind speeds.

speeds at different latitudes. In each case, as the wind speed increases, the Coriolis force increases; hence *the stronger the wind speed, the greater the deflection.* Also, note that the Coriolis force increases for all wind speeds from a value of zero at the equator to a maximum at the poles. We can see this latitude effect better by examining Fig. 12.14.

Imagine in Fig. 12.14 that there are three planes, each at a different latitude and each flying along a straight-line path, with no external forces acting on them. The destination of each aircraft is due east and is marked on the diagram (Fig. 12.14a). Each plane travels in a straight path relative to an observer positioned at a fixed spot in space. The earth rotates beneath the moving planes, causing the destination points at latitudes 30° and 60° to change direction slightly (to the observer in space) (Fig. 12.14b). To an observer standing on the earth, however, it is the plane that appears to deviate. The amount of deviation is greatest toward the pole and nonexistent at the equator. Therefore, the Coriolis force has a far greater effect on the plane at high latitudes (large deviation) than on the plane at low latitudes (small deviation). On the equator, it has no effect at all. The same is true of its effect on winds.

In summary, to an observer on the earth, moving objects deflect from their intended path. The amount of deflection depends upon:

1. the rotation of the earth
2. the latitude
3. the object's speed*

In addition, the Coriolis force acts at right angles to the wind, only influencing wind direction and never wind speed.

The Coriolis "force" behaves as a real force, constantly tending to "pull" the wind to its right in the Northern Hemisphere and to its left in the Southern Hemisphere. Moreover, this effect is present in all motions relative to the earth's surface. However, in most of our everyday experiences, the Coriolis force is so small that it is negligible and, contrary to popular belief, does not cause water to turn clockwise or counterclockwise when draining from a sink (the shape of the sink actually plays a much larger role). The Coriolis force is also minimal on small-scale winds, such as those that blow inland along coasts in summer. Only where winds blow over vast regions is the effect significant.

Wind Flow Aloft

We now know that the *PGF* is the driving force behind the wind and that the Coriolis force influences wind direction only. Let's examine these two factors to see how they produce the observed flow of air aloft. We will look at the wind flow above the *friction layer*—a level typically 1000 m (3300 ft) above the ground. (The friction layer is discussed in detail in Chapter 13.)

Geostrophic Wind   Figure 12.15 shows a map of the Northern Hemisphere, with horizontal pressure variations at an altitude of about 1 km. The evenly spaced isobars indicate a constant *PGF* directed from south toward north as indicated by

*These three factors are grouped together and shown in the expression

Coriolis force = $2\Omega V \sin \phi$,

where Ω is the earth's angular rate of spin (a constant), V is the speed of the object, and ϕ is the latitude.

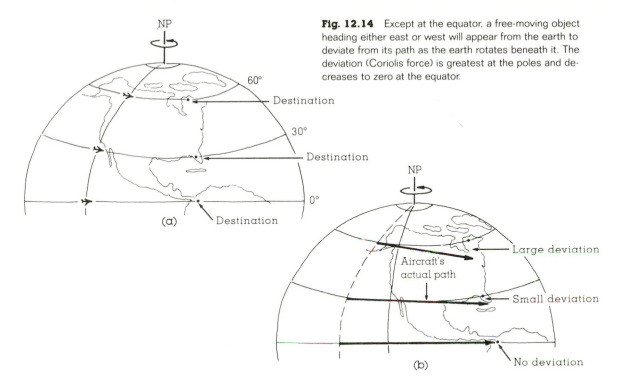

Fig. 12.14 Except at the equator, a free-moving object heading either east or west will appear from the earth to deviate from its path as the earth rotates beneath it. The deviation (Coriolis force) is greatest at the poles and decreases to zero at the equator.

the arrow at the left. Why, then, does the map show a west wind? We can answer this question by placing a parcel of air at position 1 in the diagram and watching its behavior.

At position 1, the *PGF* acts immediately upon the parcel, accelerating it northward toward lower pressure. However, the instant the air begins to move, the Coriolis force deflects the air towards its right, curving its path. As the parcel of air increases in speed (positions 2, 3, and 4), the magnitude of the Coriolis force increases (as shown by the longer arrows), bending the wind more and more to its right. Eventually, the wind speed increases to a point where the Coriolis force just balances the *PGF*. At this point (position 5), the wind no longer accelerates because the net force is zero. Here the wind flows in a straight path, parallel to the isobars at a constant speed.* This flow of air is called a **geostrophic** (*geo*: earth;

strophic: turning) **wind**. Notice that the geostrophic wind blows in the Northern Hemisphere with lower pressure to its left and higher pressure to its right.

When the flow of air is purely geostrophic, the isobars are straight and evenly spaced, and the wind speed is constant. In the atmosphere, isobars are rarely straight or evenly spaced, and the wind normally changes speed as it flows along. So, the geostrophic wind is usually only an ap-

*At first, it may seem odd that the wind blows at a constant speed with no net force acting on it. But when we remember that the net force is necessary only to accelerate ($F = ma$) the wind, it makes more sense. For example, it takes a considerable net force to push a car and get it rolling from rest. But once the car is moving, it only takes a force large enough to counterbalance friction to keep it going. There is no net force acting on the car, yet it rolls along at a constant speed.

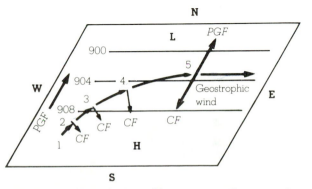

Fig. 12.15 Above the level of friction, air initially at rest will accelerate until it flows parallel to the isobars at a steady speed with the pressure gradient force (*PGF*) balanced by the Coriolis force (*CF*). Wind blowing under these conditions is called *geostrophic*.

FOCUS ON A SPECIAL TOPIC

A Closer Look at the Geostrophic Wind

We know from an earlier discussion that the geostrophic wind gives us a good approximation of the real wind above the level of friction, about 500 to 1000 m above the earth's surface. Above the friction layer, the winds blow more or less parallel to the isobars, or contours. We know that, for any given latitude, the speed of the geostrophic wind is proportional to the pressure gradient. This may be represented as:

$$V_g \sim \frac{\Delta p}{d},$$

where V_g is the geostrophic wind and Δp is the pressure difference between two places some horizontal distance (d) apart. From this, we can see that the greater the pressure gradient, the stronger the geostrophic wind.

When we consider a unit mass of moving air, we must take into account the air density (mass per unit volume) expressed by the symbol ρ. The geostrophic wind is now directly proportional to the pressure gradient force:

$$V_g \sim \frac{1}{\rho} \frac{\Delta p}{d}.$$

We can see from this expression that, with the same pressure gradient (at the same latitude), the geostrophic wind will increase with increasing elevation because air density decreases with height.

In a previous section, we saw that the geostrophic wind represents a balance of forces be-

tween the Coriolis force and the pressure gradient force. Here, it should be noted that the Coriolis force can be expressed as

Coriolis force = $2\Omega V \sin\phi$,

where Ω is the earth's angular spin (a constant), V is the speed of the wind, and ϕ is the latitude. The $\sin\phi$ is a trigonometric function that takes into account the variation of the Coriolis force with latitude. At the equator (0°) $\sin\phi$ is 0, at 30° latitude $\sin\phi$ is 0.5, and, at the poles (90°), $\sin\phi$ is 1.

This balance between the Coriolis force and the pressure gradient force can be written as

$$CF = PGF$$

$$2\Omega V_g \sin\phi = \frac{1}{\rho} \frac{\Delta p}{d}.$$

proximation of the real wind. However, the approximation is generally close enough to help us more clearly understand the behavior of the winds aloft.

The speed of the geostrophic wind is directly related to the pressure gradient. In Fig. 12.16, we can see that a geostrophic wind flowing parallel to the isobars is similar to water in a stream flowing parallel to its banks. At position 1, the geostrophic wind is blowing at a low speed; at position

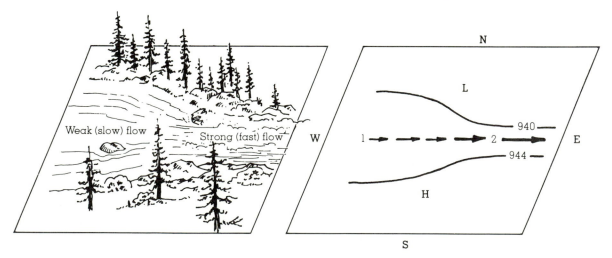

Fig. 12.16   The isobars and contours on an upper-level chart are like the banks along a flowing stream. When they are widely spaced, the flow is weak; when they are narrowly spaced, the flow is stronger.

FOCUS ON A SPECIAL TOPIC, continued

Solving for V_g, the geostrophic wind, the equation becomes

$$V_g = \frac{1}{2\Omega \sin\phi\rho} \frac{\Delta p}{d}. \quad (1)$$

Customarily, the rotational (2Ω) and latitudinal $(\sin\phi)$ factors are combined into a single value f, called the *Coriolis parameter*. Thus, we have the geostrophic wind equation written as

$$V_g = \frac{1}{f\rho} \frac{\Delta p}{d}, \quad (2)$$

Suppose we compute the geostrophic wind for the example given in Fig. 3. Here the wind is blowing parallel to the isobars in the Northern Hemisphere at latitude 40°. The spacing between the isobars is 200 km and the pressure difference is 4 mb. The altitude is 5600 m above sea

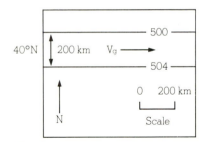

Fig. 3 A portion of an upper-air chart for part of the Northern Hemisphere at an altitude of 5600 m above sea level. The air temperature is −25°C and the air density is 0.70 kg/m³.

level, where the air temperature is −25°C (−13°F) and the air density is 0.70 kg/m³. First, we list our data and put them in the proper units:

$\Delta p = 4$ mb $= 400$ newton/m²
$d = 200$ km $= 2 \times 10^5$ m
$\sin\phi = \sin(40°) = 0.64$

$\rho = 0.70$ kg/m³
$2\Omega^* = 14.6 \times 10^{-5}$ radians/sec.

When we use equation (1) to compute the geostrophic wind, we obtain

$$V_g = \frac{1}{2\Omega \sin\phi\rho} \frac{\Delta p}{d}.$$

$$V_g = \frac{400}{14.6 \times 10^{-5} \times 0.64 \times 0.70 \times 2 \times 10^5}$$

$V_g = 30.6$ m/sec, or 59.4 knots.

*The rate of the earth's rotation (Ω) is 360° in one day, actually a sidereal day consisting of 23 hr, 56 min, 4 sec, or 86,164 seconds. This gives a rate of rotation of 4.18×10^{-3} degrees per second. Most often Ω is given in radians, where 2π radians equals 360° ($\pi = 3.14$). Therefore, the rate of the earth's rotation can be expressed as 2π radians/86,164 sec, or 7.29×10^{-5} radians/sec, and the constant 2Ω becomes 14.6 $\times 10^{-5}$ radians/sec.

2, the gradient increases and the wind speed picks up. Therefore, strong pressure gradients correspond to strong geostrophic winds; weak pressure gradients to weak geostrophic winds; and where there is no pressure gradient there is no wind.

In Fig. 12.17, we can see that the geostrophic wind direction can be determined by studying the orientation of the isobars; its speed can be estimated from the spacing of the isobars. On an isobaric chart, the geostrophic wind direction and speed are related in a similar way to the contour lines. Therefore, if we know the isobar or contour patterns on an upper-level chart, we also know the direction and relative speed of the geostrophic wind, even for regions where no direct wind measurements have been made. Similarly, if we know the geostrophic wind direction and speed, we can estimate the orientation and spacing of the isobars, even if we don't have a current weather map.

Thus far, we have seen that the winds aloft do not always blow in a straight line; frequently, they curve and bend into meandering loops as they tend to follow the patterns of the isobars. In the

Northern Hemisphere, winds blow counterclockwise around lows and clockwise around highs. The next section explains why.

Winds around Lows and Highs Look at the wind flow around the upper-level low (Northern Hemisphere) in Fig. 12.18a. At first it appears as though the wind is defying the Coriolis force by bending to the left as it moves counterclockwise

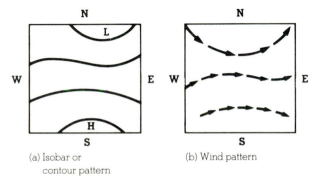

(a) Isobar or contour pattern

(b) Wind pattern

Fig. 12.17 By observing the orientation and spacing of the isobars (or contours) in diagram a, the geostrophic wind direction and speed can be determined for diagram b.

FOCUS ON OBSERVATION

Estimating Wind Direction and Pressure Patterns Aloft

Both the geostrophic wind direction and the orientation of the isobars aloft can be estimated by observing middle- and high-level clouds from the earth's surface. Suppose, for example, we are in the Northern Hemisphere watching clouds directly above us move from southwest to northeast at an elevation of 3000 m (see Fig. 4a). This indicates that the geostrophic wind at this level is southwesterly. Looking downwind, the geostrophic wind blows parallel to the isobars with lower pressure on the left and higher pressure on the right. Thus, if we stand with our backs to the direction from which the clouds are moving, lower pressure aloft will always be to our left and higher pressure to our right. From this observation, we can draw a rough upper-level chart (Fig. 4b), which shows isobars and wind direction for an elevation of approximately 3000 m.

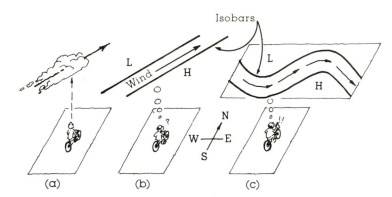

Fig. 4 This drawing of a simplified upper-level chart is based on cloud observations. Upper-level clouds moving from the southwest (a) indicate isobars and winds aloft (b). When extended horizontally, the upper-level chart appears as in (c), where lower pressure is to the northwest and higher pressure is to the southeast.

The isobars aloft will not continue in a southwest-northeast direction indefinitely; rather, they will often bend into wavy patterns. We may carry our observation one step farther, then, by assuming a bending of the lines (Fig. 4c). Thus, with a southwesterly geostrophic wind aloft, a trough of low pressure will be found to our west and a ridge of high pressure to our east. What would be the pressure pattern if the winds aloft were blowing *from* the northwest? Answer: A trough would be to the east and a ridge to the west.

---

around the system. Let's see why the wind blows in this manner.

Suppose we consider a parcel of air initially at rest at position 1. The *PGF* accelerates the air inward toward the center of the low and the Cor-

iolis force deflects the moving air to its right, until the air is moving parallel to the isobars at position 2. If the wind were geostrophic, at position 3 the air would move northward parallel to straight-line isobars at a constant speed. The wind is blowing

Fig. 12.18 Winds and related forces around areas of low and high pressure above the friction level in the Northern Hemisphere.

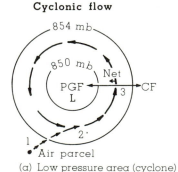

(a) Low pressure area (cyclone)

(b) High pressure area (anticyclone)

at a constant speed, but parallel to curved isobars. A wind that blows at a constant speed parallel to *curved isobars* above the level of frictional influence is termed a **gradient wind**.

Earlier in this chapter we learned that an object accelerates when there is a change in its speed or direction (or both). Therefore, the gradient wind blowing *around* the low-pressure center is constantly accelerating because it is constantly changing direction. This acceleration, called the **centripetal acceleration**, is directed at right angles to the wind, inward toward the low center.

Remember from Newton's second law that, if an object is accelerating, there must be a net force acting on it. In this case, the net force acting on the wind must be directed toward the center of the low, so that the air will keep moving in a circular path. This inward-directed force is called the **centripetal force*** (*centri*: center; *petal*: to push toward) and results from an imbalance between the Coriolis force and the pressure gradient force.†

Again, look closely at position 3 (Fig. 12.18a) and observe that the inward directed *PGF* is greater than the outward-directed Coriolis force (*CF*). The difference between these two forces—the net force—is the inward-directed centripetal force. In Fig. 12.18b, the wind around the high blows clockwise around its center. To keep the wind blowing in a circle, the inward-directed Coriolis force must be greater in magnitude than the outward-directed

*The magnitude of the centripetal force is related to the wind velocity (*V*) and the radius of the wind's path (*r*) by the formula

Centripetal force $= \dfrac{V^2}{r}$.

Where wind speeds are light and there is little curvature (large radius), the centripetal force is weak and, compared to other forces, may be considered insignificant. However, where the wind is strong and blows in a tight curve (small radius), as in the case of tornadoes and tropical hurricanes, the centripetal force is large, and becomes quite important.

†In some cases, it is more convenient to express the centripetal force (and the centripetal acceleration) as the *centrifugal force*, an apparent force that is equal in magnitude to the centripetal force, but directed outward from the center of rotation. The gradient wind is then described as a balance of forces between the centrifugal force $\dfrac{V^2}{r}$, the pressure gradient force $\dfrac{1}{\rho}\dfrac{\Delta p}{d}$ and the Coriolis force $2\Omega V \sin\phi$. Under these conditions, the gradient wind equation is expressed as

$$\frac{V^2}{r} + \frac{1}{\rho}\frac{\Delta p}{d} + 2\Omega V \sin\phi = 0.$$

PGF, so that the centripetal force (again, the net force) is directed inward.

In the Southern Hemisphere, the *PGF* starts the air moving and the Coriolis force deflects it to the left, thereby causing the wind to blow clockwise around lows and counterclockwise around highs.

Near the equator, where the Coriolis force is minimum, winds may blow around intense tropical storms with the centripetal force being almost as large as the *PGF*. In this type of flow, the Coriolis force is considered negligible, and the wind is called *cyclostrophic*.

So far we have seen how winds blow in theory, but how do they appear on an actual map?

Winds on Upper-Level Charts   Figure 12.19 shows a 500-mb chart with contour lines (heavy lines), isotherms (dashed lines), and winds for September 5, 1978. As we would expect at this level, the regions of lowest elevations (lowest pressures) are associated with the coolest air and the highest elevations (highest pressures) with the warmest air. As an example, compare the cold −25°C isotherm within the low off the Washington-Oregon coast with the relatively warm −10°C isotherm within the ridge over the south Alaska coast.

The wind directions on the map are given by lines that parallel the wind. The wind speeds are indicated by barbs and flags: Half a barb ( ) indicates a 5-knot wind; a full barb ( ), a 10-knot wind, and a flag ( ), a 50-knot wind. Note that the wind generally blows parallel to the contour lines, counterclockwise around the lows and clockwise around the highs. The strength of the wind is directly related to the magnitude of the contour gradient. (This has the same meaning as the pressure gradient.) Observe the close spacing of the contour lines over the New England states and the corresponding high winds. In the high-pressure area, where gradients are weak, the winds are light.

In a region where the winds blow in a west-east direction, paralleling lines of latitude, the wind flow is termed **zonal**. The winds in Fig. 12.19 are approximately zonal from south-central Canada to the eastern Canadian coastline. Because the winds aloft generally blow from west to east, planes

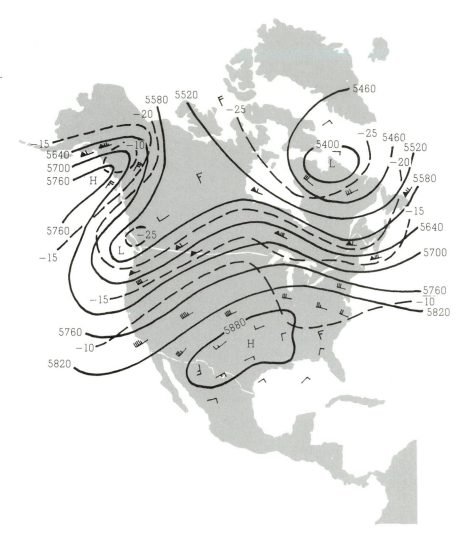

Fig. 12.19 The 500-mb chart for September 5, 1978. Solid lines are contours (meters above sea level). Dashed lines are isotherms in °C.

flying in this direction have a beneficial tail wind, explaining why a flight from San Francisco to New York City takes about 30 minutes less than the return flight. If the flow aloft is zonal, clouds, storms, and surface anticyclones tend to move rapidly from west to east.

When the wind flows in large, looping meanders, following a more north-south trajectory paralleling the meridian lines, the flow is termed **meridional**. Meridional flow can be seen in Fig. 12.19 along the west coast of North America. Here, cold polar air is brought down on the western side of the low as warm tropical air moves north ahead of it. During periods of meridional flow, surface storms tend to move slowly, often intensifying into major storm systems. (Chapter 17 details this topic.)

Moving from south to north, we can see that the contour lines decrease in elevation. We expect this, since the average air temperature is warmer to the south than it is farther north. Where horizontal temperature contrasts are large, there are also large height gradients and strong winds. We can also see that when the flow is zonal, cold air and lower heights are observed to the north of the wind. Where there is meridional flow, however, there is a gradient of height as well as temperature that runs in a west-east direction. Notice the change in height and temperature when moving eastward from the Washington coast to the tip of northern Idaho. In general, the north-south temperature gradient is strongest in winter. This is why the winds aloft are usually much stronger in winter than in summer.

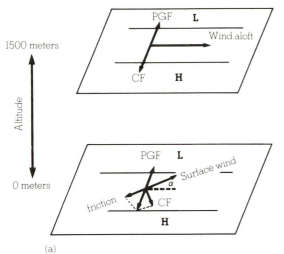

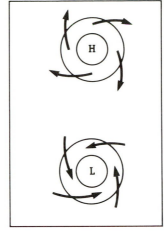

(a)

(b)   Surface map
Northern Hemisphere

Fig. 12.20 The effect of surface friction is to slow down the wind so that, near the ground, the wind crosses the isobars and blows toward lower pressure. This phenomenon produces an outflow of air around a high and an inflow around a low.

Surface Winds

Winds on a surface weather map do not blow exactly parallel to the isobars; instead, they cross the isobars, moving from higher to lower pressure. The reason for this behavior is due to friction.

In Fig. 12.20a, the wind is blowing at a level above the frictional influence of the ground. At this level (typically above 1 km), the wind is approximately geostrophic and blows parallel to the isobars, with the *PGF* on its left balanced by the Coriolis force on its right. At the earth's surface, the same pressure gradient will not produce the same wind speed, and the wind will not blow in the same direction.

Near the surface, *friction reduces the wind speed, which in turn reduces the Coriolis force.* Consequently, the weaker Coriolis force no longer balances the *PGF*, and the wind blows across the isobars toward lower pressure. The *PGF* is now balanced by the sum of the frictional force and the Coriolis force. Therefore, in the Northern Hemisphere, we find surface winds blowing counterclockwise and *into* a low; they flow clockwise and *out* of a high (Fig. 12.20b).

In the Southern Hemisphere, winds blow clockwise and inward around surface lows; counterclockwise and outward around surface highs. See the surface weather map and the general wind flow pattern on December 20, 1971, for South America (Fig. 12.21).

In Fig. 12.20a, the angle (α) at which the wind crosses the isobars to a large degree depends upon the roughness of the terrain. Everything else being equal, the rougher the surface, the larger the angle. Over hilly land, for example, the angle might average between 35° and 40°, while over an open body of relatively smooth water it may

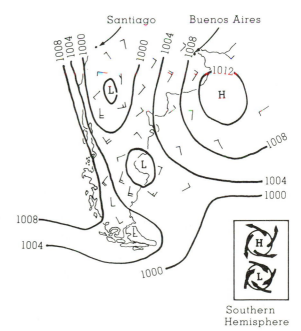

Southern
Hemisphere

Fig. 12.21 Surface weather map showing isobars and winds on December 20, 1971, in South America. The insert shows the idealized flow around surface pressure systems in the Southern Hemisphere.

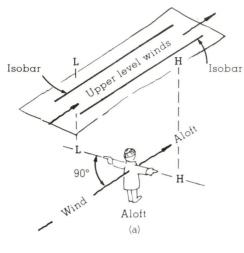

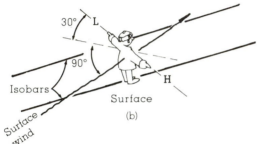

Fig. 12.22 In the Northern Hemisphere, if you stand with the wind aloft at your back, lower pressure aloft will be to your left and higher pressure to your right (a). At the surface, the same relationship holds if, with your back to the surface wind, you turn clockwise about 30° (b).

average between 10° and 15°. Taking into account all types of surfaces, the average is near 30°. This angle also depends on the wind speed. Typically, the angle is smallest for high winds and largest for gentle breezes. As we move upward through the friction layer, the wind becomes more and more parallel to the isobars.

So far, we have seen that, because of friction, surface winds move more slowly than geostrophic winds with the same pressure gradient. Surface winds also blow across the isobars toward lower pressure. The angle at which the winds cross the isobars depends upon surface friction, wind speed, and the height above the surface. As we have seen, if we stand with the wind aloft to our backs, lower pressure will be to our left and higher pressure to our right in the Northern Hemisphere. The

same rule applies to the surface wind but with a slight modification due to the fact that here the wind crosses the isobars. Look at Fig. 12.22b and notice that, at the surface, if we stand with our backs to the wind, then turn clockwise about 30°, lower pressure will be to our left. (In the Southern Hemisphere, if we stand with our backs to the wind, then turn counterclockwise about 30°, lower pressure will be to our right.) This relationship between wind and pressure is often called **Buys-Ballot's Law**, after the Dutch meteorologist Christoph Buys-Ballot (1817–1890) who formulated it.

Summary

This chapter has given us a broad view of how and why the wind blows. We examined constant pressure charts and found that low heights correspond to low pressure and high heights to high pressure. In regions where the air aloft is cold, the air pressure is normally lower than average; where the air aloft is warm, the air pressure is normally higher than average. Where horizontal variations in temperature exist, there is a corresponding horizontal change in pressure. The difference in pressure establishes a force, the pressure gradient force (PGF), which starts the air moving from higher toward lower pressure.

Once the air is set in motion, the Coriolis force bends the moving air to the right of its intended path in the Northern Hemisphere and to the left in the Southern Hemisphere. Above the level of surface friction, the wind is bent enough so that it blows nearly parallel to the isobars, or contours. Where the wind blows in a straight-line path, and a balance exists between the PGF and the Coriolis force, the wind is termed geostrophic. Where the wind blows parallel to curved isobars (or contours), the centripetal acceleration becomes important, and the wind is called a gradient wind.

The interaction of the forces causes the winds aloft in the Northern Hemisphere to blow clockwise around regions of high pressure and counterclockwise around areas of low pressure. In the Southern Hemisphere, the winds aloft blow counterclockwise around highs and clockwise around lows. The effect of surface friction is to slow down

FOCUS ON A SPECIAL TOPIC

Why Doesn't Air Rush Off Into Space?

We know that air moves in response to pressure differences. Because air pressure decreases rapidly with increasing height above the surface, there is always a strong *PGF* directed upward. Why, then, doesn't the air rush off into space?

Air does not rush off into space because the upward-directed *PGF* is nearly always exactly balanced by the downward force of gravity. When these two forces are in exact balance, the air is said to be in **hydrostatic equilibrium**.

When air is in hydrostatic equilibrium, there is no net vertical force acting on it, and so there is no net vertical acceleration. When air is not in hydrostatic balance (such as in violent thunderstorms and tornadoes), it shows appreciable vertical accelerations, but these occur over relatively small vertical distances, considering the total ver

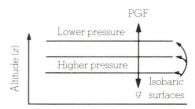

Fig. 5   When the vertical pressure gradient force (*PGF*) is in balance with the force of gravity (*g*), the air is in hydrostatic equilibrium.

tical extent of the atmosphere.

Figure 5 shows air in hydrostatic equilibrium. Since there is no net vertical force acting on the air, there is no net vertical acceleration, and the sum of the forces is equal to zero. This is represented by:

$$PGF_{vertical} + g = 0$$
$$\frac{1}{\rho}\frac{\Delta p}{\Delta z} + g = 0$$

where ρ is the air density, Δp is the decrease in pressure along

a small change in height (Δz), and g is the force of gravity. This expression is usually given as

$$\frac{\Delta p}{\Delta z} = -\rho g.$$

This equation is called the **hydrostatic equation**. The hydrostatic equation tells us that the rate at which the pressure decreases with height is equal to the air density times the acceleration of gravity (where ρg is actually the force of gravity per unit volume). The minus sign indicates that, as the air pressure decreases, the height increases. When the hydrostatic equation is given as

$$\Delta p = -\rho g\Delta z,$$

it tells us something important about the atmosphere that we learned earlier: The air pressure decreases more rapidly with height in cold (more dense) air than it does in warm (less dense) air.

---

the wind. This causes the surface air to blow across the isobars from higher pressure towards lower pressure. Consequently, in both hemispheres, surface winds blow outward, away from the center of a high, and inward, toward the center of a low.

Questions for Review

1. Explain how sea level pressure differs from station pressure.

2. Why must pressure measurements be made at the same time of day if they are to be compared?

3. What do contour lines on a constant pressure chart represent?

4. Explain why:
(a) in the Northern Hemisphere, the average height of contour lines on an upper-level isobaric chart tend to decrease northward.
(b) upper-level winds generally blow from the west.

5. What initially sets the air in motion?

6. Explain how each of the following influences the Coriolis force: (a) rotation of the earth; (b) wind speed; (c) latitude.

7. Explain how the wind aloft can be strong when the horizontal pressure gradient force just balances the Coriolis force.

8. What is a geostrophic wind?

9. Why would you *not* expect to observe a geostrophic wind at the equator?

10. How does gradient flow differ from geostrophic flow?

11. Describe how the wind blows around highs and lows aloft and near the surface (a) in the Northern Hemisphere and (b) in the Southern Hemisphere.

12. What are the forces that affect the horizontal movement of air?

13. What factors influence the angle at which surface winds cross the isobars?

14. How does Buys-Ballot's Law help to locate regions of high and low pressure aloft and at the surface?

15. The change in pressure over a given vertical distance near the surface is often 10,000 times greater than the horizontal pressure changes over the same distance. Yet, the air seldom accelerates upward. Why not?

Questions for Thought

1. A station 300 m above sea level reports a station pressure of 994 mb. What would be the sea level pressure for this station, assuming standard atmospheric conditions? If the observation were taken on a hot, summer afternoon, would the sea level pressure be greater or less than that obtained during standard conditions? Explain.

2. Pilots often use the expression "high to low, look out below." In terms of upper-level temperature and pressure, explain what this can mean.

3. Suppose an aircraft using a pressure altimeter flies along a constant pressure surface from standard temperature into warmer-than-standard air without any corrections. Explain why the altimeter would indicate an altitude lower than the aircraft's true altitude.

4. With the aid of a diagram, explain this relationship: Aloft, high and low altitudes on a constant pressure chart represent high and low pressures on a constant height chart.

5. If the earth were not rotating, how would the wind blow with respect to centers of high and low pressure?

6. Why are surface winds that blow over the ocean closer to being geostrophic than those that blow over the land?

7. If the wind aloft is blowing parallel to curved isobars, with the horizontal pressure gradient force being of greater magnitude than the Coriolis force, would the wind flow be cyclonic or anticyclonic? In this example, what would be the relative magnitude of the centripetal force, and how would it be directed?

8. With the present outside surface wind, use Buys-Ballot's Law to determine where regions of surface high- and low-pressure areas are located. If clouds are moving overhead, use the relationship to locate regions of higher and lower pressure aloft.

9. If you live in the Northern Hemisphere and a region of surface low pressure is directly west of you, what would probably be the surface wind direction at your home? If an upper-level low is also directly west of your location, describe the probable wind direction aloft and the direction in which middle-type clouds would move. How would the wind direction and speed change from the surface to where the middle clouds are located?

10. In the Northern Hemisphere, you observe surface winds shift from N to NE to E, then to SE. From this observation, you determine that a west-to-east moving high-pressure area (anticyclone) has passed north of your location. Describe how you were able to come to this conclusion.

11. The Coriolis force causes winds to deflect to the right of their intended path in the Northern Hemisphere, yet around a surface low-pressure area winds blow counterclockwise, appearing to bend to their left. Explain why.

12. Why is it that, on the equator, winds may blow either counterclockwise or clockwise with respect to an area of low pressure?

Problems and Exercises

1. Suppose you are in the Northern Hemisphere watching altocumulus clouds 4000 m (13,000 ft) above you drift from the northeast. Draw the orientation of the isobars above you. Locate and mark regions of lowest and highest pressure on this map. Finish the map by drawing isobars and the upper-level wind flow pattern hundreds of kilometers in all directions from your position. Would this type of flow be zonal or meridional? Explain.

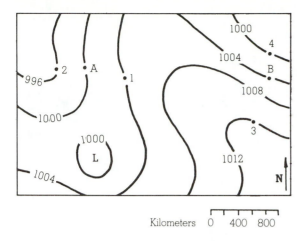

Kilometers 0   400 800

2. The map you see here is a sea level pressure chart (Northern Hemisphere), with isobars drawn for every 4 mb. Answer the following questions, which refer to this map.

(a) What is the lowest possible pressure in whole millibars that there can be in the center of the closed low? What is the highest pressure possible?

(b) Place a dashed line through the ridge and a dotted line through the trough.

(c) What would be the wind direction at point A and at point B?

(d) Where would the stronger wind be blowing, at point A or B? Explain.

(e) Compute the pressure gradient between points 1 and 2, and between points 3 and 4.

How do the computed pressure gradients relate to the pressure gradient force?

(f) If point A and point B are located at 30°N, and if the air density is 1.2 kg/m³, compute the geostrophic wind at point A and point B. (Hint: Be sure to convert km to m and mb to newton/m², where 1 mb = 100 N/m².)

(g) Would the actual winds at point A and point B be greater than, less than, or equal to the wind speeds computed in problem f? Explain.

3. (a) Suppose the atmospheric pressure at the bottom of a deep air column 5.6 km thick is 1000 mb. If the average air density of the column is 0.91 kg/m³, and the acceleration of gravity is 9.8 m/sec², use the hydrostatic equation to determine the atmospheric pressure at the top of the column. (Hint: Be sure to convert km to m and mb to newton/m², where 1 mb = 100 N/m².)

(b) If the air in the column of problem a becomes much colder than average, would the atmospheric pressure at the top of the new column be greater than, less than, or equal to the pressure computed in problem a? Explain.

(c) Determine the atmospheric pressure at the top of the air column in problem a if the air in the column is quite cold and has an average density of 0.97 kg/m³.

Windswept dunes cast their evening shadow at Great Sand Dunes National Monument,
Colorado. (Photo by author)

CHAPTER 13

Wind: Small-Scale Interactions with the Environment

The air in motion—what we commonly call *wind*—is invisible, yet we see evidence of it nearly everywhere we look. It sculptures rocks, moves leaves, blows smoke, and lifts water vapor upward to where it can condense into clouds. The wind is with us wherever we go. On a hot day, it can cool us off; on a cold day, it can make us shiver. A breeze can sharpen our appetite when it blows the aroma from the local bakery in our direction. The wind is a powerful element. The workhorse of weather, it moves storms and large fair-weather systems around the globe. It transports heat, moisture, dust, insects, bacteria, and pollens from one area to another.

Circulations of all sizes exist within the atmosphere. Little whirls form inside bigger whirls, which encompass even larger whirls—one huge mass of turbulent, twisting *eddies*. For clarity, meteorologists arrange circulations according to their size. This hierarchy of motion from tiny gusts to giant storms is called the *scales of motion*.

Contents

Scales of Motion

Consider smoke rising into the otherwise clean air from a chimney in the industrial section of a large city (Fig. 13.1a). Within the smoke, small chaotic motions—tiny eddies—cause it to tumble and turn. These eddies constitute the smallest scale of motion—the **microscale**. At the microscale, eddies with diameters of a few meters or less not only disperse smoke, they also sway branches and swirl dust and papers into the air. They form by convection, or by the wind blowing past obstructions and are usually short-lived, lasting only a few minutes at best.

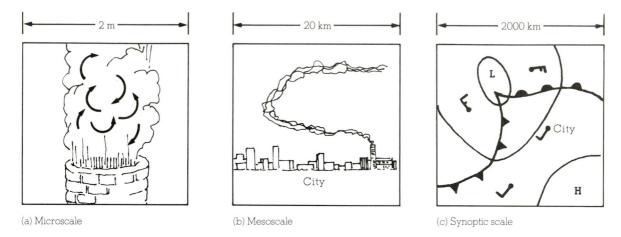

(a) Microscale   (b) Mesoscale   (c) Synoptic scale

Fig. 13.1   Scales of atmospheric motion. The tiny microscale motions constitute a part of the larger mesoscale motions, which, in turn, are part of the much larger synoptic scale. Notice that as the scale becomes larger, motions observed at the smaller scale are no longer visible.

In Fig. 13.1b observe that, as the smoke rises, it drifts toward the center of town. Here the smoke rises even higher and is carried back toward the industrial section. This circulation of city air constitutes the next larger scale—the **mesoscale** (meaning middle scale). Typical mesoscale winds range from a few kilometers to about a hundred kilometers in diameter. Generally, they last longer than microscale motions, often many minutes, hours, or in some cases as long as a day. Mesoscale circulations include local winds (which form along shorelines and mountains), as well as thunderstorms, tornadoes, and small tropical storms.

When we look at the smoke stack on a surface weather map (Fig. 13.1c), neither the smoke stack nor the circulation of city air shows up. All that we see are the circulations around high- and low-pressure areas. We are now looking at the **synoptic scale**, or weather map scale. Circulations of this magnitude dominate regions of hundreds to even thousands of square kilometers and, although the life spans of these features vary, they typically last for days and sometimes weeks.

The largest wind patterns are seen at the *planetary* or **global scale**. Here, we have wind patterns ranging over the entire earth. Sometimes, the synoptic and global scales are combined and referred to as the **macroscale**.

In this chapter, we will concentrate primarily on microscale winds and the effect they have on our environment.

Friction and Turbulence

We are all familiar with friction. If we rub our hand over the top of a table, friction tends to slow its movement because of irregularities in the table's surface. On a microscopic level, friction arises as atoms and molecules of the two surfaces seem to adhere, then snap apart as the hand slides over the table. Friction is not restricted to solid objects; it occurs in moving fluids as well. Consider, for example, a steady flow of water in a stream. When a paddle is placed in the stream, turbulent whirls called eddies form behind it. These eddies create fluid friction by draining energy from the main stream flow, slowing it down. Let's examine the idea of fluid friction in more detail.

The friction of fluid flow is called **viscosity**. When the slowing of a fluid—such as air—is due to the random motion of the gas molecules, the viscosity is referred to as **molecular viscosity**. Consider a mass of air gliding horizontally and smoothly (*laminar flow*) over a stationary mass of air. Even though the molecules in the stationary air are not moving horizontally, they are darting about and colliding with each other. At the boundary separating the air layers, there is a constant exchange of molecules between the stationary air and flowing air. The overall effect of this molecular exchange is to slow down the moving air. If molecular viscosity were the only type of friction acting on moving air,

Fig. 13.2 Winds flowing past an obstacle. In stable air, light winds produce small eddies and little vertical mixing (a). Greater winds in unstable air create deep, vertically mixing eddies that produce strong gusty surface winds (b).

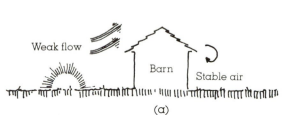

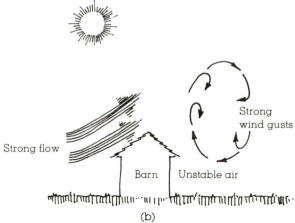

(a)   (b)

the effect of friction would disappear in a thin layer just above the surface. There is, however, another frictional effect that is far more important in reducing wind speeds.

When laminar flow gives way to irregular turbulent motion, there is an effect similar to molecular viscosity, but which occurs throughout a much larger portion of the moving air. The internal friction produced by turbulent whirling eddies is called **eddy viscosity**. Near the surface, it is related to the roughness of the ground. As wind blows over a landscape dotted with trees and buildings, it breaks into a series of irregular, twisting eddies that can influence the air flow for hundreds of meters above the surface. Within each eddy, the wind speed and direction fluctuate rapidly, producing the irregular air motion known as *wind gusts*. Eddy motions created by obstructions are commonly referred to as **mechanical turbulence**. Mechanical turbulence creates a drag on the flow of air, one far greater than that caused by molecular viscosity.

The frictional drag of the ground normally decreases as we move away from the earth's surface. Because of this, wind speeds tend to increase with height above the ground. In fact, at a height of only 10 m (33 ft), the wind is often moving twice as fast as at the surface. The atmospheric layer near the surface that is influenced by friction (turbulence) is called the **friction layer**. The top of the friction layer is usually near 1000 m (3300 ft), but this may vary somewhat since both strong winds and rough terrain extend the region of frictional influence.

Surface heating and instability also cause turbulence to extend to greater altitudes. As the earth's surface heats, thermals rise and convection cells form. The resulting vertical motion creates **thermal turbulence**, which increases with the intensity of surface heating and the degree of atmospheric instability. During the early morning, when the air is most stable, thermal turbulence is normally at a minimum. As surface heating increases, instability is induced and thermal turbulence becomes more intense. If this heating produces convective clouds that rise to great heights, there may be turbulence from the earth's surface to the base of the stratosphere.

Although we have treated thermal and mechanical turbulence separately, they occur together in the atmosphere—each magnifying the influence of the other. Let's consider a simple example: the eddy forming behind the barn in Fig. 13.2. In stable air with weak winds, the eddy is nonexistent or small. As wind speed and surface heating increase, instability develops, and the eddy becomes larger and extends through a greater depth. The rising side of the eddy carries slow moving surface air upward, causing a frictional drag on the faster flow of air aloft. Some of the faster moving air is brought down with the descending part of the eddy, producing a momentary gust of wind. Because of the increased depth of circulating eddies in unstable air, strong, gusty surface winds are more likely to occur when the atmosphere is unstable. Greater instability also leads to a greater exchange of faster moving air from upper levels with slower moving air at lower lev-

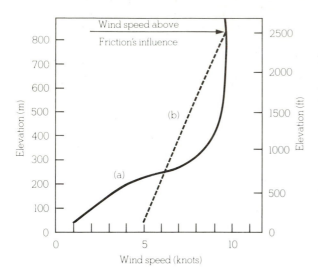

Fig. 13.3   When the air is stable and the terrain fairly smooth (a), vertical mixing is at a minimum, and the effect of surface friction only extends upward a relatively short distance above the surface. When the air is unstable and the terrain rough (b), vertical mixing is at a maximum, and the effect of surface friction extends upward through a much greater depth of atmosphere. Within the region of frictional influence, vertical mixing increases the wind speed near the ground and decreases it aloft. (Wind at the surface is measured at 10 m above the surface.)

els. In general, this exchange increases the average wind speed near the surface and decreases it aloft, producing the distribution of wind speed with height shown in Fig. 13.3.

We can now see why surface winds are usually stronger in the afternoon. Vertical mixing during the middle of the day links surface air with the faster moving air aloft. The result is that the surface air is pulled along more quickly. At night, when convection is reduced, the interchange between the air at the surface and the air aloft is at a minimum. Hence, the wind near the ground is less affected by the faster wind flow above, and so it blows more slowly.

In summary, the friction of air flow (viscosity) is a result of the exchange of air molecules moving at different speeds. The exchange brought about by random molecular motions (molecular viscosity) is quite small in comparison with the exchange brought about by turbulent motions (eddy viscosity). Therefore, the frictional effect of the surface on moving air depends largely upon mechanical

and thermal turbulent mixing. The depth of mixing and, hence, frictional influence (the friction layer) depend primarily upon three factors:

1. surface heating—producing steep lapse rate and strong thermal turbulence
2. strong wind speeds—producing strong mechanical turbulent motions
3. rough or hilly landscape—producing strong mechanical turbulence

When these three factors occur simultaneously, the frictional effect of the ground is transferred upward to considerable heights, and the wind at the surface is typically strong and gusty.

Eddies—Big and Small

When the wind encounters a solid object, an eddy forms on the object's leeward side. The size and shape of the eddy depends upon the size and shape of the obstacle and on the speed of the wind. Light winds produce small stationary eddies. Wind moving past trees, shrubs, and even your body produces small eddies. (You may have had the experience of dropping a piece of paper on a windy day only to have it carried away by a swirling eddy as you bend down to pick it up.) Air flowing over a building produces larger eddies that will, at best, be about the size of the building. Strong winds blowing past an open sports stadium can produce eddies that may rotate in such a way as to create surface winds on the playing field. These surface winds may move in a direction opposite to the wind flow above the stadium. Wind blowing over a fairly smooth surface produces few eddies, but when the surface is rough, many eddies form.

The eddies that form downwind from obstacles can produce a variety of interesting effects. For instance, wind moving over a mountain range in stable air with a speed greater than 40 knots usually produces waves and eddies, such as those shown in Fig. 13.4. We can see that eddies form both close to the mountain and beneath each wave crest. These are called *roll eddies*, or **rotors**, and have violent vertical motions that produce extreme turbulence and hazardous flying conditions. On a much smaller scale, the howling of

FOCUS ON AN APPLICATION

Eddies and ''Air Pockets''

Turbulent eddies form aloft as well as near the surface. Turbulence aloft can occur suddenly and unexpectedly, especially where the wind changes its speed or direction (or both) abruptly. Such a change is called a **wind shear**. The shearing creates forces that produce eddies along a mixing zone. To better understand these eddies, imagine that high in the atmosphere there is a stable layer of air having vertical wind speed shear as shown in Fig. 1. The top half of the layer slowly slides over the bottom half and the relative speed of both halves is low. As long as the wind shear between the top and bottom of the layer is small, few if any eddies form. However, if the shear and the corresponding relative speed of these layers increase, wavelike undulations may form. When the shearing exceeds a certain value, the waves break into large swirls, with significant ver-

tical movement. (See billow clouds, Fig. 10.13.) Eddies such as these often form in the upper troposphere near jet streams, where large wind speed shears exist. They also occur in conjunction with mountain waves, which may extend upward into the stratosphere. These huge eddies often develop in perfectly clear air; hence, this form of turbulence is referred to as **clear air turbulence (CAT)**.

The eddies that form in clear air may have diameters ranging from a couple of meters to several hundred meters. An unsuspecting aircraft entering such a region may be in for more than just a bumpy ride. If an aircraft flies into a zone of descending air, it may drop suddenly, producing the sensation that there is no air to support the wings. Consequently, these regions have come to be known as *air pockets*.

Commercial aircraft entering

an air pocket have dropped hundreds of meters, injuring passengers and flight attendants not strapped into their seats. For example, on April 4, 1981, a *DC-10* jetliner flying at 11,300 m (37,000 ft) over central Illinois encountered a region of severe clear air turbulence and reportedly plunged about 600 m (2000 ft) toward the earth before stabilizing. Twenty-one of the 154 people aboard were injured; one person sustained a fractured hip and another person, after hitting the ceiling, jabbed himself in the nose with a fork then landed in the seat in front of him. Clear air turbulence has occasionally caused structural damage to aircraft by breaking off vertical stabilizers and tail structures. Fortunately, the effects are usually not this dramatic.

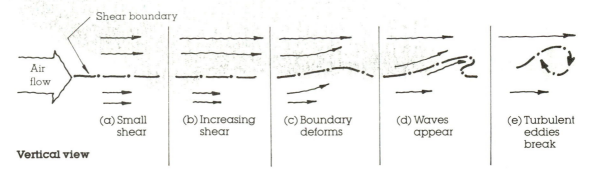

Vertical view

(a) Small shear
(b) Increasing shear
(c) Boundary deforms
(d) Waves appear
(e) Turbulent eddies break

Fig. 1 The formation of clear air turbulence (CAT) along a boundary of increasing wind speed shear.

wind on a blustery night is believed to be caused by eddies that are constantly being shed around obstructions, such as chimneys and roof corners. Whirling eddies create small areas of high air den-

sity. These tightly packed molecules act like a tiny pulse of compressed air, which pushes on the surrounding air molecules. These, in turn, push on other molecules and so on until the pulse of

Fig. 13.4 Under stable conditions, air flowing past a mountain range can create eddies many kilometers downwind from the mountain itself.

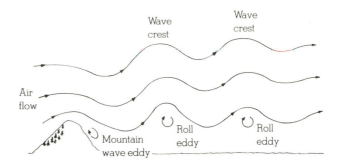

compressed air reaches your eardrums, producing a sound. On a windy night, one eddy after another forms around sharp corners with such regularity that the wind sounds as if it were howling.

The largest atmospheric eddies form as the flow of air becomes organized into huge spiraling whirls—the cyclones and anticyclones of middle latitudes—which can have diameters greater than 1000 km (600 mi). Since it is these migrating systems that make our middle latitude weather so changeable, we will examine the formation and movement of these systems in Chapters 16 and 17.

The Force of the Wind

Because a small increase in wind speed can greatly increase the wind force on an object, strong winds may blow down trees, overturn mobile homes, and even move railroad cars. Here are just a few examples to illustrate how powerful the wind really is. On November 17, 1869, in Columbia County, New York, high winds lifted a moving train from the tracks and hurled it down a 6 m embankment. In Nebraska, in 1894, strong winds propelled six

fully loaded coal cars over 160 km (100 mi) in a little more than three hours. Similarly, in February, 1965, the wind presented people in North Dakota with a "ghost train" as it pushed five railroad cars from Portal to Minot (about 125 km) without a locomotive. And, while the people in Mount Pleasant, Iowa, awaited the 1979 Fourth of July celebration, a phenomenally strong wind—estimated at 90 knots—blew the Goodyear blimp from its mooring and rolled it 300 meters into a corn field, where it came to rest in ruins.

Wind blowing with sufficient force to demolish the Goodyear blimp is uncommon. However, wind blowing with enough force to move an automobile is very common, especially when the automobile is exposed to a strong crosswind. On a normal road, the force of a crosswind is usually insufficient to move a car sideways because of the reduced wind flow near the ground. However, when the car crosses a high bridge, where the frictional influence of the ground is reduced, the increased wind speed can be felt by the driver. Near the top of a high bridge, where the wind flow is typically strongest, complicated eddies pound against the car's side as the air moves past obstructions, such as guard railings and posts. In a strong wind, these eddies may even break into extremely turbulent whirls that buffet the car, causing difficult handling as it moves from side to side. If there is a wall on the bridge, the wind may swirl around and strike the car from the side opposite the wind direction. A similar effect occurs where the wind moves over low hills paralleling a highway. (See Fig. 13.5.) When the vehicle moves by the obstruction, a wind gust from the opposite direction can suddenly and without warning push it to the opposite side. This wind hazard is a special problem for trucks, campers, and trailers, and highway signs warning of gusty wind areas are often posted.

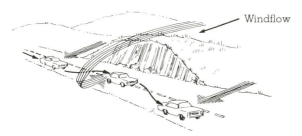

Fig. 13.5 Strong winds flowing past an obstruction can produce a reverse flow of air that strikes an object from the side opposite the general wind direction.

FOCUS ON A SPECIAL TOPIC

Pedaling into the Wind

Anyone who rides a bicycle knows that it is much easier to pedal with the wind than against it. The reason is obvious: As the wind blows against an object, it exerts a force upon it. The amount of force exerted by the wind over an area increases as the square of the wind velocity. This is shown by the relationship

$$F \sim V^2,$$

where F is the wind force and V is the wind velocity. From this, we can see that, if the wind velocity doubles, the force goes up by a factor of 2^2, or 4, which means that pedaling into a 40-knot wind requires 4 times as much effort as pedaling into a 20-knot wind.

Wind striking an object exerts a pressure on it. The amount of pressure depends upon the object's shape and size, as well as on the amount of reduced pressure that exists on the object's downwind side. Without concern for all the complications, we can approximate the wind pressure on an object with a simple formula. For example, if the wind velocity (V) is in miles per hour, and the wind force (F) is in pounds, and the object's surface area (A) is measured in square feet, the wind pressure (P), in pounds per square feet, is

$$\frac{F}{A} = P = 0.004\ V^2.$$

We can look at a practical example of this expression if we consider a bicycle rider going 10 mi/hr into a head wind of 40 mi/hr. With the total velocity of the wind against the rider (wind speed plus bicycle speed) being 50 mi/hr, the pressure of the wind is

$$P = 0.004\ V^2$$
$$P = 0.004\ (50^2)$$
$$P = 10\ \text{lb/ft}^2.$$

If the rider has a surface body area of 5 ft², the total force exerted by the wind becomes

$$F = P \times A$$
$$F = 10\ \text{lb/ft}^2 \times 5\ \text{ft}^2$$
$$F = 50\ \text{lb}.$$

This is enough force to make pedaling into the wind extremely difficult. To remedy this adverse effect, cyclists—especially racers—bend forward as low as possible in order to expose a minimum surface area to the wind.

Up to now we have seen that, when the wind meets a barrier, it exerts a force upon it. If the barrier doesn't move, the wind moves around, up, and over it. When the barrier is long and low like a water wave, the slight updrafts created on the windward side support the wings of birds, allowing them to skim the water in search of food without having to flap their wings. Elongated hills and cliffs that face into the wind create upward air motions that can support a hang-glider in the air for a long time. (The cliffs in the Hawaiian Islands and along the California coast with their steep escarpments are especially fine hang-gliding areas.) Wind speeds greater than about 15 knots blowing over a smooth yet moderately sloping ridge may provide excellent *ridge-soaring* for the sailplane enthusiast.

As stable air flows over a ridge, it increases in speed. Thus, winds blowing over mountains tend to be stronger than winds blowing at the same level on either side. In fact, the greatest wind speed ever recorded near the ground occurred at the summit of Mt. Washington, New Hampshire, elevation 1909 m (6262 ft), where the wind gusted to 200 knots on April 12, 1934. A similar increase in wind speed occurs where air accelerates as it funnels through a narrow constriction, such as a low pass or saddle in a mountain crest.

Microscale Winds Blowing Over the Earth's Surface We have already seen that winds can move objects on the ground. However, strange as it may seem, winds capable of moving large boulders are not capable of lifting fine clay and dust particles.* The reason is that, near the ground in a thin layer less than 0.1 mm thick, the wind speed is practically zero even when high wind speeds

*Diameters of clay- and dust-size particles are typically less than 0.004 mm.

Fig. 13.6 Ventifacts are rocks pitted, faceted, and polished by windblown sand.

exist only a few meters higher. How, then, do these very fine particles become airborne?

Sand grains large enough to protrude into the region of wind flow are pushed along when the wind speeds are sufficiently strong. Particles with diameters larger than 0.2 mm begin to move in a 13-knot wind, and larger sand particles whose diameters are about 2.0 mm are put in motion by a 26-knot wind. Once the grains begin to move, they roll and slide over the ground, with the smaller ones rolling faster than the larger ones. After moving a short distance, a grain may collide with a larger, stationary particle. This causes the smaller particle to jump into the air to a height of between 1 mm and several meters, depending upon the wind speed and the weight of the particle. This bouncing and skipping motion of sand grains is called **saltation**. Once airborne, the particles are swept along even more quickly by the wind. Eventually, they drop back to the ground, where they dislodge other particles, some being fine clay and dust too small to be picked up by the wind alone. If these tiny particles are caught in the wind-stream, they remain suspended for a long time and, in some cases, are carried over great distances.

Wind and Exposed Soil   Where the wind blows over exposed soil, it takes an active role in shaping the landscape. This is especially noticeable in deserts. As tiny, loose particles of sand, silt, and

dust are lifted from the surface and carried away by the wind, the ground level gradually lowers. The removal of this fine material leaves the surface covered with gravel and pebbles, which are too large to be transported by the wind. Such a landscape is termed **desert pavement**.

Because sand grains usually travel close to the ground, strong winds armed with sand can erode and shape rocks lying near the surface. If the rocks are soft and the winds are exceptionally strong (and prevail from one direction), the constant impact of hard quartz sand grains can wear solid rocks, notching the base of a cliff or a boulder. The constant sandblasting of rock fragments may cause a flattened and pitted surface on the windward side. Rocks shaped by the abrasive action of the wind are called **ventifacts** (Fig. 13.6). Two or more facets commonly form on a single rock if the wind blows from more than one direction or the rock is overturned, exposing another side. Even telephone poles have no immunity to desert winds and are often tapered near the base, where windblown sand has cut away the wood. And, finally, a motorist traveling through a wind-driven sandstorm knows all too well the disastrous effects of sandblasting on a car's paint job and windshield.

Blowing sand eventually comes to rest *behind* obstacles, which can be anything from a rock to a clump of vegetation. As the sand grains accumulate, they pile into a larger heap that, when high enough, acts as an obstacle itself. If the wind speed is strong and continues to blow in the same direction for a sufficient time, the sand piles up higher and eventually becomes a **sand dune**. On the dune's surface, the sand rolls, slides, and gradually creeps along, producing wavelike patterns called *sand ripples*. Each ripple forms perpendicular to the wind direction, with a gentle slope on the windward side and a steeper slope on the leeward side. (If the wind direction frequently changes, the ripple becomes more symmetric.) On a larger scale, the dune itself may take on this shape. Sand is carried forward and up the dune until it reaches the top. Here, the air flow is strongest, and the sand continues its forward movement and cascades down the backside of the dune into quieter air. The effect of this migration is to create a dune whose windward

Top view   Side view

Fig. 13.7 Windblown sand accumulating around an obstruction may grow into a barchan dune, which points away from the prevailing wind direction.

slope is more gentle than its leeward slope. Therefore, the shape of a sand dune reveals the prevailing wind direction during its formation.

Where there is a limited amount of sand—when the ground is flat and the wind direction fairly constant—crescent-shaped *barchan dunes* will form. In Fig. 13.7 notice how the tips of the barchan point away from the prevailing wind. In areas where sand is more abundant, barchan dunes may coalesce into *transverse dunes*, forming long, wavy ridges perpendicular to the prevailing wind.

Wind and Snow Surfaces Wind blowing over a snow-covered landscape may also create wavelike patterns several centimeters high, and oriented at right angles to the wind. These *snow ripples* are similar to sand ripples. On a larger scale, winds may create *snow dunes*, which are quite similar to sand dunes. Irregularities at the surface can cause a strong wind (40 knots) to break into turbulent eddies. If the snow on the ground is moist and sticky, some of it may be picked up by the wind and sent rolling. As it rolls along, it collects more snow and grows bigger. If the wind is sufficiently strong, the moving clump of snow becomes cylindrical, often with a hole extending through it lengthwise (Fig. 13.8). These **snow rollers** range from the size of eggs to that of small barrels. The tracks they make in the snow are typically less than 1 cm deep and several meters long. Snow rollers are rare, but, when they occur, they create a striking winter scene. In populated areas, they may escape notice as they are often mistaken as being made by children rather than by nature.

Strong winds blowing over a vast region of open plains can alter the landscape in a different way. Consider, for example, a light snowfall several centimeters deep covering a large portion of central South Dakota. After the snow stops falling, strong winds may whip it into the air, leaving fields barren of snow. The cold, dry wind also robs the soil of any remaining moisture and freezes it solid. Meanwhile, the snow settles out of the air when the wind encounters obstacles. Since the greatest density of such obstructions is normally in towns, municipal snowfall measurements may show an accumulation of many centimeters, while the surrounding countryside, which may desperately need the snow, has practically none.

To help remedy this situation, *snow fences* are constructed in open spaces (Fig. 13.9). Behind the snow fence, the wind speed is reduced because the air is broken into small eddies, which allow the snow to settle to the ground. Added snow cover is important for open areas because it acts like an insulating blanket that protects the ground from the bitter cold air, which often follows

Fig. 13.8 A snow roller.

Fig. 13.9 Snow accumulating behind snow fences in Wyoming.

in the wake of a major snowstorm. In regions of low rainfall, moisture from the spring snowmelt can be a critical factor during long, dry summers. Snow fences are also built to protect major highways in these areas. Hopefully, the snow will accumulate behind the fence rather than in huge drifts on the road.

Wind and Vegetation We have already examined how wind affects the landscape in regions

where the amount of exposed vegetation is limited. What can happen to vegetation in regions of high winds? Strong, prevailing winds can bend and twist the branches of conifers toward the leeward side, producing *wind sculptured trees* (Fig. 13.10). Storm winds can unexpectedly prune trees by breaking off their branches. Armed with sand, strong winds can damage or destroy tender new vegetation, decreasing crop productivity. Most plants increase their rate of transpiration as wind speed increases.* This leads to rapid water loss, especially in warmer areas having low humidities, and may actually dry out plants. If sustained, this drying-out effect may stunt plant growth, and, in some windy, dry regions, mature trees that should be many meters tall grow only to the height of a small shrub.

Wind-dried vegetation can result in an area of high fire danger. If a fire should begin here, any additional wind helps it along, directing its movement, adding oxygen for combustion, and carrying burning embers elsewhere to start new fires. On the open plains, where the wind blows practically unimpeded, wind-whipped prairie fires can imperil

Fig. 13.10 In the high country, trees standing unprotected from the wind are often sculpted into "flag" trees.

*This effect actually stops above a certain wind speed and varies greatly among plant species.

homes and livestock and the fires burn out of control over large areas.

Wind erosion is greatly reduced by a continuous cover of vegetation. The vegetation screens the surface from the direct force of the wind and anchors the soil. Soil moisture also helps to resist wind erosion by cementing particles together, which increases their cohesiveness. From this, we can see that land where natural vegetation has been removed for farming purposes—followed by several years of drought—is ripe for wind erosion. This happened in parts of the Great Plains in the middle 1930s, when winds carried millions of tons of dust into the air, creating vast duststorms that buried whole farmhouses, reduced millions of acres to unproductive wasteland, and financially ruined thousands of families. Because of these disastrous effects of the wind, portions of the western plains became known as the "dust bowl."

To protect crops and soil, *windbreaks*—commonly called **shelterbelts**—are planted. Shelterbelts usually consist of a series of mixed conifer and deciduous trees or shrubs planted in rows perpendicular to the prevailing wind flow. They greatly reduce the wind speed behind them. (See Fig. 13.11.) This reduction in wind flow depends both on the wind speed and the height of the barrier. Behind a shelterbelt, wind near the ground can be reduced for a distance up to 25 times the height of the barrier. The minimum wind speed usually occurs at a distance between 3 and 4 times the height of the belt. Upwind from the belt, wind flow is reduced for a distance up to 6 times the height of the barrier. As air filters through the belt, the flow is broken into small eddies, which have little mixing effect on the air near the surface. However, if trees are planted too close together, several unwanted effects may result. For one thing, the air moving past the belt may be broken into larger, more turbulent eddies, which swirl soil about. Furthermore, in high winds, strong downdrafts may damage the crops.

The use of properly designed shelterbelts has benefited agriculture. In some parts of the Central Plains, these belts have stabilized the soil and increased wheat yield. Despite their advantages, many of the shelterbelts planted during the mid-1930s drought years are now being removed. Some are economically unfeasible because they occupy valuable crop land. Others interfere with the large

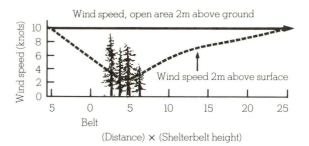

Fig. 13.11   A properly designed shelterbelt can reduce the air flow downwind for a distance of 25 times the height of the belt. The minimum wind flow behind the belt is typically measured downwind at a distance of about 4 times the belt's height.

center pivot sprinkler systems now in use. At any rate, one wonders how the absence of all the shelterbelts would affect this region if it were struck by a drought similar to that experienced in the 1930s.

Wind and Waves   The impact of the wind on the earth's surface is not limited to land; wind also influences water—it makes waves. Waves forming through wind blowing over the surface of the water are known as **wind waves**. Just as air blowing over the top of a water-filled pan creates tiny ripples, so waves are created as the frictional drag of the wind transfers energy to the water. In general, the greater the wind speed, the greater the amount of energy added, and the higher will be the waves. Actually, the amount of energy transferred to the water (and thus the height to which a wave can build) depends upon three factors:

1. the wind speed
2. the length of time that the wind blows over the water
3. the **fetch**, or distance, of deep water over which the wind blows

A sustained 50-knot wind blowing steadily for nearly three days over a minimum distance of 2600 km (1600 mi) can generate waves with an average height of 15 m (49 ft). Thus, a stationary storm system centered somewhere over the open sea is capable of creating large waves. In fact, the largest wave ever recorded in the open sea was encountered by a navy tanker during a Pacific storm with winds of over 60 knots on February 7, 1933. During this storm, the height of the largest

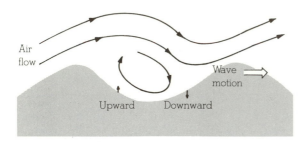

Fig. 13.12 Wind blowing over a wave creates a small eddy of air that helps to reinforce the up and down motion of the water.

wave was estimated at 34 m (112 ft). Observing the wave crest from down in the trough would be like looking up at the top of an 11-story building from a parking lot!

Microscale winds actually help waves grow taller. Consider, for example, the wind blowing over the small wave, as shown in Fig. 13.12. Observe that both the wind and the wave are moving in the same direction, and that the wave crest deflects the wind upward, producing an undulation in the airflow just above the water. This looping air motion establishes a small eddy of air between the two crests. The upward and downward motion of the eddy reinforces the upward and downward motion of the water. Consequently, the eddy helps the wave to build in height.

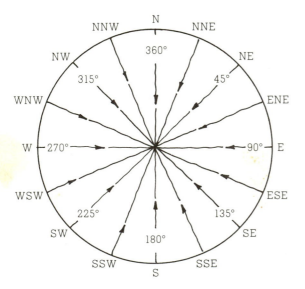

Fig. 13.13 Wind direction can be expressed in degrees about a circle or as compass points.

Traveling in the open ocean, waves represent a form of energy. As they move into a region of weaker winds, they gradually change: Their crests become lower and more rounded, forming what are commonly called *swells*. When waves reach a shoreline they transfer their energy—sometimes catastrophically—to the coast and structures along it. High, storm-induced waves can hurl thousands of tons of water against the shore. If this happens during an unusually high tide, resort homes overlooking the ocean can be pounded into a twisted mass of board and nails by the surf. Bear in mind that the storms creating these waves may be thousands of kilometers away and, in fact, may never reach the shore. Some of the largest, most damaging waves ever to strike the beach communities of Southern California arrived on what was described as "one of the clearest days imaginable." On the more positive side, in the Hawaiian Islands these high waves are excellent for surfboarding.

Determining Wind Direction and Speed

Wind is characterized by its direction, speed, and gustiness. We already know that wind direction is given as the direction from which it is blowing—a north wind blows from the north toward the south. However, near large bodies of water and in hilly regions, wind direction is expressed differently. For example, wind blowing from the water onto the land is referred to as an **onshore breeze**, while wind blowing from land to water is called an **offshore breeze**. Air moving uphill is an *upslope wind*; air moving downhill is a *downslope wind*. The wind direction may also be given as degrees about a 360° circle. These directions are expressed by the numbers shown in Fig. 13.13. For example: A wind direction of 360° is a north wind; an east wind is 90°; a south wind is 180°; and calm is expressed as zero. It is also common practice to express the wind direction in terms of compass points, such as N, NW, NE, and so on.

The Influence of Prevailing Winds At many locations, the wind blows more frequently from one direction than from any other. The **prevailing**

FOCUS ON AN ISSUE

Wind Power

Thousands of small windmills, their arms spinning in a stiff breeze, have pumped water, sawed wood, and even supplemented the electrical needs of small farms for many decades. It was not until the energy crisis of the early 1970s, however, that we seriously considered wind-driven turbines to run generators. In 1980, wind energy got a boost from Congress in the form of a bill—the Wind Energy Research, Demonstration, and Utilization Act—that authorized $900 million to define the nation's wind-energy potential and provide support for the research and development of wind generators.

Presently, several wind turbines are being built to generate electricity. The more conventional-looking horizontal-axis model has a turret that keeps the blades facing with the wind (Fig. 2). The blades themselves turn behind the tower, so that they cannot be blown against it during a windstorm. The horizontal-axis wind turbine built jointly by the U.S. Department of Energy and the National Aeronautics and Space Administration (NASA) is currently operating in Clayton, New Mexico. Monitoring equipment starts the rotor turning when the wind speed reaches 10 knots and shuts it off at 35 knots. Since the average wind speed at Clayton is 13 knots, the wind turbine can generate about 200,000 watts of electricity, enough to satisfy the power needs of 60 homes— roughly 15 percent of the town's needs. A vertical-axis wind turbine, developed by Sandia Laboratories in Albuquerque, New

Fig. 2 A horizontal-axis wind turbine in Clayton, New Mexico.

Mexico, for the U.S. Department of Energy, looks like an upside-down egg beater whose blades spin about a vertical drive shaft (Fig. 3). This type of wind machine has one definite advan-

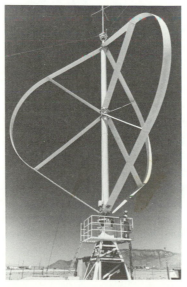

Fig. 3 A vertical-axis wind turbine.

tage: It requires no additional equipment to keep it facing into the wind.

Wind power seems an attractive way of producing energy— it is nonpolluting and, unlike solar power, is not restricted to daytime use. It does, however, pose some problems. Large wind machines may interfere with television reception in their immediate vicinity. Also, the financial aspects are another problem. The Clayton wind machine cost over $1 million to build. In addition, a region dotted with large wind machines is unaesthetic. (Probably, though, it would be no more of an eyesore than the parades of huge electrical towers marching across many open areas today.) Most important, if the wind turbine is to produce electricity, there must be wind, not just any wind, but a steady flow of air neither too weak nor too strong. A slight breeze will not turn the blades, and a powerful wind gust could severely damage the machine. Thus, regions with the greatest potential for wind-generated power would have moderate, steady winds. Examples of such areas include the central Great Plains, high mountain passes, and the coastal areas of New England and the Pacific Northwest.

In a steady wind of 17 knots, large wind machines can convert between 30 and 40 percent of the available wind energy to electricity. Their efficiency increases with both the wind speed and the size of the rotor blade. The available energy is

(continued on next page)

proportional to the cube of the wind speed. A wind speed of 20 knots, for example, can provide 8 times more energy than a wind of 10 knots, meaning that a small increase in wind speed can result in a large increase in electrical energy. Large-blade wind turbines are more efficient than small-blade models because the power output of the machine increases as the square of the blade diameter. Hence, a wind turbine with a 90 m blade is capable of producing 4 times as much electricity as one with a 45 m blade.

With fossil fuels rapidly diminishing, the wind can provide a pollution-free alternative form of energy. For example, a 1000 kilowatt (1 megawatt) wind turbine can supply power for approximately 400 average-size American homes. In California alone, there are over 2000 wind turbines, many of which are on *wind farms*—clusters of 50 or more wind turbines (Fig. 4). During 1983, one California power company purchased 31 million

Fig. 4 A portion of a wind farm near the summit of Altamount Pass, California.

kilowatt hours of electricity from wind farms, for a savings of an equivalent 52,000 barrels of oil and enough electricity to meet the annual needs of over 5200 residential customers. Present estimates are that wind power may be able to furnish between 2 and 4 percent of the nation's total energy needs by the year 2000.

---

wind is the name given to the wind direction most often observed during a given time period. Prevailing winds can greatly affect the climate of a region. For example, where the prevailing winds are upslope, clouds, fog, and precipitation are more likely than where the winds are downslope. Prevailing onshore winds in summer carry moisture, cool air, and fog into coastal regions, while prevailing offshore breezes carry warmer and drier air into the same locations. In city planning, the prevailing wind can help decide where industrial centers, factories, and city dumps should be built. All of these, of course, must be located so that the wind will not carry pollutants into populated areas. Sewage disposal plants must be situated downwind from large housing developments, and

major runways at airports must be aligned with the prevailing wind to assist aircraft in taking off or landing.

The prevailing wind can even be a significant factor in building an individual home. In the northeastern half of the United States, the prevailing wind in winter is northwest and in summer it is southwest. Thus, houses built in the northeastern United States should have windows facing southwest to provide summertime ventilation and few, if any, windows facing the cold winter winds from the northwest. The northwest side of the house should be thoroughly insulated and even protected by a windbreak.

From the prevailing wind, biologists can predict the direction disease-carrying insects and plant

Fig. 13.14 Which way was the wind blowing when this cinder cone erupted?

spores will move and hence, how a disease may spread. Geologists use the prevailing wind to predict where ejected debris from potentially active volcanoes will land.

Many local ground and landscape features show the effect of a prevailing wind. For example, smoke particles from an industrial stack settle to the ground on its downwind side. From the air, the prevailing wind direction can be seen as a discolored landscape on the leeward side of the stack Wind blowing over surfaces of snow and sand produces ripples with a more gentle slope facing into the wind. As previously mentioned, barchan sand dunes have similar shapes and, thus, show the prevailing wind direction. Trees with branches twisted and broken on their windward side appear gnarled as the remaining branches point away from the prevailing wind. Look at Fig. 13.14 and see if you can determine the prevailing wind when this cinder cone erupted.*

The prevailing wind can be represented by a **wind rose**, which indicates the percentage of time the wind blows from different directions. Extensions from the center of a circle point to the wind direction, and the length of each extension indicates the percentage of time the wind blew from

that direction. Figure 13.15 shows a wind rose for a city averaged over a period of 10 years during the month of January. Observe that the longest extension points toward the northwest and that the wind blew from this direction 25 percent of the time. This is the prevailing wind for this time period. Of course, a wind rose can be made for any time of the day, and it can represent the wind direction for any month or season of the year.

The prevailing wind in a town does not always represent the prevailing wind of an entire region. In mountainous regions, the wind is usually guided by topography and is often deflected by obstruc-

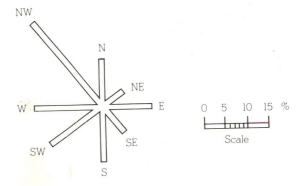

Fig. 13.15 This wind rose represents the percent of time the wind blew from different directions at a given site during the month of January for the past 10 years. The prevailing wind is NW and the wind direction of least frequency is NE.

*During eruption, the prevailing wind was from left to right. We can tell this by the volcano's shape. Material ejected from the volcano was blown by the wind to the right, where it accumulated, producing a more gentle slope.

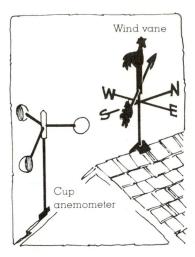

Fig. 13.16 A wind vane and a cup anemometer.

tions that cause its direction to change abruptly. Within this region, the wind may be blowing from one direction on one side of a valley and from an entirely different direction on the other side.

Wind Instruments A very old, yet reliable, weather instrument for determining wind direction is the **wind vane**. Most wind vanes consist of a long arrow with a tail, which is allowed to move freely about a vertical post (Fig. 13.16). The arrow always points into the wind and, hence, always gives the wind direction. Wind vanes can be made of almost any material. At airports, a cone-shaped bag opened at both ends so that it extends horizontally as the wind blows through it sits near the runway. This form of wind vane, called a *wind sock*, enables pilots to tell the surface wind direction when landing.

The instrument that measures wind speed is the **anemometer**. The oldest type of anemometer is the *pressure plate anemometer* developed by Robert Hooke in 1667. It consists of a rectangular metal plate, which is free to swing in the vertical. As the speed of the wind increases, the force of the wind on the plate pushes it outward at a greater angle. The wind speed is read from a scale mounted adjacent to the arm of the swinging plate. Most anemometers today consist of three (or more) hemispherical cups (*cup anemometer*) mounted on a vertical shaft. The difference in wind pressure from one side of a cup to the other causes the cups to spin about the shaft. The rate at which

they rotate is directly proportional to the speed of the wind. The spinning of the cups is usually translated into wind speed through a system of gears, and may be read from a dial or transmitted to a recorder.

The **aerovane** is an instrument commonly used to indicate both wind speed and direction. It consists of a three-bladed propeller that rotates at a rate proportional to the wind speed. Its streamlined shape and a vertical fin (Fig. 13.17) keep the blades facing into the wind. When attached to a recorder, a continuous record of both wind speed and direction is obtained.

The wind-measuring instruments described thus far are "ground-based," and only give wind speed or direction at a particular fixed location. But the wind is influenced by local conditions, such as buildings, trees, and so on. Also, wind speed normally increases rapidly with height above the ground. Thus, wind instruments should be exposed to freely flowing air at a height of at least 10 m above the surface and well above the roofs of buildings. In practice, unfortunately, anemometers are placed at various levels; the result, then, is often erratic wind observations.

A simple way to obtain winds above the surface is with a **pilot balloon**. A small balloon filled with gas is released from the surface. The balloon rises at a known rate, but drifts freely with the wind. It is manually tracked with a small telescope called a **theodolite**. Every minute (or half minute), the balloon's vertical angle (height) and horizontal angle (direction) are measured. The data from the observations are plotted on a special board, and the wind speed and direction are computed at specific intervals—usually every 300 m—above the surface.

The pilot balloon principle can be used to obtain wind information during a radiosonde observation. As a balloon carrying a radiosonde (an instrument package designed to measure the vertical profile of temperature, pressure, and humidity) rises from the surface, equipment located on the ground constantly tracks it, measuring its vertical and horizontal angles, as well as its distance from the observing station. From this information, a computer determines and prints the vertical profile of wind from the surface up to where the balloon normally pops, typically in the stratosphere near 30 km (19 mi). The observation of winds using a radiosonde balloon is called a **rawinsonde observation**.

Above an altitude of 30 km, rockets and radar can inform us about the wind flow. One type of rocket ejects an instrument attached to a parachute that drifts with the wind as it slowly falls to earth. While descending, a ground-based radar unit tracks the instrument and determines wind information for that region of the atmosphere. Other rockets eject metal strips at some desired level. Again, radar tracks these drifting pieces of chaff, which provide valuable wind speed and direction data for elevations outside the normal radiosonde range.

A device similar to radar called **lidar** (*light detection and ranging*) uses infrared or visible light in the form of a laser beam to determine wind information. Basically, it sends out a narrow beam of light that is reflected from particles, such as smoke or dust—it measures wind velocity by measuring the movement of these particles.

Another instrument presently under study as a possible wind information gatherer is the *infrared radiometer*. Radiometers do not measure wind directly; instead, they measure the intensity of infrared radiation emitted from an object. Recent studies show that strong variations in wind speed and wind direction are accompanied by sharp variations in air temperature, especially near the underside of a severe thunderstorm, where a cold downdraft pours out of the cloud. Infrared radiometers can detect these temperature variations and relate them to the wind field. More recently, studies have shown that the amount of water vapor also changes drastically in regions of strong wind shear related to clear air turbulence. Because water vapor is an excellent emitter of infrared radiation, radiometers can measure these sudden changes in radiation intensity. Installed in an airplane, this instrument could warn pilots of possible wind turbulence as far as 100 km (60 mi) in front of the aircraft.

In remote regions of the world where upper-air observations are lacking, satellites have been used to obtain wind speed and direction. So far, the most reliable wind data have come from geostationary satellites positioned above a particular location. From this vantage point, the satellite shows the movement of clouds. The direction of cloud movement indicates wind direction, and the hori-

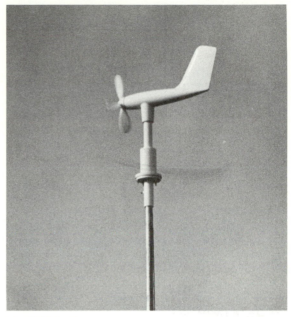

Fig. 13.17 The aerovane.

zontal distance the cloud moves during a given time period indicates the wind speed.

Summary

In this chapter, we have considered both how our environment influences the wind and how the wind influences our environment. We saw that the friction of air flow—viscosity—can be brought about by the random motion of air molecules (molecular viscosity) or by turbulent whirling eddies of air (eddy viscosity). The depth of the atmospheric layer near the surface that is influenced by surface friction (the friction layer) depends upon atmospheric stability, the wind speed, and the roughness of the terrain. Although it may vary, the top of the friction layer is usually near 1000 m.

Winds blowing past obstructions can produce a number of effects, from gusty winds at a sports stadium to howling winds on a blustery night. Aloft, winds blowing over a mountain range may generate hazardous rotors downwind of the range. And the eddies that form in a region of strong wind shear, especially in the vicinity of a jet stream, can produce extreme turbulence, even in clear air.

We have seen that the wind is a powerful weather element that affects our environment in many ways.

It can shape the landscape, influence crop production, transport material from one area to another, generate waves, and turn a wind turbine to produce electricity. Wind blowing over the earth's surface can create a variety of features. In deserts, we see sand dunes, ventifacts, and desert pavement. Over a snow surface, the wind produces snow ripples and snow rollers. Where high winds blow over a ridge, trees may be sculpted into "flag" trees. In unprotected areas, shelterbelts are planted to protect crops and soil from damaging winds.

All in all, this chapter has given us an overview of the many ways small-scale winds interact with our environment.

Questions for Review

1. Describe the various scales of motion, and give an example of each.

2. How does the earth's surface influence the flow of air above it?

3. Distinguish between molecular viscosity and eddy viscosity.

4. What causes wind gusts?

5. Why are winds near the surface typically stronger and more gusty in the afternoon?

6. A friend has just returned from a trans-Atlantic jet aircraft flight and reported that the plane dropped about 1000 m when it entered an "air pocket." Explain to your friend what apparently happened to cause this.

7. What is wind shear?

8. Explain why the car in the diagram below may experience a west wind as it travels past the wall.

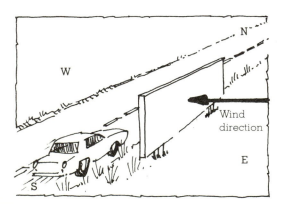

9. Strong winds can pick up large sand, but not clay and dust particles. Why?

10. Explain the various ways the wind affects vegetation.

11. What effect does the wind have on a desert environment?

12. How do snow rollers and snow ripples form?

13. Explain how shelterbelts protect sensitive crops from wind damage.

14. With the same wind speed, explain why a camper is more easily moved by the wind than a car.

15. What are the necessary conditions for the development of large wind waves?

16. How can a coastal area have heavy waves on a clear, nonstormy day?

17. If you are standing directly south of a smoke stack and the wind from the stack is blowing over your head, what would be the wind direction?

18. An upper wind direction is reported as 315°. From what compass direction is the wind blowing?

19. List as many ways as you can of determining the wind direction. The wind speed.

20. Below is a list of instruments. Describe how each one is able to measure wind speed, wind direction, or both.

    (a) wind vane   (e) radiosonde
    (b) cup anemometer   (f) lidar
    (c) aerovane   (g) satellite
    (d) pilot balloon   (h) radar

Questions for Thought

1. A pilot enters the weather service office and wants to know what time of the day she can expect to encounter the least turbulent winds at 760 m (2500 ft) above central Kansas. If you were the weather forecaster, what would you tell her?

2. Why is it dangerous during hang-gliding to enter the leeward side of the hill when the wind speed is strong?

3. After a winter snowstorm, Cheyenne, Wyoming, reports a total snow accumulation of 48 cm (19 in.), while the maximum depth in the surrounding countryside is only 28 cm (11 in.). If the storm's intensity and duration were practically the same for a radius of 50 km around Cheyenne, explain why Cheyenne received so much more snow.

4. Why is the difference in surface wind speed between morning and afternoon typically greater on a clear, sunny day than on a cloudy, overcast day?

5. Look at a map of the Pacific Ocean and locate the Hawaiian Islands. Where would be the best position (NE, N, or NW) for a large storm to provide huge surfing waves for the beaches along the north shores? Explain.

6. Average annual wind speed information in knots is given below for two cities located on the Great Plains. Which city would probably be the best site for a wind turbine? Why?

	TIME								**Average Annual Wind Speed (knots)**
	Mid-night	**3**	**6**	**9**	**Noon**	**3**	**6**	**9**	
City A:	12	7	8	13	15	18	14	13	12.5
City B:	8	6	6	13	20	22	15	10	12.5

7. Which of the sites (A, B, or C) would probably be the best place to construct a wind turbine? Which would be the worst? Explain.

Problems and Exercises

1. A model city is to be constructed in the middle of an uninhabited region. The wind rose below shows the annual frequency of wind directions for this region. With the aid of the wind rose, on a square piece of paper determine where the following should be located:

   (a) industry
   (b) parks
   (c) schools
   (d) shopping centers
   (e) sewage disposal plants
   (f) housing development
   (g) an airport with two runways

2. What would be the total force exerted on a camper 15 feet long and 8 feet high, if a wind of 40 mi/hr blows perpendicular to one of its sides?

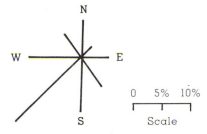

Surface heating and lifting of air along a sea breeze combine to form these clouds almost daily during the summer in southern Florida. (Photo: courtesy of T. Ansel Toney and M. Elsea)

CHAPTER 14

Local Wind Systems

Every summer, millions of people flock to the New Jersey shore, hoping to escape the oppressive heat and humidity of the inland region. On hot, humid days, these travelers often encounter thunderstorms about 30 km or so from the ocean, thunderstorms that invariably last for only a few minutes. In fact, by the time the vacationers arrive at the beach, skies are generally clear and air temperatures are much lower, as cool ocean breezes greet them. If the travelers return home in the afternoon, these "mysterious" showers occur at just about the same location as before.

The showers are not really mysterious. Actually, they are caused by a local wind system—the sea breeze. As cooler ocean air pours inland, it forces the warmer, unstable humid air to rise and condense, producing majestic clouds and rainshowers along a line where the wind systems meet.

Many such examples of local wind systems occur the world over. In this chapter, we will concentrate on mesoscale and synoptic scale winds. First, we will see how the heating and cooling of the surface and the air above produce thermal circulations. Then, we will examine sea breezes and other local winds, describing how they form and what type of weather they generally bring.

Contents

Thermal Circulations

Consider the vertical distribution of pressure shown in Fig. 14.1a. The isobaric surfaces all lie parallel to the earth's surface; thus, there is no horizontal variation in pressure (or temperature), and there is no pressure gradient and no wind. Suppose the atmosphere is cooled to the north and warmed to the south (Fig. 14.1b). In the cold, dense air above the surface, the isobars bunch closer together,

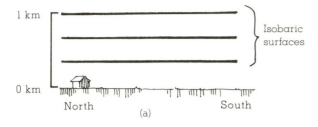

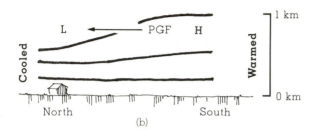

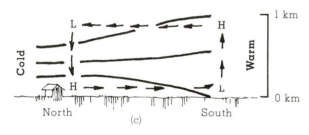

Fig. 14.1 A thermal circulation produced by the heating and cooling of the atmosphere near the ground. (The Hs and Ls refer to atmospheric pressure.)

while in the warm, less dense air, they spread farther apart. This dipping of the isobars produces a horizontal pressure gradient force aloft that causes the air to move from higher pressure (warm air) toward lower pressure (cold air).

At the surface, the air pressure remains unchanged until the air aloft begins to move. As the

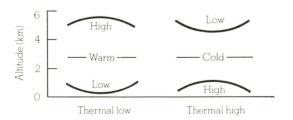

Fig. 14.2 The vertical distribution of pressure with thermal highs and thermal lows.

air aloft moves from south to north, air leaves the southern area and "piles up" above the northern area. This redistribution of air reduces the surface air pressure to the south and raises it to the north. Consequently, a pressure gradient force is established at the earth's surface from north to south and, hence, surface winds begin to blow from north to south.

We now have a distribution of pressure and temperature and a circulation of air, as shown in Fig. 14.1c. As the cool surface air flows southward, it warms and becomes less dense. In the region of surface low pressure, the warm air slowly rises, expands, cools, and flows out the top at an elevation of about 1 km above the surface. At this level, the air flows horizontally northward toward lower pressure, where it completes the circulation by slowly sinking and flowing out the bottom of the surface high. Circulations brought on by changes in air temperature, in which warmer air rises and colder air sinks, are termed **thermal circulations**.

The regions of surface high and low atmospheric pressure created as the atmosphere either cools or warms are called **thermal highs** and **thermal lows**. In general, they are shallow systems, usually extending no more than a few kilometers above the ground. These systems weaken with height. For example, at the surface, atmospheric pressure is lowest in the center of the warm thermal low in Fig. 14.2. In the warm air above the low, the isobars spread apart, and, at some intermediate level, the thermal low disappears and actually changes into a high. A similar phenomenon happens above the cold thermal high. The surface pressure is greatest in its center, but because the isobars aloft are crowded together due to the cold, dense air, the surface thermal high becomes a low a kilometer or so above the ground. We can summarize the typical characteristics of thermal pressure systems as being shallow, weakening with height, and being maintained, for the most part, by local heating and cooling.

Sea and Land Breezes

Anyone who has gone to the ocean on a hot, summer day and felt the cooling breeze blow in off the water knows what a **sea breeze** is. The uneven heating rates of land and water (described in Chapter 5) cause these mesoscale coastal winds. During the day, the land heats more quickly than the adjacent water, and the intensive heating of the air above produces a shallow thermal low. The air over the water remains cooler than the air over the land; hence, a shallow thermal high exists above the water. The overall effect of this pressure distribution is a surface breeze that blows from the sea (Fig. 14.3a). Since the strongest gradients of temperature and pressure occur near the land-water boundary, the strongest winds typically occur right near the beach and diminish inland. Further, since the greatest contrast in temperature between land and water usually occurs in the afternoon, sea breezes are strongest at this time. (The same type of breeze that develops along the shore of a large lake is called a **lake breeze**.)

At night, the land cools more quickly than the water. The air above the land becomes cooler than the air over the water, producing a distribution of pressure, such as the one shown in Fig. 14.3b. With higher surface pressure now over the land, the wind reverses itself and becomes a **land breeze**—a breeze that flows from the land toward the water. Temperature contrasts between land and water are generally much smaller at night; hence, land breezes are usually weaker than their daytime counterpart, the sea breeze.

Look at Fig. 14.3 again and observe that the rising air is over the land during the day and over the water during the night. Therefore, along the humid east coast, daytime clouds tend to form over land and nighttime clouds over water. This explains why, at night, distant lightning flashes are sometimes seen over the ocean.

Influence of a Sea Breeze   Sea breezes are best developed where large temperature differences exist between land and water. Such conditions prevail year-round in many tropical regions. In middle latitudes, however, sea breezes are invariably spring and summer phenomena.

During the summer, a sea breeze usually sets in about midmorning after the land has been warmed. By early afternoon, the breeze has increased in strength and depth. By late afternoon, the cool ocean air may reach a depth of more than

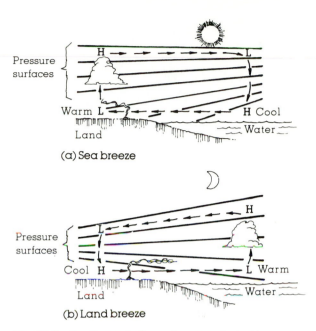

Fig. 14.3   Development of a sea breeze and a land breeze.

300 m (1000 ft) and extend inland for more than 20 km (12 mi).

The leading edge of the sea breeze is called the **sea breeze front**. As the front moves inland, a rapid drop in temperature occurs just behind it. In some locations, this temperature change may be 5°C (9°F) or more during the first hours—a refreshing experience on a hot, sultry day. Since cities near the ocean usually experience the sea breeze by noon, their highest temperature usually occurs much earlier than in inland cities. Along the east coast, the passage of the sea breeze front is marked by a wind shift, usually from west to east. In the cool ocean air, the relative humidity rises as the temperature drops. If the relative humidity increases to above 70 percent, water vapor begins to condense upon particles of sea salt or industrial smoke, producing haze. When the ocean air is highly concentrated with pollutants, the sea breeze front may meet relative clear air and thus appear as a *smoke front*, or a **smog front** (Fig. 14.4). If the ocean air becomes saturated, a mass of low clouds and fog will mark the leading edge of the marine air.

When there is a sharp contrast in air tempera-

Fig. 14.4 The leading edge of a sea breeze, marked by a distinct smog front, moves into Riverside, California.

Fig. 14.5 The convergence of two lake breezes and their influence on the maximum temperature during July in upper Michigan.

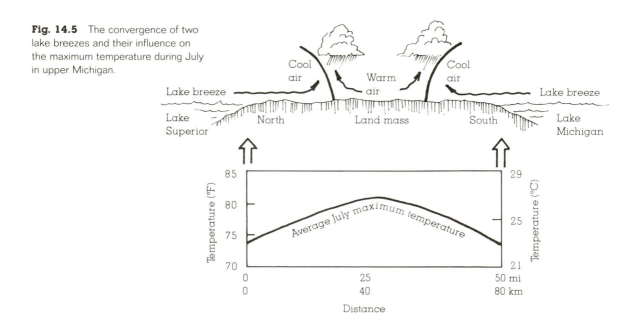

ture across the frontal boundary, the warmer, lighter air will converge and rise. In many regions, this makes for good sea breeze glider soaring. If this rising air is sufficiently moist, a line of cumulus clouds will form along the sea breeze front, and, if the air is also unstable, thunderstorms may form. As previously mentioned, on a hot, humid day one can drive toward the shore, encounter heavy showers several kilometers from the ocean, and arrive at the beach to find it sunny with a steady onshore breeze.

A sea breeze moving over a forest fire can be dangerous. First of all, gusty surface winds often make the fire difficult to control. Another problem is the return flow aloft. Along the sea breeze frontal boundary, air can rise to elevations where it becomes part of the return flow. Should burning embers drift seaward with this flow and drop to the ground behind the fire, they could start new fires. Flames from these fires pushed on by surface winds can trap firefighters between the two blazes.

When cool, dense, stable marine air encounters an obstacle, such as a row of hills, the heavy air tends to flow around them rather than over them. When the opposing breezes meet on the opposite side of the obstruction, they form what is called a *sea breeze convergence zone*. Such conditions are common along the Pacific Coast.

Sea breezes in Florida help produce that state's abundant summertime rainfall. On the Atlantic side of the state, the sea breeze blows in from the east; on the Gulf shore, it moves in from the west. The convergence of these two moist wind systems, coupled with daytime convection, produces cloudy conditions and showery weather over the land. Over the water (where cooler, more stable air lies close to the surface), skies often remain cloud-free. (See color plate 32.)

Convergence of coastal breezes is not restricted to ocean areas. Both Lake Michigan and Lake Superior are capable of producing well-defined lake breezes. In upper Michigan, these large bodies of water are separated by a narrow strip of land about 80 km (50 mi) wide. As seen in Fig. 14.5, the two breezes push inland and converge near the center of the peninsula, creating afternoon clouds and showers, while the lakeshore area remains sunny, pleasantly cool, and dry.

Local Winds and Water

Frequently, local winds will change speed and direction as they cross a large body of water. Figure 14.6 shows the wind speed and direction as air flows over a large lake. At position A, on the upwind side, the wind is blowing at 10 knots from the WNW; at position B, the wind speed is 15 knots from the NW; at position C, the wind is again blowing at 10 knots from the WNW. Why does the wind blow faster and from a slightly different direction in the center of the lake? As the air moves from the rough land over the relatively smooth lake, friction with the surface lessens, and the wind speed increases. The increase in wind speed, however, increases the Coriolis force, which turns the wind flow to the right, as shown by the report at position B. When the air reaches the opposite side of the lake, it again encounters rough land, and its speed slows. This reduces the Coriolis force, and the wind responds by shifting to a more westerly direction, as shown by the report at position C.

Changes in wind speed along the shore of a large lake can inhibit cloud formation on one side and enhance it on the other. Suppose warm, moist air flows over a lake, as illustrated in Fig. 14.7. Observe that clouds are forming on the downwind side, but not on the upwind side. The lake is slightly cooler than the air. Consequently, by the time the air reaches the lee side of the lake, it will be cooler, denser, and less likely to rise. Why, then, are clouds forming on this side of the lake? As air moves from the land over the water, it

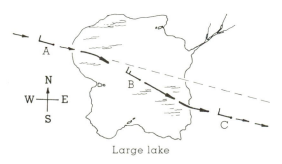

Large lake

Fig. 14.6 Wind can change in both speed and direction when crossing a large lake.

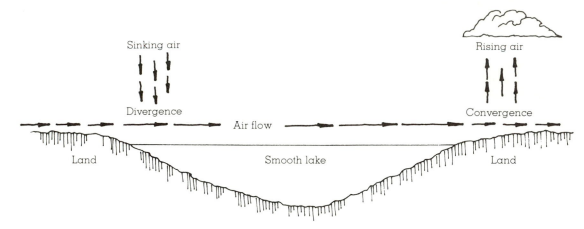

Fig. 14.7 Sinking air develops where surface winds move off shore, speed up, and diverge. Rising air develops as surface winds move on shore, slow down, and converge.

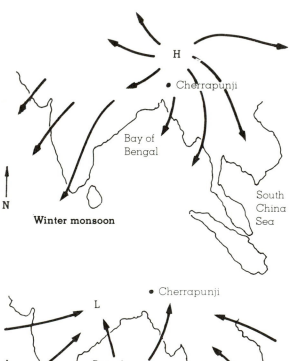

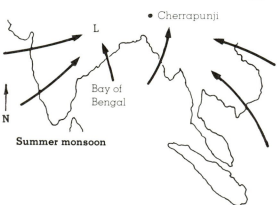

Fig. 14.8 Changing annual wind-flow patterns associated with the winter and summer Asian monsoon.

travels from a region of greater friction into a region of less friction, so it increases in speed. This causes the surface air to diverge—to spread apart. Such spreading of air forces air from above to slowly sink, which, of course, inhibits the formation of clouds. Hence, there are no clouds on the upwind side of the lake. Out over the lake, the separation between air temperature and dew point lessens. As this nearly saturated air moves on shore, friction with the rougher ground slows it down, causing it to "bunch up" or converge (which forces the air upward). This slight upward motion coupled with surface heating is often sufficient to initiate the formation of clouds along the downwind side of the lake.

Seasonally Changing Winds— The Monsoon

The word *monsoon* derives from the Arabic *mausim*, which means seasons. A **monsoon wind system** is one that *changes direction seasonally*, blowing from one direction in summer and from the opposite direction in winter. This seasonal reversal of winds is especially well developed in eastern and southern Asia.

In some ways, the monsoon is similar to a huge sea breeze. During the winter, the air over the continent becomes much colder than the air over the ocean. A large, shallow high-pressure area develops over continental Siberia, producing a *clockwise* circulation of air that flows out over the Indian Ocean and South China Sea (Fig. 14.8). Subsiding air of the anticyclone and the downslope movement of northeasterly winds from the inland plateau provide eastern and southern Asia with generally fair weather and the dry season. Hence, the winter monsoon means clear skies, with winds that blow from land to sea.

In summer, the wind-flow pattern reverses itself as air over the continents becomes much warmer than air above the water. A shallow thermal low develops over the continental interior. The heated air within the low rises, and the surrounding air responds by flowing *counterclockwise* into the low center. This results in moisture-bearing winds sweeping into the continent from the ocean. The humid air converges with a drier westerly flow, causing it to rise; further lifting is provided by hills and mountains. Lifting cools the air to its saturation point, resulting in heavy showers and thunderstorms. Thus, the summer monsoon of southeastern Asia means wet, rainy weather (wet season) with winds that blow from sea to land.

Many factors help create the monsoon wind system. The latent heat given off during condensation aids in the warming of the air over the continent and strengthens the summer monsoon circulation. Rainfall is enhanced by weak westward-moving low-pressure areas called *monsoon depressions*. The formation of these depressions is aided by an upper-level jet stream. Where winds in the jet diverge, surface pressures drop, the monsoon depressions intensify, and surface winds increase. The greater inflow of moist air supplies larger quantities of latent heat, which, in turn, intensifies the summer monsoon circulation. Preliminary findings of the Monsoon Experiment (MONEX), a joint research project of the United States and other nations, suggests that surges of rainfall over western India might be related to a strong, moist, low-level jet stream that flows along the eastern hills of Kenya then swings out over the Indian Ocean, where it feeds the monsoon winds. Hence, by monitoring the low-level jet stream, it might be possible to predict the intensity and duration of monsoon rains.

Summer monsoon rains over southern Asia can reach record amounts. Located about 300 km inland on the southern slopes of the Khasi Hills in northeastern India, Cherrapunji receives an average of 1080 cm (425 in.) of rainfall each year, most of it during the summer monsoon between April and October. The summer monsoon rains are essential to the agriculture of that part of the world. With a population of over 600 million people, India depends heavily on the summer rains, so that food crops will grow. Unfortunately, the monsoon can be unreliable in both duration and intensity. In 1973 and 1974, it did not fully develop, and left India with only small amounts of rain and a poor crop yield. The situation did not become normal until 1975, when the rains returned, providing India with a record grain yield. Since the monsoon is vital to the survival of so many people, it is no wonder that meteorologists have investigated it extensively. They have tried to develop methods of accurately forecasting the intensity and duration of the monsoon. The outlook has been hopeful, but the results are not yet impressive.

Monsoon wind systems exist in other regions of the world, where large contrasts in temperature develop between oceans and continents. Usually, however, these systems are not as pronounced as in southeast Asia. Monsoon winds can, for example, be found in Australia, Africa, and South America. In the United States, a weak monsoon-like flow occurs in the Mississippi Valley, as winter winds tend to be cold, dry, and more northerly and summer winds warm, humid, and southerly. Along the Pacific Coast, the strong onshore surface flow that begins in spring and lasts until fall is so persistent it is called the *Pacific monsoon*.

A monsoon-like circulation also exists in the southwestern United States, especially in Arizona and New Mexico, where spring and early summer are normally dry, as warm westerly winds sweep over the region. By mid-July, however, moist southerly winds are more common, and so are afternoon showers and thunderstorms.

Mountain and Valley Breezes

Mountain and valley breezes develop along mountain slopes. Observe in Fig. 14.9 that, during the

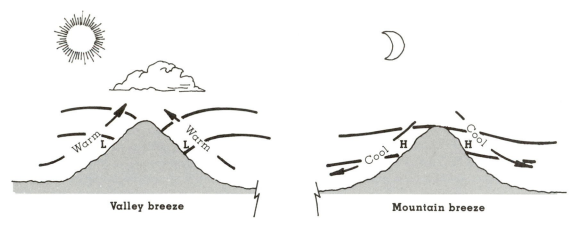

Valley breeze   Mountain breeze

Fig. 14.9 Valley breezes blow uphill during the day; mountain breezes blow downhill at night. (The Ls and Hs represent pressure, while the solid lines are pressure surfaces.)

day, sunlight warms the valley walls, which in turn warm the air in contact with them. The heated air, being less dense than the air of the same altitude above the valley, rises as a gentle upslope wind known as a **valley breeze**. At night, the flow reverses. The mountain slopes cool quickly, chilling the air in contact with them. The cooler, more dense air glides downslope into the valley, providing a **mountain breeze**. (Because gravity is the force that directs these winds downhill, they are also referred to as *gravity winds*, or *drainage*

winds.) This daily cycle of wind flow is best developed in clear, summer weather when prevailing winds are light.

In many areas, the upslope winds begin early in the morning, reach a peak speed of about 6 knots by midday, and reverse direction by late evening. The downslope mountain breeze increases in intensity, reaching its peak in the early morning hours, usually just before sunrise. In the Northern Hemisphere, valley breezes are particularly well developed on south-facing slopes, where sunlight

Fig. 14.10 As mountain slopes warm during the day, air rises and often condenses into cumuliform clouds, such as these.

is most intense. On partially shaded north-facing slopes, the upslope breeze may be weak or absent. Since upslope winds begin soon after the sun's rays strike a hill, valley breezes typically begin first on the hill's east-facing side. In the late afternoon, this side of the mountain goes into shade first, producing the onset of downslope winds at an earlier time than experienced on west-facing slopes. Hence, it is possible for campfire smoke to drift downslope on one side of a mountain and upslope on the other side.

When the upslope winds are well developed and have sufficient moisture, they can reveal themselves as building cumulus clouds above mountain summits (Fig. 14.10). Since valley breezes usually reach their maximum strength in the early afternoon, cloudiness, showers, and even thunderstorms are common over mountains during the warmest part of the day—a fact well known to climbers, hikers, and seasoned mountain picnickers.

Downslope Winds

Any wind that blows downslope is called a **katabatic wind**. These range in speed from gentle mountain breezes to winds that rush down elevated slopes at hurricane speeds.

An especially strong, cold katabatic wind is the **fall wind**, which flows from elevated plateaus usually surrounded by mountains. When winter snows accumulate on the plateau, the overlying air grows extremely cold and a shallow dome of high pressure forms near the surface (Fig. 14.11). Along the edge of the plateau, the horizontal pressure gradient force is usually strong enough to cause the cold air to flow across the isobars through gaps and saddles in the hills. Along the slopes of the plateau, the wind continues downhill as a gentle or moderate cold breeze. If the horizontal pressure gradient increases substantially, such as when a storm approaches, or if the wind is confined to a narrow canyon or channel, the flow of air can increase, often destructively, as cold air rushes downslope like water flowing over a fall.

Winds such as these are observed in different parts of the world. In North America, for example,

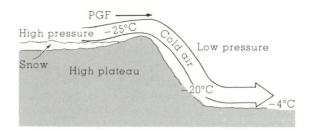

Fig. 14.11   The formation of a fall wind.

when cold air accumulates over the Columbia plateau, it may flow westward through the Columbia River Gorge as a strong, gusty, and sometimes violent, wind. Even though the sinking air warms by compression, it is so cold to begin with that it reaches the ocean side of the Cascade Mountains much colder than the marine air it replaces. The *Columbia Gorge wind* is often the harbinger of a prolonged cold spell.

Strong downslope winds funneled through a mountain canyon can do extensive damage. For example, during January, 1984, a ferocious downslope wind blew through Yosemite National Park at speeds estimated at 100 knots. The wind toppled trees and, unfortunately, caused a fatality when a tree fell on a park employee sleeping in a tent.

Along the northern Adriatic coast in Yugoslavia, a polar invasion of cold air from the Soviet Union descends the slopes from a high plateau and reaches the lowlands as the *bora*—a cold, gusty, northeasterly wind with speeds sometimes in excess of 100 knots. A similar, but often less violent, cold wind known as the *mistral* descends the western mountains into the Rhone Valley of France, and then out over the Mediterranean Sea. It frequently causes frost damage to exposed vineyards, and makes people bundle up in the otherwise mild climate along the Riviera. Strong, cold winds also blow downslope off the icecaps in Greenland and Antarctica occasionally, with speeds greater than hurricane force. Not all katabatic winds are cold. The chinook wind is a good example of a warm downslope wind.

Chinook Winds   The **chinook wind** is a warm, dry wind that descends the eastern slope of the

FOCUS ON A SPECIAL TOPIC

Snow Eaters and Rapid Temperature Changes

Chinooks are thirsty winds. As they move over a heavy snow cover, they can melt and evaporate a foot of snow in less than a day. This has led to some tall tales about these so-called "snow eaters." Canadian folklore has it that a sled-driving traveler once tried to outrun a chinook. During the entire ordeal his front runners were in snow while his back runners were on bare soil.

Actually, the chinook is important economically. It not only brings relief from the winter cold, but it uncovers prairie grass, so that livestock can graze on the open range. Also, these warm winds have kept railroad tracks clear of snow, so that trains can keep running. On the other hand, the drying effect of a chinook can create an extreme fire hazard. And when a chinook follows spring planting, the seeds may die in the parched soil. Along with the dry air comes a buildup of static electricity, making a simple handshake a shocking experience. These warm, dry winds have sometimes adversely affected human behavior. During periods of chinook winds some people feel ir-

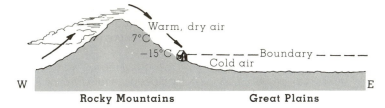

Fig. 1 Cities near the warm air–cold air boundary can experience sharp temperature changes, if cold air should rock up and down like water in a bowl.

ritable and depressed and others become ill. The exact reason for this phenomenon is not clearly understood.

Chinook winds have been associated with rapid temperature changes. Figure 1 shows a shallow layer of extremely cold air resting against the Rocky Mountains. In the cold air, temperatures are near −15°C (5°F), while just a short distance up the mountain a warm chinook wind raises the air temperature to 7°C (45°F). The cold air behaves just as any fluid, and, in some cases, atmospheric conditions may cause the air to move up and down much like water does when a bowl is rocked back and forth. This can cause extreme temperature variations for

cities located at the base of the hills along the periphery of the cold air–warm air boundary, as they are alternately in and then out of the cold air. Such a situation is held to be responsible for the unbelievable two-minute temperature change of 27°C (49°F) recorded at Spearfish, South Dakota, during the morning of January 22, 1943. On the same morning, in nearby Rapid City, the temperature fluctuated from −20°C (−4°F) at 5:30 A.M. to 12°C (54°F) at 9:40 A.M., then down to −12°C (11°F) at 10:30 A.M. and up to 13°C (55°F) just 15 minutes later. At nearby cities, the undulating cold air produced similar temperature variations that lasted for several hours.

Rocky Mountains. The region of the chinook is rather narrow (only several hundred kilometers wide) and extends from northeastern New Mexico northward into Canada. Similar winds occur along the leeward slopes of mountains in other regions of the world. In the Alps, for example, such a wind is called a **foehn** and, in Argentina, a *zonda*. When these winds move through an area, the temperature rises sharply, sometimes over 20°C (36°F) in

one hour, and a corresponding sharp drop in the relative humidity occurs, occasionally to less than 5 percent.

What is it that makes these winds so warm and dry? The answer is illustrated in Fig. 14.12. Suppose the air at the foot of the mountain (Fig. 14.12a) has a temperature of 10°C (50°F). Upon rising on the windward side, the air expands and cools at the dry adiabatic rate (10°C/km) to a

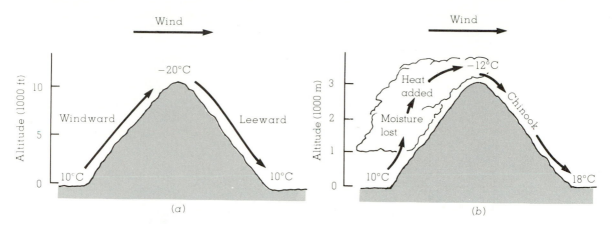

Fig. 14.12 Chinooks form as winds blowing downslope heat by compression. The moisture lost and latent heat added on the windward side enhance their warmth and dryness.

temperature of −20°C (−4°F) at the 3000 m (10,000 ft) summit. Upon descending the leeward side, the compressed air warms at the dry adiabatic rate, producing essentially the same temperature at each elevation as on the windward side. However, if warm air moves in aloft (warm advection) above the mountain peak, the sinking air will become warmer on the leeward side. If clouds form on the windward side, the air will be even warmer and drier.

Suppose the cloud in Fig. 14.12b forms at an altitude of 1000 m (about 3000 ft) above the surface. Up to this level, the rising air cools at the dry adiabatic rate. Within the cloud, latent heat is added to the rising air, as water vapor condenses into liquid cloud droplets. The addition of latent heat retards the cooling rate, so the air now cools at the moist adiabatic rate (6°C/km). By the time the air reaches the summit, it has a temperature of −12°C (10°F), some 8°C warmer than if the clouds were not present. Assuming that the cloud remains on the windward side, the descending air will warm at the dry adiabatic rate all the way down the leeward side. By the time the air reaches the base of the mountain, it will have a temperature of 18°C (64°F), which is warmer than when it started its upward journey. The air will also be drier since a good deal of its moisture was removed as rain, snow, and clouds on the windward side. (Although windward precipitation enhances the chinook, it is not a necessary ingredient.) In summary, the source of warmth for a chinook is compressional heating, which is supplemented by the release of latent heat on the windward side; the main source of dryness is the loss of moisture on the upwind slopes.

Along the front range of the Rockies, a bank of clouds forming over the mountains is a telltale sign of an impending chinook. This *chinook wall cloud* usually remains stationary as air rises, condenses, and then rapidly descends the leeward slopes, often causing strong winds in foothill communities. Figure 14.13 shows how a chinook wall cloud appears as one looks west toward the Rockies from the Colorado plains. The photograph was taken on a winter afternoon with the air temperature about −7°C (20°F). That evening, the chinook moved downslope at high speeds through foothill valleys, picking up sand and pebbles (which dented cars and cracked windshields). The chinook spread out over the plains like a warm blanket, raising the air temperature the following day to a mild 15°C (59°F). The chinook and its wall of clouds remained for several days, bringing with it a welcomed break from the cold grasp of winter.

Certain atmospheric conditions are necessary for the onset of a strong chinook. In winter, cold arctic air must move southward out of Canada and cover a large portion of the Great Plains. When the cold, shallow air mass stalls along the eastern flank of the Rockies and a trough of low

Fig. 14.13 A chinook wall cloud forming over the Colorado Rockies (viewed from the plains).

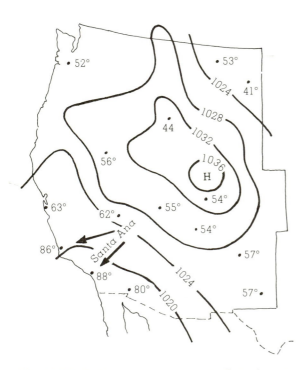

Fig. 14.14 Surface weather map showing Santa Ana conditions for January 17, 1976. Maximum temperatures for this day are given in °F. Observe that the downslope winds blowing into Southern California raised temperatures into the upper 80s, while elsewhere temperature readings were much lower.

pressure forms to the east of this region, strong westerly winds carry Pacific air up and over the range toward the region of lower pressure. As the descending warm air replaces the cold Canadian air, temperatures may jump more than 22°C (40°F) in less than an hour.

Santa Ana Winds Similar to the chinook is the **Santa Ana wind**, a warm, dry wind that blows from the east or northeast into Southern California. As the air descends from the elevated desert plateau, it funnels through mountain canyons in the San Gabriel and San Bernardino mountains, finally spreading over the Los Angeles basin and San Fernando Valley. The wind often blows with exceptional speed in the Santa Ana Canyon (the canyon from which it derives its name).

These warm, dry winds develop as a region of high pressure builds over the Great Basin. The clockwise circulation around the anticyclone forces air downslope from the high plateau. There is no precipitation to add latent heat to the air, so the compressional heating provides the primary source of warming. The air is dry, since it originated in the desert, and it dries out even more as it is heated. Figure 14.14 shows a typical wintertime Santa Ana situation.

As the wind rushes through canyon passes, it

lifts dust and sand and dries out vegetation. This sets the stage for serious brush fires, especially in autumn, when chaparral-covered hills are already parched from the dry summer.* One such fire in November of 1961—the infamous Bel Air fire—burned for three days, destroying 484 homes and causing over $25 million in damage. During October, 1977, Santa Ana-driven flames scorched 25,000 acres along a 10-mile-wide swath across the Santa Monica Mountains, destroying 91 homes collectively valued at several million dollars. To the north in Santa Barbara, a Santa Ana-like wind was responsible for the disastrous fire of July 26 and 27, 1977. In a canyon east of the town, a box kite burst into flames after touching a power line. The burning kite fell to the ground, and an east wind fanned and spread the fire over the hills. In just two days, over 300 buildings were totally destroyed, with an estimated loss exceeding $50 million.

With the protective vegetation cover removed, the land is ripe for erosion, as winter rains wash away topsoil and, in some areas, create serious mudslides. The adverse effects of a wind-driven Santa Ana fire may be felt throughout the year.

A similar wind called a *California norther* can produce unbearably high temperatures in the northern half of California's Central Valley. On August 8, 1978, for example, a ridge of high pressure formed to the north of this region, while a thermal low was well entrenched to the south. This pressure pattern produced a north wind in the area. A summertime north wind in most parts of the country means cooler weather and a welcome relief from a hot spell, but not in Red Bluff, California, where the winds moved downslope off the mountains. Heated by compression, these winds increased the air temperature in Red Bluff to an unbelievable 48°C (119°F) for two consecutive days—amazing when you realize that Red Bluff is located at about the same latitude as Philadelphia!

Other Local Winds of Interest

Up to now, we have examined the causes of some mesoscale wind systems recognized more than

*Chaparral denotes a shrubby environment, in which many of the plant species contain highly flammable oils.

Fig. 14.15 This satellite picture shows the effects of a Santa Ana wind. As a forest fire rages in the hills east of Los Angeles, a strong Santa Ana wind blows smoke from the fire first over the city, then out to sea.

just locally. Other local winds are known by various names in different locales. Let's look at some examples.

In winter, when an intense storm tracks east across the Great Plains of North America, cold northerly winds often plunge southward behind it. As the cold air moves through Texas, it may drop temperatures tens of degrees in a few hours. Such a cold wind is called a **Texas norther**, or *blue norther*, especially if accompanied by snow. If the cold air penetrates into Central America, it is known as a *norte*. Meanwhile, if the cold, gusty winds over the plains states are accompanied by drifting, blowing, and falling snow, the term *blizzard* is applied to this weather situation.

When a storm develops or intensifies off the eastern seaboard of North America, and moves

Table 14.1   Local Winds of the World

NAME	DESCRIPTION
Cold Winds	
buran	a strong, cold wind that blows over the Soviet Union and central Asia
purga	a buran accompanied by strong winds and blowing snow
pampero	a cold wind blowing from the south over Argentina, Uruguay, and into the Amazon Basin
burga	a cold northeasterly wind in Alaska usually accompanied by snow; similar to the buran and purga of the Soviet Union
bise	generally a cold north or northeast wind that blows over southern France; often brings damaging spring frosts
papagayo	a cold northeasterly wind along the Pacific coast of Nicaragua and Guatemala; occurs when a cold air mass overrides the mountains of Central America
tehuantepecer	a strong wind from the north or northwest funneled through the gap between the Mexican and Guatemalan mountains and out into the Gulf of Tehuantepec
Mild Winds	
levanter	a mild, humid, and often rainy east or northeast wind that blows across southern Spain
harmettan	a dry, dusty but mild wind from the northeast or east that originates over the cool Sahara in winter and blows over the west coast of Africa; brings relief from the hot, humid weather along the coastal region
Hot Winds	
simoom	a strong, dry, and dusty desert wind that blows over the African and Arabian deserts; name means "poison wind" because it is often accompanied by temperatures in excess of 52°C (125°F), which may cause heat stroke

northeastward along the coast, it causes strong northeasterly winds along coastal areas. These **northeasters** (or nor'easters) usually bring heavy rain, snow, or sleet and gale force winds, which frequently attain maximum intensity off the coast of New England. The northeaster of March, 1984, produced strong northeasterly winds from Virginia to Maine. (See Fig. 14.16.) Huge waves reaching heights of 6 m (20 ft) accompanied by winds gusting to 80 knots pounded the shoreline, causing extensive damage to beaches, beach front homes, seawalls, and boardwalks.

Local winds that also deserve attention are the strong down-mountain winds that occasionally accompany a chinook. Such winds are especially notorious in winter in Boulder, Colorado, where the average yearly windstorm damage is about $1 million. These *Boulder winds* have been recorded at over 100 knots, damaging roofs, uprooting trees, overturning mobile homes and trucks, and sandblasting car windows. Although the causes of these high winds are not completely understood, some meteorologists believe that they may be associated with large vertically oriented spinning whirls of air called *mountainadoes*. How these rapidly rotating vortices form is presently being investigated.

Table 14.1 lists winds of local significance observed in other regions of the world.

Desert Winds   Dust storms form in dry regions, where strong winds are able to lift and fill the air with particles of fine dust. In desert areas where loose sand is more prevalent, *sandstorms* develop, as high winds enhanced by surface heating rapidly carry sand particles close to the ground. A spectacular example of a storm composed of dust or sand is the **haboob** (from Arabic *hebbe*: blown). The haboob forms as cold downdrafts along the leading edge of a thunderstorm lift dust or sand into a huge, tumbling dark cloud that may extend horizontally for over 150 km and rise vertically to the base of the thunderstorm (Fig. 14.17). Spinning whirlwinds of dust frequently form along the turbulent cold air boundary, giving rise to sightings of huge *dust devils* and even tornadoes. Haboobs are most common in the African Sudan (where about 24 occur each year) and in the desert southwest of the United States, especially in southern Arizona.

The spinning vortices so commonly seen on hot days in dry areas are called **whirlwinds**, or **dust devils**. (In Australia, the Aboriginal word *willy-willy* is used to refer to a dust devil.) Generally, dust devils form on clear, hot days, as warm air rises above a heated surface. Wind, often deflected by small topographic barriers, flows into this region, rotating the rising air. Depending on the nature of the topographic feature, the spin of a dust devil around its central eye may be cyclonic or anticyclonic, and both directions occur with about equal frequency.

Having diameters of only a few meters and heights of less than a hundred meters, most dust devils are small and last only a short time. (See Fig. 14.18.) There are, however, some dust devils of sizable dimension, extending upward from the surface for many hundreds of meters. Such whirlwinds are capable of considerable damage; winds exceeding 75 knots may overturn mobile homes and tear the roofs off buildings. Fortunately, the majority of dust devils are small.

There are other desert winds that deserve mentioning. Winds originating over the Sahara Desert are given local names as they move into different regions. For example, the normal flow of surface air over North Africa is from the north; however, when a storm system is located west of Africa or southern Spain (position 1, Fig. 14.19), a hot, dry,

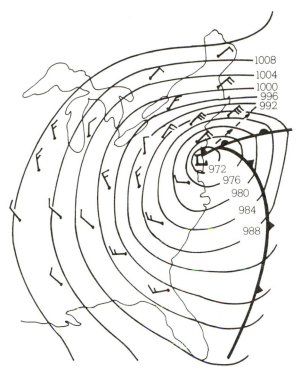

Fig. 14.16 The surface weather map for 7:00 A.M. (E.S.T.) March 29, 1984, shows an intense low-pressure area (central pressure below 972 mb, or 28.70 in.), which is generating strong northeasterly winds from the mid-Atlantic states into New England. This northeaster devastated a wide area of the eastern seaboard, causing almost $1 billion in damage. (Dashed line with arrows shows the storm's path.)

Fig. 14.17 A haboob approaching Phoenix, Arizona, from the southeast shortly after 6:00 P.M. on July 23, 1972. The height of the dust cloud is rising about 450 m (1475 ft) above the valley floor.

Fig. 14.18 A dust devil forming on a clear, hot summer day just south of Phoenix, Arizona.

and dusty easterly or southeasterly wind—the *leste*—blows over Morocco and out into the Atlantic. If the wind crosses the Mediterranean, it becomes the *leveche* when it enters southern Spain. Because of the short time it is over water, the leveche remains hot and dry.

When a low-pressure center is located at position 2, a warm, dry, dust-laden south or south-east wind originates over the Sahara Desert and blows across North Africa. This wind is known as the **sirocco**. As it moves over the Mediterranean, it picks up moisture and arrives in Sicily and southern Italy as a warm but more humid wind.

A storm located still farther to the east (position 3) can cause a dry, hot southerly wind—the **khamsin**—to blow over Egypt, the Red Sea, and Saudi

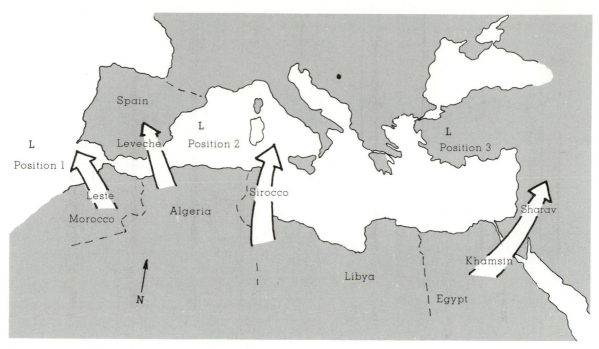

Fig. 14.19   Local winds that originate over North Africa.

Arabia. In Israel, this wind is called the *sharav*. These winds are exceedingly hot and can raise the air temperature to 50°C (122°F), while lowering the relative humidity to less than 10 percent. Because storm systems are not common over the Mediterranean in summer, scorching breezes such as these generally occur in spring or fall.

Summary

In this chapter, we have looked at winds of a more local nature. Land and sea breezes are true mesoscale winds that blow in response to local pressure differences created by the uneven heating and cooling rates of land and water. When winds move across a large body of water, they often change in speed and direction. The reduced friction of the water causes the wind to increase in speed, and the increased speed causes the Coriolis force to bend it more to the right (Northern Hemisphere). Where the winds change direction seasonally, they are termed monsoon winds. Monsoon winds exist in many parts of the world, including North America, Asia, Australia, and Africa.

Local winds that blow uphill during the day are called valley breezes and those that blow downhill at night, mountain breezes. Actually, any downslope wind is called a katabatic wind. A strong, cold katabatic wind that blows downslope from an elevated plateau is called a fall wind. Examples of fall winds are the bora in Yugoslavia and the mistral in France.

A warm katabatic wind is the chinook wind, a warm, dry wind that descends the eastern side of the Rocky Mountains. The same type of wind in the Alps is called a foehn. A warm, dry downslope wind that blows into Southern California from the east or northeast is the Santa Ana wind.

Local intense heating of the surface can produce small rotating winds, such as the dust devil, while downdrafts in a thunderstorm are responsible for the desert haboob. Some winds, such as the blizzard, are snow-bearing, while others, such as the sirocco, are dust-bearing. Finally, the wind's name may express the direction from which it blows (the Texas norther), or it may represent the region that it blows from (the Santa Ana).

Questions for Review

1. Using a diagram, explain how a thermal circulation develops.

2. Why do winds usually change direction and speed when moving over a large body of water?

3. Discuss the factors that contribute to the formation of the summer and winter monsoon in India.

4. You are fly fishing in a mountain stream during the early morning; would you expect the wind to be blowing upstream or downstream? Explain.

5. Use a diagram to explain why chinook winds are both warm and dry.

6. Name some of the benefits of a chinook wind.

7. What atmospheric conditions contribute to the development of a strong Santa Ana condition?

8. How do fall winds form?

9. Explain how a Texas norther (or blue norther) differs from a northeaster.

10. Why are haboobs more prevalent in Arizona than in Oklahoma?

11. In what part of the world would you expect to encounter each of the following winds, and what type of weather would each wind bring?

    (a) bora   (e) buran
    (b) mistral   (f) bise
    (c) blizzard   (g) sirocco
    (d) zonda   (h) leveche

Questions for Thought

1. Explain why cities near large bodies of cold water in summer experience well-developed sea breezes, but only poorly developed land breezes.

2. Why do clouds tend to form over land with a sea breeze and over water with a land breeze?

3. The convergence of two sea breezes in Florida frequently produces rain showers; the convergence of two sea breezes in California does not. Explain.

4. If campfire smoke is blowing uphill along the east-facing side of the hill and downhill along the west-facing side of the same hill, are the fires cooking breakfast or dinner? From the drift of the smoke, how were you able to tell?

5. Why don't chinook winds form on the east side of the Appalachians?

6. Why aren't dust devils more common in humid climates?

7. Explain why chinook winds can be warm and dry even when condensation does not occur on the windward side of a mountain.

8. Show, with the aid of a diagram, what atmospheric and topographic conditions are necessary for an area in the Northern Hemisphere to experience hot summer breezes from the north.

9. The prevailing winds in southern Florida are northeasterly. Knowing this, would you expect the strongest sea breezes to be along the east or west coast of southern Florida? What about the strongest land breezes?

Problems and Exercises

1. On the map of the United States, label where each of the following winds might be observed, then show with arrows the general direction of air flow that occurs with each of the winds.

    (a) Santa Ana wind
    (b) Chinook wind
    (c) California norther
    (d) Northeaster
    (e) Columbia Gorge wind (downslope)
    (f) Texas norther (blue norther)
    (g) sea breeze along the New Jersey shore
    (h) sea breeze in Los Angeles, California

2. On the accompanying map of the United States, show where the centers of atmospheric pressure should be located in order to produce the following winds. Place a large L on the map for the center of low pressure and a large H for the center of high pressure. (Be sure to place the letter representing the wind next to the L or the H.)

(a) high pressure area for a Santa Ana wind

(b) low-pressure area for a chinook wind

(c) high pressure area for a California norther

(d) low pressure area for a northeaster

(e) high pressure area for a Columbia Gorge wind (downslope wind)

(f) high- and low-pressure areas for a Texas norther (blue norther)

(g) high-pressure area for a sea breeze along the New Jersey shore

(h) low-pressure area for a sea breeze in Los Angeles, California

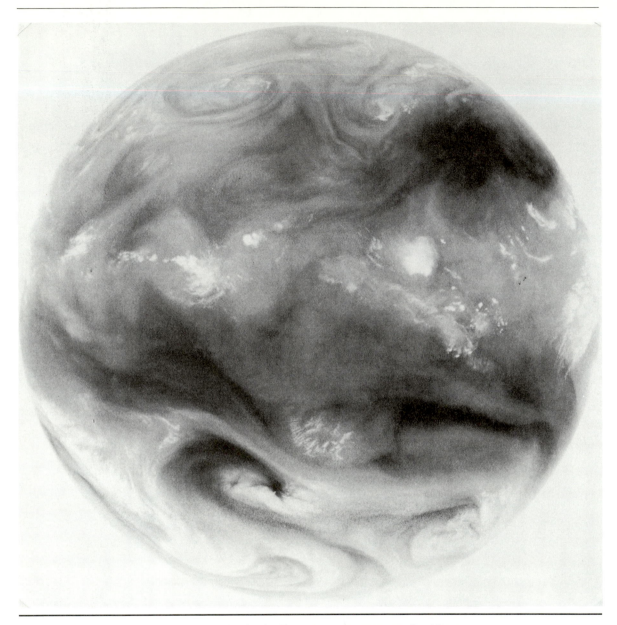

An enhanced infrared image of water vapor taken by the European geostationary satellite *Meteo-sat* on July 15, 1983. Images such as these reveal global circulation patterns with an accuracy far greater than ever before. (Photo: European Space Agency)

CHAPTER 15

Global Wind Systems

In Chapter 14, we learned that local winds vary considerably from day to day and from season to season. As you may suspect, these winds are part of a much larger circulation—the little whirls within larger whirls that we spoke of before. Indeed, if the rotating high- and low-pressure areas are like spinning eddies in a huge river, then the flow of air around the globe is like the meandering river itself. When winds throughout the world are averaged over a long period, the local wind patterns vanish, and what we see is a picture of the winds on a global scale—what is commonly called the **general circulation of the atmosphere**. Just as the eddies in a river are carried along by the overall flow of water, so the highs and lows in the atmosphere are swept along by the general circulation. We will examine this large-scale circulation of air, its effects and its features, in this chapter.

General Circulation of the Atmosphere

Before we study the general circulation, we must remember that it only represents the *average* air flow around the world. Actual winds at any one place and at any given time may vary considerably from this average. Nevertheless, the average can answer why and how the winds blow around the world the way they do—why, for example, prevailing surface winds are northeasterly in Honolulu and westerly in New York City. The average can also give a picture of the driving mechanism behind these winds, as well as a model of how heat and momentum are transported from equatorial regions poleward, keeping the climate in middle latitudes tolerable.

Contents

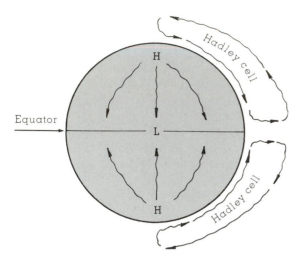

Fig. 15.1 The general circulation of air on a nonrotating earth uniformly covered with water (with the sun directly above the equator).

The underlying cause of the general circulation is the unequal heating of the earth's surface. We know from Chapter 5 that, averaged over the entire earth, incoming solar radiation is roughly equal to outgoing earth radiation. However, we also know that this energy balance is not maintained for each latitude, since the tropics experience a net gain in energy, while polar regions suffer a net loss. To balance these inequities, the atmosphere transports warm air poleward and cool air equatorward. Although seemingly simple, the actual flow of air is complex; certainly not everything is known about it. In order to better understand it, we will first look at some models (that is, artificially constructed analogies) that eliminate some of the complexities of the general circulation.

Single-Cell Model The first model is the single-cell model, in which we assume that the earth's surface is uniformly covered with water, so that differential heating between land and water does not come into play. We will further assume that the sun is always directly over the equator, so that the winds will not shift seasonally. Finally, we assume that the earth does not rotate, so that the only force we need deal with is the pressure gradient force. With these assumptions, the general circulation of the atmosphere would look much

like Fig. 15.1, a huge thermally driven convection cell in each hemisphere.

This is the **Hadley cell** (named after the eighteenth-century English meteorologist George Hadley, who first proposed the idea). It is driven by energy from the sun. Excessive heating of the equatorial area produces a broad region of surface low pressure, while at the poles excessive cooling creates a region of surface high pressure. In response to the horizontal pressure gradient, cold surface polar air flows equatorward, while at higher levels air flows toward the poles. The entire circulation consists of rising air near the equator, sinking air over the poles, an equatorward flow of air near the surface, with a return flow aloft. In this manner, some of the excess energy of the tropics is transported as sensible and latent heat to the regions of energy deficit at the poles.

Such a simple cellular circulation as this does not actually exist on the earth. For one thing, the earth rotates, so the Coriolis force would deflect the southward-moving surface air in the Northern Hemisphere to the right, producing easterly surface winds at practically all latitudes. These winds would be moving in a direction opposite to that of the earth's rotation and, due to friction with the surface, would slow down the earth's spin. We know that this does not happen and that prevailing winds in middle latitudes actually blow from the west. Therefore, observations alone tell us that a closed circulation of air between the equator and the poles is not the proper model for a rotating earth. (Models that simulate air flow around the globe have also verified this.) How, then, does the wind blow on a rotating planet? To answer, we will keep our model simple by retaining our first two assumptions; that is, that the earth is covered with water and that the sun is always directly above the equator.

Three-Cell Model If we allow the earth to spin, the simple convection system breaks into a series of rotating cells as shown in Fig. 15.2a. Although this model is considerably more complex than the single-cell model, there are some similarities. The tropical regions still receive an excess of heat and the poles a deficit. In each hemisphere, three cells instead of one have the task of energy redistribution. A surface high-pressure area is located at the poles, and a broad trough of surface low pressure still exists at the equator. From the equator to latitude 30°, the circulation closely resembles that of a Hadley cell. Let's look at this model more closely by examining what happens to the air above the equator. (Refer to Fig. 15.2 as you read the following section.)

Over equatorial waters, the air is warm, horizontal pressure gradients are weak, and winds are light. This region is referred to as the **doldrums**. (The monotony of the weather in this area has given rise to the expression "down in the doldrums.") Here, warm air rises, often condensing into huge cumulus clouds and thunderstorms called *convective "hot" towers* because of the enormous amount of latent heat they liberate. This heat makes the air more buoyant and provides energy to drive the Hadley cell. The rising air reaches the tropopause, which acts like a barrier, causing the air to move laterally toward the poles. The Coriolis force deflects this poleward flow toward the right in the Northern Hemisphere and to the left in the Southern Hemisphere, providing westerly winds aloft in both hemispheres. (We will see later that these westerly winds reach maximum velocity and produce jet streams near 30° and 60° latitudes.)

Air moving poleward from the tropics becomes more dense as it constantly cools by radiation. It also begins to converge, especially as it approaches the middle latitudes.* Recall from Chapter 2 that the convergence of air aloft produces high pressure at the surface. Hence, at latitudes near 30°, the convergence of air aloft produces belts of high pressure called **subtropical highs** (or anticyclones). As the converging air above the highs slowly descends, it becomes warmer and drier. This downward motion produces generally clear skies and warm surface temperatures; hence, it is here that we find the major deserts of the world. Over the ocean, the weak pressure gradients in the center of the high produce only weak winds. According to legend, sailing ships traveling to the New World were frequently becalmed in

*You can see why the air converges, if you have a globe of the world. Put your fingers on meridian lines at the equator and then follow the meridians poleward. Notice how the lines and your fingers bunch together in the middle latitudes.

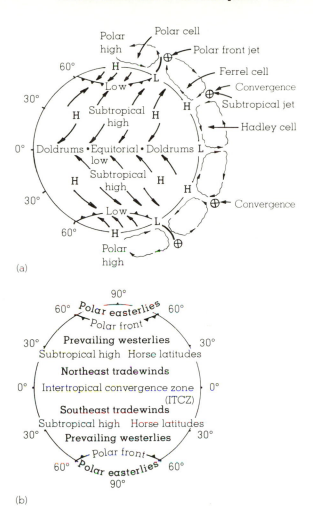

(a)

(b)

Fig. 15.2 Diagram (a) shows the wind and surface pressure distribution over a uniformly water-covered rotating earth. Diagram (b) gives the names of surface winds and pressure systems over a uniformly water-covered rotating earth.

this region; and, as food and supplies dwindled, horses were either thrown overboard or eaten. As a consequence, this region is now known as the **horse latitudes**.

From the horse latitudes, some of the surface air moves back toward the equator. It does not flow straight back, however, because the Coriolis force deflects the air, causing it to blow from the northeast in the Northern Hemisphere and from the southeast in the Southern Hemisphere. These steady winds provided sailing ships with an ocean route to the New World; hence, these winds are called the **trade winds**. Near the equator, the *northeast trades* converge with the *southeast trades* along a boundary called the **intertropical convergence zone** (ITCZ). In this region of surface convergence, air rises and continues its cellular journey.

Meanwhile, at latitude 30°, not all of the surface air moves equatorward. Some air moves toward the poles and deflects toward the east, resulting in a more or less westerly air flow in both hemispheres. Consequently, from Texas northward into Canada, it is much more common to experience winds blowing out of the west than from the east. The westerly flow is not constant; migrating areas of high and low pressure break up the surface flow pattern from time to time.

As this mild air travels poleward, it encounters cold air moving down from the poles. These two air masses of contrasting temperature do not readily mix. They are separated by a boundary called the **polar front**, a zone of low pressure where surface air converges and rises, often causing storms. Some of the rising air returns at high levels to the horse latitudes, where it sinks back to the surface in the vicinity of the subtropical high. This middle cell (called the **Ferrel cell**, after the American meteorologist William Ferrel) is completed when surface air from the horse latitudes flows poleward toward the polar front.

Behind the polar front, the cold air from the poles is deflected by the Coriolis force, so that the general flow of air is northeasterly. Hence, this is the region of the **polar easterlies**. In winter, the polar front with its cold air can move into middle and subtropical latitudes, producing a cold polar outbreak. Along the front, a portion of the rising air moves poleward, and the Coriolis force deflects the air into a westerly wind at high levels. Air aloft eventually reaches the poles, slowly sinks to the surface, and flows back toward the polar front, completing the weak polar cell.

We can summarize all of this by referring back to Fig. 15.2 and noting that, at the surface, there are two major areas of high pressure and two major areas of low pressure. Areas of high pressure exist near latitude 30° and the poles; areas of low pressure exist over the equator and near 60° latitude in the vicinity of the polar front. By knowing the way the winds blow around these systems, we have a generalized picture of surface winds throughout the world. The trade winds extend from the subtropical high to the equator; the westerlies from the subtropical high to the polar front, and the polar easterlies from the poles to the polar front.

How does this three-cell model compare with actual observations of winds and pressure? We know, for example, that upper-level winds at middle latitudes generally blow from the west. The middle Ferrel cell, however, suggests an east wind aloft as air flows equatorward. Hence, discrepancies exist between this model and atmospheric observations. This model does, however, agree closely with the winds and pressure distribution at the *surface*, and so we will examine this next.

Average Surface Winds and Pressure: The Real World When we examine the real world with its continents and oceans, mountains and ice fields, we obtain an average distribution of sea level pressure and winds for January and July, as shown in Figs. 15.3 and 15.4, respectively. Even though these data are based on sparse observations, especially in unpopulated areas, we can see that there are regions where pressure systems appear to persist throughout the year. These systems are referred to as *semipermanent highs and lows* because they move only slightly during the course of a year.

In Fig. 15.3, we can see that there are four semipermanent pressure systems in the Northern Hemisphere during January. In the eastern Atlantic, between latitudes 25° and 35°N is the **Bermuda-Azores high**, and, in the Pacific Ocean, its counterpart, the **Pacific high**. These are the subtropical anticyclones that develop in response to the convergence of air aloft near an upper-level jet stream. Since surface winds blow clockwise around these systems, we find the trade winds to the south and the prevailing westerlies to the north. In the Southern Hemisphere, where there is less contrast between land and water, the subtropical highs show up as well-developed systems with a clearly defined circulation.

Where we would expect to observe the polar front (between latitudes 40° and 65°), there are two semipermanent subpolar lows. In the North Atlantic, there is the **Icelandic low**, which covers Iceland and southern Greenland, while the **Aleu-**

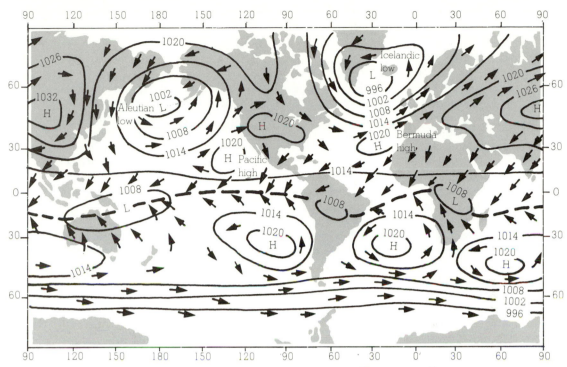

Fig. 15.3 Average January sea level pressure distribution and surface wind-flow patterns. The heavy dashed line represents the position of the ITCZ.

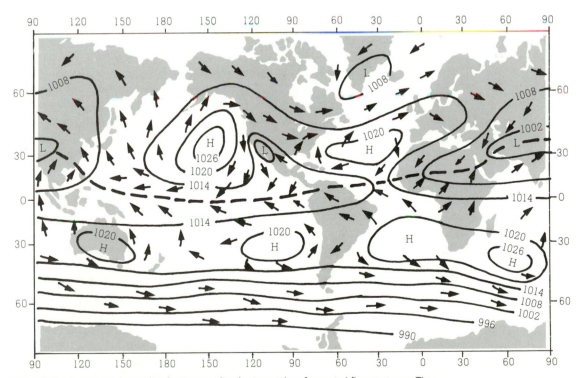

Fig. 15.4 Average July sea level pressure distribution and surface wind-flow patterns. The heavy dashed line represents the position of the ITCZ.

Wet and Dry Regions

The position of the major features of the general circulation and their latitudinal displacement (which annually averages about 10° to 15°) strongly influence the climate of many areas. For example, on the global scale, we would expect abundant rainfall where the air rises and very little where the air sinks. Consequently, areas of high rainfall exist in the tropics, where humid air rises in conjunction with the ITCZ, and between 40° and 55° latitude, where middle latitude storms and the polar front force air upward. Areas of low rainfall are found near 30° latitude in the vicinity of the subtropical highs

and in polar regions where the air is cold and dry. Poleward of the equator, between the doldrums and the horse latitudes, the area is influenced by both the ITCZ and the subtropical high. In summer (high sun period), the subtropical high moves poleward and the ITCZ invades this area, bringing with it ample rainfall. In winter (low sun period), the subtropical high moves equatorward, bringing with it clear, dry weather.

During the summer, the Pacific high drifts northward to a position off the California coast. Strong subsidence of air on its eastern side tends to keep summer weather along the west

coast relatively dry. The rainy season typically occurs in winter when the high moves south and storms are able to penetrate the region. Along the east coast, the clockwise circulation of winds around the Bermuda high brings warm, tropical air northward into the United States and southern Canada from the Gulf of Mexico. Because subsidence is not as well developed on this side of the high, the humid air can rise and condense into towering cumulus clouds and thunderstorms. So, it is the air motions associated with the subtropical highs that keep summer weather dry in California and moist in Georgia.

---

tian low sits over the Aleutian Islands in the North Pacific. These zones of cyclonic activity actually represent regions where numerous storms, having traveled eastward, tend to converge, especially in winter. In the Southern Hemisphere, the subpolar low forms a continuous trough that completely encircles the globe.

On the January map, there are other pressure systems, which are not semipermanent in nature. Over Asia, for example, there is a huge (but shallow) anticyclone called the **Siberian high**, which forms because of the intense cooling of the land. South of this system, the winter monsoon shows up clearly, as air flows away from the high across Asia and out over the ocean. A similar (but less intense) anticyclone is evident over North America.

As summer approaches, the land warms and the cold, shallow highs disappear. In some regions, areas of surface low pressure replace areas of high pressure. The lows that form over warm land are the *thermal low* and the *monsoon low*. On the July map (Fig. 15.4), warm thermal lows are found over the desert southwest of the United

States and over the plateau of Iran. Notice that these systems are located at the same latitudes as the subtropical highs. We can understand why they form when we realize that, during the summer, the subtropical high-pressure belt girdles the world aloft near 30° latitude (Fig. 15.6). Within this system, the air subsides and warms, producing clear skies (which allow intense surface heating by the sun). This air near the ground warms rapidly, rises only slightly, then flows laterally several hundred meters above the surface. The outflow lowers the surface pressure and, as we saw in Chapter 14, a shallow thermal low forms. The monsoon low over India usually begins as a thermal low when the continent of Asia warms. As the low intensifies, warm, moist air from the ocean is drawn into it, producing the wet summer monsoon so characteristic of India and Southeast Asia. Where these surface winds converge with the general westerly flow, rather weak monsoon depressions form. These enhance the position of the monsoon low on the July map.

When we compare the January and July maps,

we can see several changes in the semipermanent pressure systems. The strong, subpolar lows so well developed in January over the Northern Hemisphere are hardly discernible on the July map. The subtropical highs, however, remain dominant in both seasons. Because the sun is overhead in the Northern Hemisphere in July and overhead in the Southern Hemisphere in January, the zone of maximum surface heating shifts seasonally. In response to this, the major pressure systems, wind belts, and ITCZ (heavy dashed line) shift toward the north in July and toward the south in January.

To see what effect the annual shifting of the major features of the general circulation has on global rainfall patterns, read the focus section entitled "Wet and Dry Regions." (See p. 298.)

Average Wind Flow and Pressure Patterns Aloft Figures 15.5 and 15.6 are average global 500-mb charts for the months of January and July, respectively. Look at both charts carefully and observe that some of the surface features of the general circulation are reflected on these upper-air charts. On the January map, for example, both the Icelandic low and Aleutian low are located to the west of their surface counterparts. On the July map, the subtropical high pressure areas of the Northern Hemisphere appear as belts of high height (pressure) that tend to circle the globe south of 30°N. In both hemispheres, the air is warmer over low latitudes and colder over high latitudes. This horizontal temperature gradient establishes a horizontal pressure (contour) gradient that causes the winds to blow from the west, especially in middle and high latitudes.* Notice that the temperature gradients and the contour gradients are steeper in winter than in summer. Consequently, the winds aloft are stronger in winter than in summer. The westerly winds, however, do not extend all the way to the equator as easterly winds appear on the equatorward side of the upper-level subtropical highs.

In middle and high latitudes, the westerly winds continue to increase in speed above the 500-mb level. We already know that the wind speed increases up through the friction layer, but why should

*Remember that, at this level (about 5600 m or 18,000 ft above sea level), the winds are approximately geostrophic, and tend to blow more or less parallel to the contour lines.

it continue to increase at higher levels? You may remember from Chapter 12 that the geostrophic wind at any latitude is directly related to the pressure gradient and inversely related to the air density. Therefore, a greater pressure gradient will result in stronger winds, and so will a decrease in air density. Owing to the fact that air density decreases with height, the same pressure gradient will produce stronger winds at higher levels. Actually, the winds increase in speed up to the tropopause. Above the tropopause, the temperature gradients reverse. This changes the pressure gradients and reduces the strength of the westerly winds. Where strong winds tend to concentrate into narrow bands at the tropopause, we find the jet stream.

Jet Streams

Atmospheric **jet streams** are swiftly flowing air currents thousands of kilometers long, hundreds of kilometers wide, and only a few kilometers thick. Wind speeds in the central core of a jet stream often exceed 100 knots and occasionally exceed 250 knots. Jet streams are usually found at the tropopause at elevations between 10 and 15 km (6 and 9 mi), although they may occur at both higher and lower altitudes.

Jet streams were first encountered by high-flying military aircraft during World War II, but their existence was suspected before the war. Ground-based observations of fast-moving cirrus clouds had revealed that westerly winds aloft must be moving rapidly indeed.

Figure 15.7 illustrates the average position of the jet streams, tropopause, and general air flow for the Northern Hemisphere in winter. From this diagram, we can see that there are two jet streams, both located in tropopause gaps, where mixing between tropospheric and stratospheric air takes place. The jet stream situated at nearly 13 km (43,000 ft) above the subtropical high is the **subtropical jet**. The jet stream situated at about 10 km (33,000 ft) near the polar front is known as the **polar front jet**. Since both are found at the tropopause, they are called *tropopause jets*.

In Fig. 15.7, the wind in the jet core would be flowing as a westerly wind away from the viewer.

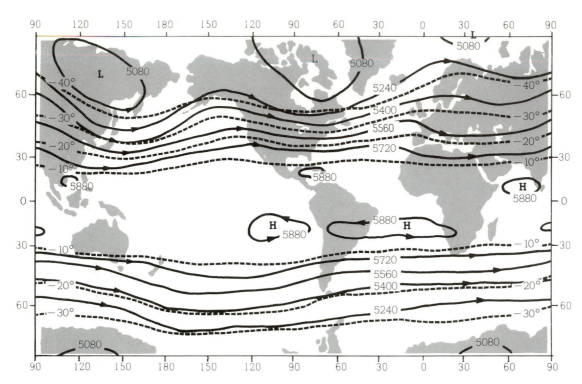

Fig. 15.5 Average 500-mb chart for the month of January. Solid lines are contour lines in meters. Dashed lines are isotherms in °C.

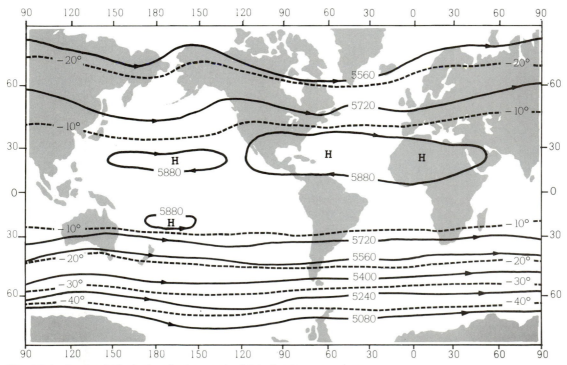

Fig. 15.6 Average 500-mb chart for the month of July. Solid lines are contour lines in meters. Dashed lines are isotherms in °C.

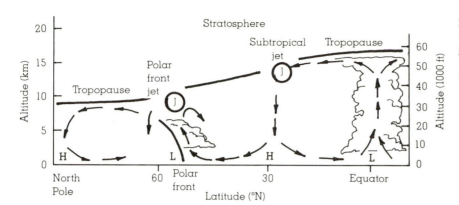

Fig. 15.7 Average position of the polar front and subtropical jet streams, with respect to a model of the general circulation in winter.

This, of course, is only an average, as jet streams often meander into broad loops that sweep north and south. This is especially true of the polar front jet, which, while sweeping into this pattern, may even merge with the subtropical jet.

We can see the looping pattern of the jet by studying Fig. 15.8. This diagram shows the position of the jet stream at the 300-mb level (near 9 km or 30,000 ft) on April 18, 1979. The airflow pattern and jet core are given by the heavy dark arrow; the dashed lines represent lines of equal wind speed (*isotachs*). Since the wind flow at the 300-mb level pretty much parallels the contour lines, troughs of low pressure exist over the western states and along the east coast, while a ridge of high pressure covers the Plains states. Note that the strongest winds are located in the troughs. This region of strong wind is called a **jet maximum** or **jet streak**. In Chapter 17, we will see how this region of high winds is an important factor in developing and intensifying surface storm systems.

An air particle moving in the jet core over Los Angeles would move northward into Canada then loop around and head southeastward, eventually moving off the coast of Virginia. This meridional pattern illustrates an important function of jet streams. On the eastern side of the trough, swiftly moving air carries warmer air poleward, while, on the western side, the more northerly flow brings cold air equatorward. Jet streams are, therefore, significant in the global transfer of heat. Since jet streams tend to meander around the world, we can also see how radioactive debris from aboveground nuclear explosions set off in western China could end up in the United States and elsewhere.

As you might guess, the fallout from the 1976 Chinese nuclear explosion was first detected in the western states and several days later along the east coast.

The ultimate cause of jet streams is the energy imbalance that exists between high and low latitudes. How, then, do jet streams actually form?

The Formation of the Polar Front Jet and the Subtropical Jet Horizontal variations in temperature and pressure offer clues to the existence of the polar front jet. Figure 15.9 shows a side profile of the atmosphere in the region of the polar front. Since the polar front is a boundary separating the cold polar air to the north from the warm subtropical air to the south, the greatest contrast in air temperature occurs along the frontal zone. We can see this as the isotherm dips sharply crossing the front. This rapid change in temperature produces a rapid change in pressure (as shown by the sharp bending of the constant pressure surface as it passes through the front). *The sudden change in pressure sets up a steep pressure gradient, which intensifies the wind speed and causes the jet stream.* Because the north-south temperature contrast along the front is strongest in winter and weakest in summer, the polar front jet shows seasonal variations. In winter, the winds blow stronger and the jet moves farther south as the leading edge of the cold air extends into subtropical regions. In summer, the jet is weaker and is usually found over more northerly latitudes.

The subtropical jet stream, which is usually strongest near the 200-mb level, tends to form along the poleward side of the Hadley cell (Fig.

Fig. 15.8 Position of the jet stream at the 300-mb level during the morning of April 18, 1979. Dashed lines are lines of equal wind speed (isotachs) in knots. Heavy dark line is the jet stream axis.

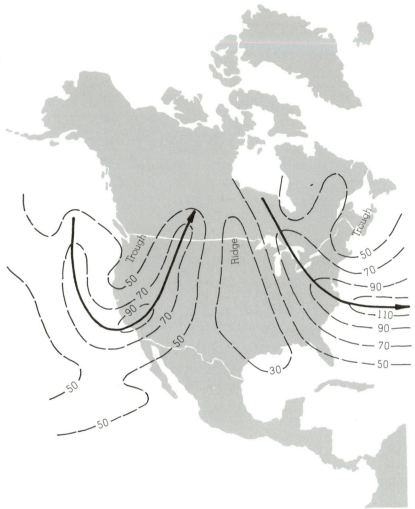

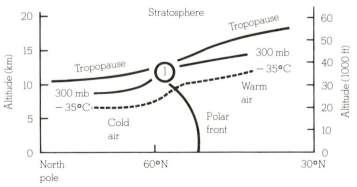

Fig. 15.9 Vertical view of the polar front jet stream in association with a sharply dipping pressure surface (solid line) and isotherm (dashed line).

15.7). Here, warm air carried poleward by the Hadley cell produces sharp temperature contrasts along a boundary called the **subtropical front**. In the vicinity of the subtropical front, sharp contrasts in temperature produce sharp contrasts in pressure and strong winds.

When we examine jet streams carefully, we see that another mechanism (other than a steep temperature gradient) causes a strong westerly flow aloft. The cause appears to be the same as that which makes an ice skater spin faster when the arms are pulled in close to the body—the **conservation of angular momentum**.

At the equator, the earth rotates toward the east at a speed close to 1000 knots. On a windless day, the air above moves eastward at the same speed. If somehow the earth should suddenly stop rotating, the air above would continue to move eastward until friction with the surface brought it to a halt; the air keeps moving because it has momentum.

Straight-line momentum—called *linear momentum*—is the product of the mass of the object times its velocity. An increase in either the mass or the velocity (or both) produces an increase in momentum. Air on a spinning planet moves about an axis in a circular path and has angular momentum. Along with the mass and the speed, angular momentum depends upon the distance (*r*) between the mass of air and the axis about which it rotates. *Angular momentum* is defined as the product of the mass (*m*) times the velocity (*v*) times the radial distance (*r*):

Angular momentum = *mvr*.

As long as there are no external twisting forces (torques) acting on the rotating system, the angular momentum of the system does not change. We say that angular momentum is *conserved*; that is, the product of the quantity *mvr* at one time will equal the numerical quantity *mvr* at some later time. Hence, a decrease in radius must produce an increase in speed and vice versa. An ice skater, for instance, with arms fully extended rotates quite slowly. As the arms are drawn in close to the body, the radius of the circular path (*r*) decreases, which causes an increase in rotational velocity (*v*), and the skater spins faster. As arms become fully extended again, the skater's speed decreases. The conservation of angular momentum, when

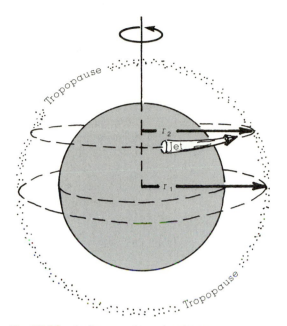

Fig. 15.10 Air flowing poleward at the tropopause moves closer to the rotational axis of the earth (r_2 is less than r_1). This decrease in radius is compensated for by an increase in velocity and the formation of a jet stream.

applied to moving air, will help us to understand the formation of a jet stream.

Consider heated air parcels rising from the equatorial surface on a calm day. As the parcels approach the tropopause, they spread laterally and begin to move poleward. If we follow the air moving northward (Fig. 15.10) we see that, because of the curvature of the earth, air constantly moves closer to its axis of rotation (*r* decreases). Because angular momentum is conserved (and since the mass of air is unchanged), the decrease in radius must be compensated for by an increase in speed. The air must, therefore, move faster to the east than a point on the earth's surface does. To an observer, this is a west wind. Hence, the conservation of angular momentum of northward-flowing air leads to the generation of strong westerly winds and the formation of a jet stream.

Other Jet Streams There is another jet stream that forms in summer near the tropopause above Southeast Asia, India, and Africa. Here, the altitude of the summer tropopause and the jet stream is near 15 km. Because the jet forms on the equatorward side of the upper-level subtropical high, its winds are easterly and hence it is known

Momentum—A Case of Give and Take

We know that energy from the sun drives the atmospheric circulation. Although the circulation pattern is complex, the general flow aloft is westerly. However, the westerly flow is not constant, for it breaks into eddies, cyclones, and anticyclones that transfer heat and momentum poleward. This transfer process feeds energy to the jet stream and maintains the westerly winds aloft. Therefore, this transfer mechanism is responsible for maintaining the general circulation of the atmosphere. How, then, does the momentum exchange take place?

There is a constant exchange of momentum between the earth and the atmosphere. Since momentum is simply mass times velocity, the momentum of an object whose mass is 1 is represented by velocity only. For such an object (a mass of air, for example), a change in momentum represents a change in velocity.

Consider the earth to be a rotating globe and the atmosphere to be your hand. When your hand rests on the globe and rotates with it at the same speed, there is, for all practical purposes, no transfer of momentum. But if your hand is on the equator and moves more slowly to the east than the rotating globe, then the friction between your hand and the globe will reduce the globe's rate of spin. Hence, there is a transfer of momentum from your hand to the globe. Because the actual winds in the tropics are easterly, there is a transfer of momentum from

the moving air to the earth, which should slow down the earth's rotation in a similar manner. But the earth's spin does not slow because of the westerly winds in middle latitudes.

To see what effect westerly winds have on the earth, place your hand on a rotating globe, then move it faster to the east than the globe rotates. Momentum transfer from your hand to the globe will cause the globe to spin faster. Similarly, the westerly winds of middle latitudes should increase the earth's rate of spin, but they do not because they are compensated for by the easterly winds in the tropics.

The rotating earth affects the momentum of the atmosphere as well. We know that the northeast trades blow from about 30°N latitude toward the equator. As the air moves closer to the equator, it also moves farther away from the earth's axis of rotation (r increases). Therefore, to conserve *angular momentum*, the increase in r must be offset by a decrease in velocity (v). However, the slower the air moves eastward on the rotating earth, the faster the wind appears to be blowing from the east to an observer on the earth's surface. (For a calm wind, the air is moving eastward at the same rate that the earth spins.) As a consequence, the trades should be strong easterly winds near the equator. In fact, however, the trades are fairly steady, but weak. The reason is that friction with the earth drags the air along more rapidly toward the east; thus, the apparent west-

ward motion of the air decreases. The net result of this frictional interaction is that the earth imparts some of its momentum to the tropical air above. So, in low latitudes, the atmosphere gains momentum from the earth.

Meanwhile, in middle latitudes, the prevailing westerlies curve slightly northward away from the subtropical high. As they move northward, they also move closer to the earth's axis of rotation (r decreases). The conservation of angular momentum requires that, as r decreases, surface wind velocity should increase and eventually reach the speed of a fast-flowing westerly jet. Surface winds do not blow that fast; again, the reason is surface friction, which slows the westerlies and reduces the air's angular momentum. Therefore, in middle latitudes, the air near the surface loses momentum to the earth.

If the atmosphere were to continually lose momentum in middle latitudes and gain momentum in low latitudes, both the prevailing westerlies and northeast trades would slow until the air is calm. Since we know this does not happen, there must be a net transfer of momentum from low latitudes toward high latitudes in order to maintain our wind systems. It is the large low- and high-pressure areas, the cyclones and anticyclones, of the middle latitudes that are primarily responsible for this transfer of momentum. We can see how this is accomplished by considering a laboratory experi-

FOCUS ON A SPECIAL TOPIC, continued

ment that simulates the circulation of the atmosphere.

The "*dishpan experiment*" consists of a flat pan filled with water. Around the edge of the pan, a heating coil supplies heat to the pan's "equator." In the center of the pan, a cooling cylinder represents the "pole," and ice water is continually supplied here. The temperature difference between "equator" and "pole" produces a thermally driven circulation that transports heat poleward; when the pan is rotated, there is a transfer of angular momentum as well.

Aluminum powder is sprinkled on the water, so that motions of the fluid can be seen. If the pan rotates slowly, the flow of water is symmetric with the polar cylinder. This would represent a zonal (west to east) wind in the atmosphere (see Fig. 1a). As the pan rotates faster, the flow increases and waves develop. Eventually, the waves break into a series of waves and rotating eddies similar to those shown in Fig. 1b. The flow develops waves and eddies because, without them, the circulation alone is unable to effectively transport heat and momentum poleward.

The atmospheric counterpart of these eddies are the cyclones and anticyclones of the middle latitudes. At the earth's surface, they occur as winds circulating around centers of low and high pressure; aloft, however, they usually form elongated troughs and ridges. For instance, a storm in the upper troposphere might look like a trough similar to the one seen in Fig. 2. Notice that it

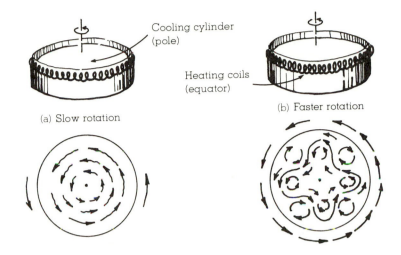

Fig. 1 (a) A slowly rotating pan produces a flow that is symmetric with the poles. (b) A pan rotating at a speed that corresponds to the rotation of the earth produces troughs, ridges, and eddies, which appear very similar to the patterns we see on an upper-level chart.

has an asymmetric shape and tilts in a southwest-northeast direction. The winds on the east side of the trough are generally stronger than the winds on the west side. Also, the northward-moving winds on the east side have a stronger west-to-east component than do the southward-moving winds on the west side. This means that there is a net west-to-east transfer of momentum from lower latitudes to the middle latitude westerlies. Therefore, the next time a storm causes you some discomfort, remember that, without such storms, the atmosphere could not maintain its circulation.

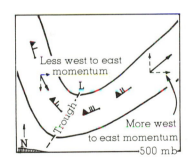

Fig. 2 A well-developed surface storm usually shows up as a wave with a tilted trough (dashed line) on a 500-mb chart. The wave transports westerly momentum poleward because the winds east of the trough have a greater westerly component than do the winds west of the trough.

as the **tropical easterly jet stream**. The formation of the tropical easterly jet stream appears to be related to the warming of the air over large elevated land masses, such as Tibet. During the summer, the air above this region (even at high elevations) is warmer than the air above the ocean to the south. This contrast in temperature produces a north to south pressure gradient and strong easterly winds that reach a maximum speed near 15°N latitude.

Not all jet streams form at the tropopause. For example, there is a jet stream that forms near the top of the stratosphere over polar latitudes. Because little, if any, sunlight reaches the polar region during the winter, air in the upper stratosphere is able to cool to low temperatures. By comparison, in equatorial regions, sunlight prevails all year long, allowing stratospheric ozone to absorb solar energy and warm the air. The horizontal temperature gradients between the cold poles and the warm tropics create steep horizontal pressure gradients, and a strong westerly jet forms above the poles at an elevation near 50 km (30 mi). Because this wind maximum occurs in the stratosphere during the dark polar winter, it is known as the **stratospheric polar night jet stream**.

In summer, the polar regions experience more hours of sunlight than tropical areas. Stratospheric temperatures over the poles increase more than at the same altitude above the equator, which causes the horizontal temperature gradient to reverse itself. The jet stream disappears, and, in its place, are weaker easterly winds.

Jet streams are not confined to the stratosphere; they also form in the upper mesosphere and in the thermosphere. Not much is known about the winds at these high levels, but they are probably related to the onslaught of charged particles that constantly bombard this region of the atmosphere.

Jet streams form near the earth's surface as well. One such jet develops over the Central Plains of the United States, where it occasionally attains speeds of 60 knots several hundred meters above the surface. This wind speed maximum, which usually flows from the south or southwest, is known as a **low-level jet**. It typically forms at night above a temperature inversion. Apparently, the stable air

reduces the interaction between the air within the inversion and the air directly above. Consequently, the air in the vicinity of the jet is able to flow faster because it is not being slowed by the lighter winds below. Also, the north-south trending Rocky Mountains tend to funnel the air northward. However, these factors alone are unable to account for the strong winds in the jet core.

Another important element contributing to the formation of the low-level jet is the downward sloping of the land from the Rockies to the Mississippi Valley, which causes nighttime air above regions to the west to be cooler than air at the same elevation to the east. This horizontal contrast in temperature causes pressure surfaces to dip toward the west. The dipping of pressure surfaces produces strong pressure gradient forces directed from east to west which, in turn, cause strong southerly winds.

During the summer, these strong southerly winds carry moist air from the Gulf of Mexico into the Central Plains. This moisture, coupled with converging-rising air of the low-level jet, enhances thunderstorm formation. Therefore, on warm, moist, summer nights, when the low-level jet is present, it is common to have nighttime thunderstorms over the plains.

Atmosphere—Ocean Interactions

Although scientific understanding of all the interactions between the oceans and the atmosphere is far from complete, there are some relationships that deserve mentioning here.

Global Wind Patterns and Surface Ocean Currents As the wind blows over the oceans, it causes the surface water to drift along with it. The moving water gradually piles up, creating pressure differences within the water itself. This leads to further motion several hundreds of meters down into the water. In this manner, the general wind flow around the globe starts the major surface ocean currents moving. The relationship between the general circulation and ocean currents can be seen by comparing Figs. 15.3 and 15.4 with Fig. 15.11.

Because of the larger frictional drag in water, ocean currents move more slowly than the prevailing wind. Typically, these currents range in speed from several kilometers per day to several kilometers per hour. As we can see in Fig. 15.11, major ocean currents do not follow the wind pattern exactly; rather, they spiral in semiclosed circular whirls called *gyres*. In the North Atlantic, the prevailing winds blow clockwise and outward from the subtropical high, while the ocean currents move in a more or less circular, but still clockwise, pattern. As the water moves beneath the wind, the Coriolis force deflects the water to the right in the Northern Hemisphere and to the left in the Southern Hemisphere. This causes the surface water to move at an angle between 20° and 45° to the direction of the wind. Hence, surface water tends to move in a circular pattern as winds blow outward away from the center of the subtropical high.

Important interactions between the atmosphere and the ocean can be seen by examining the huge gyre in the North Atlantic. Flowing northward along the east coast of the United States is a tremendous warm water current called the *Gulf Stream*.

Table 15.1 MAJOR OCEAN CURRENTS

1. Gulf Stream	9. South Equatorial Current	17. Peru or Humbolt Current
2. North Atlantic Drift	10. South Equatorial Countercurrent	18. Brazil Current
3. Labrador Current	11. Equatorial Countercurrent	19. Falkland Current
4. West Greenland Drift	12. Kuroshio Current	20. Benguela Current
5. East Greenland Drift	13. North Pacific Drift	21. Agulhas Current
6. Canary Current	14. Alaska Current	22. West Wind Drift
7. North Equatorial Current	15. Oyashio Current	
8. North Equatorial Countercurrent	16. California Current	

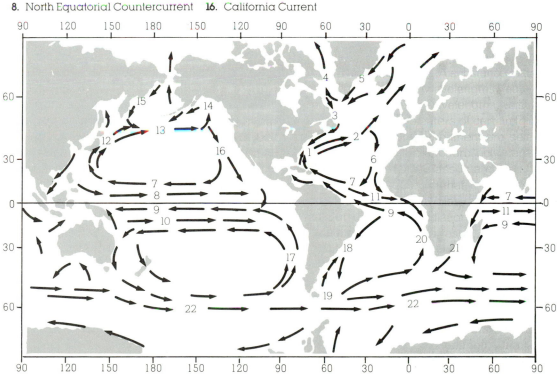

Fig. 15.11 Average position and extent of the major surface ocean currents.

The Gulf Stream carries vast quantities of warm, tropical water into higher latitudes. To the north, on the western side of the smaller subpolar gyre, cold water moves southward along the Atlantic coast of North America. This *Labrador Current* brings cold water as far south as Massachusetts in summer and North Carolina in winter. In the vicinity of the Grand Banks of Newfoundland, where the two opposing currents flow side by side, there is a sharp temperature gradient. When warm Gulf Stream air blows over the cold Labrador Current water, the stage is set for the formation of the fog so common to this region.

Meanwhile, steered by the prevailing westerlies, the Gulf Stream swings away from the coast of North America and moves eastward toward Europe. Gradually, it widens and slows as it merges into the broader *North Atlantic Drift*. As this current approaches Europe, it divides into two currents. A portion flows northward along the coasts of Great Britain and Norway, bringing with it warm water (which helps keep winter temperatures much warmer than one would expect this far north). The other part of the North Atlantic Drift flows southward as the *Canary Current* which transports cool northern water equatorward. Eventually, the Atlantic gyre is completed as the Canary Current merges with the westward-moving *North Equatorial Current*, which derives its energy from the northeast trades.

The ocean circulation in the North Pacific is similar to that in the North Atlantic. On the western side of the ocean is the Gulf Stream's counterpart, the warm, northward-flowing *Kuroshio Current*, which gradually merges into the slower-moving *North Pacific Drift*. A portion of this current flows southward along the coastline of the western United States as the cool *California Current*. In the Southern Hemisphere, surface ocean circulations are much the same except that the gyres move counterclockwise in response to the winds around the subtropical highs. In the Indian Ocean, monsoon circulations tend to complicate the general pattern of ocean currents.

Up to now, we have seen that atmospheric circulations and ocean circulations are closely linked; wind blowing over the oceans produces surface ocean currents. The currents, along with the wind, transfer heat from tropical areas, where there is a surplus of energy, to polar regions, where there is a deficit. This helps to equalize the latitudinal energy imbalance with about 40 percent of the total heat transport in the Northern Hemisphere coming from surface ocean currents. The environmental implications of this heat transfer are tremendous. If the energy imbalance were to go unchecked, yearly temperature differences between low and high latitudes would increase greatly, and the climate would gradually change.

Satellite pictures reveal that distinct temperature gradients exist along the boundaries of surface ocean currents. For example, off the east coast of the United States, where the warm Gulf Stream meets cold waters to the north, sharp temperature contrasts often exist. The boundary separating the two masses of water with contrasting temperatures and densities is called an **oceanic front**. Along this frontal boundary, a portion of the meandering Gulf Stream occasionally breaks away and develops into a closed circulation of either cold or warm water—a whirling eddy (Fig. 15.12). Because these eddies transport heat and momentum from one region to another, they may have a far-reaching effect upon the climate and a more immediate impact upon coastal waters. Scientists are investigating the effects of these eddies.

Upwelling Earlier, we saw that the cool California Current flows roughly parallel to the west coast of North America. From this, we might conclude that summer surface water temperatures would be cool along the coast of Washington and gradually warm as we move south. A quick glance at the water temperatures along the west coast of the United States during August (Fig. 15.13) quickly dispels that notion. The coldest water is observed along the northern California coast near Cape Mendocino. Why there? To answer this, we need to examine how the wind influences the movement of surface water.

As the wind blows over an open stretch of ocean, the surface water beneath it is set in motion. Under the influence of the Coriolis force, the water moves at an angle about 45° to the direction of the wind. If we imagine the top layer of ocean water to be broken into a series of layers, then

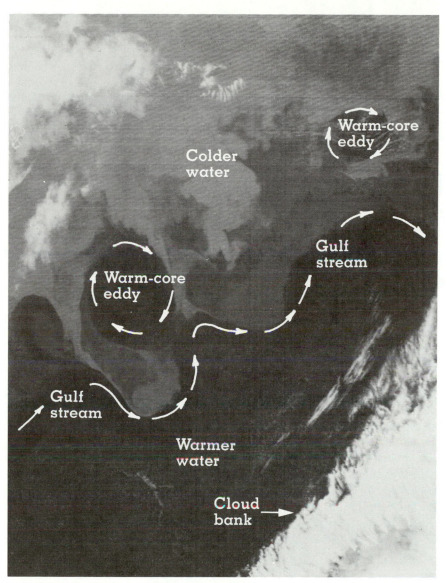

Fig. 15.12   The formation of eddies along the Gulf Stream. (Infrared satellite picture.)

each layer will exert a frictional drag on the layer below. Each successive layer will not only move a little slower than the one above, but (because of the Coriolis effect) each layer will also rotate slightly to the right of the layer above. (The rotation of each layer is to the right in the Northern Hemisphere and to the left in the Southern Hemisphere.) Consequently, descending from the surface, we would find water slowing and turning to the right until, at some depth (usually about

100 m), the water actually moves in a direction opposite to the flow of water at the surface. This turning of water with depth is known as the **Ekman Spiral**. The Ekman Spiral in Fig. 15.14 shows us that the average movement of surface water down to a depth of about 100 m is at right angles (90°) to the surface wind direction. The Ekman Spiral helps to explain why, in summer, surface water is cold along the west coast of North America.

The summertime position of the Pacific high and

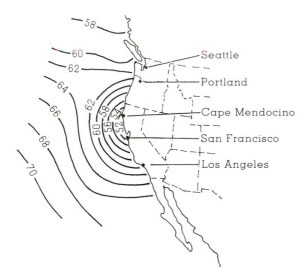

Fig. 15.13   Average sea surface temperatures (°F) along the west coast of the United States during August.

the low coastal mountains cause winds to blow parallel to the California coastline (Fig. 15.15). The net transport of surface water is at right angles to the wind, in this case, out to sea. As surface water drifts away from the coast, cold, nutrient-rich water from below rises to replace it. The rising of cold water is known as **upwelling**. Upwelling is strongest and surface water is coolest in this area because here the wind parallels the coast.

Summertime weather along the west coast often consists of low clouds and fog, as the air

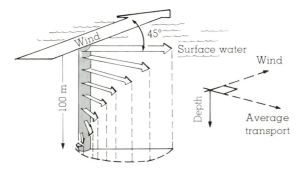

Fig. 15.14   The Ekman Spiral. Winds move the water, and the Coriolis force deflects the water to the right (Northern Hemisphere). Below the surface each successive layer of water moves more slowly and is deflected to the right of the layer above. The average transport of surface water in the Ekman layer is at right angles to the prevailing winds.

over the water is chilled to its saturation point. On the brighter side, upwelling produces good fishing, as higher concentrations of nutrients are brought to the surface (Fig. 15.16). But swimming is only for the hardiest of souls, since the average surface water temperature in summer is nearly 10°C (18°F), colder than the average coastal water temperature found at the same latitude along the Atlantic Coast.

Global Wind Patterns and Surface Ocean Temperatures   Between the ocean surface and the atmosphere, there is an exchange of heat, moisture, and momentum that depends, in part, on temperature differences between water and air. In winter, when air-water temperature contrasts are greatest, there is a substantial transfer of sensible and latent heat from the ocean surface into the atmosphere. This energy helps to maintain the global air flow. Because of the difference in heat capacity between water and air, even a relatively small change in surface ocean temperatures could modify atmospheric circulations. This could have far-reaching effects on global weather patterns.

Meteorologists have been able to correlate surface water temperatures in the Pacific Ocean with weather in the United States. Between 1971 and 1975 the surface temperatures in the central Pacific were warmer than normal, while surface temperatures were cooler than normal along the west coast of North America. During this period, the east coast experienced warm, dry winters, while the weather in the west was relatively wet. In contrast, the winter of 1976–1977 produced record-breaking cold temperatures in the east and drought in the west. At about this time, surface water temperatures in the central Pacific became cooler, while those along the west coast became warmer. During the winter of 1982–1983, the water temperatures off the west coast became much warmer than normal. Fish accustomed to colder water migrated northward. As the westerly jet stream intensified, massive storms one after another pounded the California coast, causing extensive damage and record-breaking rains. Heavy snow fell over the Rocky Mountains and widespread flooding occurred in the Gulf states. Meanwhile, the northeastern part of the nation experienced an exceptionally mild winter. The cause

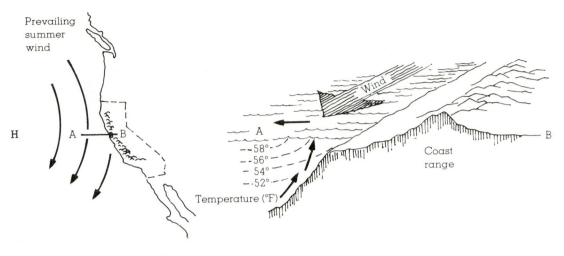

Fig. 15.15 As winds blow parallel to the west coast of North America, surface water is transported to the right (out to sea). Cold water moves up from below (upwells) to replace the surface water.

of these weather events has been linked to an atmospheric—oceanic phenomenon referred to as El Niño.

El Niño and the Southern Oscillation   Along the west coast of South America, where the cool Peru Current sweeps northward, southerly winds promote upwelling of cold, nutrient-rich water that gives rise to large fish populations, especially anchovies. The abundance of fish supports a large population of sea birds whose droppings (called *guano*) produce huge phosphate-rich deposits, which support the fertilizer industry. Near the end of each calendar year, a warm current of nutrient-poor tropical water moves southward, replacing the cold, nutrient-rich surface water. Because this condition frequently occurs around Christmas, local residents called it **El Niño** (Spanish for child), referring to the Christ child.

In most years, the warming lasts for only a few weeks to a month or more, after which weather patterns usually return to normal and fishing improves. However, when El Niño conditions last for many months, and a more extensive ocean warming occurs, the economic results can be catastrophic. (It is this extremely warm episode that scientists now refer to as El Niño.)

During a severe El Niño, large numbers of fish and marine plants may die. Dead fish and birds may litter the water and beaches of Peru; their decomposing carcasses deplete the water's oxygen supply, which leads to the bacterial production of huge amounts of smelly hydrogen sulfide. The El Niño of 1972–1973 reduced the annual Peruvian anchovy catch from 10.3 million metric tons in 1971 to 4.6 million metric tons in 1972. Since much of the harvest of this fish is converted into fishmeal and exported for use in feeding livestock and poultry, the world's fishmeal production in 1972 was greatly reduced. Countries such as the United States that rely on fishmeal for animal feed had to use soybeans as an alternative. This raised poultry prices in the United States by more than 40 percent. A less severe El Niño occurred in 1976–1977. But an extremely strong El Niño—one of the strongest ever seen—occurred during 1982–1983.

Normally, in the tropical Pacific ocean, the trades are persistent winds that blow westward from a region of higher pressure over the eastern Pacific toward a region of lower pressure centered over Indonesia. (See Fig. 15.3.) The westward-moving trades drag some of the cool water located along the South American coast with them. As this water moves westward, it is heated by sunlight and the atmosphere. Consequently, in the Pacific Ocean, surface water along the equator is cool in the east and warm in the west. In addition, the dragging of surface water raises the sea level in the western Pacific and lowers it in the eastern Pacific. This

Fig. 15.16 The enhanced infrared satellite picture shows that, due to upwelling, the coldest water (lighter tones) is observed adjacent to the coast. Filaments of light tones show where the colder upwelling water interacts with the warmer southward moving water of the California current, producing oceanic fronts. Because certain types of fish seem to concentrate near these boundaries, the location of ocean fronts is important to the commercial fishing industry.

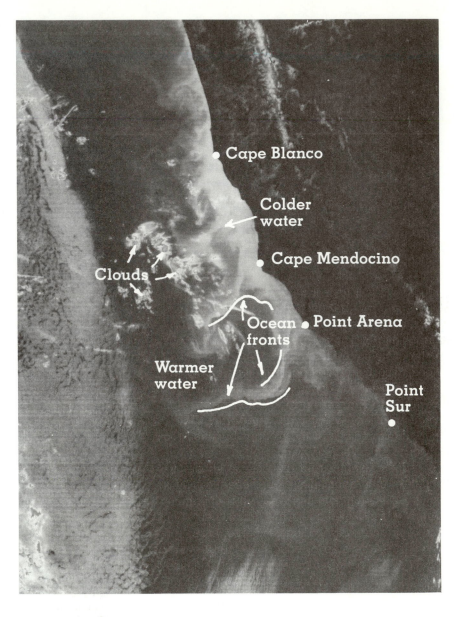

produces a thick layer of warm water over the tropical western Pacific Ocean.

Every few years, the surface atmospheric pressure patterns break down, as air pressure rises over the region of the western Pacific and falls over the eastern Pacific. This change in pressure weakens the trades, and, during strong pressure reversals, west-blowing winds are replaced by east-blowing winds. Surface water warms over a broad area of the tropical Pacific and heads eastward toward South America in a surge known as a

Kelvin wave. Toward the end of the warming period, which can last for many months, atmospheric pressure over the eastern Pacific begins to rise and, over the western Pacific, it falls. This reversal of surface air pressure at opposite ends of the Pacific Ocean is called the **Southern Oscillation**.

During the 1982–1983 El Niño, the east-blowing winds near the equator were stronger than during any previous episode. As these winds pushed eastward, they dragged surface water with them. This raised the sea level in the east and

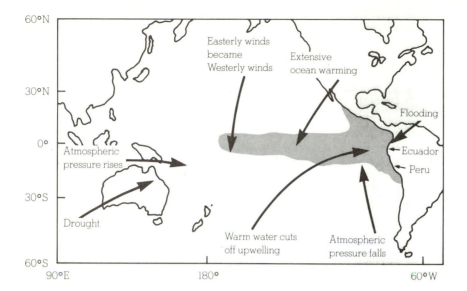

15.17 Some of the conditions that occurred during the El Niño of 1982–1983.

lowered it in the west. The eastward moving water gradually warmed under the tropical sun, becoming as much as 6°C (11°F) warmer than normal in the eastern equatorial Pacific. Gradually, a thick layer of warm water pushed into coastal areas of Ecuador and Peru, choking off the upwelling that supplies cold, nutrient-rich water to this region. The unusually warm water extended from South America's coastal region for many thousands of kilometers westward along the equator. The warm tropical water also spread northward along the west coast of North America to southern Canada.

Such a large area of abnormally warm water can have an effect on global wind patterns. The warm tropical water fuels the atmosphere with additional warmth and moisture, which the atmosphere turns into additional storminess and rainfall. The added warmth from the oceans apparently influences the westerly winds aloft in such a way that certain regions of the world experience too much rainfall, while others have too little.

Although the actual mechanism by which changes in surface ocean temperatures influence global wind patterns is not fully understood, the by-products are plain to see. For example, during the El Niño of 1982–1983, severe drought was felt in Indonesia, southern Africa, and Australia, where the 1982 production of wheat, oats, and barley was half that of the previous year. Meanwhile, record rains and flooding occurred over Ecuador and Peru, where the 1982 commercial fish catch was 50 percent of the 1981 total. In the Northern Hemisphere, an unusually strong westerly jet stream brought storms from California into the Gulf Coast states. The total damage worldwide due to flooding, winds, and drought exceeded $8 billion.

Are El Niño episodes predictable? At this time, the answer is unknown. It is known, however, that El Niño and the Southern Oscillation are part of a large-scale ocean-atmosphere interaction that can take several years to run its course. Some scientists feel that the trigger necessary to start an El Niño lies within the changing of the seasons, especially the transition periods of spring and fall. Others feel that the winter monsoon plays a major role in triggering El Niño.

It appears that El Niño and the monsoon system are intricately linked, so that a change in one brings about a change in the other. There is even speculation that the volcanic eruption of El Chichón may have played a role in the El Niño of 1982–1983. Presently, scientists, with the aid of mathematical general circulation models (GCMs), are trying to simulate atmospheric and oceanic conditions, so that El Niño and the Southern Oscillation can be anticipated. In addition, a ten-year in-depth study known as TOGA (Tropical Ocean and Global Atmosphere)—to start in 1985—will provide scientists with valuable information about the interactions that occur between the ocean and the atmosphere. The hope is that a better understanding of El Niño will provide improved long-range forecasts of weather and climate.

Summary

In this chapter, we have described the large-scale patterns of wind and pressure that persist around the world. We found that, at the surface in both hemispheres, the trade winds blow equatorward from the semipermanent high-pressure areas centered near 30° latitude. Near the equator, the trade winds converge along a boundary known as the intertropical convergence zone (ITCZ). On the poleward side of the subtropical highs, are the prevailing westerly winds. The westerlies meet cold, polar easterly winds along a boundary called the polar front, a zone of low pressure where middle latitude storms often form. The annual shifting of the major pressure areas and wind belts—northward in July and southward in January—strongly influences the annual precipitation of many regions.

Aloft, warm air (high pressure) over low latitudes and cold air (low pressure) over high latitudes produce westerly winds in both hemispheres, especially at middle and high latitudes. Near the equator, easterly winds exist. The jet streams are located where strong winds concentrate into narrow bands. The polar front jet stream forms in response to temperature contrasts along the polar front, while the subtropical jet stream forms at higher elevations above the subtropics, along a boundary called the subtropical front.

Near the surface, we examined the interaction between the atmosphere and oceans. We found the interaction to be an ongoing process where everything, in one way or another, seems to influence everything else. On a large scale, winds blowing over the surface of the water drive the major ocean currents; the oceans, in turn, release energy to the atmosphere, which helps to maintain the general circulation. Where winds and the Ekman Spiral move surface water away from a coastline, cold, nutrient-rich water upwells to replace it, creating good fishing. When atmospheric circulation patterns change, and the trade winds weaken or reverse direction, warm tropical water is able to flow eastward toward South America, where it chokes off upwelling and produces dis-

astrous economic conditions. When the warm water extends over a vast area of the tropical Pacific, the warming is called El Niño. The large-scale interaction between the atmosphere and the ocean during El Niño affects global atmospheric circulation patterns. The sweeping winds aloft provide too much rain in some areas and not enough in others. Studies now in progress are designed to determine how the interchange between atmosphere and ocean can produce such events.

Questions for Review

1. Draw a large circle. Now, place the major surface pressure and wind belts of the world at their appropriate latitudes.

2. Explain how the average surface pressure features shift from summer to winter.

3. Why is it impossible on the earth for a Hadley cell to extend from the pole to the equator?

4. Along a meridian line running from the equator to the poles, how does the general circulation help to explain zones of abundant and sparse precipitation?

5. Explain why the winds in the middle and upper troposphere tend to blow from west to east in both the Northern and the Southern Hemisphere.

6. How does the polar front influence the development of the polar front jet stream?

7. Describe how the conservation of angular momentum plays a role in the formation of a jet stream.

8. Why is the polar front jet stream stronger in winter than in summer?

9. Why does the low-level jet over the Central Plains blow mainly from the south?

10. Explain the relationship between the general circulation of air and the circulation of ocean currents.

11. List at least four important interactions that exist between the ocean and the atmosphere.

12. Describe how the Ekman Spiral forms.

13. What conditions are necessary for upwelling to occur along the west coast of North America? The east coast of North America?

14. Describe how the Southern Oscillation influences the phenomenon known as El Niño.

Questions for Thought

1. What effect would continents have on the circulation of air in the single-cell model?

2. How would the general circulation of air appear in summer and winter if the earth were tilted on its axis at an angle of 45° instead of 23½°?

3. Summer weather in the southwestern section of the United States is influenced by a subtropical high-pressure cell, yet Fig. 15.4 shows an area of low pressure at the surface. Explain.

4. Explain why icebergs tend to move at right angles to the direction of the wind.

5. Over the open ocean in the vicinity of the Pacific high, observations have indicated that ozone concentrations hundreds of meters above the surface are greater than expected. Give a possible explanation for this.

6. Give *two* reasons why pilots would prefer to fly in the core of a jet stream rather than just above or below it.

7. Why do the major ocean currents in the North Indian Ocean reverse direction between summer and winter?

8. Explain why the surface water temperature along the northern California coast is warmer in winter than it is in summer.

9. You are given an upper-level map that shows the position of two jet streams. If one is the polar front jet and the other the subtropical jet, how would you be able to tell which is which?

10. The Coriolis force deflects moving water to the right in the Northern Hemisphere and to the left in the Southern Hemisphere. Why, then, does upwelling tend to occur along the western margin of continents in both hemispheres?

Problems and Exercises

1. Locate the following cities on a world map. Then, based on the general circulation of surface winds, predict the prevailing wind for each one during July and January.

(a) Nashville, Tennessee
(b) Oklahoma City, Oklahoma
(c) Melbourne, Australia
(d) London, England
(e) Paris, France
(f) Reykjavik, Iceland
(g) Fairbanks, Alaska
(h) Seattle, Washington

2. Suppose a balloon is released over Los Angeles on the morning of April 18, 1979. (Refer to Fig. 15.8.) If the balloon rises to the 300-mb level and is "caught" by the jet stream, calculate about how long it would take the balloon to cross the Canadian border.

3. Below is a list of average weather conditions that prevail during the month of July at San Francisco, California, and Atlantic City, New Jersey. Both cities lie adjacent to an ocean at nearly the same latitude; however, the average weather conditions vary greatly. In terms of the average surface winds and pressure systems (Fig. 15.4) and the interaction between the atmosphere and ocean, explain what accounts for the variation between the two cities of each weather element.

Atlantic City, New Jersey
(latitude 39°N)

Average weather, July

Temperature maximum	84°F
minimum	66°F
Dew point	64°F
Precipitation	3.72 in.
Prevailing wind	S
Weather: Clear to partly cloudy with the possibility of afternoon showers	
Water temperature	70°F

San Francisco, California
(latitude 37°N)

Average weather, July

Temperature maximum	64°F
minimum	53°F
Dew point	53°F
Precipitation	0.01 in.
Prevailing wind	NW
Wind speed (average)	11 mph
Weather: Fog and low clouds during the night and morning with afternoon clearing	
Water temperature	53°F

An approaching cold front with a band of cumulonimbus clouds is moving southeastward across the plains of Colorado. (Photo by author)

CHAPTER 16

Air Masses and Fronts

Every so often, someone attributes the present state of the atmosphere to something called an *air mass*. Perhaps you have heard statements such as, "Our bitterly cold weather is a result of a huge polar air mass," or, "A tropical air mass is causing today's unusually high humidity," or, "The rain today is nothing more than air mass weather." What exactly is an air mass? Where and how does it form and what types of weather can it bring? We will answer these and other questions in the following sections.

Air Masses

An **air mass** is an extremely large body of air whose properties of temperature and moisture are fairly similar in any horizontal direction at any given altitude. Air masses may cover many thousands of square kilometers. In Fig. 16.1, a large, winter air mass, associated with a high pressure area, covers over half of the United States on January 9, 1976. Note that, although the surface air temperature and dew point vary somewhat, everywhere the air is cold and dry, with the exception of the zone of snow showers on the eastern shores of the Great Lakes. This cold, shallow anticyclone will drift eastward, carrying with it the temperature and moisture characteristic of the region where the air mass formed; hence, in a day or two, cold air will be located over the central Atlantic Ocean. Part of weather forecasting is, then, a matter of determining air mass characteristics, predicting how and why they change, and in what direction the systems will move.

Source Regions Regions where air masses originate are known as **source regions**. In order for a huge mass of air to develop uniform char-

Contents

Fig. 16.1 A large, cold winter air mass is dominating the weather over much of the United States on the morning of January 9, 1976. At almost all cities, the air is cold and dry. Upper number is air temperature (°F); bottom number is dew point (°F).

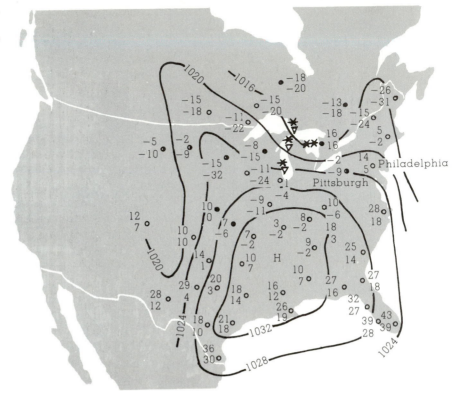

acteristics, its source region should be generally flat and of uniform composition, with light surface winds. The longer the air remains stagnant over its source region, the more likely it will acquire properties of the surface below. Consequently, ideal source regions are usually those areas dominated by high pressure. They include the ice- and snow-covered arctic plains in winter and subtropical oceans and desert regions in summer. The middle latitudes, where surface temperatures and moisture characteristics vary considerably, are not good source regions. Instead, this region is a transition zone where air masses with different physical properties move in, clash, and produce an exciting array of weather activity.

Classification Air masses are grouped into four general categories according to their source region. Air masses that originate in polar latitudes are designated by the capital letter "P"; those that form in warm tropical regions are designated by the capital letter "T." If the source region is

land, the air mass will be dry and the lowercase letter "c" (for continental) precedes the P or T. If the air mass originates over water, it will be moist—at least in the lower layers—and the lowercase letter "m" (for maritime) precedes the P or T. We can now see that polar air originating over land will be classified cP on a surface weather chart, while tropical air originating over water will be marked as mT. Table 16.1 lists the four basic air masses.*

After the air mass spends some time over its source region, it usually begins to move in response to the winds aloft. As it moves away from its source region, it encounters surfaces that may be warmer or colder than itself. The relationship between the air mass and its present environment

---

*Some classification schemes designate extremely cold air masses as arctic (A) and exceptionally hot air masses as equatorial (E). Usually, however, it is difficult to distinguish between arctic and polar air masses and equatorial and tropical air masses, and so we will not make this distinction here.

leads to an additional classification. If the air mass is colder than the surface over which it is moving, the lowercase letter "k" (for cold) is added. When the air mass is warmer than the underlying surface, the lowercase letter "w" is added. Hence, a cPk air mass is polar air that originated over land and is presently colder than the surface directly beneath it. What is a mTw air mass? (Answer: Moist, tropical air that is warmer than the surface over which it is moving.)

When the air mass is colder than the underlying surface, it is warmed from below, which results in a steeper lapse rate and instability at low levels. In this case, increased convection and turbulent mixing near the surface usually produce good visibility, cumuliform clouds, and showers of rain or snow. On the other hand, when the air mass is warmer than the surface below, the lower layers are chilled by contact with the cold earth. Warm air above cooler air produces a stable lapse rate with little vertical mixing. This causes the accumulation of dust, smoke, and pollutants, which restricts surface visibilities. In moist air, stratiform clouds accompanied by drizzle or fog may form.

Air Masses of North America   The principal air masses (with their source regions) that invade the United States are shown in Fig. 16.2. We are now in a position to study the formation and modification of each of these air masses and the variety of weather that accompanies them.

cP (Continental Polar) Air Masses   The bitterly cold weather that enters the United States in winter is associated with continental polar air masses. These originate over the ice- and snow-covered regions of northern Canada and Alaska where long, clear nights allow for strong radiational cooling of the surface. Air in contact with the surface becomes quite cold and stable. Since little moisture is added to the air, it is also quite dry, and specific humidities are often less than 0.3 g/kg. Eventually a portion of this cold air breaks away and, under the influence of the air flow aloft, moves southward as an enormous shallow high-pressure area.

As the cold air moves into the interior plains, there are no topographic barriers to restrain it, so it continues southward, bringing with it cold wave warnings and frigid temperatures. The infamous

Table 16.1   Air Mass Classification and Characteristics

SOURCE REGION	POLAR (P)	TROPICAL (T)
Land continental (c)	cP cold, dry, stable	cT hot, dry, stable— aloft; unstable— surface
Water maritime (m)	mP cool, moist, unstable	mT warm, moist; usually unstable

Texas norther is associated with cP air. As the air mass moves over warmer land to the south, the air temperature moderates slightly and becomes cPk as it is heated from below. However, even during the afternoon, when the surface air is most unstable, cumulus clouds are rare because of the extreme dryness of the air mass. At night, when the winds die down, rapid surface cooling and clear skies combine to produce low minimum temperatures. If the cold air moves as far south as central or southern Florida, the winter vegetable crop may be severely damaged.

In winter, the generally fair weather accompanying cP air is due to the stable nature of the atmosphere aloft. Sinking air develops above the large dome of high pressure. The subsiding air warms by compression and creates warmer air, which lies above colder surface air. Therefore, a strong upper-level subsidence inversion often forms. Should the anticyclone stagnate over a region for several days, the visibility gradually drops as pollutants become trapped in the cold air near the ground. Usually, however, winds aloft move the cold air mass either eastward or southeastward.

The Rockies, Sierra Nevada, and Cascades normally protect the Pacific Northwest from the onslaught of cP air, but, occasionally, cP air masses do invade these regions. When the upper-level winds over Washington and Oregon blow from the north or northeast on a trajectory beginning over northern Canada or Alaska, cold cP air can slip over the mountains and extend its icy fingers all the way to the Pacific Ocean. As the air moves off the high plateau, over the mountains, and on into the lower valleys, compressional heating of the sinking air causes its temperature to rise, so that by the time it reaches the lowlands, it is considerably warmer than it was originally. How-

Fig. 16.2 Air mass source regions and their paths.

ever, in no way would this air be considered warm. In some cases, the subfreezing temperatures slip over the Cascades and extend southward into the coastal areas of southern California.

A similar but less dramatic warming of cP air occurs along the East Coast. Air rides up and over the lower Appalachian Mountains. Turbulent mixing and compressional heating increase the air temperatures on the downwind side. Consequently, cities located to the east of the Appalachian Mountains usually do not experience temperatures as low as those on the west side. In Fig. 16.1, notice that for the same time of day— 7 A.M. EST—Philadelphia is 16°F (9°C) warmer than Pittsburgh.

Figure 16.3 shows two upper-air wind patterns that led to extremely cold outbreaks of cP air in 1977 and 1978. Upper-level winds typically blow from west to east, but, in both of these cases, the flow, as given by the heavy, dark arrows, had a strong north-south (meridional) trajectory. The H represents the positions of the cold surface anticyclones. Numbers on the map represent minimum temperatures (°F) recorded during the cold spells. In the central and eastern parts of the United States, some of the lowest temperatures ever recorded occurred on the mornings of January 18 and 19, 1977. Along the West Coast, the cP air mass caused ice to form on swimming pools in Los Angeles on the morning of December 8,

FOCUS ON A SPECIAL TOPIC

Lake-Effect Snows

During the winter, when the weather in the Midwest is dominated by clear, brisk cP air, people living on the eastern shores of the Great Lakes brace themselves for heavy snow showers. Snowstorms that form on the downwind side of one of these lakes are known as **lake-effect snows**. These storms are highly localized, extending from just a few kilometers to more than 50 km inland. The snow usually falls as a heavy shower or squall in a concentrated zone. So centralized is the region of snowfall, that one part of a city may accumulate many centimeters of snow, while, in another part, the ground is bare.

Lake-effect snows are most numerous from November to January. During these months, cP air moves over the lakes when they are relatively warm and not quite frozen. The contrast in temperature between water and air can be as much as 25°C (45°F). Studies show that the greater the contrast in temperature, the greater the potential for snow showers. In Fig. 1, we can see that, as the cold air moves over the warmer water, the air mass is quickly warmed from below, making it more buoyant and less stable. Rapidly, the air sweeps up moisture, soon becoming saturated. Out over the water, the vapor condenses into steam fog. As the air continues to warm, it rises and forms billowing cumuliform clouds, which continue to grow as the air becomes more unstable. Eventually, these clouds produce heavy showers of snow, which make the

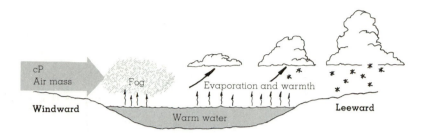

Fig. 1   The formation of lake-effect snows. Cold, dry air crossing the lake gains moisture and warmth from the water. The more buoyant air now rises, forming clouds that deposit large quantities of snow on the lake's leeward shores.

lake seem like a snow factory. Once the air and clouds reach the downwind side of the lake, additional lifting is provided by low hills and the convergence of air as it slows down over the rougher terrain. In late winter, the frequency and intensity of lake-effect snows taper off as the temperature contrast between water and air diminishes and larger portions of the lakes freeze.

Generally, the longer the stretch of water over which the air mass travels (the longer the fetch), the greater the amount of warmth and moisture derived from the lake, and the greater the potential for heavy snow showers. In fact, studies show that, for significant snowfall to occur, the air must move across 80 km (50 mi) of open water. Consequently, forecasting lake-effect snowfalls depends to a large degree on determining the trajectory of the air as it flows over the lake. Regions that experience heavy lake-effect snowfalls are shown in Fig. 2.

As the cP air moves farther east, the heavy snow showers

usually taper off; however, the western slope of the Appalachian Mountains produces further lifting, enhancing the possibility of more and heavier showers. The heat given off during condensation warms the air and, as the air descends the eastern slope, compressional heating warms it even more. Snowfall ceases, and by the time the air arrives at Philadelphia, New York, or Boston, the only remaining trace of the snow showers occurring on the other side of the mountains are the puffy cumulus clouds drifting overhead.

Fig. 2   Shaded areas show regions that experience heavy lake-effect snows.

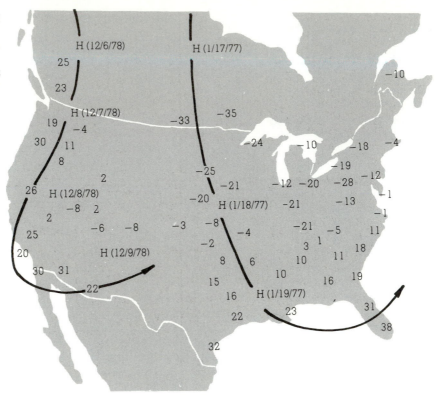

Fig. 16.3 Average upper-level wind flow (heavy arrows) and surface position of anticyclones (H) associated with two extremely cold outbreaks of cP air. Numbers on the map represent minimum temperatures (°F) measured during each cold snap.

1978. Notice in both cases how the upper-level wind directs the paths of the air masses.

The cP air that moves into the United States in summer has properties much different from its winter counterpart. The source region remains the same, but is now characterized by long summer days that melt snow and warm the land. The air is only moderately cool, and surface evaporation adds water vapor to the air. A summertime cP air mass usually brings relief from the oppressive heat in the central and eastern states, as cooler air lowers the air temperature to more comfortable levels. Daytime heating warms the lower layers, producing surface instability. With its added moisture, the rising air may condense and create a sky dotted with fair weather cumulus clouds (cumulus humilis). A typical profile of temperatures for a summer and a winter cP air mass is given in Fig. 16.4. Notice that the strong inversion so prevalent in winter is absent in summer.

When an air mass moves over a large body of water, its original properties may change considerably. For instance, cold, dry cP air moving over the Gulf of Mexico warms rapidly and gains mois-

ture. The air quickly assumes the qualities of a maritime air mass. Notice in Fig. 16.5 that rows of cumulus clouds are forming over the Gulf parallel to northerly surface winds as cP air is being warmed by the water beneath it. As the air continues its journey southward into Mexico and Central America, strong, moist northerly winds build into heavy clouds (bright area) and showers along the northern coast. Hence, a once cold, dry, and stable air mass can be modified to such an extent that its original characteristics are no longer discernible. When this happens, the air mass is given a new designation.

Notice also in Fig. 16.5 that a similar modification of cP air is occurring along the Atlantic Coast, as northwesterly winds are blowing over the mild Atlantic. When this air encounters the much warmer Gulf Stream water, it warms rapidly and becomes unstable. Vertical mixing brings down faster-flowing cold air from aloft. This creates strong, gusty surface winds and choppy seas, which can be hazardous to ships. Thus, the modification of cP air by a large body of water can adversely affect shipping.

mP (Maritime Polar) Air Masses   During the winter, cP air originating over Asia and frozen polar regions is carried eastward and southward over the Pacific Ocean by the circulation around the Aleutian low. The ocean water modifies the cP air by adding warmth and moisture to it. Since this air has a fetch of many hundreds or even thousands of kilometers, it gradually changes into mP air.

By the time this air mass reaches the Pacific Coast it is cool, moist, and unstable. The ocean's effect is to keep air near the surface warmer than the air aloft. Temperature readings in the 40s and 50s (°F) are common near the surface, while air at an altitude of about a kilometer or so may be at the freezing point. Within this colder air, characteristics of the original cP air mass may still prevail. As the air moves inland, coastal mountains force it to rise, and much of its water vapor condenses into rain-producing clouds. In the colder

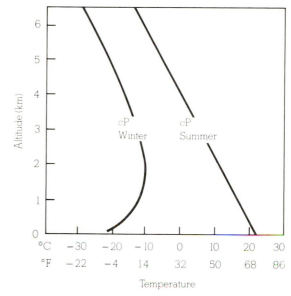

Fig. 16.4   Vertical temperature profile for a summer and a winter cP air mass.

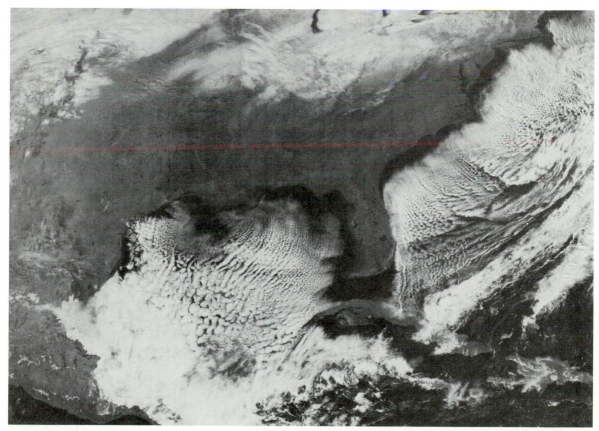

Fig. 16.5   Visible satellite picture showing the modification of cP air as it moves over the warmer Gulf of Mexico and the Atlantic Ocean.

FOCUS ON A SPECIAL TOPIC

The Return of the Siberian Express

The winter of 1983–1984 was one of the coldest on record across the United States. Unseasonably cold weather arrived in December, which, for much of the country, was one of the coldest Decembers since records have been kept. During the first part of the month, continental polar air covered most of the northern and central plains. As the cold air moderated slightly, far to the north a huge mass of bitter cold arctic air was forming over the frozen reaches of the Canadian Northwest territories.

By midmonth, the frigid air, associated with a massive high pressure area, covered all of northwest Canada. Meanwhile, an upper-level ridge was forming over Alaska. On the eastern side of the ridge, strong northerly winds associated with the jet stream directed the frigid air southward over the prairie provinces of Canada. A portion of the cold air broke away, and, like a large swirling bubble, it moved as a cold, shallow anticyclone over the Great Plains of the United States. Along the southwestern margin of the cold air, a weak storm developed and moved eastward into the Great Lakes. Behind the storm, strong, upper-level northerly winds directed the leading edge of a massive surface anticyclone and extraordinarily cold air southward into the United States. Be-

cause the frigid air was accompanied in some regions by winds gusting to 45 knots, at least one news reporter dubbed the onslaught of this arctic blast, "the Siberian Express."

The express dropped temperatures to some of the lowest readings ever recorded during the month of December. On December 22, Elk Park, Montana, recorded an unofficial low of −53°C (−64°F), only 4°C higher than the all-time low of −57°C (−70°F) for the nation (excluding Alaska) recorded at Rogers Pass, Montana, on January 20, 1954.

The center of the massive anticyclone gradually pushed southward out of Canada. By December 24, its center was over eastern Montana (Fig. 3), where the sea level pressure at Miles City reached an incredible 1064 mb (31.40 in.)—a new United States record that topped the old mark of 1063 mb set in Helena, Montana, on January 10, 1962. An enormous ridge of high pressure stretched from the Canadian arctic coast to the Gulf of Mexico. On the east side of the ridge, cold westerly winds brought lake-effect snows to the eastern shores of the Great Lakes. To the south of the high-pressure center, cold easterly winds, rising along the elevated plains, brought light amounts of *upslope snow* to sections of the

Rocky Mountain states. Notice in Fig. 3 that, on Christmas Eve, arctic air covered almost 90 percent of the United States. As the cold air swept eastward and southward, a hard freeze caused hundreds of millions of dollars in damage to the fruit and vegetable crops in Texas, Louisiana, and Florida. On Christmas Day, 125 record low temperature readings were set in twenty-four states. That afternoon, at 1:00 P.M., it was actually colder in Atlanta, Georgia, at −13°C than it was in Fairbanks, Alaska (−12°C). The worst cold wave to occur in December during this century continued through the week, as many new record lows were established in the deep south from Texas to Louisiana.

By January 1, the extreme cold had moderated, as the upper-level winds became more westerly. These winds brought milder mP Pacific air eastward into the Great Plains. The warmer pattern continued until about January 10, when the Siberian Express decided to make a return visit. Driven by strong upper-level northerly winds, impulse after impulse of arctic air from Canada swept across the United States. On January 18, an all time record low of −54°C (−65°F) was recorded for the state of Utah at Middle Sinks. On January 19, temperatures plum-

air aloft, the rain changes to snow, with heavy amounts accumulating in mountain regions. A typical upper-level wind flow pattern that brings mP air onto the west coast of North America is shown in Fig. 16.6.

When the mP air moves inland, it loses much of its moisture as it crosses a series of mountain ranges. Beyond these mountains, it travels over a cold, elevated plateau that chills the surface air and slowly transforms it into dry, stable cP air.

FOCUS ON A SPECIAL TOPIC, continued

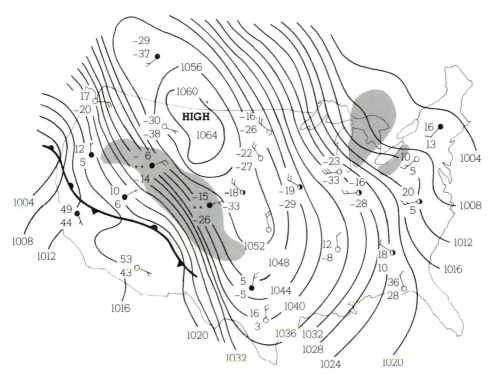

Fig. 3 Surface weather map for 7 A.M., EST, December 24, 1983. Solid lines are isobars. Shaded area represents precipitation. An extremely cold air mass covers nearly 90 percent of the United States.

meted to a new low of −22°C (−7°F) for the airports in Philadelphia and Baltimore. Toward the end of the month, the upper-level winds once again became more westerly. Over much of the nation, the cold air moderated. But the express was to return at least one more time.

The beginning of February saw relatively warm air covering much of the nation from California to the Atlantic coast. On February 4, an arctic outbreak of cP air spread southward and eastward across the nation. Although freezing air extended southward into central Florida, the express ran out of steam, and a February heat wave soon engulfed most of the United States east of the Rocky Mountains. Maritime tropical air from the Gulf of Mexico brought record warmth to much of the eastern two-thirds of the nation. Near the middle of the month, Louisville, Kentucky, reported 23°C (73°F) and Columbus, Ohio, 21°C (69°F). Even though February was one of the warmest months on record over parts of the United States, the winter of 1983–1984 (December, January, and February) will go down in the record books as the sixth coldest winter for the United States as a whole since reliable record keeping began in 1931.

East of the Rockies this air mass often brings fair weather and temperatures not nearly as cold as the frigid cP air, which invades this region from northern Canada. In fact, when mP air from the west replaces retreating cP air from the north, chinook winds often develop. Furthermore, when the modified mP air replaces moist tropical air, storms can form along the boundary separating the two air masses.

Along the east coast, mP air originates in the

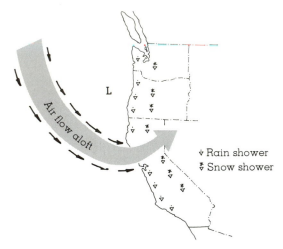

Fig. 16.6   A winter upper-air pattern that brings mP air into the west coast of North America. The large arrow represents the upper-level flow. Note the trough of low pressure along the coast. The small arrows show the trajectory of the mP air at the surface. Regions that normally experience precipitation under these conditions are also shown on the map.

North Atlantic as cP air moves southward some distance off the Atlantic Coast. It then swings southwestward toward the northeastern states. Because the water of the North Atlantic is very cold and the fetch is short, wintertime Atlantic mP air masses are usually much colder than their Pacific counterparts. Because the prevailing winds aloft are westerly, Atlantic mP air masses are also much less common.

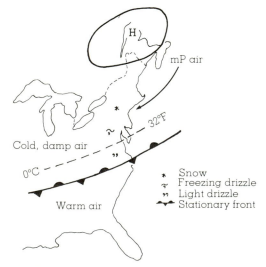

Fig. 16.7   Winter and early spring surface weather patterns that usually prevail during the invasion of mP air into the mid-Atlantic and New England states.

Figure 16.7 illustrates a typical late winter or early spring surface weather pattern that carries mP air from the Atlantic into the New England and middle Atlantic states. A slow-moving, cold anticyclone drifting to the east (north of New England) causes a northeasterly flow of mP air to the south. The boundary separating this invading colder air from warmer air even farther south is marked by a stationary front. North of this front, northeasterly winds provide generally undesirable weather, consisting of damp air and low, thick clouds from which light precipitation falls in the form of rain, drizzle, or snow. When upper atmospheric conditions are right, storms may develop along the stationary front, move eastward, and intensify into Atlantic northeasters near the shores of Cape Hatteras. (Such developing storms will be treated in detail in the next chapter.)

mT (Maritime Tropical) Air Masses   The wintertime source region for Pacific mT air is the subtropical east Pacific Ocean. Air from this region must travel over 1600 km of water before it reaches the southern California coast. Consequently, these air masses are very warm and moist by the time they arrive along the West Coast. The warm air produces heavy precipitation usually in the form of rain, even at high elevations. Melting snow and rain quickly fill rivers, which overflow into the low-lying valleys. The rapid snowmelt leaves local ski slopes barren, and the heavy rain can cause disastrous mudslides in the steep canyons. Fortunately, the invasion of a real mT air mass into latitudes north of southern California is rare.

The mT air that influences much of the weather east of the Rockies originates over the Gulf of Mexico and Caribbean Sea. In winter, cold polar air tends to dominate the continental weather scene, so mT air is usually confined to the Gulf and extreme southern states. Occasionally, a slow-moving storm system over the Central Plains draws mT air northward. Gentle, moist south or southwesterly winds blow into the central and eastern parts of the nation in advance of the system. Since the land is still extremely cold, air near the surface is chilled to its dew point. Fog and low clouds form in the early morning, dissipate by midday, and reform in the evening. This mild winter weather in the Mississippi and Ohio valleys lasts, at best, only a few days. Soon cold polar air will move down from the north behind the eastward-moving storm system. Along the boundary between the two air masses, the mT air is lifted above the more dense cP air. This often leads to heavy and widespread precipitation and storminess.

When a storm system stalls over the Central Plains, a constant supply of mT air from the Gulf of Mexico can bring record-breaking maximum temperatures to the eastern half of the country. Sometimes the air temperatures are higher in the mid-Atlantic states than they are in the deep south, as compressional heating warms the air even more as it moves downslope after crossing the Appalachian Mountains. Figure 16.8 shows a surface weather map and the associated upper air flow (heavy arrow) that brought unseasonably warm

mT air into the central and eastern states during April 1976. A large high centered off the southeast coast coupled with a strong southwesterly flow aloft carried warm, moist air into the Midwest and East, causing a record-breaking April heat wave. The flow aloft prevented the surface low and the cP air behind it from making much eastward progress, so that the warm spell lasted for five days. Providence, Rhode Island, experienced a record-breaking high temperature for April of 36°C (96°F). Note that, on the west side of the surface low, the winds aloft funneled cold cP air from the north into the western states, creating unseasonably cold weather from California to the Rockies. Hence, while people in the Southwest were huddled around heaters, others several

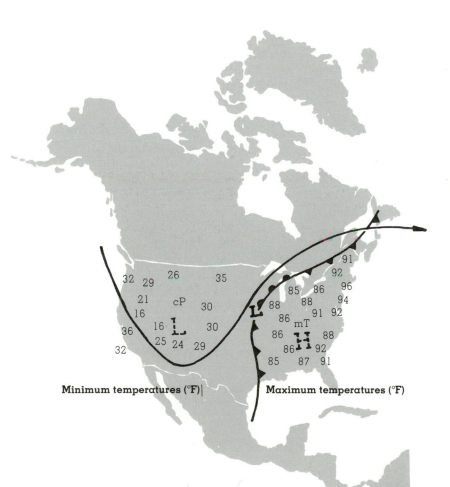

Fig. 16.8 Weather conditions during an unseasonably hot spell in the East that occurred between the 15th and 20th of April, 1976. The surface low-pressure area and fronts are shown for April 17. Numbers to the east of the low are maximum temperatures recorded during the hot spell, while those to the west of the low are minimums reached during the same time period. The heavy arrow is the average upper-level flow during the period. The dashed L and H show average positions of the upper-level trough and ridge.

Minimum temperatures (°F)   Maximum temperatures (°F)

thousand kilometers away in the Northeast were turning on air conditioners. We can see that it is the upper-level meridional flow, directing cP air southward and mT air northward, that makes these contrasts in temperature possible.

As maritime air moves inland over the hot continent, it warms, rises, and frequently causes cumuliform clouds, which produce afternoon showers and thunderstorms. You can almost count on thunderstorms developing along the Gulf Coast each afternoon in summer. As evening approaches, thunderstorm activity typically dies off. Nighttime cooling lowers the air temperature and, if the air becomes saturated, fog or low clouds form. These, of course, dissipate by late morning as surface heating warms the air again.

A weak, but often persistent, flow around an upper-level anticyclone will spread mT air from the Gulf of Mexico into the southern and central Rockies, where it causes afternoon thunderstorms. Occasionally, this easterly flow may work its way even farther west, producing shower activity in the otherwise dry southwestern desert.

Normally, mT air originating in the tropical eastern Pacific remains far south of California. Occasionally, a weak upper-level southerly flow will spread this humid air northward into the southwestern United States, most often Arizona, Nevada, and southern California. In many places, the moist, unstable air aloft only shows up as middle and high cloudiness, especially altocumulus and cirrocumulus castellanus. However, where the moist flow meets a mountain barrier, it usually rises and condenses into towering shower-producing clouds.

cT (Continental Tropical) Air Masses The only real source region for hot, dry cT air in North America is northern Mexico and the adjacent arid southwestern United States. During the summer, the air mass is hot, dry, and unstable at low levels with frequent dust devils forming during the day. Because of the low relative humidity (typically less than 10 percent during the afternoon), air must rise over 3000 m (10,000 ft) before condensation begins. Furthermore, an upper-level ridge usually produces weak subsidence over the region, tending to make the air aloft rather stable and the surface air even warmer. Consequently, skies are generally clear, the weather is hot, and rainfall is

practically nonexistent where cT air masses prevail. If this air mass moves outside its source region and into the Great Plains and stagnates over that region for any length of time, a severe drought may result. Figure 16.9 shows a weather map situation where cT air covers a large portion of the western United States and produces hot, dry weather northward to Canada.

So far, we have examined the various air masses that enter North America annually. The characteristics of each depend upon the air mass source region and the type of surface over which the air mass moves. The winds aloft determine the trajectories of these air masses. Occasionally, an air mass will control the weather in a region for some time. These persistent weather conditions are known as *air mass weather*.

Air mass weather is especially common in the southeastern United States during summer as, day after day, mT air from the Gulf brings sultry conditions and afternoon thunderstorms. It is also common in the Pacific Northwest in winter when unstable, cool mP air accompanied by widely scattered showers dominates the weather for several days or more. The real weather action, however, usually occurs not within air masses but at their margins, where air masses with sharply contrasting properties meet—in the zone marked by weather fronts.*

Fronts

Although we briefly looked at fronts in Chapter 1, we are now in a position to study them in-depth, which will aid us in forecasting the weather. We will now learn about the general nature of fronts—how they move and what weather patterns are associated with them.

A **front** is the transition zone between two air masses of different densities. Since density differences are most often caused by temperature differences, fronts usually separate air masses with contrasting temperatures. Often, they separate air masses with different humidities as well.

*The word "front" is used to denote the clashing of two air masses, probably because it resembles the fighting in Western Europe during World War I, when the term originated.

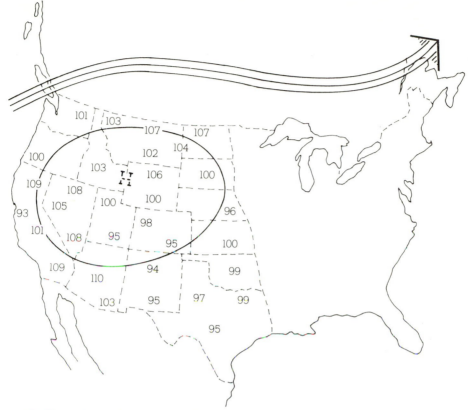

Fig. 16.9 During August 6th and 7th, 1983, cT air covered a large area of the western United States. Numbers on the map represent maximum temperatures (°F) during this period. The dashed H with the contour line shows the position of a subtropical high at 500 mb. Sinking air associated with the high contributed to the hot weather. Winds aloft were weak with the main flow (heavy line) positioned over Canada.

Remember that air masses have both horizontal and vertical extent; consequently, the upward extension of a front is referred to as a *frontal surface*, or a *frontal zone*.

Figure 16.10 shows a simplified weather map illustrating four different fronts. As we move from west to east across the map, the fronts appear in the following order: a stationary front between points A and B; a cold front between points B and C; a warm front between points C and D; and an occluded front between points C and L. Let's examine the properties of each of these fronts.

Stationary Fronts A stationary front has essentially no movement. On a colored weather map, it is drawn as an alternating red and blue line. Semicircles face toward colder air on the red line and triangles point toward warmer air on the blue line. The stationary front between points A and B in Fig. 16.10 marks the boundary where cold, dense cP air from Canada butts up against the north-south trending Rocky Mountains. Unable to cross the barrier, the cold air shows little or no westward movement. The stationary front is drawn along a line separating the cP from the milder mP air to the west. Notice that the surface winds tend to blow parallel to the front, but in opposite directions on either side of it.

The weather along the front is clear to partly cloudy, with much colder air lying on its eastern side. Because both air masses are dry, there is no precipitation. This is not, however, always the case. When warm, moist air rides up and over the cold air, widespread cloudiness with light precipi-

Fig. 16.10 A simplified weather map showing surface pressure systems, air masses, and fronts.

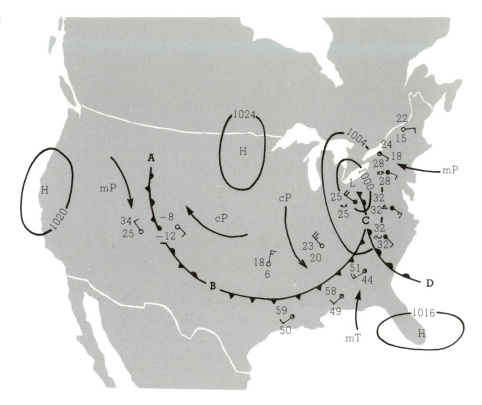

tation can cover a vast area. These are the conditions that prevail north of the east-west running stationary front in Fig. 16.7.

If the warmer air to the west begins to move and replace the colder air to the east, the front in Fig. 16.10 will no longer remain stationary; it will become a warm front. If, on the other hand, the colder air slides up over the mountain and replaces the warmer air on the other side, the front will become a cold front.

Cold Fronts The cold front between points B and C on the surface weather map (Fig. 16.10) represents a zone where cold, dry, stable polar air is replacing warm, moist unstable tropical air. The direction of movement of the cold front is given by the triangles along the front. On a weather map in color, the front is drawn as a solid blue line. How did the meteorologist know to draw the front at that location? A closer examination of the situation will give us the answer.

The weather in the immediate vicinity of this cold front in the southeastern United States is shown in Fig. 16.11. The data plotted on the map

represent the current weather at selected cities. The station model used to represent the data at each reporting station is a simplified one that shows temperature, dew point, present weather, cloud cover, sea level pressure, wind direction and speed. The little line in the lower right-hand corner of each station gives the pressure tendency during the last three hours. With all of this information, the front can be properly located. (Appendix B explains the weather symbols and the station model more completely.)

The following criteria are used to locate a front on a surface weather map:

1. sharp temperature changes over a relatively short distance
2. changes in the air's moisture content (as shown by marked changes in the dew point)
3. shifts in wind direction
4. pressure changes
5. clouds and precipitation patterns

In Fig. 16.11, we can see a large contrast in air temperature and dew point on either side of the front. There is also a wind shift from southwesterly ahead of the front, to northwesterly behind it. Notice that each isobar kinks as it crosses the front, forming an elongated area of low pressure—a trough—which accounts for the wind shift. Since surface winds normally blow across the isobars toward lower pressure, we find winds with a southerly component ahead of the front and winds with a northerly component behind it.

Since the cold front is a trough of low pressure, sharp changes in pressure can be significant in locating the front's position. One important fact to remember is that the lowest pressure usually occurs just as the front passes a station. Notice that, as you move toward the front, the pressure drops, and, as you move away from it, the pressure rises. This is clearly shown by the pressure tendencies for each station on the map. Just before the front passes, the pressure tendency shows a falling barometer (\), while just behind the front, the barometer is now beginning to rise (√), and farther behind the front, the barometer is rising steadily (/).

The cloud and precipitation patterns are better seen in a side view of the front along the line X–X' (Fig. 16.12). We can see from Fig. 16.12 that, at the front, the cold, dense air wedges under the

warm air, forcing it upwards. As the moist, unstable air rises, it condenses into a series of cumuliform clouds. Strong, upper-level westerly winds blow the delicate ice crystals (which form near the top of the cumulonimbus) into cirrostratus and cirrus. These clouds usually appear far in advance of the approaching front. At the front itself, a relatively narrow band of thunderstorms produces heavy showers with gusty winds. Behind the front, the air cools quickly. (Notice how the freezing level dips as it crosses the front.) The winds shift from southwesterly to northwesterly, pressure rises, and precipitation ends. As the air dries out, the skies clear, except for a few lingering fair weather cumulus clouds.

Observe that the leading edge of the front is steep. The steepness is due to friction, which slows the air flow near the ground. The air aloft pushes forward, blunting the frontal surface. If we could walk from where the front touches the surface back into the cold air, a distance of 50 km, the front would be about 1 km above us. Thus, the slope of the front—the ratio of vertical rise to horizontal distance—is 1:50. This is typical for a fast-moving cold front—those that move about 25 knots. In a slower-moving cold front—one that

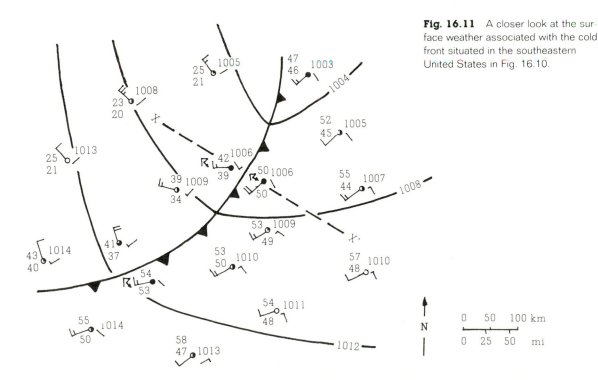

Fig. 16.11 A closer look at the surface weather associated with the cold front situated in the southeastern United States in Fig. 16.10.

Fig. 16.12 A vertical view of the weather across the cold front in Fig. 16.11 along the line X–X'.

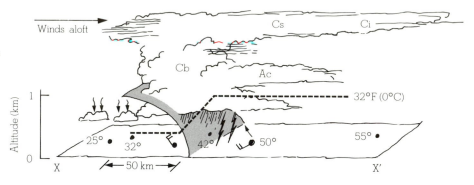

moves about 15 knots—the slope is much more gentle, averaging about 1:100.

With slow-moving cold fronts, clouds and precipitation usually cover a broad area behind the front. When the ascending warm air is stable, stratiform clouds, such as nimbostratus, become the predominate cloud type and even fog may develop in the rainy area. Occasionally, out ahead of a fast-moving front, a line of active showers and thunderstorms, called a *squall line*, develop parallel to and ahead of the advancing front. A squall line precedes the front by 100 to 300 km (60 to 180 mi), producing heavy precipitation and strong, gusty winds. (We will examine squall lines more closely in Chapter 19.)

As the temperature contrast across a front lessens, the front will often weaken and dissipate. Such a condition is known as **frontolysis**. On the other hand, an increase in the temperature contrast across a front can cause it to strengthen and regenerate into a more vigorous frontal system, a condition called **frontogenesis**.

An example of a regenerated front is shown in the infrared satellite pictures, Fig. 16.13. The cold front in Fig. 16.13a is weak as indicated by the low clouds (gray tones) along the front. As the front moves offshore, over the warm Gulf Stream (Fig. 16.13b), it intensifies into a more vigorous frontal system as surface air becomes unstable and convective activity develops. Notice that extensive cloudiness and thunderstorms are forming along the frontal zone as indicated by the bright areas in the infrared picture.

So far, we have considered the general weather patterns of "typical" cold fronts. There are, of course, exceptions. For example, if the rising warm air is dry and stable, scattered clouds are all that form, and there is no precipitation. In extremely dry weather, a marked change in the dew point, accompanied by a slight wind shift, may be the only clue to a passing front. During the winter, a series of cold polar outbreaks may travel across the United States so quickly that warm air is unable to develop ahead of the front. In this case, frigid arctic air usually replaces cold polar air, and a drop in temperature is the only indication that a cold front has moved through your area. Along the west coast, the Pacific Ocean modifies the air so much that cold fronts, such as those described in the previous section, are never seen. In fact, as a cold front moves inland from the Pacific Ocean, the surface temperature contrast across the front may be quite small. Topographic features usually distort the wind pattern so much that locating the position of the front and the time of its passage are exceedingly difficult. In this case, the pressure tendency is the most reliable indication of a frontal passage.

Cold fronts usually move southward or eastward, but sometimes they move westward. In New England, this occurs when northeasterly surface winds, blowing clockwise around an anticyclone centered to the north over Canada, push a cold front southwestward often as far south as Boston. Because the cold front moves in from the east, it is known as a "*back door*" cold front. As the front passes, westerly surface winds usually shift to easterly or northeasterly and temperatures drop as mP air flows in off the Atlantic Ocean.

Even though cold-front weather patterns have many exceptions, learning these patterns can be to your advantage if you live in the eastern two-

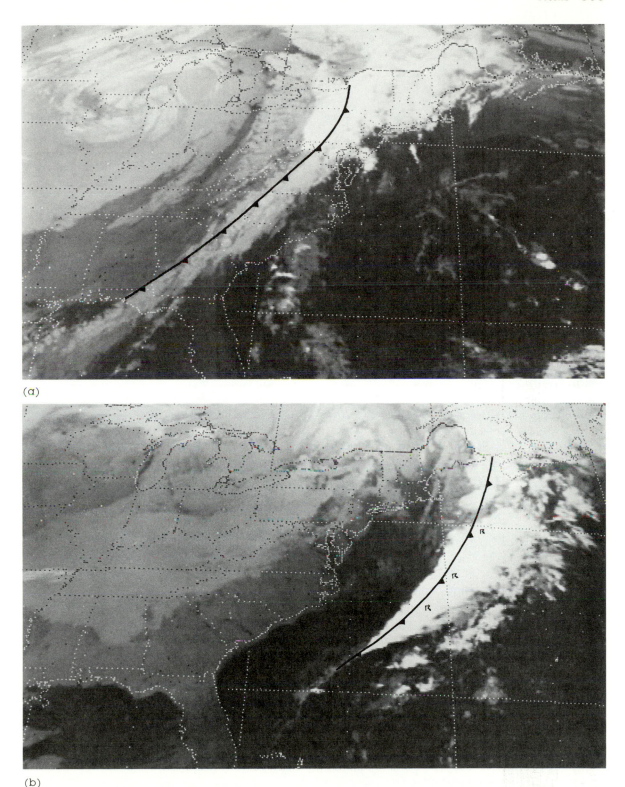

(a)

(b)

Fig. 16.13 (a) The weakening cold front over land on the morning of November 21, 1975 intensifies into (b) a vigorous front over water on the morning of November 22, 1975. (Infrared satellite pictures.)

Table 16.2 Typical Weather Conditions Associated with a Cold Front

WEATHER ELEMENT	BEFORE PASSING	WHILE PASSING	AFTER PASSING
Winds	south-southwest	gusty, shifting	west-northwest
Temperature	warm	sudden drop	colder
Pressure	falling steadily	sharp rise	rising steadily
Clouds	increasing Ci, Cs, then either Tcu or Cb	Tcu or Cb	often Cu
Precipitation	short period of showers	heavy showers of rain or snow sometimes with hail, thunder, and lightning	decreasing intensity of showers, then clearing
Visibility	fair to poor in haze	poor, followed by improving	good except in showers
Dew point	high; remains steady	sharp drop	lowering

thirds of the United States, where well-defined cold fronts are experienced. Knowing them improves your own ability to make short-range weather forecasts. For your reference, Table 16.2 summarizes idealized cold-frontal weather.

Warm Fronts In Fig. 16.10, a warm front is drawn along the line running from points C to D. Here, the leading edge of advancing mTw air from the Gulf of Mexico replaces the retreating cold mP from the North Atlantic. The direction of frontal movement is given by the half circles, which point into the cold air; this front is heading toward the northeast. When the map is in color, the warm front shows as a solid red line. The average speed of a warm front is about 10 knots, or about half that of an average cold front. During the day, as mixing occurs on both sides of the front, its movement may be much faster. Warm fronts often move in a series of rapid jumps, which show up on successive weather maps. At night, however, radiational cooling creates cool, dense surface air behind the front. This inhibits both lifting and the front's forward progress. When the forward surface edge of the warm front passes a station, the wind shifts, the temperature rises, and the overall weather conditions improve. To see why, we will examine the weather commonly associated with the warm front both at the surface and aloft.

Look at Figs. 16.14 and 16.15 closely and observe that the warmer, less-dense air rides up and over the colder, more-dense surface air. This rising of warm air over cold, called **overrunning**, produces clouds and precipitation well in advance of

the front's surface boundary. The warm front that separates the two air masses has an average slope of about 1:150, a much more gentle slope than that of a typical cold front. Warm air overriding the cold air creates a stable atmosphere (see the vertical temperature measurements in Fig. 16.15b). Notice that a temperature inversion exists in the region of the upper-level front. Another fact to notice is that the wind veers (shifts clockwise) with altitude, so that the southeasterly surface winds become southwesterly and westerly aloft.

Suppose we are standing at the position marked P′ in Figs. 16.14 and 16.15. Note that we are over 1200 km (750 mi) ahead of the surface front. Here, the surface winds are light and variable. The air is cold and about the only indication of an approaching warm front is the high cirrus clouds overhead. We know the front is moving slowly toward us and that within a day or so it will pass our area. Suppose that, instead of waiting for the front to pass us, we drive toward it, observing the weather as we go.

Heading toward the front, we notice that the cirrus clouds gradually thicken into a thin, white veil of cirrostratus whose ice crystals cast a halo around the sun.* Almost imperceptibly, the clouds thicken and lower, becoming altocumulus and altostratus through which the sun shows only as a faint spot against an overcast gray sky. Snowflakes begin to fall, and we are still over 500 km

*If the warm air is relatively unstable, ripples or waves of cirrocumulus clouds will appear as a "mackerel sky."

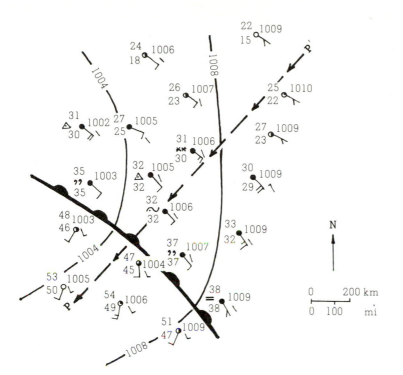

Fig. 16.14 Surface weather associated with a typical warm front.

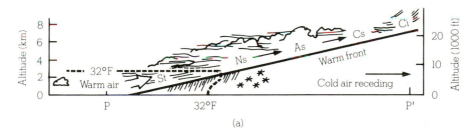

(a)

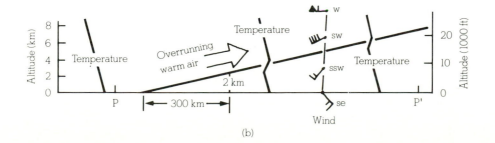

(b)

Fig. 16.15 Vertical view of (a) the clouds and precipitation and (b) the temperature and winds across the warm front in Fig. 16.14 along the line P–P'.

(300 mi) from the surface front. The snow increases, and the clouds thicken into a sheetlike covering of nimbostratus. The winds become brisk and out of the southeast, while the barometer slowly falls. Within 300 km (180 mi) of the front, the cold surface air mass is now quite shallow. The surface air temperature moderates and, as we approach the front, the light snow changes first into sleet. It then becomes freezing rain and finally rain and drizzle as the air temperature climbs above freezing. Overall, the precipitation remains light or moderate but covers a broad area. Moving still closer to the front, warm raindrops partially evaporate in the cool air, saturating it and creating ragged wind-blown stratus and fog. (Thus, flying in the vicinity of a warm front is quite hazardous.)

Finally, after a trip of over 1200 km, we reach the warm front's surface boundary. As we cross the front, the weather changes are noticeable, but much less pronounced than those experienced with the cold front; they show up more as a gradual transition rather than a sharp change. On the warm side of the front, the air temperature and dew point rise, the wind shifts from southeast to south or southwest, and the barometer stops falling. The light rain ends and, except for a few stratocumulus, the fog and low clouds vanish.

This scenario of an approaching warm front represents average warm-front weather. In some instances, the weather can differ from this dramat-

ically. For example, if the overrunning warm air is relatively dry and stable, only high and middle clouds will form, and no precipitation will occur. On the other hand, if the warm air is relatively moist and unstable (as is often the case during the summer), heavy showers can develop as thunderstorms become embedded in the cloud mass.

Along the west coast, the Pacific Ocean significantly modifies the surface air so that warm fronts are difficult to locate on a surface weather map. Also, not all warm fronts move northward or northeastward. On rare occasions, a front will move into the eastern seaboard from the Atlantic Ocean as the front spins all the way around a deep storm positioned off the coast. Cold northeasterly winds ahead of the front usually become warm northeasterly winds behind it. Even with these exceptions, knowing the normal sequence of warm-front weather will be useful, especially if you live east of the Rockies, where warm fronts become well developed. You can look for certain cloud and weather patterns and make reasonably accurate short-range forecasts of your own. Table 16.3 summarizes typical warm-front weather.

Occluded Fronts   When a cold front catches up to and overtakes a warm front, the frontal boundary created between the two air masses is called an **occluded front**, or, simply, an **occlusion**.

Table 16.3   Typical Weather Associated with a Warm Front

WEATHER ELEMENT	BEFORE PASSING	WHILE PASSING	AFTER PASSING
Winds	south-southeast	variable	south-southwest
Temperature	cool–cold	steady rise	warmer
Pressure	usually falling	leveling off	slight rise, followed by fall
Clouds	in this order: Ci, Cs, As, Ns, St and fog; occasionally Cb in summer	stratus-type	clearing with scattered Sc; occasionally Cb in summer
Precipitation	light to moderate rain, snow, sleet, or drizzle	drizzle	usually none; sometimes light rain or showers
Visibility	poor	poor, but improving	fair in haze
Dew point	steady	slow rise	rise, then steady

On the surface weather map, it is represented as alternating cold-front triangles and warm-front half circles; both symbols point in the direction toward which the front is moving. In color, an occluded front is shown by a solid purple line. In Fig. 16.10, notice that the air behind the front is colder than the air ahead of it. This is known as a *cold-type occluded front*, or **cold occlusion**. Let's see how this front develops.

The development of a cold occlusion is shown in Fig. 16.16. Along line *A—A'*, the cold front is rapidly approaching the slower-moving warm front. Along line *B–B'*, the cold front overtakes the warm front, and, as we can see in the vertical view across *C–C'*, lifts both the warm front and the warm air mass off the ground. As a cold-occluded front approaches, the weather sequence is similar to that of a warm front with high clouds lowering and thickening into middle and low clouds, with precipitation forming well in advance of the surface front. Since the front represents a trough of low pressure, southeasterly winds and falling barometers occur ahead of it. The frontal passage, however, brings weather similar to that of a cold front: heavy, often showery precipitation with winds shifting to west or northwest. After a period of wet weather, the sky begins to clear, barometers rise, and the air turns colder. The most violent weather usually occurs where the cold front is just overtaking the warm front, at the point of occlusion, where the greatest contrast in temperatures occurs. Cold occlusions are the most prevalent type of front that moves into the Pacific coastal states. Occluded fronts frequently form over the North Pacific and North Atlantic, as well as in the vicinity of the Great Lakes.

Continental polar air over eastern Washington and Oregon may be much colder than milder maritime polar air moving inland from the Pacific Ocean. Figure 16.17 illustrates this situation. Observe that the air ahead of the warm front is colder than the air behind the cold front. Consequently, when the cold front catches up to and overtakes the warm front, the milder, lighter air behind the cold front is unable to lift the colder, heavier air off the ground. As a result, the cold front rides "piggyback" along the sloping warm front. This produces a *warm-type occluded front*, or a **warm**

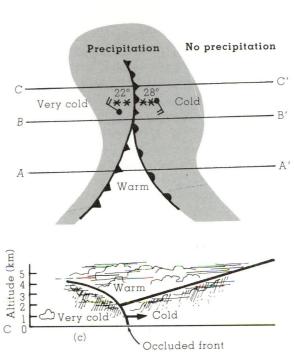

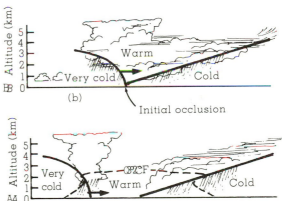

Fig. 16.16 The formation of a cold occluded front. The faster-moving cold front (a) catches up to the slower-moving warm front in (b) and forces it to rise off the ground in (c).

occlusion. The surface weather associated with a warm occlusion is similar to that of a warm front.

Contrast Fig. 16.16 to Fig. 16.17. Note that the primary difference between the warm- and cold-type occluded front is the location of the upper-level front. In a warm occlusion, the upper-level cold front *precedes* the surface occluded front by as much as 325 km (200 mi), whereas in a cold

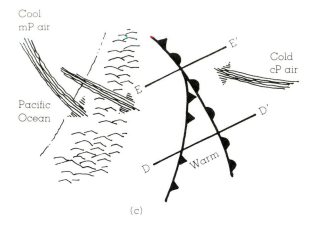

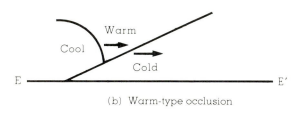

(b) Warm-type occlusion

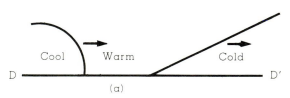

Fig. 16.17 The formation of a warm-type occluded front. The faster-moving cold front in (a) overtakes the slower-moving warm front in (b). The lighter air behind the cold front rises up and over the denser air ahead of the warm front. Diagram (c) shows a surface map of the situation.

occlusion the upper warm front *follows* the surface occluded front, usually by about 50 km (30 mi).

In the world of weather fronts, occluded fronts are the mavericks. In our discussion, we treated occluded fronts as forming when a cold front overtakes a warm front. Some do form in this manner, but others apparently form as new fronts, which develop when a surface storm intensifies in a

region of cold air after its trailing cold and warm fronts have broken away and moved eastward. The new occluded front shows up on a surface chart as a trough of low pressure separating two cold air masses. Because of this, locating and defining occluded fronts at the surface is often difficult for the meteorologist. Similarly, you too may find it hard to recognize an occlusion. In spite of this, we will assume that the weather associated with occluded fronts behaves in a similar way to that shown in Table 16.4.

The frontal systems described in this chapter are actually part of a much larger storm system—the middle latitude cyclone. Chapter 17 details these storms, explaining where, why, and how they form.

Summary

In this chapter, we have considered the different types of air masses and the various weather each brings to a particular region. Continental polar air masses are responsible for very cold, dry weather in winter and cool, pleasant weather in summer. Maritime polar air, having traveled over an ocean for a considerable distance, brings cool, moist weather to an area. The hot, dry weather of summer is associated with continental tropical air masses, while warm humid conditions are due to maritime tropical air masses. Where air masses with sharply contrasting properties meet, we find weather fronts.

A front is a boundary between two air masses of different densities. Stationary fronts have essentially no movement with cold air on one side and warm air on the other. Winds tend to blow parallel to the front, but in opposite directions on either side of it. Along the leading edge of a cold front, where colder air replaces warmer air, showers are prevalent, especially if the warmer air is moist and unstable. Along a warm front, warmer air rides up and over colder surface air, producing widespread cloudiness and precipitation that can cover thousands of square kilometers. Cold fronts typically move faster and are more steeply sloped

Table 16.4 Summary of Typical Weather Associated with Occluded Fronts

WEATHER ELEMENT	BEFORE PASSING	WHILE PASSING	AFTER PASSING
Winds	southeast-south	variable	west-to-northwest
Temperature			
Cold type	cold–cool	dropping	colder
Warm type	cold	rising	milder
Pressure	usually falling	low point	usually rising
Clouds	in this order: Ci, Cs, As, Ns	Ns, sometimes Tcu and Cb	Ns, As, or scattered Cu
Precipitation	light, moderate, or heavy precipitation	light, moderate, or heavy continuous precipitation or showers	light-to-moderate precipitation followed by general clearing
Visibility	poor in precipitation	poor in precipitation	improving
Dew point	steady	usually slight drop, especially if cold-occluded	slight drop, although may rise a bit if warm-occluded

than warm fronts. Occluded fronts, which are often difficult to locate and define on a surface weather map, may have characteristics of both cold and warm fronts.

Questions for Review

1. What is an air mass?

2. If an area is described as a ''good air mass source region'' what information can you give about it?

3. It is summer. What type of afternoon weather would you expect from an air mass designated as mTk? Explain.

4. Why is cP air an unwelcome visitor to the Great Lakes in winter and yet very welcome in summer?

5. What is meant by the term *air mass weather*?

6. Explain why the central United States is not a good air mass source region.

7. Why do air temperatures tend to be a little higher on the eastern side of the Appalachian Mountains than on the western side, even though the same winter cP air mass dominates both areas?

8. Explain the effect of the upper-level winds on surface air masses.

9. List the temperature and moisture characteristics of each of the major air mass types.

10. What are lake-effect snows and how do they form?

11. Why are mP air masses along the east coast of the United States usually colder than those along the nation's west coast? Why are they also *less* prevalent?

12. Why are the boundaries between air masses more strongly developed in winter than in summer?

13. What type of air mass would be responsible for the weather conditions listed below?
  (a) heavy snow showers and low temperatures at Buffalo, New York
  (b) hot, muggy summer weather in the Midwest and the East
  (c) daily afternoon thunderstorms along the Gulf Coast
  (d) heavy snow showers along the western slope of the Rockies
  (e) refreshing, cool, dry breezes after a long summer hot spell on the Central Plains
  (f) heavy summer rainshowers in southern Arizona
  (g) drought with high temperatures over the Great Plains
  (h) persistent cold, damp weather with drizzle along the east coast
  (i) summer afternoon thunderstorms forming along the eastern slopes of the Sierra Nevada Mountains

14. On a surface weather map, what do you

know about a region where the word *frontogenesis* is marked?

15. Explain why barometric pressure usually falls with the approach of a cold or occluded front.

16. Based on the following weather forecasts, what type of front will most likely pass the area?

(a) Light rain and cold today, with temperatures just above freezing. Southeasterly winds shifting to westerly tonight. Turning colder with rain becoming heavy and possibly changing to snow.

(b) Cool today with rain becoming heavy at times by this afternoon. Warmer tomorrow. Winds southeasterly becoming westerly by tomorrow morning.

(c) Increasing cloudiness and warm today, with the possibility of showers by evening. Turning much colder tonight. Winds southwesterly, becoming gusty and shifting to northwesterly by tonight.

(d) Increasing high cloudiness and cold this morning. Clouds increasing and lowering this afternoon, with a chance of snow or rain tonight. Precipitation ending tomorrow morning. Turning much warmer. Winds light easterly today, becoming southeasterly tonight and southwesterly tomorrow.

17. Draw side views of a typical cold front, warm front, and cold-occluded front. Include in each diagram, cloud types and patterns, areas of precipitation, and relative temperatures on each side of the front.

Questions for Thought

1. Suppose an mP air mass moving eastward from the Pacific Ocean travels across the United States. Describe all of the modifications that could take place as this air mass moves eastward in winter. In summer.

2. Explain how an anticyclone during the autumn can bring record-breaking low temperatures and cP air to the southeastern states, and only several days later very high temperatures and mT air.

3. In Fig. 16.4, there is a temperature inversion. How does this inversion differ from the two inversions illustrated in Fig. 16.15b?

4. When a very cold air mass covers half of the United States, a very warm air mass often covers the other half. Explain how this happens.

5. Explain why freezing rain more commonly occurs with warm fronts than with cold fronts.

6. In winter, cold-front weather is typically more violent than warm-front weather. Why? Explain why this is not necessarily true in summer.

7. When a cold front passes a station in the Northern Hemisphere, the wind shifts in a clockwise manner. How would the winds shift during the passage of a cold front in the Southern Hemisphere?

8. Why does the same cold front typically produce more rain over Kentucky than over western Kansas?

9. A cold front moves through your area and yet you observe freezing rain. Show with a diagram the structure of the cold front, including slope, cloud patterns, and vertical temperature profile.

10. How would a cold front, a warm front, and an area of low pressure appear on a surface weather map in the Southern Hemisphere?

Problems and Exercises

1. Make a sketch of North America and show the upper-air wind-flow pattern that would produce:

(a) very cold cP air moving into the far western states in winter

(b) cold cP air over the Central Plains in winter

(c) warm mT air over the Midwest in winter

(d) warm, moist mT air over southern California and Arizona during the summer

2. You are presently taking a weather observation. The sky is full of wispy cirrus clouds estimated to be about 6 km (20,000 ft) overhead. If a warm front is approaching from the south, about how far away is it (assuming a normal slope)? If it is moving toward you at an average warm-front speed, how long will it take before it passes your area?

3. On the surface weather map (page 341), locate the cold front, warm front, and occluded front. Draw them in. (Hint: Tables 16.2, 16.3, and 16.4 can help in locating the fronts.) Station model information is given in Appendix B.

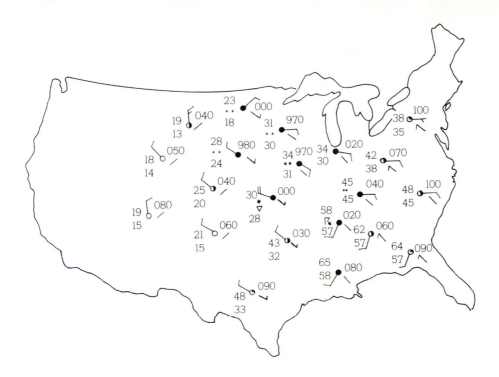

Visible satellite picture of an occluded middle latitude cyclone (Photo: National Oceanic and Atmospheric Administration)

CHAPTER 17

Middle Latitude Cyclones

Early weather forecasters were aware that precipitation generally accompanied falling barometers and areas of low pressure. However, it was not until after the turn of this century that scientists began to piece together the information that yielded the ideas of modern meteorology.

Working largely from surface observations, a group of scientists in Bergen, Norway, developed a model explaining the life cycle of an extratropical storm; that is, a storm that forms at middle and high latitudes outside of the tropics. This extraordinary group of meteorologists included Vilhelm Bjerknes, his son Jakob, Halvor Solberg, and Tor Bergeron. They published their theory shortly after World War I. It was widely acclaimed and became known as the "polar front theory of a developing wave cyclone" or, simply, the **polar front theory**. What these meteorologists gave to the world was a working model of how a midlatitude cyclone progresses through the stages of birth, growth, and decay. An important part of the model involved the development of weather along the polar front. As new information became available, the original work was modified, so that, today, it serves as a convenient way to describe the structure and weather associated with a migratory storm system.

Contents

Polar Front Theory

The development of a wave cyclone, according to the Norwegian model, begins along the polar front. You will remember (from our discussion of the general circulation in Chapter 15) that the polar front is a semicontinuous global boundary separating cold polar air from warm subtropical air. The stages of a developing wave cyclone are illus-

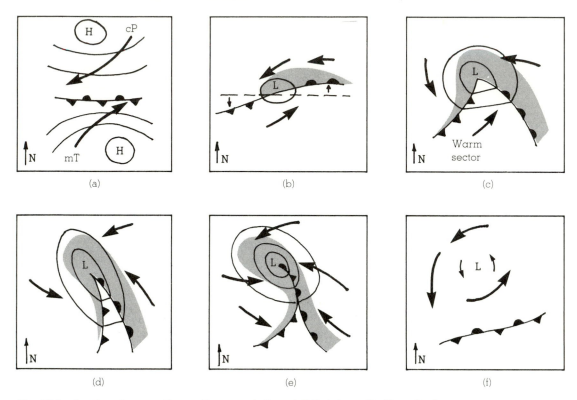

Fig. 17.1 A series of consecutive weather maps (a through f) that shows the life cycle of a wave cyclone (Northern Hemisphere) based on the polar front theory.

trated in the sequence of surface weather maps shown in Fig. 17.1.

Figure 17.1a shows a segment of the polar front as a stationary front. It represents a trough of lower pressure with higher pressure on both sides. Cold air to the north and warm air to the south flow parallel to the front, but in opposite directions. This type of flow sets up a cyclonic wind shear. You can conceptualize the shear more clearly if you place a pen between the palms of your hands and move your left hand toward your body; the pen turns counterclockwise, cyclonically.

Under the right conditions, a wavelike kink forms on the front, as shown in Fig. 17.1b. The wave that forms is known as a **frontal wave**. Watching the formation of a frontal wave on a weather map is like watching a water wave from its side as it approaches a beach: It first builds, then breaks, and finally dissipates. This is why a cyclonic storm system is known as a **wave cyclone**. Figure 17.1b shows the newly formed wave with a cold front pushing southward and a warm front moving

northward. The region of lowest pressure is at the junction of the two fronts. As the cold air displaces the warm air upward along the cold front, and as overrunning occurs ahead of the warm front, a narrow band of precipitation forms (shaded area). Steered by the winds aloft, the system typically moves east or northeastward and gradually becomes a fully developed *open wave* in 12 to 24 hours (Fig. 17.1c). The central pressure is now much lower, and several isobars encircle the wave's apex. These more tightly packed isobars create a stronger cyclonic flow, as the winds swirl counterclockwise and inward toward the low center. Precipitation forms in a wide band *ahead* of the warm front and along a narrow band *behind* the cold front. The region of warm air between the cold and warm fronts is known as the **warm sector**. Here, the weather tends to be partly cloudy, although scattered showers may develop if the air is unstable.

Energy for the storm is derived from several sources. As the air masses try to attain equilibrium, warm air rises and cold air sinks, transforming potential energy into kinetic energy. Condensation supplies energy to the system in the form of latent heat. And, as the surface air converges toward the low center, wind speeds increase, producing an increase in kinetic energy.

As the open wave moves eastward, central pressures continue to decrease, and the winds blow more vigorously. The faster-moving cold front constantly inches closer to the warm front, squeezing the warm sector into a smaller area, as shown in Fig. 17.1d. Eventually, the cold front overtakes the warm front and the system becomes occluded. At this point, the storm is usually most intense, with clouds and precipitation covering a large area. The intense storm system shown in Fig. 17.1e gradually dissipates, because cold air now lies on both sides of the occluded front. Without the supply of energy provided by the rising warm, moist air, the old storm system dies out and gradually disappears (Fig. 17.1f). Occasionally, however, a new wave will form on the westward end of the trailing cold front. We can think of the sequence of a developing wave cyclone as a whirling eddy in a stream of water that forms behind an obstacle, moves with the flow, and gradually vanishes downstream. The entire life cycle of a wave cyclone can last from a few days to over a week.

Figure 17.2 shows a series of wave cyclones at various stages of development along the polar front in winter. Such a succession of storms is known as a ''family'' of cyclones. Observe that to the north of the front are cold anticyclones; to the south over the Atlantic Ocean is the warm, semipermanent Bermuda high. The polar front itself has developed into a series of loops, and at the apex of each loop is a cyclone. The cyclone over the northern plains (Low 1) is just forming; the one along the East Coast (Low 2) is an open wave; and the system near Iceland (Low 3) is dying out. If the average rate of movement of a wave cyclone from birth to decay is 25 knots, then it is entirely possible for a storm to develop over the central part of the United States, intensify into a large storm over New England, become occluded over the ocean, and reach the coast of England in its dissipating stage less than a week after it formed.

Up to now, we have considered the polar front

model of a developing wave cyclone, which represents a rather simplified version of the stages that an extratropical storm system must go through. In fact, few (if any), storms adhere to the model exactly. Nevertheless, it serves as a good foundation for understanding the structure of storms. So keep the model in mind as you read the following sections.

Developing Middle Latitude Storms—Cyclogenesis

Any development or strengthening of a cyclone is called **cyclogenesis**. Within the United States, there are regions that show a propensity for cyclogenesis, including the eastern slopes of the Rockies, the Great Basin, the Gulf of Mexico, and the Atlantic Ocean east of the Carolinas. Near Cape Hatteras, North Carolina, for example, warm Gulf Stream water can supply moisture and warmth to the region south of a stationary front, thus increasing the contrast between air masses to a point where storms may suddenly and unexpectedly spring up along the front. These cyclones normally move northeastward along the Atlantic Coast, bringing high winds and heavy snow or rain to coastal areas. Before the age of modern communications and weather prediction such coastal

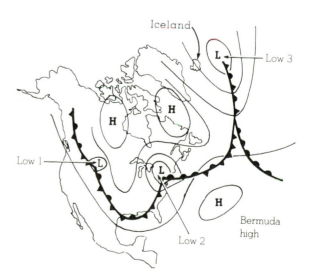

Fig. 17.2 A series of wave cyclones (''family'' of cyclones) forming along the polar front.

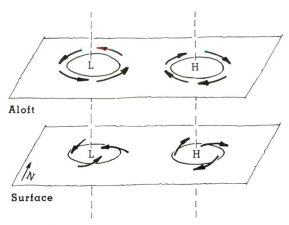

Aloft

Surface

Fig. 17.3 If lows and highs aloft were always directly above lows and highs at the surface, the surface systems would quickly dissipate.

storms would often go undetected during their formative stages; and sometimes an evening weather forecast of "fair and colder" along the eastern seaboard would have to be changed to "heavy snowfall" by morning. Fortunately, with

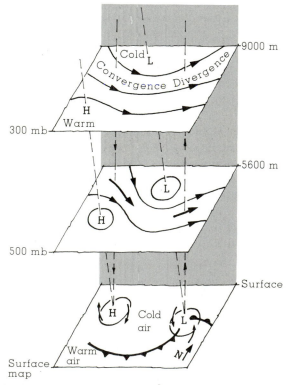

Fig. 17.4 The vertical structure of cyclones and anticyclones.

today's weather information gathering and forecasting techniques, these storms rarely strike by surprise.

When extratropical cyclones deepen rapidly, the term *explosive cyclogenesis*, or "bomb," is used to describe them. As an example, explosive cyclogenesis occurred in a storm that developed over the warm Atlantic just east of New Jersey on September 10, 1978. As the central pressure of the storm dropped nearly 60 mb (1.8 in.) in 24 hours, hurricane force winds battered the ocean liner *Queen Elizabeth II* and sank the fishing vessel *Captain Cosmo*.

Frontal waves that develop into huge storms are called **unstable waves**. These waves form suddenly, grow in size, and then slowly dissipate with the entire process taking several days to a week. Other waves, called **stable waves**, remain small and never grow into giant weather producers. Why is it that some waves develop into huge storms, whereas others simply dissipate in a day or so?

This question poses one of the real challenges in weather forecasting. The answer is complex. Indeed, there are many surface conditions that do influence the formation of a wave, including mountain ranges and land-ocean temperature contrasts. However, the real key to the development of a wave cyclone is found in the upper-wind flow, in the region of the high-level westerlies. Therefore, before we can arrive at a reasonable answer to our question, we need to see how the winds aloft influence surface pressure systems.

Vertical Structure of Deep Pressure Systems

In Chapter 14, we learned that thermal pressure systems are shallow and weaken with increasing elevation. On the other hand, developing surface storm systems are deep lows that usually intensify with height. This means that a simple surface depression associated with one of these systems will appear on an upper-level chart as either a closed low or a trough.

Suppose the upper-level low is directly above the surface low (see Fig. 17.3). Notice that only at the surface do the winds blow inward toward the low's center. As these winds converge, air density increases directly above the surface low. This increase in mass causes surface pressures to rise; gradually, the low fills and the surface low dissipates. The same reasoning can be applied to surface anticyclones. Winds blow outward away from the center of a surface high. If a closed high or ridge lies directly over the surface anticyclone, divergence at the surface will remove air from the column directly above the high. Surface pressures fall and the system weakens. So it appears that, if upper-level pressure systems were always located directly above those at the surface, cyclones and anticyclones would die out soon after they form (if they could form at all). What, then, is it that allows these systems to develop and intensify?

We know from Chapter 12 that lows aloft are generally associated with cold air. Note in Fig. 17.4 that behind the cold front there is cold air both at the surface and aloft. Consequently, the upper-level low is located *behind*, or to the *west*, of the surface low. Observe how the surface low tilts toward the northwest, showing up as a closed system at 500 mb and as a trough at 300 mb. Directly above the surface low, at 300 mb, the air spreads out and diverges (as indicated by the contour lines). This allows the converging surface air to rise and flow out of the top of the air column just below the tropopause, which acts as a constraint to vertical motions. We now have a mechanism for developing storms. When upper-level divergence is stronger than surface convergence (more air is taken out at the top than is brought in at the bottom), surface pressures drop, and the low intensifies. By the same token, when upper-level divergence is less than surface convergence (more air flows in at the bottom than is removed at the top), surface pressures rise, and the system fills.

We can also use Fig. 17.4 to explain the structure of the anticyclone. High pressures aloft are associated with warm air. At the surface and aloft, warm air lies to the southwest of the surface high. This causes the surface anticyclone to tilt toward the southwest at higher altitudes. Moving upward, we observe that the closed system becomes a ridge at the 300-mb level. Also notice that directly above the surface high at 300 mb there is convergence of air (as indicated by the contour lines).

Convergence causes an accumulation of air above the high. This allows the air to sink slowly and replace the diverging surface air. Hence, when upper-level convergence exceeds low-level divergence (inflow at top is greater than outflow near the surface), surface pressures rise, and the anticyclone builds. On the other hand, when upper-level convergence is less than low-level divergence, the anticyclone weakens as surface pressures fall.

Look at the wind direction at the 500-mb level in Fig. 17.4. Winds at this altitude tend to steer surface systems in the same direction that they are moving. Thus, the surface storm will move toward the northeast, while the surface anticyclone will move toward the southeast. These paths indicate the average movement of surface pressure systems in the eastern two-thirds of the United States (Fig. 17.5). In general, the surface storm centers travel across the United States at about 16 knots in summer and about 27 knots in winter. The faster winter velocity reflects the stronger upper-level flow during this time of year.*

So far, we have seen that deep pressure systems exist at the surface and aloft throughout much of the troposphere. When the surface pressure system does not lie directly beneath the upper-level trough but tilts toward the west, the atmosphere is able to redistribute its mass. Regions of low-level convergence are compensated for by regions of upper-level divergence and vice versa. Cyclones and anticyclones can intensify and, steered by the winds aloft, move away from their region of formation.

Since regions of strong upper-level divergence and convergence typically occur along the flanks of a jet stream, we will examine this area next.

Developing Cyclones and the Jet Stream

The polar front jet stream can provide some of the necessary ingredients for the development of

---

*As a forecasting rule of thumb, surface pressure systems tend to move in the same direction as the wind at the 500-mb level. The speed at which the surface systems move is about half the speed of the 500-mb winds.

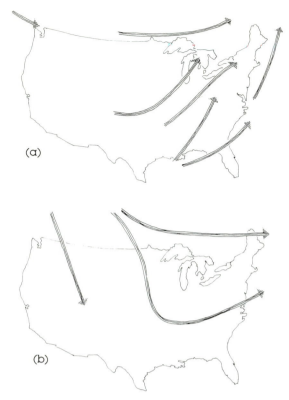

(a)

(b)

Fig. 17.5 Typical paths of (a) winter cyclones and (b) winter anticyclones.

a wave cyclone. Remember (Chapter 15) that the axis of the polar front jet pretty much coincides with the polar front. The core of the jet—the jet maximum or jet streak—represents the region of strongest winds. As winds approach the jet maximum, they speed up; when they leave, they slow down. These accelerations and decelerations, coupled with strong wind speed shears and a curving jet stream, cause regions of convergence and divergence to exist along the flanks of the jet (see Fig. 17.6).*

Figure 17.7 shows us how areas of divergence and convergence correlate with the development of surface pressure systems. The heavy line on the upper-level chart represents the jet stream with its regions of convergence and divergence; the lower chart is a surface weather map with two developing storm systems. The jet stream lies to the north of the small surface wave (A). Directly

*To better understand why these areas of convergence and divergence form, read the focus section entitled "A Closer Look at Convergence and Divergence" (p. 350).

above the wave is a region of divergence that pumps surface air upward to the jet. The jet then quickly sweeps this air downstream, causing surface pressures to drop rapidly. As surface pressure gradients increase, the wind speeds pick up, and the fronts move more quickly as surface air circulates counterclockwise around the depression. As the flow aloft steers the storm along, the surface low intensifies, and the jet becomes distorted, usually developing into a large, looping wave. Notice that, as the storm occludes (B), the jet crosses the occluded front, providing an area of divergence above the surface low. Also observe that above the anticyclones, regions of convergence feed air downward into these surface systems. Hence, we find jet streams supplying air to the surface anticyclones and removing air above the surface cyclones.

The polar front jet stream is strongest and moves farther south in winter. Consequently, midlatitude storms across the United States are usually better developed and tend to occur more frequently during the coldest months. During the summer, when the polar jet shifts northward, developing midlatitude storm activity occurs principally over Alberta, Canada, and the Northwest Territories.

Since the polar front jet stream forms along the polar front, why isn't there always a line of cyclones developing on the front? As we have just seen, regions of divergence and convergence along the flanks of the jet are found only at specific zones near the jet maximum. Furthermore, studies of developing cyclones show that most middle latitude storm systems develop when the jet stream becomes distorted; that is, when it starts to bend. A bending jet indicates that deep troughs and ridges—waves—exist in the flow aloft. The next section examines these waves and their influence on a developing storm.

Upper-Level Waves and Surface Storms

If we look at an upper-level chart of the Northern (or Southern) Hemisphere similar to the one in Fig. 17.8, we see a series of ridges and troughs,

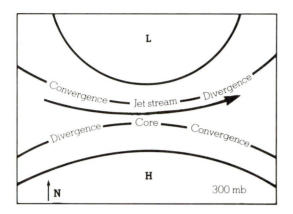

Fig. 17.6 Regions of divergence and convergence around the core of the polar front jet stream on an upper-level chart.

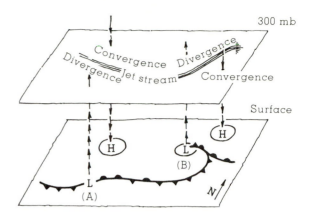

Fig. 17.7 Areas around the polar front jet stream supply the necessary regions of divergence and convergence for surface pressure systems.

known as **longwaves**, encircling the globe. At any given time, there are between four and six of these waves looping around the earth. Since mountain ranges tend to disturb the upper-level wind flow, these waves are often found to the east of such topographic barriers as the Rockies and the Tibetan Plateau. The wavelength of longwaves varies between 4000 and 8000 km (2500 to 5000 mi). Longwaves are also known as **Rossby waves**, after C. G. Rossby, a famous meteorologist who carefully studied their motion. Imbedded in longwaves are **shortwaves**, which are small disturbances, or ripples.

Rossby found that the shorter the wavelength of a particular wave, the faster it moved downstream. Shortwaves tend to move eastward at a speed proportional to the average wind flow near the 700-mb level, about 3 km above sea level. Longwaves, on the other hand, often remain stationary, move eastward very slowly at less than 4° of longitude per day, or even move westward (retrograde). We can obtain a better idea of this wave movement if we think of longwaves as being huge meanders (loops) in a swiftly flowing stream of water. Water moves through the loops quickly, while the loops themselves move eastward very slowly, as the fast-flowing water cuts away at one bank and deposits material on the other. Suppose debris tumbles into the stream, disturbing the flow. The disturbed flow appears as a small wrinkle that travels downstream through the loops, at a speed near the average stream flow. This wrinkle in the flow is analogous to a shortwave in the atmosphere.

Atmospheric shortwave troughs, often simply called *shortwaves*, move quickly around the longwaves. Shortwaves usually deepen when they approach a longwave trough and weaken when they move through a ridge. Figure 17.9 illustrates shortwaves moving around longwaves on an upper-level chart.

Over a large area of the chart (Fig. 17.9), the isotherms (dashed lines) roughly parallel the isobars (solid lines). In this region, winds do not cross the isotherms, and there is no temperature ad-

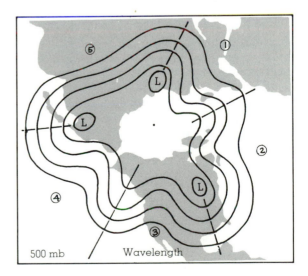

Fig. 17.8 Map of the Northern Hemisphere showing five longwaves on a 500-mb chart. Note that the wavelength of wave number 3 is greater than the width of the United States.

FOCUS ON A SPECIAL TOPIC

A Closer Look at Convergence and Divergence

Convergence is the piling up of air above a region, while *divergence* is the spreading out of air above some region. We know that, at the surface, winds diverge about the center of a high-pressure area, while aloft—directly above the high—the winds converge. When upper-level convergence exceeds low-level divergence, the surface pressure increases, and we say that the high-pressure area is *building*. By the same token, we know that surface winds converge about the center of a surface low-pressure area, and aloft—directly above the low—the winds diverge. When upper-level divergence exceeds low-level convergence, surface pressure decreases, and we say that the storm system is *intensifying*, or *deepening*.

Convergence and divergence of air may result from changes in wind direction and wind

speed. For example, convergence occurs when moving air is funneled into an area, much in the way cars converge when they enter a crowded freeway. Divergence occurs when moving air spreads apart, much as cars spread out when a congested two-lane freeway becomes three lanes. On an upper-level chart, this type of convergence (also called *confluence*) occurs when contour lines move closer together, as a steady wind flows parallel to them. On the same chart, this type of divergence (also called *diffluence*) occurs when the contour lines move apart as a steady wind flows parallel to them. (In Fig. 17.4, observe the contour lines and the regions of convergence and divergence on the 300-mb chart.)

Convergence and divergence may also result from changes in wind speed. Convergence oc-

curs when the wind slows down as it moves along, while divergence occurs when the wind speeds up. We can grasp these relationships more clearly if we imagine air molecules to be marching in a band. When the marchers in front slow down, the rest of the band members squeeze together, causing convergence; when the marchers in front start to run, the band members spread apart, or diverge. In summary, *speed convergence* takes place when the wind speed decreases downwind and *speed divergence* takes place when the wind speed increases downwind.

Earlier, we saw that areas of convergence and divergence form at specific regions around the jet stream. To understand why, consider air moving through a jet stream as depicted in Fig. 1. The heavy shaded area is the jet maximum, or jet

---

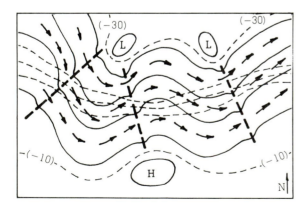

Fig. 17.9 Upper-level chart showing shortwaves (heavy dashed line) moving around several longwaves. Solid lines are isobars. Dashed lines are isotherms. Arrows show wind direction.

vection.* Where these conditions exist, the atmosphere is said to be **barotropic**. By comparison, there are regions (especially toward the middle of the map) where isotherms are close together. Here, temperature and pressure gradients are steep, isotherms cross the isobars, and strong winds produce temperature advection. Where these conditions prevail, the atmosphere is said to be **baroclinic**. Note that the baroclinic atmosphere exists only in a narrow zone. Below this zone lies the polar front; above it flows the polar front jet

*The warming of air produced by the wind is called *warm advection*. The cooling of air produced by the wind is called *cold advection*. For warm advection to occur, the wind must blow across the isotherms from warmer to colder regions. For cold advection, the wind must blow across the isotherms from colder to warmer regions.

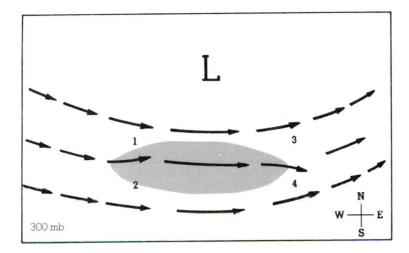

Fig. 1 Changing air motions within the jet stream maximum (shaded area) cause convergence of air at points 1 and 4 and divergence at points 2 and 3.

gradient force. The greater force temporarily exceeds the Coriolis force, and the air swings slightly to the north across the contour lines. This causes convergence of air at point 1 and divergence at point 2. Toward the middle of the jet maximum, the increase in wind speed causes the Coriolis force to increase and the wind to become nearly geostrophic again. However, as the air leaves the jet maximum, the pressure gradient force is reduced and the Coriolis force temporarily exceeds it, causing the air to swing slightly to the south. This causes convergence of air at point 4 and divergence at point 3. Downwind, the air returns to a nearly geostrophic balance. At the surface beneath the area of strongest divergence (point 3) is the deepening storm, while beneath the area of strongest convergence (point 1) is the building anticyclone.

streak—the region where the winds are strongest. As air enters the jet maximum from the west, it increases in speed; as it exits the maximum, it decreases in speed. Remember from Chapter 12 that at this elevation the wind flow is nearly in geostrophic balance with the pressure gradient force (directed north) and the Coriolis force (directed south). As the air enters the jet maximum, it increases in speed due to an increase in the pressure

---

stream. Notice how the shortwaves disturb the air flow and accentuate the region of baroclinicity. This disturbed flow can amplify the wave and develop or intensify a surface storm system. The theory explaining how this occurs is known as the *baroclinic wave theory of developing cyclones.*

The Necessary Ingredients for a Developing Wave Cyclone

To better understand why wave cyclones develop and intensify into huge storms, we need to examine atmospheric conditions at the surface and aloft. Suppose that a portion of a longwave trough at the 500-mb level lies directly above a surface

stationary front (see Fig. 17.10a). On the 500-mb chart, contour lines (solid lines) and isotherms (dashed lines) parallel each other and are crowded close together. Colder air is located in the northern half of the map, while warmer air is located to the south. Winds are blowing at fairly high velocities, which produce a sharp change in wind speed—a strong wind speed shear—from the surface up to this level. Suppose a shortwave moves through this region, disturbing the flow (see Fig. 17.10b). This sets up a kind of instability in the flow (as warmer air rises and colder air sinks) known as **baroclinic instability**.

With the onset of baroclinic instability, horizontal and vertical air motions begin to enhance the formation of cyclonic storms. As the flow aloft becomes disturbed, a region of converging air

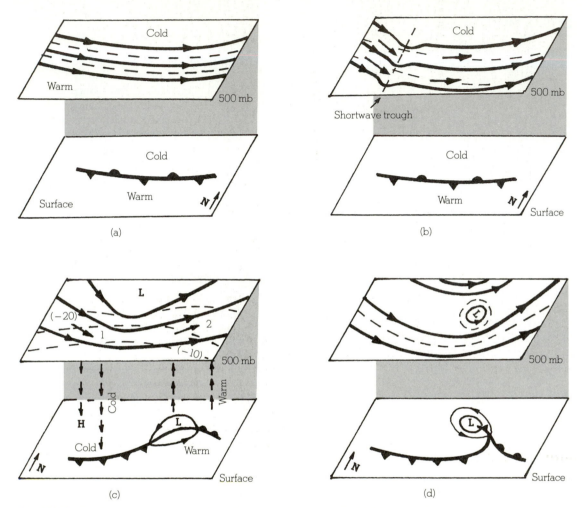

Fig. 17.10 The formation of a wave cyclone during baroclinic instability. (a) A longwave trough at 500 mb lies parallel to and directly above a surface stationary front. (b) A shortwave (heavy dashed line) disturbs the flow aloft. This initiates temperature advection. (c) The upper trough intensifies and provides the necessary vertical motions for the development of the surface cyclone. (d) The surface storm occludes, and a cold pool of air remains above it.

forms behind the shortwave trough (to the left of the trough in Fig. 17.10b). Meanwhile, a region of diverging air forms ahead of the trough (to the right in Fig. 17.10b). Beneath these zones, surface pressures change and wind speeds increase. If initially there are no fronts on the surface map, they may begin to form where air masses with contrasting properties are brought together, as surface air rises and the surrounding air flows inward to replace it. As the converging surface air develops cyclonic spin, cold air flows southward and warm air northward. We can see in Fig. 17.10c that the western half of the stationary front is now

a cold front and the eastern half a warm front. Cold air moves in behind the cold front, while warm air slides up along the warm front. These regions of cold and warm advection occur all the way up to the 500-mb level.

On the 500-mb chart in Fig. 17.10c, cold advection is occurring at position 1 as the wind crosses the isotherms, bringing cold air into the trough. The cold advection makes the air more dense and lowers the height of the air column from the surface up to the 500-mb level. (Recall that, on a 500-mb chart, lower heights mean the same as lower pressures; consequently, the pressure in the trough lowers and the trough deepens.) Meanwhile, at position 2, warm advection is taking place, which has the effect of raising the height of a column of air; here, the 500-mb heights increase and a ridge builds (strengthens). Therefore, *the overall effect of differential temperature advection is to intensify the wave*.

Regions of cold and warm advection are associated with vertical motions. Where there is cold advection, some of the cold, heavy air sinks; where there is warm advection, some of the warm, light air rises. Hence, air must be sinking in the vicinity of position 1 and rising in the vicinity of position 2. The sinking of cold air and the rising of warm air provide energy for a developing cyclone, as potential energy is transformed into kinetic energy. Further, condensation in the ascending air releases latent heat, which strengthens the system even more. So, we now have a full-fledged middle latitude cyclone with all of the necessary ingredients for its development.

Eventually, the storm system occludes (Fig. 17.10d) and dies, as the supply of warm surface air is cut off. Often, an upper-level pool of cold air (which has broken away from the main flow) lies almost directly above the surface low. The isotherms around this **cut-off low** tend to parallel the contour lines, which indicates that no significant temperature advection is occurring. Without the necessary energy transformation, the surface system gradually dissipates. The upper-level low, however, may remain stationary for many days. If air is forced to ascend into this cold pocket, widespread clouds and precipitation may persist for some time, even though the surface storm system itself has moved on east out of the picture.

In general, we now have a fairly good concept as to why some wave cyclones intensify into huge storms while others do not. For a storm system to intensify, there must be an upper-level counterpart that lies to the west of the surface depression. As shortwaves disturb the flow aloft, they cause regions of differential temperature advection to appear, leading to an intensification of the upper-level trough. Zones of converging and diverging air develop both aloft and at the surface. These regions are accompanied by vertical motions, which provide the proper energy conversions for the storm's growth. With this atmospheric situation, storms may form even where there are

no fronts. In regions where the upper-level flow is not disturbed by shortwaves or where no upper trough exists, the necessary vertical and horizontal motions are insufficient to enhance cyclonic storm development, even where a surface front exists. The horizontal and vertical motions, cloud patterns, and weather that typically occur with a developing open wave cyclone are summarized in Fig. 17.11.

Vorticity, Longwaves, and Weather Systems

Earlier in this chapter, we saw that the flow aloft frequently forms into longwaves and that storms often develop downwind of a mountain range. To see why these occur, we need to investigate an important concept in meteorology called **vorticity**.

When something spins, it has vorticity. The faster it spins, the greater its vorticity. In meteorology, vorticity is a measure of the spin of small air parcels. Although the spin can be in any direction, our concern will be with the spin of horizontally flowing air about a vertical axis, much like an ice skater spins about an imaginary vertical axis. Because it is such a complex topic, we will begin our study of vorticity by examining large-scale air flow. Our goal is to see how vorticity relates vertical motions to the divergence and convergence of air. Before beginning, we must give vorticity some quantitative value.

Let us define air that spins cyclonically (counterclockwise) as having *positive* vorticity and air that spins anticyclonically (clockwise) as having *negative* vorticity. Figure 17.12a shows a shallow column of air rotating cyclonically. This weak circulation has positive vorticity. The ice skater inside the column will help us to illustrate the column's rate of spin.

Suppose the air flow aloft begins to diverge above the column. The divergence aloft is compensated for by convergence near the surface. If we assume that the total mass of the air in the column does not change, the column must stretch vertically, producing upward air motions. Notice that, as the column stretches, the ice skater's arms are pulled in close, and the skater spins much faster. Therefore, the rate at which air flows

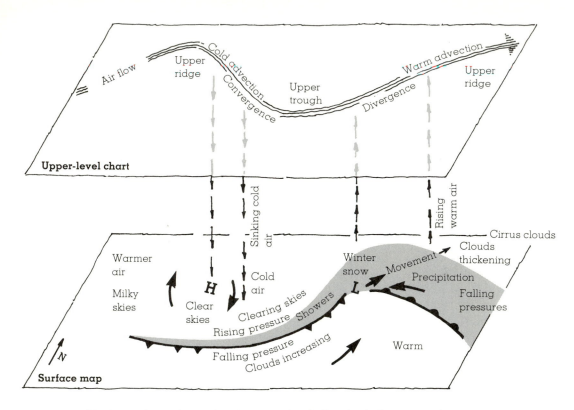

Fig. 17.11   Clouds, weather, and vertical motions associated with a developing wave cyclone.

around the center of the column must also increase, which means that the vorticity of the column increases, becoming more positive.

Figure 17.12b illustrates what happens to the original column when convergence of air occurs above the column. Upper-level convergence creates low-level divergence, and the column shrinks vertically, producing downward air motions. As the column shrinks vertically, it spreads out horizontally, and the ice skater's spin rate decreases, as the arms move outward away from the skater's sides. Hence, the column's spin rate decreases

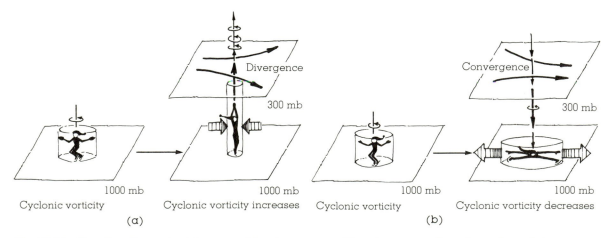

Fig. 17.12   The effect of upper-level divergence (a) and convergence (b) on the vorticity of a surface column of air.

FOCUS ON A SPECIAL TOPIC

Lee-Side Lows

Vorticity helps to explain the development of storms on the leeward side of mountains. We can see why when we examine the flow of air over a mountain barrier (see Fig. 2). Assume that the wind is westerly and there is no shear, so that the relative vorticity of the air column as it approaches the mountain will be zero. Theoretical studies show that the absolute vorticity of the column divided by its depth is equal to a constant. Thus,

$$\frac{\text{Earth's vorticity} + \text{relative vorticity}}{\text{Depth of column}} = \text{constant}$$

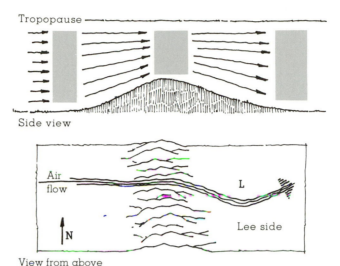

Fig. 2 Due to changes in the air's vorticity, low-pressure areas (lee-side lows) tend to form on the leeward side of mountain ranges.

As the column moves up the side of the mountain, the tropopause acts as a lid on the top of the column. This causes the column to shrink vertically and expand horizontally. To compensate for the column's decreasing depth, there must be a similar decrease in absolute vorticity. Hence, the air flow starts to curve anticyclonically toward the southeast. Once the column begins to move downslope, it stretches vertically. Its depth in-

creases, and so must its absolute vorticity. The relative vorticity of the southeastward moving air is zero, but the air is moving into a region of decreasing earth vorticity. Consequently, there must be a bending of the air flow in the cyclonic sense to increase the column's absolute vorticity. The air flow now swings toward

the northeast, which creates a trough of low pressure called a **lee-side low** on the downwind side of the barrier. Now we can see why so many wave cyclones form along the flanks of the Rocky Mountains, especially in eastern Colorado, where the crest of the mountains reaches its maximum elevation.

and its vorticity decreases, becoming less positive. Aloft, however, where there is horizontal convergence, an air layer's spin rate would increase, which would increase its cyclonic vorticity. By the same token, where there is horizontal divergence aloft, an air layer would lose cyclonic vorticity. Therefore, it follows that *a gain (or loss) of cyclonic vorticity in the lower troposphere is counterbalanced by a loss (or gain) of cyclonic vorticity in the upper troposphere.*

Before we consider how vorticity ties in with the development of weather systems, we need to

examine two important types of this phenomenon: the earth's vorticity and relative vorticity.

Because the earth spins, it has vorticity. In the Northern Hemisphere, the *earth's vorticity* is always positive because the earth spins counterclockwise about its vertical North Pole axis. The amount of earth vorticity imparted to any object—even to those that are not moving relative to the earth's surface—depends upon the latitude. In Fig. 17.13, an observer standing on the equator would not spin about his or her own *vertical axis*; further north, the observer would spin very slowly,

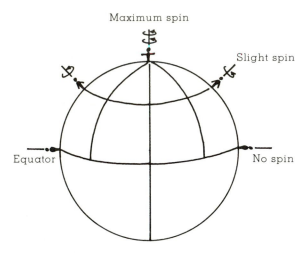

Fig. 17.13 Due to the rotation of the earth, the rate of spin of observers about their vertical axes increases from zero at the equator to a maximum at the poles.

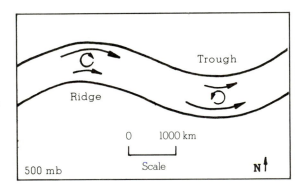

Fig. 17.14 Because the contour lines curve, air moving through a ridge spins clockwise and gains anticyclone relative vorticity. In the trough, the air spins counterclockwise and gains cyclonic relative vorticity.

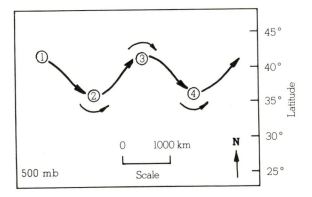

Fig. 17.15 The wavy path of air aloft due to the conservation of absolute vorticity. (See text for explanation.)

while at the North Pole the observer would spin at a maximum rate of one revolution per day. It is now apparent that any object on the earth has vorticity simply because the earth is spinning, and the amount of this earth vorticity increases from zero at the equator to a maximum at the poles.*

Moving air will generally have additional vorticity relative to the earth's surface. This type of vorticity, called *relative vorticity*, is the sum of two effects: the curving of the air flow (*curvature*) and the changing of the wind speed over a horizontal distance (*shear*). Figure 17.14 illustrates vorticity due to curvature. Air moving through the trough tends to spin cyclonically, increasing its relative vorticity. In the ridge, the spin tends to be anticyclonic, and the relative vorticity of the air increases in the negative direction. Whenever the wind blows faster on one side of an air parcel than on the other, a shear force is imparted to the parcel and it will spin and gain relative vorticity.

The sum of the earth's vorticity and the relative vorticity is called the **absolute vorticity**. We can use the concept of absolute vorticity to explain why the westerly flow aloft tends to form into waves. To explain the wave motion, we will assume that there is no divergence or convergence of air; columns of air may not stretch or contract. Where this condition prevails, the absolute vorticity of air will be conserved. This means that the numerical value of the sum of the earth's vorticity and the relative vorticity will not change with time:

Absolute vorticity = earth's vorticity +
relative vorticity = constant.

Hence, any decrease in the earth's vorticity must be compensated for by an increase in the relative vorticity and vice versa.

Consider, for example, that air in Fig. 17.15 is moving horizontally at a constant speed at the 500-mb level. At this level, divergence is usually near zero, so our initial assumption approaches a real situation. Since there is no wind speed shear, any change in the relative vorticity will be due to curvature. At position 1, the air is flowing southeastward. Heading equatorward, the air moves into a region of decreasing earth vorticity. To keep

*The earth's vorticity at any latitude is equal to the product of twice the earth's angular rate of spin (2Ω) and the sine of the latitude (ϕ)—$2\Omega \sin(\phi)$. This expression is referred to as the *Coriolis parameter*.

the absolute vorticity of the air constant, there must be a corresponding increase in the relative vorticity. Since increasing relative vorticity implies cyclonic curvature, the air turns counterclockwise at position 2 and heads northeastward. But now the air is moving into a region where the earth's vorticity steadily increases. To offset this increase, the relative vorticity must decrease. The relative vorticity will decrease if the curvature becomes anticyclonic, so at position 3 air turns clockwise and heads toward the equator once again. This again brings the air into a region where the earth's vorticity decreases. To compensate, the air must now turn cyclonically at position 4, and so on. In this manner, a series of upper-level longwaves may develop, encircling the entire earth.

We are now in a position to see how vorticity ties in with the development of middle latitude storms. Earlier we saw that, in the Northern Hemisphere, anticyclonic flow—the clockwise flow around a high-pressure area—produces negative relative vorticity. However, the earth's positive vorticity usually exceeds the anticyclone's negative relative vorticity. Consequently, in the middle and upper troposphere, we usually find middle latitude cyclones and anticyclones associated with regions of high and low (positive) absolute vorticity, about the size of the state of Iowa. Between these regions are insignificant amounts of vorticity. The regions of high and low absolute vorticity are swept along by the flow aloft without any change in value. As a consequence, a region of positive absolute vorticity aloft moves with the wind and is known as an area of **positive vorticity advection (PVA)**.

Regions of PVA are extremely valuable in weather forecasting because they link divergence aloft, vertical motions, and surface storm development. For instance, sophisticated mathematical equations show that downwind from a vorticity maximum (to the east of the region of PVA in Fig. 17.16) there is upper-level divergence and low-level convergence. This means rising air and the possibility of clouds, precipitation, and cyclonic storm development. It follows that when a region of positive vorticity moves over a surface stationary front, a wave will form along the front and a storm will develop. Even in the absence of fronts, a zone of organized clouds with precipitation may form in conjunction with an area of PVA. On the other hand, upwind from a vorticity maximum (to

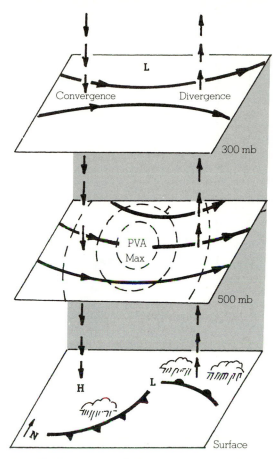

Fig. 17.16 An area of PVA (positive vorticity advection) on a 500-mb chart means divergence aloft and ascending air motions on its east side, and convergence aloft with descending motions on its west side.

the west of the region of PVA in Fig. 17.16) or in a region of very low vorticity, there is upper-level convergence, low-level divergence, slowly sinking air, and the generally fair weather that we associate with surface anticyclones.

Because of the gaps in observational data, scientists often found it difficult to locate centers of positive vorticity on satellite pictures, especially in regions where there were no clouds. Now, however, certain geostationary satellites are equipped to observe atmospheric water vapor. The swirling patterns of moisture clearly identify the position of vorticity centers. In Fig. 17.17 observe the cyclonic swirl of water vapor associated with a region of maximum vorticity just off the coast of Northern California.

Fig. 17.17 Infrared water vapor image taken by *GOES West* on August 30, 1982. The white band of clouds (blobs) north of the equator is the intertropical convergence zone. The cyclonic swirl of water vapor off the California coast defines a region of maximum vorticity. The stretched-out band of moisture in the Southern Hemisphere is the jet stream. To the south of the jet is another vorticity maximum.

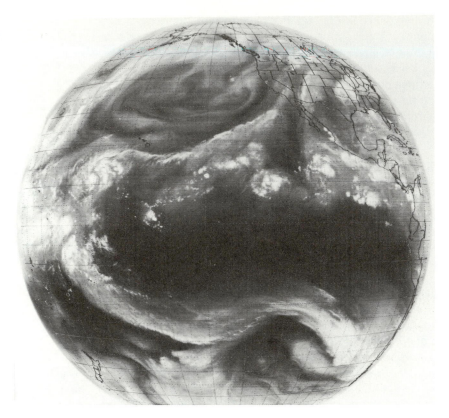

Summary

In this chapter, we discussed where, why, and how a wave cyclone forms. We began by examining the early polar front theory proposed by Norwegian scientists after World War I. We saw the important effect that the upper-air flow, including the jet stream, has on the intensification and movement of surface high- and low-pressure areas. Along the flanks of a jet maximum, regions of convergence supply air to surface anticyclones and regions of divergence remove air above surface midlatitude cyclones. The curving nature of the jet stream tends to direct anticyclones southeastward and midlatitude cyclones northeastward.

When the jet stream bends, waves in the form of deep troughs and ridges exist in the flow aloft. When an upper-level trough lies to the west of a surface depression and when a shortwave disturbs the flow aloft, horizontal and vertical air motions begin to enhance the formation of the surface storm. The rising of warm air and the sinking of cold air provide the proper energy conversions for the storm's growth, as potential energy is transformed into kinetic energy.

Finally, we looked at the concept of vorticity and how it relates to developing cyclones, longwaves, and lee-side lows. We found that, as an area of positive absolute vorticity approaches a region, it is often accompanied by diverging air aloft, converging air at the surface, and cyclonic storm development. Where air moves over a mountain range, changes in the absolute vorticity of the air column cause low-pressure areas to either intensify or form on the leeward side of the mountain.

Questions for Review

1. Why do wave cyclones "die out" after they become occluded?

2. List four regions in North America where wave cyclones tend to develop.

3. What are the necessary ingredients for a wave cyclone to develop into a huge storm system?

4. Explain this fact: Without upper-level divergence, a surface open wave would probably persist for less than a day.

5. Why do middle latitude surface pressure systems tilt westward with increasing height?

6. What are unstable waves in the atmosphere? What are stable waves?

7. How does the polar front jet stream influence the formation of a wave cyclone?

8. Explain why, even though the polar front jet stream coincides with the polar front, some surface regions are more favorable for the development of wave cyclones than others.

9. How are longwaves in the atmosphere different from shortwaves?

10. Using a diagram, explain why a surface high-pressure area over North Dakota will typically move southeastward while, at the same time, a deep storm system over the Great Lakes will generally move northeastward.

11. How would you be able to tell where cold advection and warm advection are occurring by examining a 500-mb chart?

12. How does the conservation of absolute vorticity help to explain the formation of Rossby waves?

13. Why does cyclogenesis often occur in the region just east of the Rockies?

Questions for Thought

1. An English friend of yours says that last night's rain over northern Great Britain was caused by a storm that originally formed east of the Colorado Rockies. Explain how this could happen.

2. Would a wave cyclone intensify or dissipate if the upper trough were located to the *east* of the surface disturbance? Explain your answer with the aid of a diagram.

3. Explain why, at 500 mb when cold advection is occurring, the air temperature does not drop as fast as it should. (Hint: What type of vertical air motions are also occurring?)

4. Baroclinic waves seldom form in the tropics. Why?

5. Suppose that the earth stops rotating. How would this affect the earth's vorticity? What would

happen to the absolute vorticity of a moving air parcel? If the parcel were initially moving southwestward, how would its direction change, if at all?

6. Why do Pacific storms often redevelop on the eastern side of the Sierra Nevada Mountains?

7. If you only had isotherms on an upper-level chart, how would a cut-off low appear?

8. How could you tell a barotropic atmosphere from a baroclinic atmosphere by examining an upper-level chart?

9. Why does baroclinic instability tend to occur near the polar front?

Problems and Exercises

1. With the aid of several diagrams, describe the different stages of a developing wave cyclone using the polar front theory.

2. On the 300-mb chart below (Fig. 17.18), suppose the winds are blowing at a constant speed parallel to the contour lines.

(a) On the chart, mark where regions of convergence and divergence are occurring.

(b) Put an "L" on the chart below where you might expect to observe a developing midlatitude storm.

(c) Put an "H" on the chart below where you might expect to observe an anticyclone.

(d) In which directions would the midlatitude cyclone and anticyclone most likely move?

(e) In terms of convergence and divergence, what are the necessary conditions for the intensification of the surface midlatitude storm? For the building of the anticyclone?

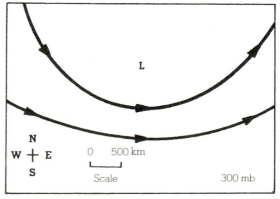

Fig. 17.18 Chart for exercise 2.

A watchful eye and a little weather information are important when making a local short-range weather forecast. These altocumulus clouds, southwesterly surface winds, a falling barometer with a sea level pressure of 1016 mb (30.00 in.), and the instant weather forecast chart (Appendix E) all suggest that air temperatures will rise and that precipitation is possible within 12 hours. (Photo: Ross DePaola)

CHAPTER 18

Weather Forecasting

Knowing what the weather will be like in the future is vital to many human activities. For example, a summer forecast of extended heavy rain and cool weather would have construction supervisors planning work under protective cover, department stores advertising umbrellas instead of bathing suits, and ice cream vendors vacationing as their business dropped off. The forecast would alert farmers to harvest their crops before their fields become too soggy to support the heavy machinery needed for the job. And the commuter? Well, the commuter knows that prolonged rain could mean clogged gutters, flooded highways, stalled traffic, and late dinners.

On the other side of the coin, a forecast calling for extended high temperatures with low humidity has an entirely different effect. As ice cream vendors prepare for record sales, the dairy farmer anticipates a decrease in milk and egg production. The forest ranger prepares warnings of fire danger in parched timber and grasslands. The construction worker is on the job outside once again, but the workday begins in the early morning and ends by early afternoon to avoid the oppressive heat. And the commuter prepares for increased traffic stalls due to overheated car engines.

Put yourself in the shoes of a weather forecaster: It is your responsibility to predict the weather accurately so that thousands (possibly millions) of people will know whether to carry an umbrella, wear an overcoat, or prepare for a winter storm. Since weather forecasting is not an exact science, your predictions will occasionally be incorrect. If your erroneous forecast misleads many people, you may become the target of jokes, insults, and even anger. There are even people who expect you to be able to predict the unpredictable. For example, on Monday you may be asked whether

Contents

next Saturday will be a nice day for a picnic. And, of course, what about next winter? Will it be bitterly cold?

Unfortunately, accurate answers to such questions are beyond meteorology's present technical capabilities. Will forecasters ever be able to answer such questions confidently? If so, what steps are being taken to improve the forecasting art? How are forecasts made, and why do they go wrong? These are just a few of the questions we will address in this chapter.

Assembling the Data

Weather forecasting basically entails predicting how the present state of the atmosphere will change. Consequently, if we wish to make a weather forecast, present weather conditions over a large area must be known. To obtain this information, a network of observing stations are located throughout the world. Over 10,000 land-based stations and hundreds of ships provide surface weather information four times a day.* Most airports observe conditions hourly. Additional information, especially upper-air data, is supplied by radiosondes, aircraft, and satellites. Radiosonde data are usually available only at 0000 and 1200 GMT, while aircraft and satellite observations may occur throughout the day.

A United Nations agency—the World Meteorological Organization (WMO)—consists of over 130 nations. WMO is responsible for the international exchange of weather data and certifies that the observation procedures do not vary among nations, an extremely important task, since the observations must be comparable.

After an observation is taken, it is immediately sent to a communication substation by electronic means, usually teletype, telephone, or satellite relay. From there, the data collected at many observation stations are sent to World Meteorological Centers (located in Melbourne, Australia; Moscow, U.S.S.R., and Washington, D.C.). Worldwide weather information is then sent to the

*Observations are usually taken at 0000, 0600, 1200, and 1800 Greenwich Mean Time (GMT)—local time at the Greenwich Observatory in England. To convert from GMT to your local time, see Appendix C.

National Meteorological Center (NMC) located in Camp Springs, Maryland, just outside Washington, D.C. Here, the massive job of analyzing the data, preparing weather maps and charts, and predicting the weather on a global and national basis begins. From NMC, this information is transmitted to public and private agencies worldwide.

The compiled charts, maps, and forecasts are sent electronically to Weather Service Forecast Offices (WSFO). These stations use the data for preparing regional weather forecasts, as well as for advisories and warnings of impending severe weather. The region serviced by one of these offices is a state or a large portion of a state. To supply the forecast needs of a smaller region, such as a metropolitan area, over 200 Weather Service Offices (WSO) issue local forecasts generally adapted from the original area forecasts of the WSFOs.

The public hears weather forecasts over radio or television. Many stations hire professional meteorologists to make their own forecasts aided by NMC material or to modify a WSO forecast. Other stations hire meteorologically untrained announcers who usually read the forecasts of the National Weather Service word for word, sometimes without properly accrediting the originating agency.

Since we now know how weather information is obtained, let's examine some of the methods and procedures used in making a forecast.

Methods of Weather Forecasting

Probably the easiest weather forecast to make is a **persistence forecast**, which is simply a prediction that future weather will be the same as present weather. If it is snowing today, a persistence forecast would call for snow through tomorrow. Such forecasts are most accurate for time periods of several hours and become less and less accurate after that.

Another method of forecasting is the **steady-state**, or **trend method**. The principle involved here is that surface weather systems tend to move in the same direction and at approximately the same speed as they have been moving, providing no evidence exists to indicate otherwise. Suppose, for example, that a cold front is moving eastward at an average speed of 30 km (19 mi) per hour and it is 90 km (56 mi) west of your home. Using the steady-state method, we might predict that the front should pass through your area in three hours.

The **analogue method** is yet another form of weather forecasting. Basically, this method relies on the fact that existing features on a weather chart (or a series of charts) may strongly resemble features that produced certain weather conditions sometime in the past. Prior weather events can then be utilized as a guide to the future. The problem here is that, even though weather situations may appear similar, they are never *exactly the same*. There are always sufficient differences in the variables to make applying this method a challenge.

Predicting the weather by **weather types** employs the analogue method. In general, weather patterns are categorized into similar groups or "types," using such criteria as the position of the subtropical highs, the upper-level flow, and the prevailing storm track. As an example, when the Pacific high is weak or depressed southward and the flow aloft is zonal (west-to-east), surface storms tend to travel rapidly eastward across the Pacific Ocean and into the United States without developing into deep systems. But when the Pacific high is to the north of its normal position and the upper-air flow is meridional (north-south), looping waves form in the flow with surface lows usually developing into huge storms. As we saw in Chapter 17, these upper-level longwaves move slowly, usually remaining almost stationary for perhaps a few days to a week or more. Consequently, the particular surface weather at different positions around the wave is likely to persist for some time. Figure 18.1 presents an example of weather conditions most likely to prevail with a meridional weather type.

Weather types can be used as an approach to *long-range* (a month or more in advance) *weather forecasting*. Typically, the upper-air circulation changes gradually from zonal to meridional over four to six weeks. As this slow change occurs in the upper air, the surface weather may repeat itself at specific intervals. For instance, winter cold fronts may sweep into New England every four days or so, bringing showers and below-normal temperatures. By projecting trends such as

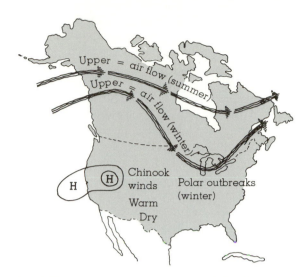

Fig. 18.1 Weather type showing upper-air flow (heavy arrows), surface position of Pacific high, and general weather conditions that should prevail.

these, and assuming that the atmosphere's behavior will not change radically (an assumption not always valid), *extended weather forecasts* can be made. At best, these forecasts only show the broad-scale weather features. They do not adequately predict specific weather elements.

Despite the apparently unsuccessful nature of long-range forecasts, the National Weather Service currently issues an extended forecast of 6 to 10 days, as well as a 30-day outlook for the coming month, and a 90-day seasonal outlook. These are not forecasts in the strict sense, but rather an overview of how average precipitation and temperature patterns may compare with normal conditions. The 30-day outlook is based on the relationship between the projected average flow at the 700-mb level and the surface weather conditions that this type of flow will create. Figure 18.2 gives a typical 30-day outlook. (The 90-day outlooks are based more on persistence statistics that carry over the general weather patterns from immediately preceding months, seasons, and years.)

A forecast based on the climatology (average weather) of a particular region is known as a **climatological forecast**. Anyone who has lived in Los Angeles for a while knows that July and August are practically rain-free. In fact, rainfall data for the summer months taken over many years reveal that rainfall amounts of more than a trace occur in Los Angeles about 1 day in every 90, or only about 1 percent of the time. Therefore, if we predict that it will not rain on some day next year during July or August in Los Angeles, our chances are nearly 99 percent that the forecast will be correct based on past records. Since it is unlikely that this pattern will significantly change in the near future, we can confidently make the same forecast for the year 2000.

Climatological forecasts can also be used to predict other weather elements, such as the maximum temperature. Suppose that in New York City the average maximum temperature on a particular date for the past 30 years is 10°C (50°F). By statistically relating the maximum temperatures on this date to other weather elements—such as the wind, cloud cover, and humidity—a relationship between these variables and maximum temperature can be drawn. By comparing these relationships with current weather information, the forecaster can predict the maximum temperature for the day.

When the Weather Service issues a forecast calling for rain, it is usually followed by a probability. For example: "The chance of rain is 60 percent." Does this mean (a) that it will rain on 60 percent of the forecast area or (b) that there is a 60 percent chance that it will rain within the forecast area? Neither one! The expression means that there is a 60 percent chance that any random place in the forecast area, such as your home, will receive measurable rainfall. Looking at the forecast in another way, if the forecast on 10 days calls for a 60 percent chance of rain, it should rain where you live on 6 of those days.

An example of a **probability forecast** using climatological data is given in Fig. 18.3. The map shows the probability of a "White Christmas"— 1 inch or more of snow on the ground—across the United States. The map is based on the average of 30 years of data, and gives the likelihood of snow in terms of a probability. For instance, the chances are 90 percent (9 Christmases out of 10) that portions of northern Minnesota, Michigan, and Maine will experience a White Christmas. In Chicago, it is 50 percent; and in Washington, D.C., about 20 percent. Many places in the far west and south have probabilities less than 5 percent, but nowhere is the probability exactly 0, for there is always some chance (no matter how small) that a mantle of white will cover the ground on Christmas day.

The Computer and Weather Forecasts: Numerical Weather Prediction

As recently as the mid-1950s, all weather maps and charts were plotted by hand and analyzed by individuals. Meteorologists predicted the weather

Fig. 18.2 The 30-day outlook is an estimate of the average temperature and precipitation for the next 30 days.

Temperature
A = Above normal
B = Below normal
N = Near normal

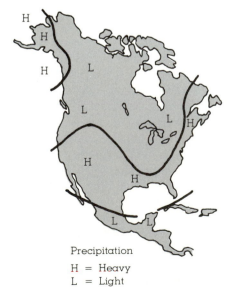

Precipitation
H = Heavy
L = Light

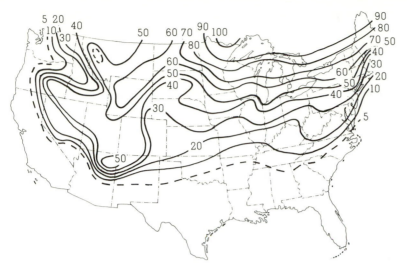

Fig. 18.3 Probability of a "White Christmas"—1 inch or more of snow on the ground—based on a 30-year average from 1940 to 1970.

using certain rules that related to the particular weather system in question. For short-range forecasts of six hours or less, the steady-state technique provided (and still provides) a good method for moving surface fronts and pressure systems on a surface weather map. Upper-air charts, especially the 500-mb analysis, were used to predict where surface storms would develop and where pressure systems aloft would intensify or weaken. The predicted positions of these systems were extrapolated into the future using current maps. In many cases, these forecasts turned out to be amazingly accurate. They were good but, with the advent of the modern computers, today's forecasts are even better.

Modern electronic computers can analyze large quantities of data extremely fast. Each day the many thousands of observations transmitted to NMC are fed into a high-speed computer, which plots these data on a chart. An instrument called a *curve plotter* then draws lines that interpret the weather patterns. The final chart is referred to as an **analysis**.

The computer not only plots and analyzes data, it also predicts the weather. The routine daily forecasting of weather by the computer has come to be known as **numerical weather prediction**.

A computer cannot make a forecast unless it is correctly programmed. Because the many weather variables are constantly changing, meteorologists have devised *atmospheric models* that describe the present state of the atmosphere. These are not physical models that paint a picture of a de-veloping storm; they are, rather, mathematical models consisting of between six and eight mathematical equations that describe how atmospheric temperature, pressure, and moisture will change with time. Actually, the models do not represent the real atmosphere but are approximations formulated to retain the most important aspects of the atmosphere's behavior.

The models are programmed into the computer, and surface and upper-air observations of temperature, pressure, moisture, winds, and air density are fed into the equations. To determine how each of these variables will change, each equation is solved for a small increment of future time—say 5 minutes—for a large number of locations called *grid points*, each situated about 200 km apart.* In addition each equation is solved for as many as 10 levels in the atmosphere. The results of these computations are then fed back into the original equations. The computer again solves the equations with the new "data," thus predicting weather over the following 5 minutes. This procedure is done repeatedly until it reaches some desired time in the future, usually 12, 24, 36, or 48 hours. The computer then prints this informa-

*Some models have a grid spacing as small as 50 km, while the spacing in others exceeds 350 km. A model in operation at NMC since 1980, called the *spectral model*, actually describes the atmosphere, using a set of mathematical equations with wavelike characteristics rather than a set of discrete numbers associated with grid points. In the spectral model, the resolution is defined by the number of waves used rather than the size of the grid mesh.

tion, analyzes it, and draws the projected positions of pressure systems with their contour lines. The final forecast chart representing the atmosphere at a specified future time is called a **prognostic chart**, or, simply, a **prog**. Computer-drawn progs have come to be known as "machine-made" forecasts.

The computer solves the equations more quickly and efficiently than could be done by hand. For example, just to produce a 24-hour forecast chart for the Northern Hemisphere requires many hundreds of millions of mathematical calculations. It would, therefore, take a group of meteorologists working full time with hand calculators years to produce a single chart; by the time the forecast was available the weather for that day would already be ancient history.

The forecaster uses the progs as a guide to predicting the weather. At present, there are a variety of models (and, hence, progs) from which to choose, each producing a slightly different interpretation of the weather for the same projected time and atmospheric level. (See Fig. 18.11, p. 381.) The differences between progs result from the way the models use the equations, the distance between grid points, or—in the case of the spectral model—the number of waves used. Some models predict some features better than others: One model works best in predicting the position of troughs at the 500-mb level, while another forecasts the position of surface lows quite well. A good forecaster knows the idiosyncracies of each model and carefully scrutinizes all the progs. The forecaster then makes a prediction based on the *guidance* of the computer, a personalized practical interpretation of the weather situation and any local geographic features that influence the weather within the specific forecast area.

Since the forecaster has access to over 200 maps produced at NMC each day, why is it that forecasts are sometimes wrong?

Problems in Weather Prediction

Computer-forecast models idealize the real atmosphere, meaning that each model makes certain assumptions about the atmosphere. These assumptions may be on target for some weather

situations and be way off for others. Consequently, the computer may produce a prog that on one day comes quite close to describing the actual state of the atmosphere and not so close on another. A forecaster who bases a prediction on an "off day" computer prog may find a forecast of "rain and windy" turning out to be a day of "clear and colder."

Even though many thousands of weather observations are taken worldwide each day, there are still regions where observations are sparse, particularly in the upper air over the oceans. Temperature data recorded by satellites have helped to alleviate this problem, although not completely because correlating temperatures measured by satellites with those measured by radiosondes has not been an easy task. Since the computer's forecast is only as good as the data fed into it, and since observing stations may be so far apart that they miss certain weather features, a denser network of observations is needed, especially in remote areas, to ensure better forecasts in the future.

Earlier in this chapter, we saw that the computer solves the equations that represent the atmosphere at many locations called grid points, each about 200 km apart. As a consequence, weather systems larger than this distance, such as extensive midlatitude cyclones and anticyclones, show up on computer progs, while systems much smaller than this distance, such as thunderstorms, do not. The computer models are, therefore, better at predicting the widespread precipitation associated with a large cyclonic storm than local showers and thunderstorms. In summer, when much of the precipitation falls as local showers, a computer prog may have indicated good weather, while outside it is pouring rain. To capture the smaller-scale weather features, the distance between grid points on some models is being reduced. However, as the horizontal spacing between grid points decreases, the number of computations increases. When the distance is halved, there are eight times as many computations to perform, and the time required to run the model goes up by a factor of 16.

Another forecasting problem is that computer models cannot adequately interpret many of the factors that influence surface weather, such as the interactions of water, ice, and local terrain on

A Word about Accuracy and Skill

In spite of the complexity and everchanging nature of the atmosphere, forecasts made for between 12 and 24 hours are usually quite accurate. Those made for between 1 and 2 days are fairly good. Beyond about 3 days, however, forecast accuracy falls off rapidly. Although weather predictions made for up to 2 days are by no means perfect, they are far better than simply flipping a coin. But how accurate are they?

One problem with determining forecast accuracy is deciding what constitutes a right or wrong forecast. Suppose tomorrow's forecast calls for a minimum temperature of 5°C. If the official minimum turns out to be 6°C, is the forecast incorrect? Is it as incorrect as one 10 degrees off? By the same token, what about a forecast for snow over a large city, and the snow line cuts the city in half with the southern portion receiving heavy amounts and the northern portion none? Is the forecast right or wrong? At present, there is no clear-cut answer to the question of determining forecast accuracy.

How does forecast accuracy compare with forecast skill? Suppose you are forecasting the daily summertime weather in Los Angeles. It is not raining today and your forecast for tomorrow calls for "no rain." Suppose that tomorrow it doesn't rain. You made an accurate forecast, but did you show any skill in so doing? In a previous section, we saw that the chance of measurable rain in Los Angeles on any summer day is very small indeed; chances are good that day after day it will not rain. For a forecast to show skill, it should be more than one based solely on the current weather (persistence) or on the "normal" weather (climatology) for a given region. Therefore, during the summer in Los Angeles, a forecaster will have many accurate forecasts calling for "no measurable rain," but will need skill to predict correctly on which summer days it will rain.

Meteorological forecasts, then, show skill when they are more accurate than a forecast utilizing only persistence or climatology. Persistence forecasts are usually difficult to improve upon for a period of time of several hours or less. Weather forecasts ranging from 12 hours to a few days generally show much more skill than those of persistence. However, as the range of the forecast period increases, the skill drops quickly. For periods up to 5 days, the only elements showing moderate skill are daily temperature forecasts. The 6 to 10 day outlooks are somewhat better than climatology. Beyond 10 days, specific forecasts are generally about as good as those based on climatology The greatest improvement in forecasting skill during the past 25 years has been made in the area of severe storm warnings for hurricanes and tornadoes. Despite large population increases in areas generally threatened by these storms, there has been a decrease over the years in the number of lives lost because of them.

---

weather systems. Some models do take large geographic features (such as mountain chains and oceans) into account, while ignoring smaller features (such as hills and lakes). These smaller features can have a marked influence on the local weather. Given the effect of local terrain, as well as the impact of some of the other problems previously mentioned, computer forecasts presently do an inadequate job of predicting local weather conditions, such as surface temperatures, winds, and precipitation.

In summary, imperfect numerical weather predictions may result from a sparcity of data or from inadequate representation of the many pertinent processes and interactions that take place within the atmosphere.

Improving Weather Forecasts

Is there any hope that forecasts, and especially extended forecasts, will be better in the future? If so, what efforts are presently being made to improve them?

In search of better and more accurate forecasts, meteorologists have launched extensive research programs requiring international scientific cooperation. The *Global Atmospheric Research Program* (GARP) is a large-scale research effort designed to develop a better understanding of worldwide weather systems, so that large-scale forecasts using numerical models can be pushed to their theoretical limit of a week or more.

To provide the meteorologist with knowledge of global weather system interaction, a field research program known as GATE (GARP Atlantic Tropical Experiment) was conducted during the summer of 1974 over a comparatively small area in the tropical Atlantic just west of Africa. The main purpose of GATE was to determine the role that tropical cumulus cloud clusters play in the atmosphere's general circulation. Thousands of people from 70 nations with dozens of ships and aircraft, satellites, radar, and ocean buoys amassed literally tons of data. While these data were being analyzed, another research effort began.

Designed to observe the entire atmosphere for a whole year, the *First GARP Global Experiment* (FGGE) got under way in the late 1970s. Using weather buoys, satellites, and other observing tools, the project set out to gain a better understanding of atmospheric motions on a global scale, so that improved numerical models could be developed for extended forecasts. How effective these efforts are in improving forecasts will be seen as new models are fully developed and used to predict the weather a week or more into the future.

Although scientists may never be able to skillfully predict the weather beyond about 10 days using available observations, the prediction of climatic trends appears to be more promising. Whereas individual weather systems vary greatly and are difficult to forecast very far in advance, global-scale patterns of winds and pressure frequently show a high degree of persistence and predictable change over periods of a few weeks to a month or more. With the latest generation of high-speed computers, such as the CRAY-1 at the National Center for Atmospheric Research, general circulation models (GCMs)—numerical computer models that simulate global patterns of wind, pressure, and temperature and how they change with time—are doing a far better job at predicting large-scale atmospheric behavior than did the earlier models. In fact, the new GCMs are able to simulate a number of global patterns quite

well, such as *blocking highs* that can cause precipitation and temperature patterns to deviate considerably from average conditions. As new knowledge and methods of modeling are fed into the GCMs, it is hoped that they become a reliable tool in the forecasting of weather and climate. (In Chapter 22, we will examine in more detail the climatic predictions based on numerical models.)

To help the forecaster at Weather Service Forecast Offices, new high-speed data modeling systems using minicomputers have been installed at selected stations. This new and improved communication of weather information is known as AFOS (Automation of Field Operations and Services). A push of a button displays weather information from any desired region on a television screen. The AFOS system is shown in Fig. 18.4.

Predicting the Weather from Local Signs

Because the weather affects every aspect of our daily lives, attempts to predict it accurately have been made for centuries. One of the earliest attempts was undertaken by Theophrastus, a pupil of Aristotle, who in 300 B.C. compiled all sorts of weather indicators in his *Book of Signs*. A dominant influence in the field of weather forecasting for 2000 years, this work consists of ways to foretell the weather by examining natural signs, such as the color and shape of clouds, and the intensity at which a fly bites. Some of these signs have validity and are a part of our own weather folklore—''a halo around the moon portends rain'' is one of these. Today, we realize that the halo is caused by the bending of light as it passes through ice crystals and that ice crystal-type clouds (cirrostratus) are often the forerunners of an approaching storm.

Weather predictions can be made by observing the sky and using a little weather wisdom. If you keep your eyes open and your senses keenly tuned to your environment, you should, with a little practice, be able to make fairly good short-range local weather forecasts by interpreting the messages written in the weather elements.

To help you forecast the weather, the instant weather forecast chart (Appendix E) has been prepared by considering the relationship that the pressure and wind have to various weather sys-

Fig. 18.4 New AFOS all-electronic computerized system displays weather information on TV-type consoles at the push of a button.

tems. While the chart is applicable to much of the United States, local influences, such as mountains and large bodies of water, can affect the local weather to such an extent that the large-scale weather patterns on which the chart is based do not always show up clearly. (The chart works best during the fall, winter, and spring when the weather systems are active.)

The chart is simply a guide; in no way is it infallible. For example, rising pressures in most cases indicate improving weather conditions. On the Great Plains, however, a snowstorm may begin with a rising barometer as the storm system passes south or east of an area and upslope conditions prevail. Along the eastern seaboard, a slowly rising barometer combined with northwesterly winds could spell heavy precipitation as a storm drifts northeastward parallel to the coast.

By adding such information as cloud type, temperature, and the wind aloft (as indicated by cloud movement), each forecast on the chart can be improved upon. For example, suppose that the barometer reading is 1016 mb, the pressure is falling, and the surface wind is blowing from the southeast. The instant weather forecast chart predicts that precipitation is on the way. This is a reasonable forecast, for Buys-Ballot's law (Chapter 12) suggests that the location of the surface storm is to the west. But how far west?

Suppose that the air temperature is near 10°C (50°F), and that overhead a layer of altostratus (at about 4 km, or 13,000 ft) is moving from southwest to northeast. Again, using Buys-Ballot's law (but this time for the winds aloft), we determine that there is an upper trough of low pressure to the west. As a general rule of thumb, the lower the stratiform cloud layer, the closer you are to the storm or frontal system. High cirrostratus clouds usually indicate that an approaching frontal system is 1 or 2 days away, while a layer of low stratus (especially with fog and drizzle) usually means that the front is close indeed. Altostratus clouds suggest that the approaching system is probably between 500 and 700 km to the west. The falling barometer, southeasterly surface winds, and southwesterly winds aloft suggest that the frontal system is moving toward us and that precipitation (probably rain) is likely within the next 6 to 8 hours. (If you have forgotten the typical weather associated with fronts, review Tables 16.2, 16.3, and 16.4, as well as the cyclone model, Fig. 17.11.)

Weather Forecasting Using Surface Charts

We are now in a position to forecast the weather, utilizing more sophisticated techniques. Suppose, for example, that we wish to make a short-range weather prediction and the only information available is a surface weather map. Can we make a

FOCUS ON OBSERVATION

Forecasting Temperature Advection by Watching the Clouds

A knowledge of temperature advection aloft is a valuable tool in forecasting the weather. In summer, when the surface is warm, cold advection aloft sets up instability and increases the likelihood of towering cumulus clouds and showers. On the other hand, warm advection aloft usually increases the temperature of the air, thus making it more stable. During the winter, this often leads to smoke and haze accumulating in the colder air near the surface.

By watching the movement of clouds, we get a good indication as to the wind direction at cloud level and also the type of advection. For instance, a cloud moving from the west indicates a west wind, a cloud from the south a south wind, and so on. Clouds at different levels frequently move in different directions, meaning that the wind direction is changing with height. Wind that changes direction in a clockwise sense (north to northeast to east, etc.) is a **veering wind**. Wind that changes direction in a counterclockwise sense (north to northwest to west, etc.) is a **backing wind**. There are two general rules that will help us determine

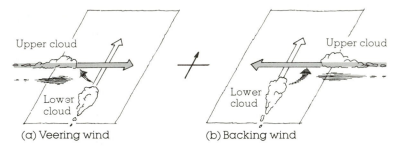

Fig. 1   (a) The wind veers with height, suggesting warm advection is occurring between the cloud layers. (b) The wind is backing with height, and cold advection is occurring between the cloud layers.

where cold or warm advection is occurring in a layer of air above us:

1. Winds that *back* with height (change counterclockwise) indicate *cold* advection.
2. Winds that *veer* with height (change clockwise) indicate *warm* advection.

As an example, suppose we observe lower clouds moving from a southerly direction (south wind) and higher clouds moving from a westerly direction (west wind). (See Fig. 1a.) The wind direction is veering with height; warm advection is occurring between the cloud layers, and the air should be getting warmer. If, on some other day, we see lower clouds moving

from a southerly direction and higher clouds moving from an easterly direction (Fig. 1b), the wind is backing with height and the atmosphere between the cloud layers is probably becoming colder.*

An example of the relationship between winds and advection is seen in the vertically shifting winds that accompany weather fronts. Figure 2b is a surface map showing a typical open wave cyclone with its accompanying warm and cold fronts. Behind the cold front, swiftly moving cumulus clouds indicate a northwesterly wind exists about a kilometer or so

*In both of these examples, we are assuming horizontal air motion only.

---

forecast from such a chart? Most definitely. And our chances of that forecast being correct improve markedly, if we have maps available from several days back. We can use these past maps to locate the previous position of surface features and predict their movement.

A simplified surface weather map is shown in Fig. 18.5. The map portrays early winter weather

conditions on Tuesday morning at 6:00 A.M. A single isobar is drawn around the pressure centers to show their positions without cluttering the map. Note that an open wave cyclone is developing over the Central Plains. The weather conforms to the cyclone model (see Fig. 17.11, p. 354), with showers forming along the cold front and light rain and snow ahead of the warm front. The dashed

FOCUS ON OBSERVATION, continued

above the surface. Ahead of the advancing warm front, stratiform clouds indicate that here southeasterly winds prevail about a kilometer above the ground. We know from Chapter 17 that warm advection takes place *ahead* of the warm front and cold advection *behind* the cold front. In both cases, the advection usually occurs in a layer from the surface up to at least the 500-mb level.

At the 500-mb level (Fig. 2a), the position of the upper trough and the region of coldest air is to the west of the surface low. The direction of wind and also cloud movement is shown by arrows. Because of the upper trough's position, the winds aloft are westerly behind the cold front and southwesterly ahead of the warm front. Figure 2c shows how the wind direction changes from the surface to the 500-mb level. Behind the cold front, the winds back from northwesterly to westerly as we move upward. Cold advection is taking place as chilling air moves in from the west. Just ahead of the approaching warm front the wind veers with height from southeasterly to southwesterly as warm air glides up and over the cool surface air.

We can use this information to

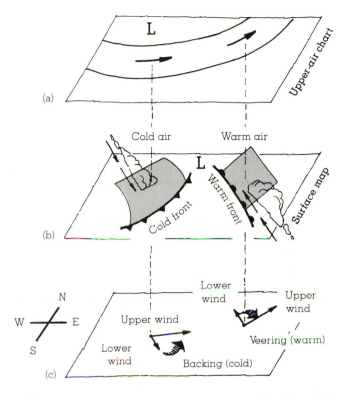

Fig. 2 Clouds, winds, and advection associated with a cold and a warm front.

improve upon a weather forecast. For instance, if you happen to be located ahead of an advancing warm front and the winds above you are veering with height, the chances are that even if precipitation begins as snow it may change to rain as warm air moves in overhead.

Behind a cold front where winds are backing with height, the influx of cold air may lower the temperature sufficiently so that rain first becomes mixed with snow, and then changes to snow before the storm moves eastward.

lines on the map represent the position of the weather systems 6 hours ago. Our first question is: How will these systems move?

Determining the Movement of Weather Systems There are several methods we can use in forecasting the movement of surface pressure systems and fronts. For short-time intervals, storms

and fronts tend to move at a steady rate; that is, they move in the same direction and at approximately the same speed as they did during the previous 6 hours (providing, of course, there is no evidence to indicate otherwise). Based on present trends, the storm center in Fig. 18.5 should move northeast. Another way to predict this movement is based on the fact that lows tend to move in a direction that parallels the isobars in the warm sector. Still another fact to consider when dealing with moving pressure systems is the pressure tendency. In general, lows tend to move toward the region of greatest pressure drop, while highs tend to move toward the region of greatest rise.

If *pressure tendencies* are plotted on our map, we could draw lines connecting points of equal pressure change. These lines, called **isallobars**, help us to visualize the regions of falling and rising pressure. The distribution of pressure change for our map might look like the one in Fig. 18.6. Drawn at 2-mb intervals, the isallobars show a broad region of falling pressure ahead of the warm front, with the largest drop occurring to the northeast of the storm. This fits with the previous observations and strengthens the prediction that the low center will move toward the northeast. The area of rising pressure immediately behind the cold front suggests that the anticyclone near the United States-Canadian border will continue to move southeastward.

Pressure tendencies not only help predict the

movement of highs and lows, they also indicate how the pressure systems are changing with time. The rapid fall in pressure in advance of the low indicates that the storm center is deepening as it moves. A deepening low means more closely spaced isobars, a greater pressure gradient, and stronger winds—something to take into account when we make our weather forecast. A drop in pressure, on the other hand, in the vicinity of an anticyclone suggests that it is weakening, while a rise in pressure means that its central pressure is increasing. Hence, the anticyclone moving out of Canada is strong (1034 mb) and will remain so, whereas the anticyclone centered off the Georgia coast is either moving eastward or weakening rapidly as indicated by the falling pressure in that area.

Before we complete our prediction about the movement of the pressure centers in Fig. 18.5, we need to look closely at the anticyclone off the Georgia coast. Strong highs, especially slow-moving ones, often retard the eastward progress of lows, deflecting them either north or south. From all indications—falling pressures and past

Fig. 18.5 Surface weather map for 6:00 A.M. Tuesday. Dashed lines indicate positions of weather features 6 hours ago. Shaded areas are receiving precipitation.

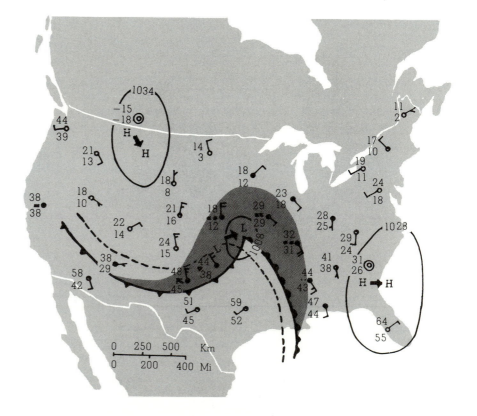

movement—this anticyclone is weakening and drifting slowly eastward. It should, therefore, pose no immediate problem to the northeastward movement of the storm center.

Even if we do not have access to pressure tendencies or previous weather maps, we can make an initial approximation of how pressure systems will move by using Fig. 17.5 (see p. 348), which shows the average tracks of lows and highs during the winter months. From this diagram, it appears that the cyclones and anticyclones in Fig. 18.5 are following rather typical trajectories.

A Forecast for Six Cities   Our objective now is to make a weather forecast for six cities. To do this, we will project the pressure systems, fronts, and current weather into the future by assuming steady-state conditions. Figure 18.7 gives the 12- and 24-hour projected positions of these features.

A word of caution before we make our forecasts. We are assuming that the pressure systems and fronts are moving at a constant rate. This may or may not occur. Storm systems, for example, tend to accelerate until they occlude,

after which their rate of movement slows. Furthermore, the direction of moving systems may change due to "blocking" highs and lows that exist in their path or because of shifting upper-level wind patterns. We will assume a constant rate of movement and forecast accordingly, always keeping in mind that the longer our forecasts extend into the future, the more susceptible they are to error.

Using Fig. 18.7 to follow the storm center eastward, we can make a basic forecast. The cold front moving into north Texas on Tuesday morning is projected to pass Dallas by that evening, so a forecast for the Dallas area would be "warm with showers, then turning colder." But we can do much better than this. Knowing the weather conditions that accompany advancing pressure areas and fronts, we can make more detailed weather forecasts that will take into account changes in temperature, pressure, humidity, cloud cover, precipitation, and winds. Our forecast will include the 24-hour period from Tuesday morning to Wednesday morning for the cities of Augusta, Georgia; Washington, D.C.; Chicago, Illinois; Memphis,

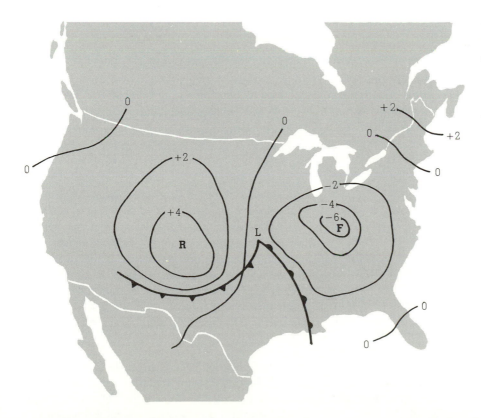

Fig. 18.6 Isallobars—lines of equal 3-hour pressure change—for 6:00 A.M. Tuesday. The "F" represents the region of greatest pressure fall, while the "R" shows the region of greatest pressure rise.

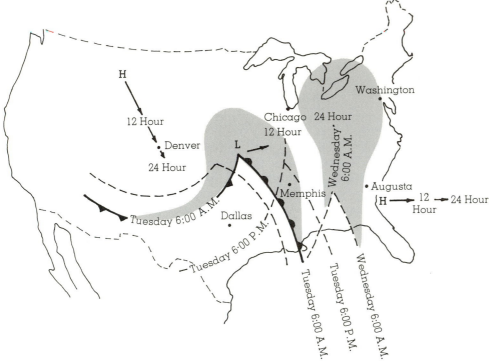

Fig. 18.7 Projected 12- and 24-hour movement of fronts, pressure systems, and precipitation (shaded area) from 6:00 A.M. Tuesday until 6:00 A.M. Wednesday.

Tennessee; Dallas, Texas; and Denver, Colorado. We will begin with Augusta.

Weather Forecast for Augusta, Georgia On Tuesday morning, cP air associated with a high-pressure center brought freezing temperatures and fair weather to the Augusta area (Fig. 18.5). Clear skies, light winds, and low humidities allowed rapid nighttime cooling so that, by morning, temperatures were in the low thirties. Now look closely at Fig. 18.7 and observe that the anticyclone is moving slowly eastward. Southerly winds on the western side of this system will bring warmer and more moist air to the region. Therefore, afternoon temperatures will be warmer than those of the day before. As the warm front approaches from the west, clouds will increase, appearing first as cirrus, then thickening and lowering into the normal sequence of warm-front clouds. Barometric pressure should fall. Clouds and high humidity should keep minimum temperatures well above freezing on Tuesday night. Note that the projected area of precipitation (shaded region) does not quite reach Augusta. With all of this in mind, our forecast might sound something like this:

Clear and cold this morning with moderating temperatures by afternoon. Increasing high clouds with skies becoming overcast by evening. Cloudy and not nearly as cold tonight and tomorrow morning. Winds will be light and out of the south or southeast. Barometric pressure will fall slowly.

Wednesday morning we discover that the weather in Augusta is foggy with temperatures in the upper 40s (°F). But fog was not in the forecast. What went wrong? We forgot to consider that the ground was still cold from the recent cold snap. The warm, moist air moving over the cold surface was chilled to its dew point, resulting in fog. Above the fog were the low clouds we predicted. The minimum temperatures remained higher than anticipated because of the release of latent heat during fog formation and the absorption of infrared energy by the fog droplets. Not bad for a start. Now we will forecast the weather for Washington, D.C.

Rain or Snow for Washington, D.C.? Look at Fig. 18.7 and observe that the storm center is slowly approaching Washington, D.C. from the west. Hence, the clear weather, light southwesterly winds, and low temperatures on Tuesday morning (Fig. 18.5) will gradually give way to increasing cloudiness, winds shifting to the southeast, and slightly higher temperatures. By Wednesday morning, the projected band of precipitation will be over the city. Will it be in the form of rain or snow? Without any data for temperatures aloft, this is difficult to determine. We can see in Fig. 18.5, however, that cities south of Washington, D.C.'s latitude are receiving snow. So a reasonable forecast would call for snow, possibly changing to rain as warm air moves in aloft in advance of the approaching fronts. A 24-hour forecast for Washington, D.C. might sound like this:

> Increasing clouds today and continued cold. Snow beginning by early Wednesday morning, possibly changing to rain. Winds will be out of the southeast. Pressures will fall.

Wednesday morning a friend in Washington, D.C., calls to tell us that the sleet began to fall but has since changed to rain. Sleet? Another fractured forecast! Well, almost. What we forgot to account for this time was the intensification of the storm. As the storm moved eastward, it deepened; central pressure lowered, pressure gradients tightened, and southeasterly winds blew stronger than anticipated. As air moved inland off the warmer Atlantic, it rode up and over the colder surface air. Snow falling into this warm layer at least partially melted; it then re-froze as it entered the colder air near ground level. The advection of warmer air from the ocean slowly raised the surface temperatures, and the sleet soon became rain. Although we did not see this possibility when we made our forecast, a forecaster more familiar with local surroundings would have. Let's move on to Chicago.

Big Snow Storm for Chicago   From Figs. 18.5 and 18.7 it appears that Chicago is in for a major snow storm. Overrunning of warm air has produced a wide area of snow which, from all indications, is heading directly for the Chicago area. Since cold air north of the low center will be over Chicago, precipitation reaching the ground should be frozen. On Tuesday morning the leading edge of precipitation is less than 6 hours away from Chicago. Based on the projected path of the storm, light snow should begin to fall around noon.

By evening, as the storm intensifies, snowfall

should become heavy. It should taper off and finally end around midnight as the storm moves on east. If it snows for a total of 12 hours—6 hours as light snow (around 1 in. every 3 hours) and 6 hours as heavy snow (around 1 in. per hour)—then the total expected accumulation will be between 6 and 10 in. As the depression moves eastward, passing south of Chicago, winds on Tuesday will gradually shift from southeasterly to easterly, then northeasterly by evening. Since the system is intensifying, it should produce strong winds that will swirl the snow into huge drifts, which may bring traffic to a crawl.

The winds will continue to shift to the north and finally become northwesterly by Wednesday morning. By then the storm center will probably be far enough east so that skies should begin to clear. Cold air advected from the northwest behind the storm will cause temperatures to drop further. Barometer readings during the storm will fall as the low center approaches and reach a low value sometime Tuesday night, after which they will begin to rise. A weather forecast for Chicago might be:

> Cloudy and cool with light rain, low clouds, and fog becoming heavy by evening and ending by Wednesday morning. Total accumulations will range between 6 and 10 in. Winds will be strong and gusty out of the east or northeast today becoming northerly tonight and northwesterly by Wednesday morning. Barometric pressure will fall sharply today and rise tomorrow.

A call Wednesday morning to a friend in Chicago reveals that our forecast was correct except that the total snow accumulation so far is 13 in. We were off in our forecast because the storm system slowed as it became occluded. We did not consider this because we moved the system by the steady-state method. At this time of year (early winter), Lake Michigan is not quite frozen over and the added moisture picked up from the lake by the strong easterly winds also helped to produce a heavier-than-predicted snowfall. Again, a knowledge of the local surroundings would have helped make a more accurate forecast. The weather 600 km south of Chicago should be much different from this.

Mixed Bag of Weather for Memphis   Observe in Fig. 18.7 that, within 24 hours, both a warm and a cold front should move past Memphis. The light rain that began Tuesday morning should saturate the cool air, creating a blanket of low clouds and fog by midday. The warm front, as it moves through sometime Tuesday afternoon, should cause temperatures to rise slightly as winds shift to the south or southwest. At night, clear to partly cloudy skies should allow the ground and air above to cool, offsetting any tendency for a rapid rise in temperature. Falling pressures should level off in the warm sector, then fall once again as the cold front approaches. According to the projection in Fig. 18.7, the cold front should arrive sometime before midnight on Tuesday, bringing with it gusty northwesterly winds, showers, the possibility of thunderstorms, rising pressures, and colder air. Taking all of this into account, our weather forecast for Memphis will be:

> Cloudy and cool with light rain, low clouds and fog early today, becoming partly cloudy and warmer by tonight. Clouds increasing with possible showers and thunderstorms later tonight or early Wednesday morning and turning colder. Winds southeasterly this morning, becoming southerly or southwesterly this evening and shifting to northwesterly by Wednesday morning. Pressures falling this morning, leveling off this evening then falling again tonight and rising by Wednesday morning.

A friend who lives near Memphis calls Wednesday to inform us that our forecast was correct except that the thunderstorms did not materialize and that Tuesday night dense fog formed in low-lying valleys, but by Wednesday morning it had dissipated. Apparently, in the warm sector, winds were not strong enough to mix the cold, moist air that had settled in the valleys with the warm air above. It's on to Dallas.

Cold Wave for Dallas   From Fig. 18.7, it appears that our weather forecast for Dallas should be straightforward, since a cold front is expected to pass the area around noon. Weather along the front is showery with a few thunderstorms developing; behind the front the air is clear but cold. By Wednesday morning it looks as if the cold front will be far to the east and south of Dallas and an anticyclone will be centered over Colorado. North or northwesterly winds on the east side of the high will bring cold continental polar air into Texas,

dropping temperatures as much as 40°F within a 24-hour period. With minimum temperatures well below freezing, Dallas will be in the grip of a cold wave. Our weather forecast should therefore sound something like this:

> Increasing cloudiness and mild this morning with the possibility of showers and thunderstorms this afternoon. Clearing and turning much colder tonight and tomorrow. Winds will be southwesterly today becoming gusty north or northwesterly this afternoon and tonight. Pressures falling this morning then rising later today.

How did our forecast turn out? A quick call to Dallas on Wednesday morning reveals that the weather there is cold but not as cold as expected, and the sky is overcast. Cloudy weather? How can this be?

The cold front moved through on schedule Tuesday afternoon, bringing showers, gusty winds, and cold weather with it. Moving southward, the front gradually slowed and became stationary along a line stretching from the Gulf of Mexico westward through southern Texas and northern Mexico. (From the surface map alone we had no way of knowing this would happen.) Along the stationary front a wave formed. This caused warm, moist Gulf air to slide northward up and over the cold surface air. Clouds formed, minimum temperatures did not go as low as expected, and we are left with a fractured forecast. Let's give Denver a try.

Clear but Cold for Denver   In Fig. 18.7, we can see that, based on our projections, the cold anticyclone will be almost directly over Denver by Wednesday morning. Sinking air aloft should keep the sky relatively free of clouds. Weak pressure gradients will produce only weak winds and this, coupled with dry air, will allow for intense radiational cooling. Minimum temperatures will probably drop to well below 0°F. Our forecast should therefore read:

> Clear and cold through tomorrow. Northerly winds today becoming light and variable by tonight. Low temperatures tomorrow morning will be below zero. Barometric pressure will continue to rise.

Almost reluctantly Wednesday morning, we inquire about the weather conditions at Denver. "Clear and very cold" is the reply. A successful forecast at last! We are told, however, that the minimum temperature did not go below zero; in fact, 13°F was as cold as it got. A downslope wind coming off the mountains to the west of Denver kept the air mixed and the minimum temperature higher than expected. Again, a forecaster familiar with the local topography of the Denver area would have foreseen the conditions that lead to such downslope winds and would have taken this into account when making the forecast.

A complete picture of the surface weather systems for 6:00 A.M. Wednesday morning is given in Fig. 18.8. By comparing this chart with Fig. 18.7, we can summarize why our forecasts did not turn out exactly as we had predicted. For one thing, the storm center near the Great Lakes moved slower and further to the north than expected. This allowed a southeasterly flow of mild Atlantic air to overrun cooler surface air ahead of the storm while, behind the low, cities remained in the snow area for a longer time. The weak wave that developed along the trailing cold front brought cloudiness and precipitation to Texas and prevented the really cold air from penetrating deep into the south. Further west, the high originally over Montana moved more southerly than south-

easterly, which set up a pressure gradient that brought westerly downslope winds to eastern Colorado.

The forecasting techniques discussed so far are those you can use when making a weather prediction. The following section describes how a meteorologist predicts the weather in a region where, to the west, surface weather features are extensively modified by a vast body of water and only scanty surface and upper-air data are available. Here, the forecaster must rely heavily on experience as well as more sophisticated tools, which include satellite data, upper-air charts, and computer progs.

A Meteorologist Makes a Prediction

It is late afternoon, and outside the weather forecast office near San Francisco the meteorologist mulls over what is going on in the sky. Overhead is a thin covering of cirrostratus; to the west, draped over the foothills, is the ever present stratus and fog. The air is cool and the winds are westerly. It is Sunday, March 25, and the forecaster's task is to predict the weather for the coastal area of central California.

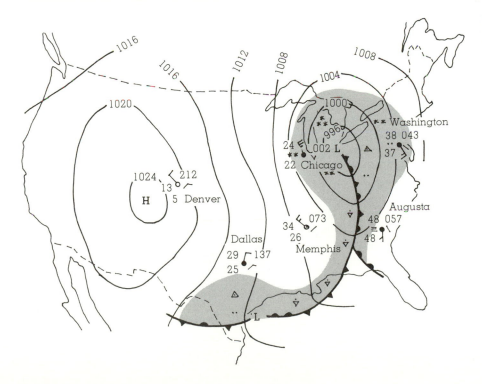

Fig. 18.8 Surface weather map for 6:00 A.M. Wednesday.

What will tomorrow's weather be like? Will it be similar to today's or will it change markedly? A slowly falling barometer of 1016 mb (30 in.) and the high clouds moving in from the west point to an approaching storm system. A forecast of persistence might be good for the next several hours, but what about tomorrow morning or tomorrow afternoon?

The late afternoon surface analysis provides little assistance with these questions. The surface map for 4:00 P.M. (PST) Sunday (Fig. 18.9) shows there are no weather fronts approaching the West Coast. In fact, the nearest front is a stationary one that has stalled over the Rockies. There is, however, a region of low pressure centered about 1100 km (700 mi) west of San Francisco, which (according to previous maps) has been there for several days. With a central pressure of only about 1012 mb (29.88 in.), the system is fairly weak. Could this weak depression be causing the increase in high cloudiness and the falling barometer? And will this pattern lead to rain tomorrow? A look at the 500-mb chart may help with these questions.

Help from the 500-mb Chart   Figure 18.10 shows the 500-mb analysis for 4:00 P.M. Sunday afternoon. While examining the chart, the mete-orologist recognizes certain clues that will aid in making the forecast. For one thing, the 5640-m height line is over northern California. When this contour line is situated here or further south, the statistical probability of receiving measurable rainfall over central California increases greatly. There are, in fact, some forecasters who base their precipitation forecasts solely on the position of that line.

West of San Francisco the flow is meridional with a cut-off warm, upper high situated just south of Alaska. To the south both east and west of the high are troughs. Because the shape of this flow around the high resembles the Greek letter omega (Ω), the high and its accompanying ridge is known as an **omega high**. The forecaster recognizes the omega high as a *blocking high*, one that tends to persist in the same geographic location for many days. This blocking pattern also tends to keep the troughs in their respective positions, which has been the case for several days now. But the chart indicates that the cold upper trough located west of San Francisco may be changing somewhat.

Observe the spacing of the contour lines around this trough. Even with a limited number of actual wind observations, the close spacing of the contours to the west and northwest of the trough and

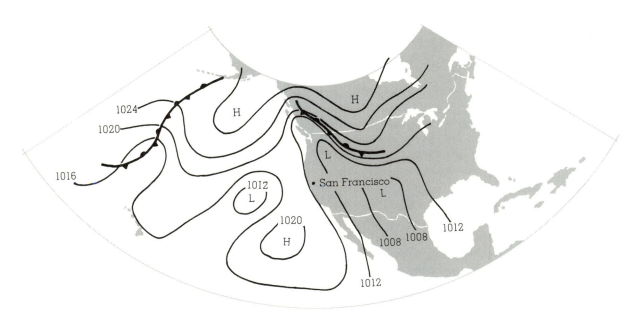

Fig. 18.9   Surface weather map for 4:00 P.M. Sunday, March 25, 1979.

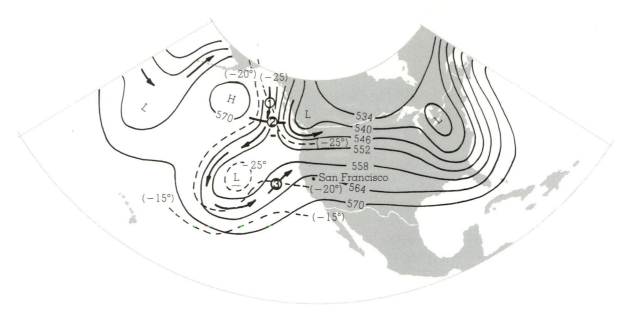

Fig. 18.10 The 500-mb chart for 4:00 P.M. Sunday, March 25, 1979. Arrows indicate wind flow. Solid lines are height contours where 564 equals 5640 m. Dashed lines are isotherms in °C. Heavy, solid line shows position of shortwave trough.

the more widely spaced contours to the east of the trough hint that stronger winds exist to the west of the trough. The forecaster knows from past experience that this usually means the trough will deepen. Also note that to the northwest of the trough (position 1) the wind crosses the −25°C isotherm, indicating that cold advection is occurring. The heavy, solid line on the chart off the Canadian coast (position 2) represents the position of a shortwave trough, which is moving rapidly southward. The injection of cold air and the shortwave into the main trough should cause it to intensify. To the east of the main trough (position 3), the wind is advecting warmer subtropical air northeastward. It is the lifting and condensing of this moist air that is producing the high clouds over San Francisco. All of these conditions—high wind speeds, regions of temperature advection, and a shortwave moving into a longwave trough—manifest themselves as a deepening of the longwave trough. As the upper trough deepens, it will be capable of providing the necessary conditions favorable for the development of the surface low into a major storm system. (You may remember from Chapter 17 that the generation of this type of dynamic instability is called baroclinic instability.)

One of the main ingredients necessary for the development and intensification of the surface depression is divergence of the air flow aloft. The forecaster knows that divergence aloft causes surface pressures to decrease beneath it. This in turn causes surface air to converge, rise, and condense into widespread cloudiness. But where will regions of divergence, convergence, and rising air be found on tomorrow's map? And how will tomorrow's map be different from today's? This is where the computer and the forecaster work together to come up with a prediction.

The Computer Provides Assistance The computer progs predict the future positions of weather systems. Some of the progs also predict where shortwave troughs will be located. It is important to know where the shortwaves will be found, because to the east of them there is usually upper-level divergence, lower-level convergence, rising air, clouds, and precipitation. Hence, predicting the position of a shortwave means predicting regions of inclement weather. (The position of the shortwaves also pretty much corresponds to the position of the vorticity maximums discussed in Chapter 17.)

Three models that predict the positions of the shortwaves, upper-level pressure systems, and flow aloft at 500 mb for 12, 24, and 36 hours into the future are shown in Fig. 18.11.* (Each prediction is made on Sunday afternoon.) Observe that each prog moves the upper trough slowly eastward and keeps it off the coast for the entire period. However, the actual positioning of the trough and the shortwaves (heavy dashed lines) differ for each model. The baroclinic and limited-area fine mesh (LFM) models move the trough eastward more quickly than does the barotropic model. Also, the baroclinic and LFM progs for Monday morning both show several shortwaves moving around the upper trough with one shortwave west of San Francisco. The barotropic model predicts that the same shortwave will be much farther to the west.

After examining each prog carefully, the forecaster must decide which model most accurately describes the future state of the atmosphere. Because this is a baroclinic situation (contour lines cross isotherms), more credence is given to the baroclinic model than the barotropic. But forecasts based on the LFM are usually better, since it uses more closely spaced grid points and a greater number of data points.

Using experience and the progs, the meteorologist sets out to predict the weather. In Fig. 18.11c, the LFM prog has a shortwave moving toward the California coast on Monday morning. The southwesterly flow above this system will steer it toward the California coast, causing the clouds to increase and thicken. According to the progs, the shortwave should move overhead sometime before noon, bringing with it cloudy skies and a chance of rain. Because the upper-level low is predicted to remain off the coast, southwesterly winds aloft will continue to pump moisture into the region. The 36-hour LFM prog shows that the second shortwave is scheduled to move through early Tuesday morning. The chance of rain should increase at this time. This same prog shows several shortwaves still off the coast, so the threat of rain will persist at least through Tuesday. Therefore, the precipitation forecast will sound like this:

*Explaining the differences among the three models is beyond the scope of this book. In very simple terms the *barotropic model* assumes that the isotherms on the upper-level chart are parallel to the contour lines; the *baroclinic model* assumes that the isotherms cross the contour lines. The *limited-area fine mesh model* (LFM) has a much closer grid-point spacing than the other two models—130 km (81 mi).

Increasing cloudiness Sunday night with rain beginning Monday morning. Periods of rain likely through Tuesday with heavy amounts falling Tuesday morning.

A Valid Forecast   By early Monday morning, the maps begin to show the changes that the computer progs predicted. The surface map for 4:00 A.M. (PST) Monday morning (Fig. 18.12) shows that the surface low in the Pacific has moved eastward and developed into a broad trough west of California. (Compare its position with Fig. 18.9.) The trough has deepened considerably as indicated by its central pressure of 1004 mb (29.65 in.). The approach of the storm is evidenced in San Francisco by thick middle clouds, southerly winds, and a falling barometer, nearly 4 mb lower than 12 hours ago. All these signs suggest that rain is on the way.

On the 500-mb chart (Fig. 18.13) for 4:00 A.M. Monday morning, we can see that the injection of cold air and the shortwave into the main upper trough have caused it to deepen. Note that the height contours are now displaced farther south and that the contour in the middle of the trough is lower than on the previous 500-mb map (Fig. 18.10). Compare Fig. 18.13 with the LFM 12-hour computer prog (Fig. 18.11c) and notice how well they match. The computer did a good job predicting the position of the upper-level pressure systems and shortwaves (heavy, dashed lines). Since the shortwaves are moving with the flow toward San Francisco, it should rain today. But at what time will the rain begin? Here is where satellite information assists the forecaster.

Assistance from the Satellite   The infrared satellite photograph taken at 6:45 A.M. Monday (Fig. 18.14) shows that the middle clouds presently over California will soon give way to a band of organized cumuliform clouds that look like a comma in the photograph. This coherent band of clouds is known as a **comma cloud**. Observe that this cloud band lies almost beneath the shortwave shown in Fig. 18.13. Also note that to the west a relatively unorganized mass of clouds is beginning to form near the second shortwave. Strong southwesterly winds aloft should carry the large comma cloud and its weather into California.

By examining the movement of the cloud mass on successive satellite photographs, the forecaster can predict its arrival time and hence when

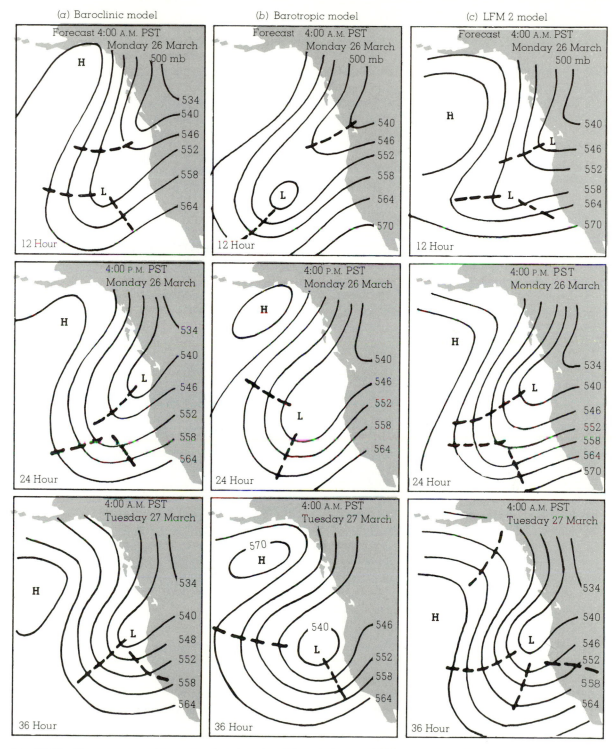

Fig. 18.11 Three computer-drawn progs that show the 12-hour, 24-hour, and 36-hour projected 500-mb chart. Solid lines are contours. Dashed lines represent projected positions of shortwaves. (These predictions were made on Sunday, March 25, 1979 at 4:00 P.M., PST.)

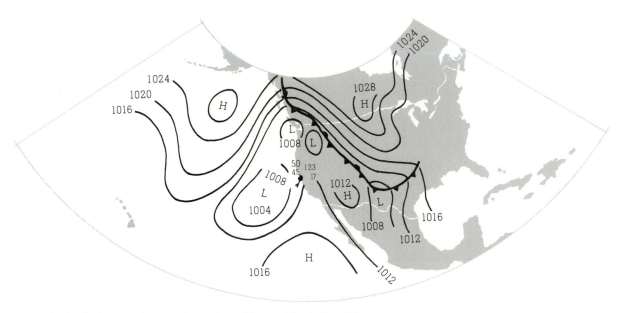

Fig. 18.12 Surface weather map for 4:00 A.M. Monday, March 26, 1979.

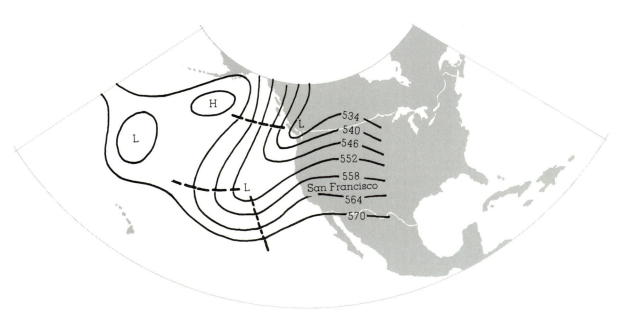

Fig. 18.13 The 500-mb chart for 4:00 A.M. Monday, March 26, 1979. Heavy, dashed lines show positions of shortwaves. (Compare with Fig. 18.11, the 12-hour prog.)

rainfall will begin. According to satellite photographs, the leading edge of the comma cloud should be just off shore by Monday afternoon. Also, radar indicates that, just off the coast, light rain is now falling from the middle cloud layer. Consequently, rainfall should begin sometime in the morning, becoming heavier by the afternoon as the cumuliform clouds move in. Because the surface low is beneath the upper-level shortwave, the surface depression will probably continue to intensify, and pressure gradients around it will increase, creating strong and gusty winds from the south as the storm approaches. An amended forecast for San Francisco might read:

Rain beginning this morning becoming heavy by this afternoon. Strong and gusty southerly winds.

Fig. 18.14 Infrared satellite picture taken at 6:45 A.M. (PST) Monday, March 26, 1979. Heavy, dashed line shows position of ''comma'' cloud.

A Day of Rain and Wind The first raindrops falling from altostratus clouds dampen city streets near the end of the morning rush hour. Quickly, the rain spreads inland and by late Monday afternoon weather radar shows that precipitation is falling throughout northern and central California as gusty southerly winds and moderate rain greet commuters on their way home.

The barometer has fallen sharply all day at San Francisco and by 4:00 P.M. the barometer reading is 1004 mb, a drop of 7 mb in just 6 hours. We can see the reason for this on the surface map for 4:00 P.M. Monday afternoon (Fig. 18.15). The low has not only moved closer to the coast, it has intensified considerably, as indicated by the drop of 11 mb in central pressure in just 12 hours. Spiraling around the low, a cloud front marks the position of the comma cloud. At first, this may seem surprising, since no fronts were drawn on the previous map. However, remember that this is a baroclinic situation with cold advection behind the low, warm advection in front of it, and divergence in the flow aloft. At the surface, air masses with contrasting temperatures are being brought together in the region of the comma cloud. Since the cold air is on the western side of the comma cloud, the meteorologist saw fit to draw in a cold front. Notice that, to the north of the low, a sta-

tionary front marks the boundary between cold mP air to the west and modified cool mP air to the east.

As the surface low intensifies, it and the spiraling band of clouds move eastward more slowly. The front will, therefore, move through later than anticipated, sometime late Monday night or early Tuesday morning. The forecaster expects that the winds will remain strong and precipitation will be heavy as the front passes.

Early Tuesday morning, the front moves on shore, bringing with it heavy rain and winds with gusts exceeding 45 knots. Billowing cumulus clouds, and in some areas thunderstorms, drench the entire Pacific Coast with rain, and with snow at higher elevations. The storm center, constantly being drained of air by strong upper-level divergence, has deepened into a furious system with a central pressure of 988 mb (29.17 in.).

The satellite photograph for 9:00 A.M. Tuesday morning (Fig. 18.16) provides us with a visual interpretation of the storm. We see superimposed on the photograph the positions of the surface low, fronts, and the winds aloft (heavy arrow). Note that a front with its heavy band of clouds stretches from Idaho southward into Nevada and Southern California, while the surface low is still positioned off the northern California coast. Mov-

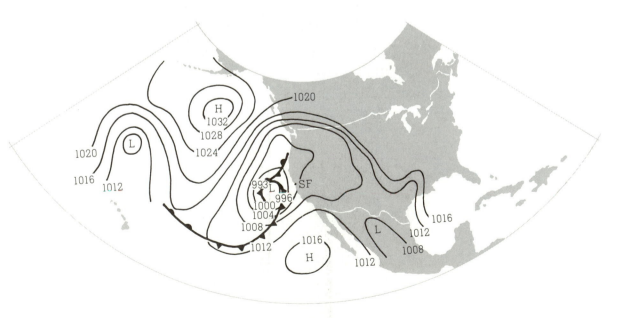

Fig. 18.15 Surface weather map for 4:00 P.M. Monday, March 26, 1979.

Fig. 18.16   Visible satellite photograph for 9:00 A.M. Tuesday, March 27, 1979. Included in the picture are the positions of surface fronts, the upper-level flow (heavy arrow), and precipitation patterns.

ing through central California is a band of clouds and showers associated with a shortwave, spinning counterclockwise around the low.

The upper flow indicates that the trough aloft is still off the coast just as Sunday's LFM computer prog had predicted. If we follow the flow southward out of Canada, we see a patch of clear weather just off shore. Here, the air pushing southward off the land is cold and dry, so clouds do not form. However, as the air moves further south, it warms and picks up moisture from the water below. The rippled cloud pattern in the photograph is cumulus clouds, which form in the unstable air. Look closely at the photograph and notice that a comma cloud is developing to the southwest of the deep surface low. If you refer back to Fig. 18.11c, you will see that the 36-hour LFM prog predicted that a shortwave would be located in this region on Tuesday morning. Is this cloud band organizing into another onslaught of high winds, heavy precipitation, and thunderstorms? If so, when will it arrive? And what about the storm off the coast? Will it deepen or fill, move inland or remain stationary? It's back to the drawing board—to the computer progs, the charts, and the satellite photographs. The challenge and anticipation of making another forecast are at hand.

Summary

Forecasting tomorrow's weather entails a variety of techniques and methods. Persistence and steady state forecasts are useful when making a short range (0–6 hour) prediction. For a longer range forecast, the current analysis, satellite data, weather typing, intuition, experience, along with guidance from the many computer progs supplied by the National Weather Service, all go into making a prediction.

Different computer progs are based upon different atmospheric models that describe the state of the atmosphere and how it will change with time. At present, computer progs are better at forecasting the position of midlatitude highs and lows than local showers and thunderstorms. To skillfully forecast smaller features, the grid spacing on the models must be reduced. This results in many more computations and a longer running time.

As new information from atmospheric research programs is fed into the latest generation of computers, it is hoped that the progs will be able to show skill in predicting the weather up to 10 days in the future. More promising at this time is the simulation of large-scale climatic trends by the most recent general circulation models.

The latter part of this chapter was not intended to make you an expert weather forecaster, nor was it designed to show you all the methods of weather prediction. It is hoped, however, that you have gained an understanding of the problems confronting anyone who attempts to predict the behavior of this churning mass of air we call our atmosphere.

Questions for Review

1. What is the function of NMC?

2. Make a persistence forecast for your area for this same time tomorrow. Did you use any skill in making this prediction? Explain.

3. Describe four methods of forecasting the weather and give an example for each one.

4. Suppose that the forecast for your area today calls for an ''80 percent chance of rain.'' What exactly does that mean?

5. In what ways has the computer assisted the meteorologist in making weather forecasts?

6. How does a prog differ from an analysis?

7. Briefly describe how the computer goes about making a weather forecast.

8. What are some of the problems associated with machine-made forecasts?

9. Describe the research programs of GATE and FGGE.

10. Would a forecast calling for a 20 percent chance of rain be sufficiently high enough for you to cancel your plans for a picnic? Explain.

11. Use the instant weather forecast chart (Appendix E) as a guide to making a short-range weather forecast when the surface weather elements indicate those shown in Table 18.1, p. 387.

12. If low clouds at an elevation of 1 km above you are moving from the southeast and clouds about 2 km higher are moving from the southwest, is cold or warm advection taking place between the cloud layers? Explain.

Table 18.1 Surface Weather Elements (to be used in answering review question 11)

TIME	AIR PRESSURE (mb)	ANY CHANGES	AIR TEMPERATURE (°F)	WIND DIRECTION	CLOUD TYPE	PRECIPITATION
morning	1017	falling slowly	29	SE	As	none
afternoon	1014	rising rapidly	36	NW	Tcu	rain showers
afternoon	1025	steady	86	NW	Ci	none
afternoon	990	falling rapidly	26	NE	Ns	snow
early morning	1020	rising	13	NW	clear	none
nighttime	1012	falling	31	SW	Sc	none
afternoon	1006	falling	52	SE	Ns	rain
nighttime	1005	rising rapidly	45	W	Cb	rain showers

13. List four methods that you could use to predict the movement of a surface open wave cyclone.

14. What is an omega high? What influence does it have on the movement of surface highs and lows?

15. Suppose that where you live the middle of January is typically several degrees warmer than the rest of the month. If you forecast this "January thaw" for the middle of next January, what type of a weather forecast will you have made?

Questions for Thought

1. From Fig. 18.3 determine the probability of a "White Christmas" for your area.

2. Suppose the chance for a "White Christmas" at your home is 10 percent. Last Christmas was a white one. If for next year you forecast a "non-white" Christmas, will you have shown any skill if your forecast turns out to be correct? Explain.

3. Suppose that it is presently warm and raining. A cold front will pass your area in 3 hours. Behind the front it is cold and snowing. Make a persistence forecast for your area 6 hours from now. Would you expect this forecast to be correct? Explain. Now, make a forecast for your area using the steady-state, or trend, method.

Problems and Exercises

1. From the newspaper or the National Weather Service, obtain a copy of next month's 30-day outlook. Based on the temperature and precipitation patterns of this forecast, draw several contour lines representing the upper-air pattern that would be necessary for such a forecast to come true.

2. Keep a notebook handy, and each day sketch the upper-air and surface weather patterns. Do this for a month or more, noting the type of weather each pattern brings. Use this information to start making your own forecasts by examining the current surface and upper-air charts.

3. When a persistent winter pattern at 500-mb appears similar to that shown in Fig. 18.17, show on the map where you would forecast the following: (a) good chance of precipitation; (b) above seasonal temperatures; (c) below seasonal temperatures; (d) generally dry weather.

4. In Fig. 18.5, p. 372, mark the position of the following cities: Cleveland, Ohio; Albuquerque, New Mexico; and New Orleans, Louisiana. Based on the projected movement of the surface weather systems in Fig. 18.7, p. 374, make a 24-hour weather forecast for each of these cities. In your forecast, include temperature, pressure, cloud cover, humidity, winds, and precipitation (if any). Compare your forecasts with the actual weather at the end of the period in Fig. 18.8, p. 377.

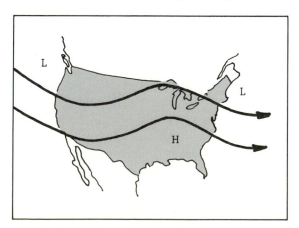

Fig. 18.17 Diagram for problem 3.

On June 8, 1966, this tornado caused almost total destruction along a path 13 kilometers (8 mi) long and 4 blocks wide through the heart of Topeka, Kansas. (Photo: *Topeka Capital Journal*)

CHAPTER 19

Thunderstorms and Tornadoes

I huddled in the bathtub with a blanket pulled over my head, waiting for the end to come. I believed I was going to die. The shrill whistling and awful rumbling grew unbearable. The roof was ripped off. The garage disappeared. I thought it was the end of the world; I thought we were gone—all of us gone. It was the most eerie sound I ever heard. It was kind of like a train, with a real shrill whistle. I don't ever want to hear anything like that again. (From "The Story of the Grand Island Tornadoes," *The Grand Island Daily Independent*, 1980.)

That was an actual account of a person's experience with one of nature's most awesome storms—the tornado. A series of these storms struck Grand Island, Nebraska, on the night of June 3, 1980, devastating a large portion of the town and killing seven persons.

Tornadoes such as these are associated with severe thunderstorms. Consequently, we will first examine the different types of thunderstorms. Later, we will focus on tornadoes, examining how and where they form, and why they are so destructive.

What Are Thunderstorms?

It probably comes as no surprise that a thunderstorm is merely a storm containing lightning and thunder. Sometimes it produces gusty surface winds with heavy rain and hail. The storm itself may be a single cumulonimbus cloud, a cluster of them, or even a line of clouds that in some cases extends for more than 100 km (62 mi).

The birth of a thunderstorm occurs when warm, humid air rises in an unstable environment. The trigger needed to start air moving upward may be the unequal heating of the surface, the effect of terrain, or the lifting of warm air along a frontal

Contents

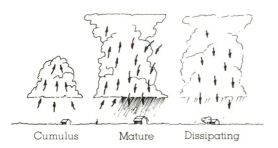

Fig. 19.1   The life cycle of an air-mass thunderstorm. (Arrows show vertical air currents.)

zone. Diverging upper-level winds also provide a favorable region for thunderstorm development, as they tend to draw air upwards beneath them. Usually, several of these mechanisms work together to generate severe thunderstorms.

Scattered thunderstorms that form in summer are often referred to as **air-mass thunderstorms** because they tend to develop in warm, maritime tropical air masses away from weather fronts. These storms are usually short-lived and rarely produce strong winds and large hail. On the other hand, the **severe thunderstorms** that form in a long line may produce high winds, flash floods, damaging hail, and even tornadoes. Let's examine the air-mass thunderstorms first.

Air-Mass Thunderstorms

Extensive studies carried out during the 1940s showed that thunderstorms go through a cycle of development from birth, to maturity, to decay. The first stage is known as the **cumulus stage**. As humid air rises, it cools and condenses into a single cumulus cloud or a cluster of clouds (Fig. 19.1). If you have ever watched a thunderstorm develop, you may have noticed that at first the cumulus clouds grow upward only a short distance, then they dissipate. This is because the cloud droplets evaporate as the drier air surrounding the cloud mixes with it. However, after the water drops evaporate, the air is more moist than before. So, the rising air is now able to condense at successively higher levels, and the cumulus cloud grows taller, often appearing as a rising dome or tower.

As the cloud builds, the transformation of water

vapor into liquid or solid cloud particles releases large quantities of latent heat. This keeps the air inside the cloud warmer than the air surrounding it. The cloud continues to grow in the unstable air as long as it is constantly fed by rising air from below. In this manner, a cumulus cloud may show extensive vertical development in just a few minutes. During the cumulus stage, rapid updrafts prevent any precipitation from falling out of the cloud. Also, there is no lightning or thunder during this stage.

As the cloud builds well above the freezing level, the precipitation particles grow larger. They also become heavier; eventually, the rising air is no longer able to keep them suspended, and they begin to fall. As they descend, they drag some of the air along with them, creating a downdraft. The downdraft is usually strengthened as drier air is drawn into the cloud, causing some of the raindrops to evaporate. This chills the air, making it both colder and heavier than the air around it. The downdraft and updraft within the cloud constitute a *cell*. In most storms, there are several cells, each of which may last for half an hour or so. The appearance of the downdraft and the storm cell marks the beginning of the **mature thunderstorm**.

During its mature stage, the thunderstorm is most intense. The top of the cloud, having reached the stable stratosphere, begins to take on the familiar anvil shape, as strong upper-level winds spread the cloud's ice crystals horizontally. The cloud itself may extend upward to an altitude of over 12 km (40,000 ft) and be several kilometers in diameter near its base. Updrafts and downdrafts reach their greatest strength in the middle of the cloud, creating severe turbulence. Lightning and thunder are also present. Heavy rain (and occasionally small hail) falls from the cloud. The rainfall may or may not reach the surface, depending on the relative humidity beneath the storm. In the dry air of the desert southwest, for example, a mature air-mass thunderstorm may look ominous and contain all of the ingredients of any other storm, except that the raindrops evaporate before reaching the ground.

About fifteen minutes to half an hour after the storm enters the mature stage, it begins to dissipate. The **dissipating stage** occurs when the falling precipitation causes downdrafts to form throughout the cloud. Deprived of its rich supply

Fig. 19.2 A dissipating thunderstorm. Most of the cloud particles in the lower half of the storm have evaporated. Only the cirrus anvil stands out as a distinguishing feature.

of warm, humid air, cloud droplets no longer form. Light precipitation now falls from the cloud, accompanied by only weak downdrafts. As the storm dies, the lower-level cloud particles evaporate rapidly, sometimes leaving only the cirrus anvil as the reminder of the once mighty presence (Fig. 19.2).

A single air-mass thunderstorm may go through these three stages in an hour or less. The reason it does not last very long is that the storm's own precipitation produces the downdrafts that cut off the storm's fuel supply by destroying the humid updrafts. Hence, the storm tends to collapse on itself.

Not only do these storms produce summer rainfall for a large portion of the United States but they also bring with them momentary cooling after an oppressively hot day. The cooling comes during the mature stage, as the downdraft reaches the surface in the form of a blast of welcome relief. Sometimes, the air temperature may lower as much as 10°C (18°F) in just a few minutes. Unfortunately, the cooling effect is short-lived, as the downdraft diminishes or the thunderstorm moves on. In fact, after the storm has ended, the air temperature usually rises; and as the moisture from the rainfall evaporates into the air, the humidity increases, sometimes to a level where it actually feels more oppressive after the storm than it did before.

Upon reaching the surface, the cold downdraft

has another effect. It may force warm, moist surface air upwards. This rising air then condenses and gradually builds into a new thunderstorm. Thus, it is entirely possible for a series of thunderstorms to grow in a line, one next to the other, each in a different stage of development. (See Fig. 19.3.) Thunderstorms that form in this manner are termed **multicell storms**. Most air-mass thunderstorms are multicell storms, as are most severe thunderstorms.

There are many thunderstorms that form in response to forced lifting along a frontal boundary and, thus, they are not air-mass thunderstorms. Most of these **frontal thunderstorms** are not severe, and their life cycle can be described in much the same way as the air-mass storms.

Severe Thunderstorms

Severe thunderstorms are capable of producing large hail, strong, gusty surface winds, flash floods, and tornadoes. Just as the air-mass thunderstorm, they form as moist air is forced to rise into unstable air. But, severe thunderstorms also form in areas with a strong vertical wind shear. This causes the updraft to tilt in the mature stage.

Figure 19.4 is a model depicting the air motions within a severe thunderstorm. The storm is mov-

Fig. 19.3   A multicell storm. This storm is composed of a series of cells in successive stages of growth. The thunderstorm in the middle is in its mature stage, with a well-defined anvil. Heavy rain is falling from its base. To the right of this cell, a thunderstorm is in its cumulus stage. To the left, a well-developed cumulus congestus cloud is about ready to become a mature thunderstorm.

ing from left to right. Strong winds aloft cause the system to tilt so that the updrafts move up and over the downdrafts. This is very important to the development of the system. When the precipitation becomes too heavy for the updrafts to support, it falls into the downdrafts rather than into the updrafts, as in the air-mass thunderstorm. Hence, the updrafts remain strong and do not dissipate. Because the updrafts flow unabated, they are capable of obtaining speeds of more than 50 knots.

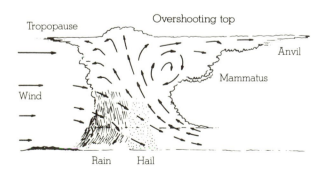

Fig. 19.4   A model describing air motions and other features associated with a severe thunderstorm.

The updrafts may be so strong that the cloud top is able to intrude well into the stable stratosphere, a condition known as *overshooting*. In some cases, the top of the cloud may extend to more than 18 km (60,000 ft) above the surface. The violent updrafts keep hailstones suspended in the cloud long enough for them to grow to considerable size. Once they are large enough, they either fall out the bottom of the cloud with the downdraft or a strong updraft may toss them out the side of the cloud, or even from the base of the anvil. Aircraft have actually encountered hail in clear air several kilometers from a storm. Also, strong downdrafts within the upper part of the cloud may produce beautiful mammatus clouds.

When we look more closely at the lower half of a severe thunderstorm (Fig. 19.5), we see that the downdraft is fed by the frictional drag of the precipitation. When the drier air surrounding the cloud is sucked into the system, some of the precipitation evaporates. This further cools the air and enhances the downdraft. The cool air reaching the ground acts like a wedge, forcing warm, moist surface air up into the system. Thus, the downdraft helps to maintain the updraft and vice versa, so that the severe thunderstorm is able to maintain itself for, in some cases, many hours.

The Gust Front and Microburst   Look at Fig. 19.5 again and notice that the downdraft spreads laterally after striking the ground. The boundary separating this cold downdraft from the warm surface air is known as a **gust front**. To an observer on the ground, the passage of the gust front resembles that of a cold front. During its passage, the wind shifts and becomes strong and gusty, with speeds occasionally exceeding 45 knots; temperatures drop sharply and, in the cold heavy air of the downdraft, the surface pressure rises. Sometimes it may jump several millibars, producing a small area of high pressure called a **meso-high** (meaning mesoscale high). The cold air may linger close to the ground for several hours, well after the thunderstorm activity has ceased.

Along the leading edge of the gust front, the air is quite turbulent. Here, strong winds can pick up loose dust and soil and lift them into a huge tumbling cloud—the haboob that we described in Chapter 14. As warm, moist air rises along the forward edge of the gust front, a **roll cloud** (also

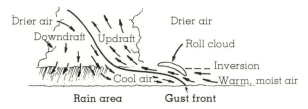

Fig. 19.5   The lower half of a severe squall-line-type thunderstorm and some of the features associated with it.

called an **arcus cloud**) may form, such as the one shown in Fig. 19.6. These clouds are especially prevalent when an inversion exists near the base of the thunderstorm. Sometimes the gust front forces warm, moist air upward, producing new thunderstorms.

Beneath a severe thunderstorm, the downdraft may become localized, so that it hits the ground and spreads horizontally in a radial burst of wind, much like water pouring from a tap and striking the sink below. Such downdrafts are called **downbursts**. When the downburst is less than about 4 km (2.5 mi) across, it is termed a **microburst**. When the microburst reaches the ground and continues as an expanding outflow, a gust front may

Fig. 19.6   A dramatic example of a roll cloud associated with a severe thunderstorm. The photograph was taken in the Philippines as the thunderstorm approached from the northwest.

Fig. 19.7 A supercell thunderstorm. This awesome storm dumped hail the size of golf balls and more than 5 centimeters (2 in.) of rain over portions of eastern Colorado.

form. Hence, a microburst can evolve into a gust front.

Microbursts are capable of blowing down trees and inflicting heavy damage upon poorly built structures. In fact, microbursts may be responsible for some damage once attributed to tornadoes. Apparently, a microburst from an intense thunderstorm slammed into a skyway gondola at Disneyland during April, 1983, stranding scores of terrified visitors on the ride as they swayed back and forth high above the amusement park.

Microbursts and their accompanying *wind shear* appear to be responsible for several airline crashes. When an aircraft flies through a microburst, it first encounters a headwind that generates extra lift. However, in a matter of seconds, the headwind is replaced by a tailwind that causes a sudden loss of lift and a subsequent decrease in the performance of the aircraft. One accident attributed to a microburst occurred at New Orleans' Moisant International Airport during July, 1982, when shortly after takeoff a Boeing 727, flying beneath a thunderstorm apparently flew directly into a microburst, lost altitude, and crashed, killing all 145 persons aboard.

Studies conducted in the early 1980s at Den-

ver's Stapleton International Airport suggest that most microbursts emanate from virga—rain falling from a thunderstorm but evaporating before reaching the ground. As the rain evaporates, it cools the air. The cooler, heavy air then plunges downward through the warmer, lighter air below. In more humid climates, other conditions—such as precipitation drag and the height at which the downdraft originates—may be more important in the development of microbursts.

Supercell and Squall-Line Thunderstorms There are two main types of severe thunderstorms: the **supercell storm** and the **squall line**. The *supercell storm* is an enormous thunderstorm whose updrafts and downdrafts are so sufficiently in balance that it is able to maintain itself as a single entity for hours on end. Figure 19.7 shows a supercell storm near Denver, Colorado. It is this type of storm that produces tornadoes and destructive hail.

The *squall line* forms as a line of thunderstorms. Sometimes they are right along a cold front, but more often they are 100 to 300 km out ahead of it. These prefrontal squall-line thunderstorms may be caused by air aloft flowing over the cold front and developing into waves, much like the waves that form downwind of mountain chains. Notice in Fig. 19.8 that the downward-moving part of the wave inhibits cloud formation, while the rising part, about 100 km from the cold front, favors uplift. It is here that clouds and thunderstorms form in unstable air.

Another possible cause of the prefrontal squall-line thunderstorm is shown in Fig. 19.9. The map shows conditions that lead to severe thunderstorms over the Central Plains, especially during the spring. A developing midlatitude cyclone forms as an open wave with a cold front, a warm front, and three distinct air masses. Behind the cold front, cold, dry cP air pushes in from the north. In the warm sector, ahead of the cold front, warm, dry cT air moves in from the southwest. Further east, warm but very moist mT air moves northward from the Gulf of Mexico. Along the cold front—where cold, dry air replaces warm, dry air—there is insufficient moisture for thunderstorms to form. However, in the warm sector, where the more-dense dry air encounters the less-dense moist air, convergence and lifting occur. It is along this boundary (called a **dry line**) that squall-line thunderstorms form. Once these thunderstorms develop, the outflow of cold air along the ground is able to initiate the uplift necessary for the generation of new, and possibly more severe, thunderstorms.

Mesoscale Convective Complexes Where conditions are favorable for convection, a number of individual thunderstorms will occasionally grow in size and organize into a large convective weather system. These convectively driven systems, called **Mesoscale Convective Complexes (MCCs)**, are quite large—they can be as much as 1000 times larger than an individual air-mass thunderstorm. In fact, they are often large enough to cover an entire state, an area in excess of 100,000 km² (Fig. 19.10).

Within the MCC, the individual thunderstorms apparently work together to generate a long-lasting weather system that moves slowly (normally less than 20 knots) and often exists for periods exceeding 12 hours. The circulation of the MCC supports the growth of new thunderstorms as well as a region of widespread precipitation. These systems are beneficial, as they provide a signifi-

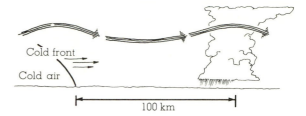

Fig. 19.8 Squall-line thunderstorms may form ahead of an advancing cold front as the upper-air flow develops waves downwind from the cold front.

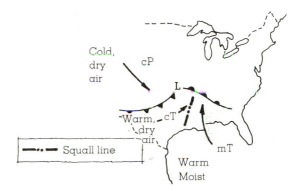

Fig. 19.9 Surface weather conditions favorable for the generation of prefrontal severe thunderstorms.

cant portion of the growing season rainfall over much of the corn and wheat belts of the United States. However, MCCs can also produce a wide variety of severe weather, including hail, high winds, destructive flash floods, and tornadoes.

Thunderstorm Movement Most thunderstorms move roughly in the direction of the winds in the middle troposphere. In some cases, however, a multicell storm over the central United States will take a peculiar path: Individual cells will move with the wind, while the thunderstorm itself will move to the right of the wind. We will be able to understand this if we take a closer look at the weather conditions necessary for the development of severe storms.

Severe thunderstorms form over the central states when warm, moist surface air streaming northward from the Gulf of Mexico lies beneath drier southwesterly (or westerly) winds aloft. This unstable air generates severe thunderstorms, such as the one forming in Fig. 19.11. Warm, humid surface air fuels the storm, so that it is able to grow into a huge multicell storm. As the storm

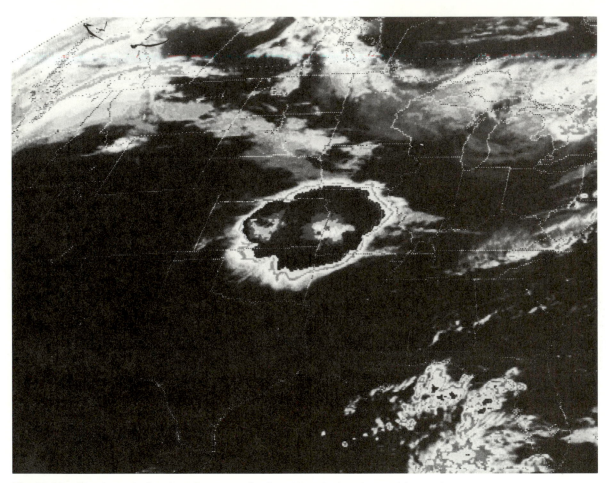

Fig. 19.10 An enhanced infrared satellite picture for June 22, 1981 that shows a Mesoscale Convective Complex extending from central Kansas across western Missouri. This organized mass of thunderstorms brought hail, heavy rain, and flooding to this area.

Fig. 19.11 The formation of multicell thunderstorms that move to the right of the upper-level winds. The developing cell is steered to the northeast by the flow aloft. Downdrafts within the storm generate a new cell to the south. The new cell, which grows into a mature thunderstorm, cuts off the supply of moisture to the old cell, and so it dissipates. This entire sequence shows cells drifting to the northeast, while the whole thunderstorm moves slightly to the right of the upper-level winds.

1 Developing cell
2 Mature thunderstorm
3 Dissipating thunderstorm

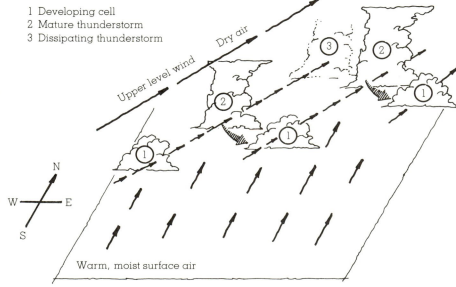

matures, it begins to deprive areas to its north of the warm, humid surface air that is necessary for new thunderstorm growth. As the upper-level winds steer the storm toward the northeast, the storm's own downdraft pushes out its southern side, forcing moist air to rise and form new cells. Meanwhile, the old cell to the north, being deprived of its moisture, gradually dissipates. Therefore, as new cells form on the southern side of the thunderstorm, old cells to the north slowly die out, and the entire multicell thunderstorm moves to the right of the general wind flow.*

Distribution of Thunderstorms

It is estimated that more than 40,000 thunderstorms occur each day throughout the world. Hence, over 14 million occur annually. The combination of warmth and moisture make equatorial land masses especially conducive to thunderstorm formation. Here, thunderstorms occur on about one out of every three days. Thunderstorms are also prevalent over water along the ITCZ, where the low-level convergence of air helps to initiate uplift. The heat energy liberated in these storms helps the earth maintain its heat balance by providing energy for the Hadley cell (see Chapter 15). Thunderstorms rarely occur in dry climates, such as the polar regions and the desert areas of the subtropical highs.

Figure 19.12 shows the average number of days each year having thunderstorms in various parts of the United States. Notice that they occur most frequently in the southeastern states along the Gulf coast with a maximum in Florida. A secondary maximum exists over the central Rockies. The region with the fewest thunderstorms is the Pacific coastal and interior valleys.

In many areas, thunderstorms form primarily in summer during the warmest part of the day when the surface air is most unstable. There are some exceptions, however. During the summer in the valleys of central and southern California, dry, sinking air produces an inversion that inhibits the development of towering cumulus clouds. In these

*Some storms move to the left of the upper-level winds. Others have actually split into two storms: one that moves to the right and the other to the left. No attempt will be made here to try and explain these different types of movement.

regions thunderstorms are most frequent in winter and spring, particularly when cold, moist, unstable air aloft moves in over moist, mild surface air. The surface air remains relatively warm because of its proximity to the ocean. Over the Central Plains states, thunderstorms tend to form more frequently at night. This may be caused by the position of the low-level southerly jet stream that we described in Chapter 15. This southerly wind, which forms at night, not only carries humid air northward but also produces areas of surface convergence, which help to trigger uplift and, hence, thunderstorms. Another cause of the nighttime thunderstorms may be the downslope flow of air from the mountains, which also causes convergence and uplift.

At this point it is interesting to compare Fig. 19.12 and Fig. 19.13. Notice that, even though the greatest frequency of thunderstorms is near the Gulf Coast, the greatest frequency of hailstorms is over the western Great Plains. One reason for this is that conditions over the Great Plains are more favorable to the development of severe thunderstorms. We also find that, in summer along the Gulf Coast, a thick layer of warm, moist air extends upward from the surface. Most hailstones falling into this layer will melt before reaching the surface. Over the plains, the warm surface layer is much shallower and drier. Falling hailstones do begin to melt, but the water around their periphery quickly evaporates in the dry air. This cools the hailstones and slows the melting rate so that many survive as ice all the way to the surface.

Now that we have looked at the development and distribution of thunderstorms, we are ready to examine an interesting, though yet not fully understood, aspect of all thunderstorms—lightning.

Lightning and Thunder

Lightning is simply a discharge of electricity, a giant spark, which occurs in mature thunderstorms. Lightning may take place within a cloud, from one cloud to another, from a cloud to the surrounding air, or from a cloud to the ground. The lightning stroke can heat the air through which it travels to an incredible 30,000°C (54,000°F). This

FOCUS ON A SPECIAL TOPIC

Terrifying Flash Flood in the Big Thompson Canyon

July 31, 1976, was like any other summer day in the Colorado Rockies, as small cumulus clouds with flat bases and dome-shaped tops began to develop over the eastern slopes near the Big Thompson and Cache La Poudre rivers. At first glance, there was nothing unusual about these clouds, as almost every summer afternoon they form along the warm mountain slopes. Normally, strong upper-level winds push them over the plains, causing rainshowers of short duration. But the cumulus clouds on this day were different. For one thing, they were much lower than usual, indicating that the southeasterly surface winds were bringing in a great deal of moisture. Also, their tops were somewhat flattened, suggesting that an inversion aloft was stunting their growth. But these harmless-looking clouds gave no clue that later that evening in the Big

Thompson Canyon more than 135 people would lose their lives in a terrible flash flood.

By late afternoon, a few of the cumulus clouds were able to puncture the inversion. Fed by moist southeasterly winds, these clouds soon developed into spectacular supercell thunderstorms with tops exceeding 18 km (60,000 ft). By early evening, these same clouds were producing incredible downpours in the mountains.

In the narrow canyon of the Big Thompson River, some places received as much as 30.5 cm (12 in.) of rain in the four hours between 6:30 P.M. and 10:30 P.M. local time. This is an incredible amount of precipitation, considering that the area normally receives about 40.5 cm (16 in.) for an entire year. The heavy downpours turned small creeks into raging torrents, and the Big Thompson River was

quickly filled to capacity. Where the canyon narrowed, the river overflowed its banks and water covered the road. The relentless pounding of water caused the road to give way.

Soon cars, tents, mobile homes, resort homes, and campgrounds were being claimed by the river. Where the debris entered a narrow constriction, it became a dam. Water backed up behind it, then broke through, causing a wall of water to rush downstream. Of the approximately 2000 people in the canyon that evening, over 135 lost their lives. Property damage exceeded $35.5 million.

Figure 1 shows the weather conditions during the evening of July 31, 1976. A cool front moved through earlier in the day and is south of Denver. The weak inversion layer associated with the front kept the cumulus clouds from building to great heights

---

extreme heating causes the air to expand explosively, thus initiating a shock wave that becomes a booming sound wave—called **thunder**—that travels outward in all directions from the flash.

What causes lightning? The normal fair weather electric field of the atmosphere is characterized by a negatively charged surface and a positively charged upper atmosphere. For lightning to occur, separate regions containing opposite electrical charges must exist within a cumulonimbus cloud. Exactly how this charge separation comes about is not totally comprehended; however, there are many theories to account for it.

Electrification of Clouds One popular theory proposes that clouds become electrified as graupel and hail fall through a region of supercooled droplets and ice crystals. As liquid droplets collide with a hailstone, they freeze on contact and release latent heat. This keeps the surface of the

hailstone warmer than that of the surrounding ice crystals. When the hailstone comes in contact with an ice crystal an important phenomenon occurs: *electrons flow from the colder object toward the warmer object*. Hence, the hailstone becomes negatively charged. The same effect occurs when supercooled droplets freeze on contact with a hailstone and tiny splinters of positively charged ice break off. These lighter, positively charged particles are then carried to the upper part of the cloud by updrafts. The larger hailstones, left with a negative charge, fall toward the bottom of the cloud. By this mechanism, the upper part of the cloud becomes positively charged, while the lower part becomes negatively charged.* (See Fig. 19.14.)

---

*Although not fully understood, the region of positive charge near the base of the cloud may be the result of ice particles becoming positively charged while melting.

FOCUS ON A SPECIAL TOPIC

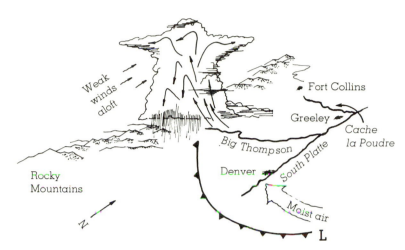

Fig. 1 Weather conditions that led to the development of severe thunderstorms that remained nearly stationary over the Big Thompson Canyon in the Colorado Rockies. The arrows within the thunderstorm represent air motions.

earlier in the afternoon. However, the strong southeasterly flow behind the cool front pushed unusually moist air upslope along the mountain range. Heated from below, the conditionally unstable air eventually punctured the inversion and developed into severe supercell thunderstorms. These remained nearly stationary for several hours due to the weak southerly winds aloft. The deluge may have deposited 19 cm (7.5 in.) of rain on the main fork of the Big Thompson River in about one hour.

The flash floods associated with such severe thunderstorms occur with unpleasant frequency. On June 9, 1972, a destructive flood occurred at Rapid City, South Dakota. The weather conditions were very similar to those that caused the Big Thompson flood. A slowly moving cold front was situated to the south of Rapid City with moist easterly winds behind it. Winds aloft were weak, so the severe thunderstorms moved slowly. Although there are other similarities between the two storms, we will not examine them here. However, a meteorologist, recognizing these conditions, would be able to alert a community to the probability of a flash flood and, hopefully, save many lives.

The Lightning Stroke Because unlike charges attract one another, the negative charge at the bottom of the cloud causes a region of the ground beneath it to become positively charged. As the thunderstorm moves along, this region of positive charge follows the cloud like a shadow. The positive charge is most dense on protruding objects, such as trees, poles, and buildings. The difference in charges causes an electric potential between the cloud and ground, which may be 10,000 volts per meter. In dry air, however, a flow of current does not occur because the air is a good electrical insulator. Gradually, the electrical potential builds, and when the electric field associated with it exceeds about 3 million volts per meter, the insulating properties of the air break down, a current flows, and lightning occurs.

Cloud-to-ground lightning begins as a flow of electrons from the middle of the cloud rushes toward the base. This discharge of electrons proceeds toward the ground in a series of steps. Each discharge covers about 50 to 100 m, then stops for about 50 millionths of a second, then occurs again over another 50 m or so. This **stepped leader** is very faint and is usually invisible to the human eye. As the tip of the stepped leader approaches the ground, the potential gradient (the voltage per meter) increases, and a current of positive charge starts upward from the ground (usually along elevated objects) to meet it. After they meet, large numbers of electrons flow to the ground and a much larger, more luminous **return stroke** several centimeters in diameter surges upward to the cloud along the path followed by the stepped leader. Even though the bright return stroke travels from the ground up to the cloud, it happens so quickly—in $1/10,000$ of a second—that our eyes cannot resolve the motion, and we see what ap-

Fig. 19.12 The average number of days each year on which thunderstorms are observed throughout the United States. (Due to the paucity of data, the number of thunderstorms is underestimated in the mountainous west.)

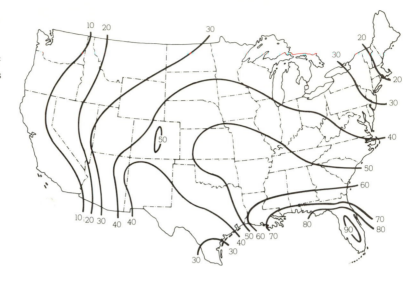

Fig. 19.13 The average number of days each year on which hail is observed throughout the United States.

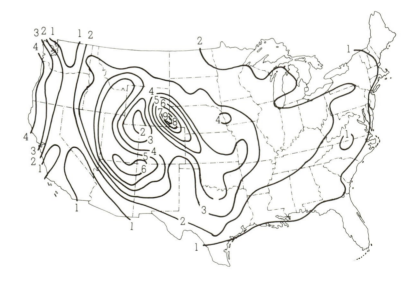

pears to be a continuous bright flash of light. (See Fig. 19.15.)

Sometimes there is only one lightning stroke, but more often the leader-and-stroke process is repeated in the same ionized channel at intervals of about a tenth of a millionth of a second. The subsequent leader, called a **dart leader**, proceeds from the cloud along the same channel as the original stepped leader; however, it proceeds downward more quickly because the electrical re-

sistance of the path is now lower. As the leader approaches the ground, normally a less energetic return stroke than the first one travels from the ground to the cloud. Typically, a lightning flash will have three or four leaders, each followed by a return stroke. A lightning flash consisting of many strokes (one photographed flash had 26 strokes) usually lasts less than a second. During this short period of time, our eyes may barely be able to perceive the individual strokes, and the flash ap-

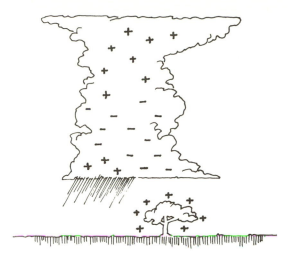

Fig. 19.14 The generalized charge distribution in a mature thunderstorm.

Fig. 19.15 A time exposure of cloud-to-ground lightning during an intense thunderstorm. The bright flashes are return strokes. The lighter forked-flashes probably are dart leaders propagating down new channels.

pears to flicker. (For additional information on the various forms of lightning, read the focus section on "Experiencing Thunder and Lightning.")

Lightning rods are placed on buildings to protect them from lightning damage. The rod is made of metal and has a pointed tip, which extends well above the structure. (See Fig. 19.16.) The positive charge concentration will be maximum on the tip of the rod, thus increasing the probability that the lightning will strike the tip and follow the metal rod harmlessly down into the ground, where the other end is deeply buried.

Lightning Detection and Suppression For many years, lightning strokes were detected primarily by visual observation. Today, cloud-to-ground lightning is located by means of an instrument called a *lightning detection-finder*. A web of these magnetic devices is a valuable tool in pinpointing lightning strokes in forested regions throughout the western United States, Canada, and Alaska. Currently, a network of these instruments is being installed along the East Coast of the United States. Lightning detection devices allow scientists to examine in detail the lightning activity inside a storm as it intensifies and moves. This gives forecasters a better idea where intense lightning strokes might be expected. In addition, when this information is correlated with satellite images, a more complete

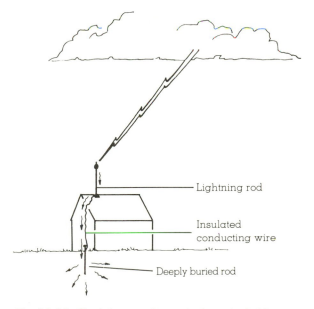

Fig. 19.16 The lightning rod extends above the building, increasing the likelihood that lightning will strike the rod rather than some other part of the structure. After lightning strikes the metal rod, it follows an insulated conducting wire harmlessly into the ground.

Don't Sit Under the Apple Tree

Because a single lightning stroke may involve a current as great as 100,000 amperes, animals and humans can be electrocuted when struck by lightning. The average yearly death toll in the United States attributed to lightning is well over 100. Many victims are struck in open places, riding on farm equipment, playing golf, or sailing in a small boat. Some live to tell about it, as did the champion golfer Lee Trevino and the former Shenandoah National Park ranger who was struck seven times and dubbed "the lightning conductor of Virginia" by the *Guinness Book of World Records*. Others are less fortunate. When you see someone struck by lightning, immediately

give CPR (cardiopulmonary resuscitation), as lightning normally leaves its victims unconscious and stops their breathing.

The largest single location of lightning fatalities is in the vicinity of relatively isolated trees. Because positive charge tends to concentrate in upward projecting objects, the upward return stroke that meets the stepped leader is most likely to originate from such objects. Clearly, sitting under a tree during an electrical storm is not wise. What *should* you do?

When caught in a thunderstorm, the best protection, of course, is to get inside a building. If no such shelter exists, be sure to avoid elevated places

and isolated trees. If you are on level ground, try to keep your head as low as possible, but do not lie down. Because large amounts of charge often concentrate at the surface near the point of a lightning strike, a surface current may travel through your body and injure or kill you. Therefore, crouch down as low as possible and minimize the contact area you have with the ground. There are some warning signs to alert you to a strike. If your hair begins to stand on end, or your skin begins to tingle and you hear clicking sounds, beware—lightning may be about to strike. And if you are standing upright, you may be acting as a lightning rod.

and precise structure of a thunderstorm is obtained. In order to monitor lightning activity continuously over a broad area of the earth, geostationary satellites are, by the late 1980s, likely to be equipped with a *lightning mapper sensor*.

Each year, approximately 9000 fires are started by lightning in the United States alone and around $50 million worth of timber is destroyed. For this reason, tests have been conducted to see whether the number of cloud-to-ground lightning discharges can be reduced. One technique that has shown some success in suppressing lightning involves seeding a cumulonimbus cloud with hair-thin pieces of aluminum about 10 cm long. The idea is that these pieces of metal will produce many tiny sparks, called *corona discharges*, and prevent the electrical potential in the cloud from building to a point where lightning occurs. While the results of this experiment are inconclusive, many forestry specialists point out that nature itself may use a similar mechanism to prevent excessive lightning damage. The long, pointed

needles of pine trees may act as tiny lightning rods, diffusing the concentration of electric charges and preventing massive lightning strokes.

Up to now, we have examined thunderstorms and their associated thunder and lightning. We are now ready to explore a product of a thunderstorm that is one of nature's most awesome phenomena: the tornado, a rapidly spiraling column of air that can strike sporadically and violently.

Tornadoes

Tornadoes are rapidly rotating winds that blow around a small area of intense low pressure. Sometimes called *twisters* or *cyclones*, tornadoes nearly always begin as a funnel-shaped cloud that looks like an elephant's trunk hanging from a large cumulonimbus cloud (Fig. 19.17). The funnel cloud is called a tornado only after it touches the ground. When viewed from above, the majority of tornadoes rotate counterclockwise. A few have been seen rotating clockwise, but these are rare.

The diameter of most tornadoes is between

Experiencing Thunder and Lightning

Light travels so fast that we see light instantly after a lightning flash. But the sound of thunder, traveling at only about 330 m/sec (1100 ft/sec), takes much longer to reach the ear. If we start counting seconds from the moment we see the lightning until we hear the thunder, we can determine how far away the stroke is. Because it takes sound about 3 seconds to travel 1 kilometer (5 seconds for each mile), if we see lightning and hear the thunder 15 seconds later, the lightning stroke occurred 5 km (3 mi) away.

When the lightning stroke is very close—100 m (330 ft) or less—thunder sounds like a clap or a crack followed immediately by a loud bang. When it is farther away, it often rumbles. The rumbling can be due to the sound emanating from different areas of the stroke. In Fig. 2 the person standing several kilometers from the lightning will hear the thunder rumble because the sound wave from the upper part of the stroke travels a slightly longer distance than the wave near the ground and, hence, it will arrive later. The rumbling is accentuated when the sound wave reaches an observer after having bounced off obstructions, such as hills and buildings.

In some instances, lightning is seen but no thunder is heard. Does this mean that thunder was not produced by the lightning? Actually, there is thunder, but the atmosphere refracts (bends) and attenuates the sound waves, making the thunder inaudible. Sound travels faster in warm air than in cold air.* Because thunderstorms form in unstable air, where the temperature normally drops rapidly with height, sound waves usually travel faster in the warm air near the surface and bend upward, away from an observer at the surface. Notice in Fig. 3 that the observer 5 km away is able to hear the thunder, while the observer 20 km away is not.

Even when a viewer is as close as several kilometers to a lightning flash, thunder may not be heard. For one thing, the complex interaction of sound waves and air molecules tends to attenuate the thunder. In addition, turbulent eddies of air less than 50 m in diameter scatter the sound waves. Consequently, when thunder from a low-energy lightning flash travels several kilometers through turbulent air, it may become inaudible.

A sound occasionally mistaken for thunder is the **sonic boom**. Sonic booms are pro-

(continued on next page)

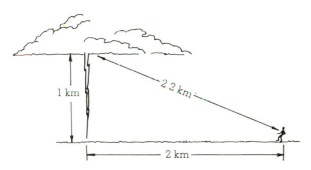

Fig. 2 Because the sound waves from the lower part of the lightning stroke reach the observer before those from the upper part of the stroke, the thunder appears to rumble.

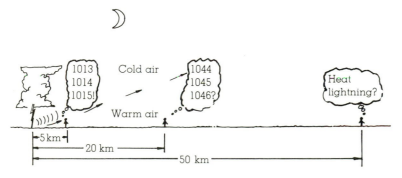

Fig. 3 When the air is warm near the surface and is cold aloft, sound waves bend upward. This bending causes thunder not to be heard when the observer is more than about 20 km (12 mi) from the lightning.

*The speed of sound in calm air is equal to $20\sqrt{T}$, where T is the air temperature in Kelvins. Also, when everything else is the same, the speed of sound increases with increasing wind speed.

FOCUS ON AN OBSERVATION, continued

duced when an aircraft exceeds the speed of sound at the altitude at which it is flying. The aircraft compresses the air, forming a shock wave that trails out as a cone behind the aircraft. Along the shock wave, the air pressure changes rapidly over a short distance. The rapid pressure change causes the distinct boom. (Exploding fireworks generate a similar shock wave and a loud bang.)

Lightning may take on a variety of shapes and forms. When a dart leader moving toward the ground deviates from the original path taken by the stepped leader, the lightning appears crooked or forked, and it is called *forked lightning. Ribbon lightning* forms when the wind moves the ionized channel between each return stroke, causing the lightning to appear as a ribbon hanging from the cloud. If the lightning channel breaks up, or appears to break up, the lightning looks like a series of

beads tied to a string. The actual cause of this *bead lightning* **is not known. However, one theory** proposes that it occurs when the lightning stroke is partially obscured by clouds or falling rain. *Ball lightning* looks like a luminous sphere that appears to float in the air or slowly dart about for several seconds. Although many theories have been proposed, the actual cause of ball lightning remains an enigma. *Sheet lightning* forms when either the lightning flash occurs inside a cloud, or intervening clouds obscure the flash, such that a portion of the cloud (or clouds) appears as a luminous white sheet. Distant lightning from thunderstorms that is seen but not heard is commonly called *heat lightning* because it frequently occurs on hot, summer nights when the overhead sky is clear. As the light from distant electrical storms is refracted through the atmosphere, air molecules and

fine dust scatter the shorter wavelengths of visible light, often causing heat lightning to appear orange to a distant observer.

As the electric field near the ground increases, a current of positive charge moves up pointed objects, such as antennas and masts of ships. However, instead of a lightning stroke, a luminous greenish or bluish halo may appear above them, as a continuous supply of sparks—a *corona discharge*—is sent into the air. This electric discharge, which can cause the top of a ship's mast to glow, is known as **St. Elmo's Fire**, named after the patron saint of sailors. St. Elmo's Fire is also seen around power lines and the wings of aircraft. When St. Elmo's Fire is visible and a thunderstorm is nearby, a lightning flash may occur in the near future, especially if the electric potential gradient of the atmosphere is increasing.

100 and 600 m, although some are just a few meters wide and others have diameters exceeding 1600 m (1 mi). Tornadoes that form ahead of an advancing cold front are often steered by southwesterly winds, and therefore tend to move from the southwest toward the northeast at speeds usually between 20 and 40 knots. However, some have been clocked at speeds greater than 70 knots. Most tornadoes last only a few minutes and have an average path length of about 7 km (4 mi). There are cases where they have traveled for hundreds of kilometers and have existed for many hours, such as the one that lasted over 7 hours and cut a path 470 km (292 mi) long through portions of Illinois and Indiana on May 26, 1917.

Each year, tornadoes take the lives of many people. The yearly average is about 100, although over 100 may die in a single day. The deadliest tornadoes are those that occur in *families*; that

is, different tornadoes spawned by the same thunderstorm. (Some thunderstorms produce a sequence of several tornadoes over two or more hours and over distances of 100 km or more.) Tornado families usually form along squall lines and often constitute what is termed a *tornado outbreak*. One of the most violent outbreaks ever recorded occurred on April 3 and 4, 1974. During a 16-hour period, 148 tornadoes cut through parts of 13 states, killing 307 people, injuring more than 6000, and causing an estimated $600 million in damage. Some of these tornadoes were among the most powerful ever witnessed. The combined path of all the tornadoes during this *superoutbreak* amounted to 4181 km (2598 mi), well over half of the total path for an average year. The greatest loss of life attributed to tornadoes occurred during the tri-state outbreak of March 18, 1925, when 746 people died as 7 tornadoes trav-

Fig. 19.17   Tornado photographed near Tracy, Minnesota, on June 13, 1968.

eled a total of 703 km (437 mi) across portions of Missouri, Illinois, and Indiana. In yet another outbreak, on Palm Sunday 1965, more than 30 tornadoes moved through 5 midwestern states, inflicting great damage and killing 256 people.

Tornado Occurrence   Tornadoes occur in many parts of the world, but no country experiences more tornadoes than the United States, which averages more than 700 annually. Although tornadoes have occurred in every state, including Alaska and Hawaii, the greatest number occur in the ''tornado belt'' of the Central Plains, which stretches from central Texas to Nebraska (Fig. 19.18). Within this region, Oklahoma has the dubious distinction of reporting more tornadoes per 10,000-square-mile area than any other state.*

The central plains region is most susceptible to tornadoes because it provides the proper atmospheric setting for the development of the severe

thunderstorms that spawn tornadoes. Here (especially in spring) warm, humid surface air is overlain by cooler, dryer air aloft, producing an unstable atmosphere. When a strong vertical wind shear exists and the surface air is forced upward, large thunderstorms capable of spawning tornadoes may form. Therefore, tornado frequency is highest during the spring and lowest during the winter when the warm surface air is absent.

About three-fourths of all tornadoes in the United States develop from March to July. The month of May normally has the greatest number of tornadoes (the average is about 4 per day) while the most violent tornadoes seem to occur in April, when horizontal and vertical temperature and moisture contrasts are greatest. Although tornadoes have occurred at all times of the day and night, they are most frequent in the late afternoon (between 4:00 P.M. and 6:00 P.M.), when the surface air is most unstable; they are least frequent in the early morning before sunrise, when the air is most stable.

Although large, destructive tornadoes are most common in the Central Plains; however, if condi-

*Many of the tornadoes that form along the Gulf Coast are generated by thunderstorms embedded within the circulation of hurricanes.

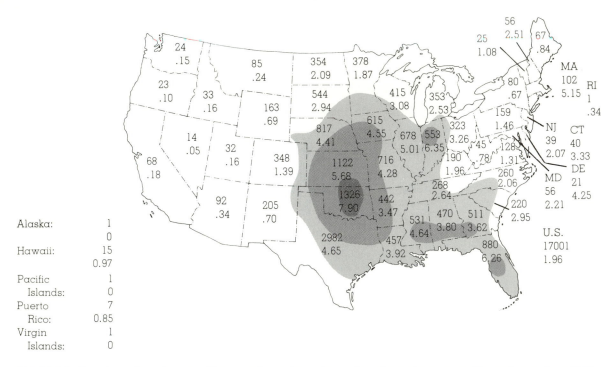

Alaska: 1
0

Hawaii: 15
0.97

Pacific
Islands: 1
0

Puerto
Rico: 7
0.85

Virgin
Islands: 1
0

Fig. 19.18 Tornado incidence by state. The upper figure shows the number of tornadoes reported by each state from 1953 through 1976. The lower figure is the average annual number of tornadoes per 10,000 square mi. The darker the shading, the greater the frequency of tornadoes.

tions are right, they can develop anywhere. For example, a series of at least 36 tornadoes, more typical of those that form over the plains, marched through North and South Carolina on March 28, 1984, claiming 59 lives and causing hundreds of millions of dollars in damage. One tornado was enormous, with a diameter of at least 4000 m (2.5 mi) and winds that exceeded 200 knots. No place is totally immune to a tornado's destructive force. On March 1, 1983, a rare tornado cut a 5-km swath of destruction through downtown Los Angeles, California, damaging more than 100 homes and businesses and injuring 33 people.

Even in the central part of the United States, the statistical chance that a tornado will strike a particular place this year is quite small. However, tornadoes can provide many exceptions to statistics. Oklahoma City, for example, has been struck by tornadoes at least 32 times in the past 90 years. And the little town of Codell, Kansas, was hit by tornadoes in 3 consecutive years—1916, 1917, and 1918—and each time on the same date: May 20! Considering the many millions of tornadoes that must have formed during the geological past, it is likely that at least one actually

moved across the land where your home is located, especially if it is in the Central Plains.

Tornado Winds The strong winds of a tornado can destroy buildings, uproot trees, and hurl all sorts of lethal missiles into the air. People, animals, and home appliances all have been picked up, carried several kilometers, then deposited. Tornadoes have accomplished some astonishing feats, such as lifting a railroad coach with its 117 passengers and dumping it in a ditch 25 m away. Showers of toads and frogs have poured out of a cloud after tornadic winds sucked them up from a nearby pond. Other oddities include chickens losing all of their feathers and pieces of straw being driven into metal pipes. Miraculous events have occurred, too. In one instance, a schoolhouse was demolished and the 85 students inside were carried over 100 m without one of them being killed.

No weather instrument has been able to withstand the furious winds of a tornado, so our knowledge of these winds comes mainly from observations of the damage done and the analysis of motion pictures. Because of the destructive nature of the tornado, it was once thought that it packed winds greater than 500 knots. However, studies conducted after 1973 reveal that even the most powerful twisters seldom have winds exceeding 220 knots, and most tornadoes probably have winds of less than 125 knots. Nevertheless, being confronted with even a small tornado can be terrifying.

The high winds of the tornado cause the most damage as walls of buildings buckle and collapse when blasted by the extreme wind force. Also, as high winds blow over a roof, lower air pressure forms above the roof. The greater air pressure inside the building then lifts the roof just high enough for the strong winds to carry it away. A similar effect occurs when the tornado's intense low-pressure center passes overhead. Because the pressure in the center of a tornado may be more than 100 mb (3 in.) lower than that of its surrounding, there is a momentary drop in outside pressure when the tornado is above the structure. The greater air pressure inside can push outward, literally causing the building to explode. To reduce this explosive pressure difference, open the windows and possibly the door, but only if there is time. Damage from tornadoes may also be inflicted on people and structures by flying debris. Hence, the wisest course to take when confronted with an approaching tornado is to seek shelter immediately.

When a tornado is approaching from the southwest, its strongest winds are on its southeast side. We can see why in Fig. 19.19. The tornado is heading northeast at 50 knots. If its rotational speed is 100 knots, then its forward speed will add 50 knots to its southeastern side (position

D) and subtract 50 knots from its northwestern side (position A). Because the most destructive and extreme winds will be on the tornado's southeastern side, it is the southwest side of the building that will receive the maximum impact of the winds. Hence, in a large building without a basement, the safest place to be is on the lowest floor in a northwest corridor.

It now appears that the most violent tornadoes (with winds exceeding 180 knots) contain smaller whirls that rotate within them. Such tornadoes are called *multi-vortex tornadoes* and the smaller whirls are called **suction vortices** (Fig. 19.20). Suction vortices are only about 10 m (30 ft) in diameter, but they rotate very fast and apparently do a great deal of damage.

In the late 1960s, Dr. T. Theodore Fujita, a noted authority on tornadoes at the University of Chicago, proposed a scale for classifying tornadoes according to their rotational wind speed and the damage done by the storm. Table 19.1 gives this scale. A study conducted in 1980 by the staff of the National Severe Storms Forecast Center in Kansas City, Missouri, showed that of all the tornadoes reported between 1950 and 1980, about two-thirds were F0 and F1 (weak tornadoes) and only 2 percent were above the F3 classification (violent). However, this study also showed that it was these latter powerful storms that accounted for over two-thirds of all tornado-related deaths.

Tornado Formation Although everything is not known about the formation of a tornado, we do know that tornadoes tend to form with severe thunderstorms and that unstable air is essential for their development. One atmospheric situation

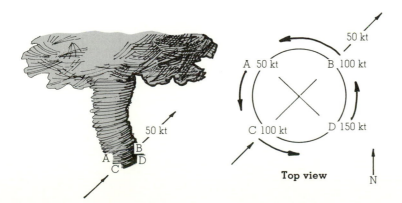

Fig. 19.19 The total wind speed of a tornado is greater on one side than on the other. When facing an on-rushing tornado, the strongest winds will be on your left side.

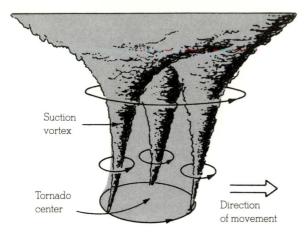

Fig. 19.20 A powerful multi-vortex tornado.

that frequently leads to severe thunderstorms with tornadoes in the spring is shown in Fig. 19.21.

At the surface, we find an open wave middle latitude cyclone with its cold front and warm front. Behind the cold front is modified mP Pacific air which, having crossed several mountain ranges, is now relatively cold and dry. (It may also be cP air penetrating southward, east of the Rockies.) Ahead of the advancing cold front, warm and humid mT air pushes northward from the Gulf of Mexico. Above this warm sector at 850 mb, a wedge of warm, moist air is streaming northward. Directly above the moist layer, between 700 mb and 500 mb, is a wedge of cold, dry air—called a *dry tongue*—moving in from the southwest. At 500 mb, a trough of low pressure exists to the west of the surface storm and at 300 mb the polar front jet stream swings over the region. At this level, the jet stream takes air away so quickly that air from below is drawn up to replace it. The stage is now set for the development of severe storms.

The position of cold air above warm air produces an unstable atmosphere. As the warm air rises from the surface, many thunderstorms should form throughout the warm sector. But often this is not the case as the atmospheric conditions that generate severe thunderstorms tend to prevent many smaller ones from forming. To see why, we will examine the vertical profile of temperature and moisture—a *sounding*—in the warm sector.

Figure 19.22 shows a typical sounding of temperature and dew point in the warm sector before tornadoes occur. From the surface up to 800 mb, the conditionally unstable air is warm and very humid. At 800 mb, a shallow inversion acts like a cap on the moist layer below. Above the inversion, the air is cold and much drier. This air is also conditionally unstable, as the temperature drops at just about the dry adiabatic rate (10°C/1000 m). The cooling of this upper layer is due, mainly, to cold advection. Cold, dry, unstable air sitting above a warm, humid layer produces convective instability, which means that the atmosphere will become even more unstable if a layer of air is somehow forced upward (see Chapter 10).

The lifting of warm surface air can occur at the frontal zones, but the air may also begin to rise anywhere in the warm sector when the surface air heats up during the day. However, in the morning, the inversion acts as a lid on rising thermals and only small cumulus clouds form. As the day progresses (and the surface air heats even more), rising air breaks through the inversion at isolated places and clouds build rapidly, sometimes explosively, as the moist air is vented upward through the opening. Thus, we can see that the inversion prevents many small thunderstorms from forming. When the surface air is finally able to puncture the inversion, the upper-level jet stream rapidly draws the moist air up into the cold, unstable air, and a large storm quickly develops. The severe

Table 19.1 Fujita Scale for Damaging Wind

SCALE	CATEGORY	MI/HR	KNOTS	EXPECTED DAMAGE
F0	Weak	40–72	35–62	light: tree branches broken, sign boards damaged
F1		73–112	63–97	moderate: trees snapped, windows broken
F2	Strong	113–157	98–136	considerable: large trees uprooted, weak structures destroyed
F3		158–206	137–179	severe: trees leveled, cars overturned, walls removed from buildings
F4	Violent	207–260	180–226	devastating: frame houses destroyed
F5		261–318	227–276	incredible: structures the size of autos moved over 100 m, steel-reinforced structures highly damaged

thunderstorm soon builds to the tropopause, and an anvil forms. However, the strong updraft inside the storm causes its top to overshoot the anvil by 2 to 4 km.

In order for the storm to spawn a tornado, the updraft must rotate. Remember from our earlier discussion that severe thunderstorms form in a region of strong vertical wind shear. In Fig. 19.21, the rapidly increasing wind speed with height (vertical wind speed shear) and the changing wind direction with height—from southerly at low levels to westerly at high levels (vertical wind direction shear)—cause the updraft inside the storm to rotate cyclonically. This rising, spinning column of air, perhaps 5 to 10 km across, is called a **mesocyclone**. Doppler radar observations have shown that rotation begins in the middle of the thunderstorm and gradually works downward. As air rushes in toward the low pressure of the mesocyclone, the rotational wind speeds increase due to the conservation of angular momentum (see Chapter 15). At the same time, the mesocyclone stretches vertically and shrinks horizontally, and the spinning air is accelerated upward. Inside the mesocyclone, which is now between 2 and 4 km wide, a spinning vortex of increasing wind speed (a tornado) may—for reasons not fully understood—appear near the midlevel of the cloud and gradually extend downward to the cloud base. At this point, the mesocyclone is called a *tornado cyclone*.

As air rushes into the low-pressure vortex from all directions, it expands, cools, and, if sufficiently moist, condenses into a visible cloud—the **funnel cloud**. As the air beneath the funnel is drawn into the core, it cools rapidly and condenses, and the funnel cloud descends toward the surface. Upon reaching the ground it is called a *tornado*. Here, it usually picks up dirt and debris, making it appear both dark and ominous. While the air along the outside of the funnel is spiraling upward, Doppler radar reveals that, within the core of violent tornadoes, the air is descending toward the extreme low pressure at the ground. As the air descends, it warms, causing the cloud droplets to evaporate. This process leaves the core free of clouds. At the surface, where the descending air in the core meets the air flowing into the funnel, suction vortices are formed as the combined flows spin and turn rapidly upward.

Observations reveal that the most strong and violent tornadoes develop near the right rear sec-

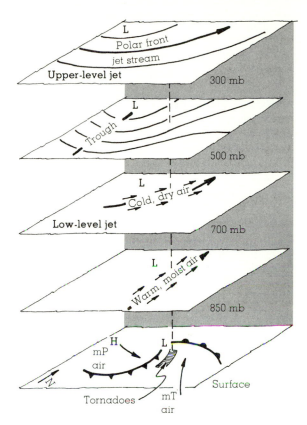

Fig. 19.21 Conditions leading to the formation of severe thunderstorms that can spawn tornadoes.

tor of a severe thunderstorm (on the southwestern side of an eastward moving storm). However, weaker tornadoes may not only develop in the main updraft, but along the gust front, where the cool downdraft forces warm inflowing air upward. Al-

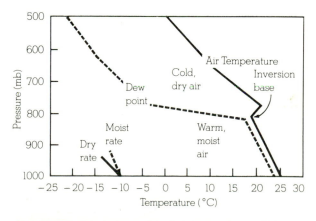

Fig. 19.22 A typical sounding of air temperature and dew point that frequently precede the development of severe thunderstorms that spawn tornadoes.

FOCUS ON INSTRUMENTS

Doppler Radar

Most of what is known about the air motions inside a tornado-generating thunderstorm has been gained by a measurement technique called *Doppler radar*. Before we investigate this remote-sensing device, we will examine how precipitation inside a severe thunderstorm appears on the scope of conventional radar.

Remember from Chapter 11 that a conventional radar transmitter sends out microwave pulses and that, when this energy strikes an object, a small fraction is scattered back to the antenna. Because precipitation particles are large enough to bounce microwaves back to the antenna, the white area on the radar scope in Fig. 4 represents precipitation inside a severe thunderstorm. Notice that the pattern is in the shape of a hook. A **hook-shape echo** such as this indicates the possible presence of a tornado. The dark area within the hook echo represents the region inside the severe thunderstorm, where strong updrafts carry cloud particles upward so rapidly that they are unable to grow large enough to

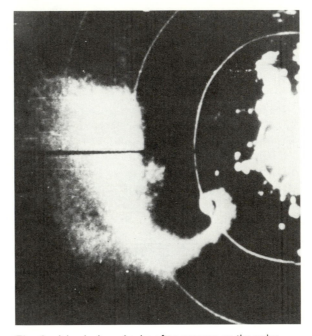

Fig. 4   A hook-shaped echo often appears on the radar screen with tornado-spawning thunderstorms.

reflect microwaves. When tornadoes form, they do so near the tip of the hook. However, many severe thunderstorms (as well as smaller ones) do not show a hook echo, but still spawn tornadoes. Sometimes, when the hook echo does appear, the tornado

is already touching the ground. Consequently, a better technique was needed in detecting tornado-producing storms. To answer this need, Doppler radar was developed.

Doppler radar is like a conventional radar in that it can de-

though it appears that most strong and violent tornadoes form within the mesocyclone, not all mesocyclones produce tornadoes. Certainly, all the processes that go into generating a tornado are not completely understood at this time.

While it is difficult to tell which thunderstorm will spawn a tornado, meteorologists can predict where tornado-generating storms are most likely to form. Notice in Fig. 19.21 that this area (the boxed-off area on the surface map) is situated where the polar front jet stream and the cold tongue of air cross the wedge of warm, moist air. Knowing this helps to explain why the region of

greatest tornado activity shifts northward from winter to summer. During the winter, tornadoes are most likely to form over the southern Gulf states. The polar front jet is above this region, and the contrast between warm and cold air masses is greatest. In spring, humid Gulf air surges northward; contrasting air masses and the jet stream also move northward and tornadoes become more prevalent from the southern Atlantic states westward into the southern Great Plains. In summer, the contrast between air masses lessens, and the jet stream is normally near the Canadian border; hence, tornado activity tends to be concentrated

FOCUS ON INSTRUMENTS

tect areas of precipitation and measure the speed of falling precipitation. But a Doppler radar can do more—it can actually measure the speed at which precipitation is moving horizontally toward or away from the radar antenna. Because precipitation particles are carried by the wind, Doppler radar can peer into a severe storm and unveil its winds.

Doppler radar works on the principle that, as precipitation moves toward or away from the antenna, the returning radar pulse will change in frequency. A similar change occurs when the high-pitched sound (high frequency) of an approaching noise source, such as a siren or train whistle, becomes lower in pitch (lower frequency) after it passes by the person hearing it. This change in frequency is called the *Doppler shift* and this, of course, is where the Doppler radar gets its name.

A single Doppler radar cannot detect winds that blow parallel to the antenna. Consequently, two or more units probing the same thunderstorm are able to give a three-dimensional picture of the winds within the storm. To help distinguish the

storm's air motions, wind velocities can be displayed in color. Color contouring the wind field gives a good picture of the storm. (See color plate 33.)

Even a single Doppler radar can uncover many of the features of a severe thunderstorm. For example, studies conducted in the 1970s revealed, for the first time, the existence of the swirling winds of the *mesocyclone* inside tornado-producing thunderstorms. Mesocyclones have a distinct image (signature) on the radar display. Tornadoes also have a distinct signature, known as the *tornado vortex signature* (*TVS*), which shows up as a region of rapidly changing wind speeds within the mesocyclone. Unfortunately, the resolution of the Doppler radar is not high enough to measure actual wind speeds of most tornadoes, whose diameters are only a few hundred meters or less. However, a new and experimental Doppler system—called *Doppler lidar*—uses a light beam (instead of microwaves) to measure the change in frequency of falling precipitation, cloud particles, and dust. Because it uses a shorter wavelength of radiation, it has a narrower beam and a

higher resolution than does Doppler radar.

By the early 1990s, the National Weather Service plans to install a network of Doppler radar units at selected weather stations. Detecting the signatures of mesocyclones and tornadoes with these units will assist forecasters in determining which severe thunderstorms will likely spawn tornadoes. In addition, they should give advanced and improved warning of an approaching tornado. More reliable warnings, of course, should cut down on the number of false alarms.

Because the Doppler radar shows air motions within a storm, it can identify the magnitude of other severe weather phenomena, such as gust fronts, microbursts, and wind shears that are dangerous to aircraft. Certainly, as Doppler radar becomes part of the major radar network, our understanding of the processes that generate severe thunderstorms will be enhanced, and hopefully there will be an even better tornado and severe storm-warning system, resulting in fewer deaths and injuries.

from the northern plains eastward to New York State.

Observing Tornadoes   The appearance of mammatus clouds at the base of a severe thunderstorm may indicate that the storm is capable of producing tornadoes. These bulging pouches suggest vertical motion (see Chapter 10) and may extend well below the cloud base, as shown in Fig. 19.23. Mammatus clouds are not funnel clouds because they do not rotate. The first sign that the thunderstorm is about to produce a tornado is the sight of *rotating* clouds at the base of the storm.

If the area of rotating clouds lowers, it becomes a **wall cloud** (Fig. 19.24). Usually within the wall cloud a smaller rapidly rotating funnel extends toward the surface. Sometimes the air is so dry that the swirling wind remains invisible until it reaches the ground and begins to pick up dust. Unfortunately, people have mistaken these "invisible tornadoes" for dust devils, only to find out (often too late) that they were not. Occasionally, the funnel cannot be seen due to falling rain, clouds of dust, or darkness. Even when not clearly visible, many tornadoes have a distinctive roar that can be heard for several kilometers. This sound, which has been described as "a roar like a thousand freight trains," appears to be loudest when the tornado is touching the surface. However, not all tornadoes make this sound and, when these storms strike, they become silent killers.

When tornadoes are likely to form during the next few hours, a **tornado watch** is issued by the National Severe Storms Forecast Center in Kansas City, Missouri, to alert the public that tornadoes may develop within a specific area during a certain time period. Many communities have trained volunteer spotters, who look for tornadoes after the watch is issued. Once a tornado is spotted—either visually or on a radar screen—a **tornado warning** is issued by the local National Weather Service Office. In some communities, sirens are sounded to alert people of the approaching storm. Radio and television stations interrupt regular programming to broadcast the warning. Although not completely effective, this warning system is apparently saving many lives. Despite the large increase in population in the tornado belt during the past twenty years, tornado-related deaths have actually shown a slight decrease.

In an attempt to unravel some of the mysteries of the tornado, several studies are underway. In one—conducted jointly by the University of Oklahoma and the National Severe Storms Laboratory in Norman, Oklahoma—still and motion pictures of tornadoes are correlated with radar data in an attempt to better estimate wind speeds. In another, a sturdy instrument package resembling a barrel with an antenna (called *TOTO*—for *To*table *To*rnado *O*bservatory) is placed in the path of an approaching mesocyclone in hope of determining surface conditions such as winds, temperature, pressure, and the electric field strength beneath a tornado cyclone. The readings are recorded in TOTO's tornado-proof shell, and read after the storm passes. On the more theoretical side, numerical cloud modeling studies are offering new insights into the formation and development of tornado-breeding thunderstorms.

Waterspouts When a tornado passes over a large body of water it is called a **tornadic waterspout**. However, **"fair weather" waterspouts** may form over water, especially above the warm, shal-

Fig. 19.23 A tornado sky. The pouches of these mammatus clouds almost touch the tops of buildings in Greeley, Colorado, on a day when several tornadoes were sighted.

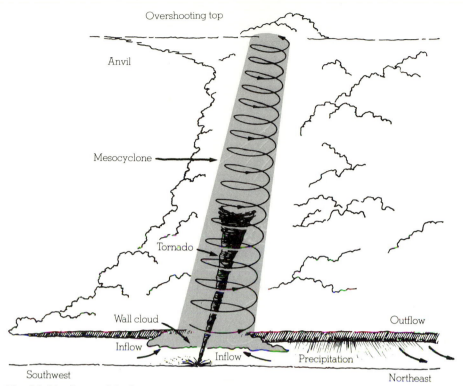

Fig. 19.24 Some of the features associated with a tornado-breeding thunderstorm as viewed from the southeast. The thunderstorm is moving to the northeast. The tornado forms in the southwest part of the thunderstorm.

low coastal waters of the Florida Keys, where almost 100 occur each month during the summer. Here, waterspouts average between 3 and 100 m in diameter. Fair-weather waterspouts are much smaller than an average tornado. They are also less intense, as their rotating winds are typically less than 45 knots. Waterspouts tend to move more slowly than tornadoes and they only last for 10 to 15 minutes, although some have existed for up to one hour.

Waterspouts tend to form when the air is unstable and clouds are developing. Unlike the tornado, they do not need a severe thunderstorm to generate them. Some form with small thunderstorms, but most form with developing cumulus congestus clouds whose tops are frequently no higher than 3600 m (12,000 ft) and do not extend to the freezing level. Apparently, the warm, humid air near the water helps to create atmospheric instability, and the updraft beneath the resulting cloud helps initiate uplift of the surface air.

The waterspout funnel is similar to the tornado funnel in that both are clouds of condensed water vapor with converging winds that rise about a central core. Contrary to popular belief, the waterspout does not draw water up into its core; however, swirling spray may be lifted several meters when the waterspout funnel touches the water. Apparently, the most destructive waterspouts are those that begin as tornadoes over land, then move over water. A photograph of a particularly well-developed and intense waterspout near the Florida Keys is shown in Fig. 19.25.

Summary

In this chapter, we have examined thunderstorms and the atmospheric conditions that produce them. The ingredients for an isolated air-mass thunderstorm are humid surface air, plenty of sunlight to heat the ground, and an unstable atmosphere. When these conditions prevail, small cumulus clouds may grow into towering clouds and thunderstorms within 20 minutes.

When conditions are ripe for thunderstorm development and a strong vertical wind shear exists,

Fig. 19.25 A waterspout over the warm waters of the Florida Keys.

the stage is set for the generation of severe thunderstorms. Supercell thunderstorms may exist for many hours, as their updrafts and downdrafts are nearly in balance. Thunderstorms that form in a line, especially ahead of an advancing cold front, are called a squall line.

Lightning is a discharge of electricity that occurs in mature thunderstorms. The lightning stroke momentarily heats the air to an incredibly high temperature. The rapidly expanding air produces a sound called thunder. Along with lightning and thunder, severe thunderstorms produce violent weather, such as destructive hail, downbursts, and the most feared of all atmospheric storms—the tornado.

Tornadoes are rapidly rotating columns of air that extend downward from the base of a thunderstorm. Most tornadoes are less than a few hundred meters wide with wind speeds less than 100 knots, although violent tornadoes may have wind speeds that exceed 250 knots. A violent tornado may actually have smaller whirls (suction vortices) rotating within it. With the aid of Doppler radar, scientists are probing tornado-spawning thunderstorms, hoping to better predict tornadoes and to better understand where, when, and how they form.

A normally small and less destructive cousin of the tornado is the "fair weather" waterspout that commonly forms above the warm waters of the Florida Keys.

Questions for Review

1. What is a thunderstorm?

2. Describe how a cumulus cloud grows into an air-mass thunderstorm.

3. How do downdrafts form in thunderstorms?

4. What is necessary for a multicell thunderstorm to form?

5. Why do air-mass thunderstorms most frequently form in the afternoon?

6. Explain why air-mass thunderstorms tend to dissipate much sooner than severe thunderstorms.

7. What is a gust front? What type of weather does it bring when it passes?

8. What is a downburst? a microburst?

9. Why are severe thunderstorms not very common in polar latitudes?

10. Give two possible explanations for the generation of prefrontal squall-line thunderstorms.

11. What is a Mesoscale Convective Complex?

12. Where does the highest frequency of thunderstorms occur in the United States? Why there?

13. Why is large hail more common in Kansas than in Florida?

14. Describe one process by which thunderstorms become electrified.

15. Explain how a cloud-to-ground lightning flash develops.

16. Why is it dangerous to seek shelter under a tree during a thunderstorm?

17. How is thunder produced?

18. If you see lightning and 10 seconds later you hear thunder, how far away is the lightning stroke?

19. What is heat lightning? sheet lightning? St. Elmo's fire?

20. What are tornadoes? Give some average statistics about tornado size, winds, and direction of movement.

21. Why is the central part of the United States more susceptible to tornadoes than any other region of the world?

22. Describe how Doppler radar measures the winds inside a severe thunderstorm. Why is Doppler radar superior to conventional radar for this purpose?

23. Describe the atmospheric conditions at the

surface and aloft that are responsible for the development of the majority of tornado-spawning thunderstorms.

24. Explain both how and why there is a shift in tornado activity from winter to summer.

25. Define the following terms: (a) suction vortice; (b) wall cloud; (c) mesocyclone; and (d) funnel cloud.

26. What conditions lead to the formation of waterspouts?

Questions for Thought

1. Why does the bottom half of a dissipating thunderstorm usually "disappear" before the top?

2. Sinking air warms, yet the downdrafts in a thunderstorm are cold. Why?

3. Explain why squall-line thunderstorms often form ahead of advancing cold fronts but seldom behind them.

4. On a house without a lightning rod, where do you feel lightning would most likely strike?

5. Why is the old adage "lightning never strikes twice in the same place" wrong?

6. Suppose while you are on a high mountain ridge a thundercloud passes overhead. What would be the wisest thing to do—stand upright? lie down? or crouch? Explain.

7. If you are confronted in an open field by a large tornado and there is no way that you could outrun it, probably the only thing that you could do would be to run and lie down in a depression. If given the choice, would you run toward your right or left as the tornado approaches? Explain your reasoning.

8. Tornadoes apparently form in the region of a strong updraft, yet they descend from the base of a cloud. Why?

9. In the Fujita scale for classifying tornadoes, into which category would the majority of waterspouts fall?

Problems and Exercises

1. On a map of the United States, place the surface weather conditions (air masses, fronts, and so on) that are necessary for the formation of most tornadoes.

2. On the surface weather map below (Fig. 19.26), at which number would you most likely observe a line of thunderstorms forming? Explain why you chose that location.

3. A multi-vortex tornado with a rotational wind speed of 125 knots is moving from southwest to northeast at 30 knots. Assume the suction vortices within this tornado have rotational winds of 100 knots:

(a) What is the maximum wind speed of this multi-vortex tornado?

(b) If you are facing the approaching tornado, on which side (northeast, northwest, southwest, or southeast) would the strongest winds be found? the weakest winds? Explain both of your answers.

(c) According to the Fujita scale (Table 19.1), how would this tornado be classified?

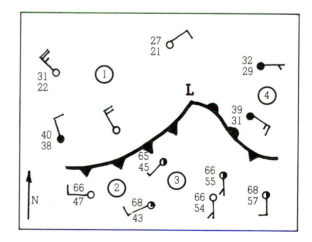

Fig. 19.26 Surface weather map for Problem 2.

Hurricane Gladys, about 150 km southwest of Tampa, Florida, as photographed from the *Apollo* 7 spacecraft. With surface winds of 80 knots and a central pressure near 986 mb, Gladys rates as a 1 on the Saffir-Simpson hurricane scale. (Photo: NASA)

CHAPTER 20

Hurricanes

Born over warm tropical waters and nurtured by a rich supply of water vapor, the *hurricane* can grow into a ferocious storm that generates enormous waves, heavy rains, and winds that may exceed 150 knots. What exactly are hurricanes? How do they form? And why do they strike the east coast of the United States more frequently than the west coast? These are some of the questions we will consider in this chapter.

Tropical Weather

In the broad belt around the earth known as the tropics—the region 23½° north and south of the equator—the weather is much different from that of the middle latitudes. In the tropics, the noon sun is always high in the sky, and so diurnal and seasonal changes in temperature are small. The daily heating of the surface and high humidity favor the development of cumulus clouds and afternoon thunderstorms. Most of these are individual thunderstorms that are not severe. However, if a jet stream should exist above the developing storm, a strong vertical wind shear can cause the updraft to tilt, producing a more violent squall-line type thunderstorm similar to those that form in middle latitudes.

As it is warm all year long in the tropics, the weather is not characterized by four seasons which, for the most part, are determined by temperature variations. Rather, most of the tropics are marked by seasonal differences in precipitation. The greatest cloudiness and precipitation occur during the high-sun period, when the intertropical convergence zone moves into the region. Even during the dry season, precipitation can be irregular, as

Contents

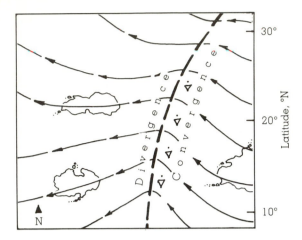

Fig. 20.1 An easterly wave as shown by the bending of streamlines. (The heavy dashed line is the axis of the trough.) The wave moves slowly westward, bringing fair weather on its western side and showers on its eastern side.

periods of heavy rain, lasting for several days, may follow an extremely dry spell.

The winds in the tropics generally blow from the east, northeast, or southeast. Because the variation of sea level pressure is normally quite small, drawing isobars on a weather map provides little useful information. Instead of isobars, *streamlines* that depict windflow are drawn. Streamlines are useful because they show where surface air converges and diverges. Occasionally, the streamlines will be disturbed by a weak trough of low pressure called an **easterly wave** (Fig. 20.1).

Easterly waves have wavelengths on the order of 2500 km (1550 mi) and travel from east to west at speeds between 10 and 20 knots. Look at Fig. 20.1 and observe that, on the western side of the trough (heavy dashed line), where easterly and northeasterly surface winds diverge, sinking air produces generally fair weather. On its eastern side, where the southeasterly winds converge, rising air generates showers and thunderstorms. Consequently, the main area of showers form *behind* the trough. Occasionally, an easterly wave will intensify and grow into a hurricane.

Anatomy of a Hurricane The **hurricane** is an intense storm of tropical origin, with sustained winds exceeding 64 knots (74 mi/hr), which forms over the warm northern Atlantic and eastern north Pacific Oceans. This same type of storm is given

different names in different regions of the world. In the western North Pacific, it is called a *typhoon*, in the Philippines a *baguio* (or a typhoon), and in India and Australia a *cyclone*. For simplicity, we will refer to all of these storms as hurricanes.

Figure 20.2 is a satellite picture of hurricane Fico situated in the eastern Pacific on July 14, 1978. The storm is approximately 600 km (375 mi) in diameter, which is about average for hurricanes. The area of broken clouds at the center is its **eye**. Fico's eye is almost 80 km (50 mi) wide, somewhat larger than the average 20 to 50 km eye diameter. Within the eye, winds are light and clouds are mainly broken. The surface air pressure is very low, nearly 955 mb.* The dark blotches in the eye are regions where the sky is clear. Notice that the clouds align themselves in bands (called *rain bands*) that spiral in toward the storm's center where they wrap themselves around the eye. Surface winds increase in speed as they blow counterclockwise and inward toward this center. Adjacent to the eye is the **eye wall**, a ring of intense thunderstorms that whirl around the storm's center and extend upward to almost 15 km above sea level. Within the eye wall, we find the heaviest precipitation and the strongest winds, which, in this storm, are 110 knots, with peak gusts of 125 knots.

If we were to venture from west to east (left to right) through the storm in Fig. 20.2, what might we experience? As we approach the hurricane, the sky becomes overcast with cirrostratus clouds; barometric pressure drops slowly at first, then more rapidly as we move closer to the center. Winds blow from the north and northwest with ever-increasing speed as we near the eye. The high winds, which generate huge waves over 10 m (33 ft) high, are accompanied by heavy rainshowers. As we move into the eye, the air temperature rises, winds slacken, rainfall ceases, and the sky brightens, as middle and high clouds appear overhead. The barometer is now at its lowest point (955 mb), some 60 mb lower than the pressure measured on the outskirts of the storm. The brief respite ends as we enter the eastern region of the eye wall. Here, we are greeted by heavy rain and strong southerly winds. As we move away

*An extreme low pressure of 870 mb (25.70 in.) was recorded in typhoon Tip during October, 1979.

Fig. 20.2 Hurricane Fico. (Visible satellite picture.)

from the eye wall, the pressure rises, the winds diminish, the heavy rain lets up, and eventually the sky begins to clear.

This brief, imaginary venture raises many unanswered questions. Why, for example, is the surface pressure lowest at the center of the storm? And why is the weather clear almost immediately outside the storm area? To help us answer such questions, we need to look at a vertical view, a profile of the hurricane along a slice that runs directly through its center. Of course, this view would be almost impossible to obtain in reality. However, a model that describes such a profile is given in Fig. 20.3.

The model shows that the hurricane is composed of an organized mass of thunderstorms that are an integral part of the storm's circulation. Near the surface, moist tropical air flows in toward the hurricane's center. Adjacent to the eye, this air rises and condenses into huge thunderstorms

Fig. 20.3 A model that shows a vertical view of air motions, clouds, and precipitation in a typical hurricane.

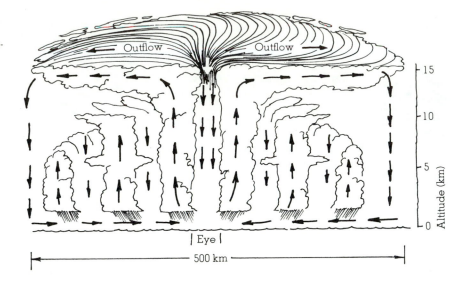

that produce heavy rainfall, as much as 25 cm (10 in.) per hour. Near the top of the thunderstorms, the relatively dry air, having lost much of its moisture, begins to flow outward away from the center. This diverging air aloft actually produces a clockwise (anticyclone) flow of air several hundred kilometers from the eye. As this outflow reaches the storm's periphery, it begins to sink and warm, inducing clear skies. In the vigorous thunderstorms of the eye wall, the air warms due to the release of large quantities of latent heat. This produces slightly higher pressures aloft, which initiate downward air motion within the eye. As the air subsides, it warms by compression. This helps to account for the warm air and the absence of thunderstorms in the center of the storm.

As surface air rushes in toward the region of much lower surface pressure, it should expand and cool, and we might expect to observe cooler air around the eye, with warmer air further away. But, apparently, so much heat is added to the air from the warm ocean surface that the surface air temperature remains fairly uniform throughout the hurricane.

We are now left with an important question: Where and how do hurricanes form? Although not everything is known about their formation, it is known that certain necessary ingredients are required before a weak tropical disturbance will develop into a full-fledged hurricane.

Hurricane Formation and Dissipation Hurricanes form over tropical waters where the winds are light and the surface water temperature is uniform over a vast area, typically 26°C (79°F) or greater. Over the tropical and subtropical North Atlantic and North Pacific Oceans these conditions prevail in summer and early fall; hence, the hurricane season normally runs from June through November.*

For a mass of unorganized thunderstorms to develop into a hurricane, the surface winds must converge. In the Northern Hemisphere, converging air spins counterclockwise. Because this type of rotation will not develop on the equator where the Coriolis force is zero (see Chapter 12), hurricanes form in subtropical regions, usually between 5° and 20° latitude. Convergence may occur along a front that has moved into the tropics from middle latitudes. Although the temperature contrast between the air on both sides of the front is gone, developing thunderstorms and converging surface winds may form, especially when the front is accompanied by a cold upper-level trough.

We know from Chapter 15 that the surface winds converge along the intertropical convergence zone (ITCZ). Occasionally, when the ITCZ is displaced away from the equator, a wave in the

*During December, 1983, hurricane Winnie formed over the still warm eastern tropical Pacific Ocean.

ITCZ forms into an area of low pressure, convection becomes organized, and the system grows into a hurricane. Weak convergence also occurs on the eastern side of an easterly wave, where hurricanes have been known to form. However, only a small fraction of all of the tropical disturbances that form over the course of a year ever grow into hurricanes.

Even when all of the surface conditions appear near perfect for the formation of a hurricane (e.g., warm water, converging winds, and so forth), the storm may not develop if the weather conditions aloft are not just right. For example, in the region of the trade winds and especially near latitude 20°, the air is often sinking due to the subtropical high. The sinking air warms and creates an inversion known as the **trade wind inversion**. When the inversion is strong it can inhibit the formation of intense thunderstorms and hurricanes. Also, hurricanes do not form where the upper-level winds are strong. Strong winds tend to disrupt the organized pattern of convection and disperse the heat, which is necessary for the growth of the storm. In fact, stronger than normal winds over the tropical North Atlantic may have attributed to the exceptionally low number of hurricanes that formed over this region during 1982 and 1983.

For hurricanes to form, the thunderstorms must become organized so that the latent heat that drives the system can be confined to a limited area. If thunderstorms start to organize along the ITCZ or along an easterly wave and if the trade wind inversion is weak, the stage may be set for the birth of a hurricane. The likelihood of hurricane development is enhanced if the air aloft is unstable. Such instability can be brought on when a cold upper-level trough from middle latitudes moves over the storm area. When this happens, the cumulonimbus clouds are able to build rapidly and grow into enormous thunderstorms. (See Fig. 20.4.)

Although the upper air is initially cold, it warms rapidly due to the huge amount of latent heat released during condensation.* As this cold air is

*Estimates are that the latent heat released in a mature hurricane in one day is equivalent to the energy released by 400 twenty-megaton hydrogen bombs. If this energy were converted to electricity, it would be enough to supply the needs of the United States for half a year.

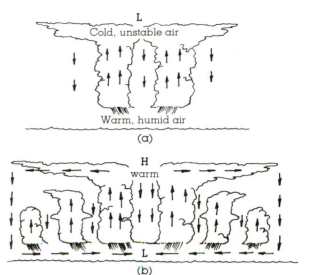

Fig. 20.4 Development of a hurricane. (a) Cold air above an organized mass of tropical thunderstorms generates unstable air and large cumulonimbus clouds. (b) The release of latent heat warms the upper troposphere, creating an area of high pressure. Upper-level winds move outward away from the high. This, coupled with the warming of the air layer, causes surface pressures to drop. As air near the surface moves toward the lower pressure, it converges, rises, and fuels more thunderstorms. Soon a chain reaction develops, and a hurricane forms.

transformed into much warmer air, the air pressure above the developing storm rises, producing an area of high pressure (see Chapter 2). Now the air aloft begins to move outward away from the region of developing thunderstorms. This diverging air aloft, coupled with warming of the air layer, causes surface pressure to drop, and a small area of surface low pressure forms. The surface air begins to spin counterclockwise and in toward the region of low pressure. As it moves inward, its speed increases due to the conservation of angular momentum (see Chapter 15). The winds then generate rough seas, which increase the friction on the moving air. This causes the winds to converge and ascend about the center of the storm. We now have a chain reaction in progress, or what meteorologists call a *feedback mechanism*. The rising air, having picked up added moisture and warmth from the choppy sea, fuels more thunderstorms and releases more heat, which causes the surface pressure to lower even more. The lower pressure near the center creates a greater friction, more convergence, more rising air, more thunderstorms, more heat, lower surface pressure, stronger winds, and so on until a full-blown hurricane is born.

As long as the upper-level outflow of air is greater than the surface inflow, the storm will intensify and the surface pressure will drop. Because the air pressure within the system is controlled to a large extent by the warmth of the air, the storm will intensify only up to a point. The controlling factors are the temperature of the water and the release of latent heat. Consequently, when the storm is literally full of thunderstorms, it will use up just about all of the available energy, so that air temperature will no longer rise and pressure will level off. Because there is a limit to how intense the storm can become, peak wind gusts seldom exceed 200 knots. When the converging surface air near the center exceeds the outflow at the top, surface pressure begins to increase, and the storm dies out.

If the hurricane remains over warm water, it may survive for a long time. Hurricane Fico traveled for thousands of kilometers over warm Pacific waters and maintained hurricane force winds for 17 days. Hurricane Ginger (1971) remained a hurricane for 20 of its 30-day life span. However, most hurricanes last for less than a week; they weaken rapidly when they travel over colder water and lose their heat source. They also dissipate rapidly over land. Here, not only is their energy source removed, but their winds decrease in strength (due to the added friction) and blow into the center, causing the central pressure to rise.

Hurricanes go through a set of stages from birth to death. Initially, the mass of thunderstorms with only a slight wind circulation is known as a **tropical disturbance**. The tropical disturbance becomes a **tropical depression** when the winds increase to between 20 and 34 knots and several closed isobars appear about its center on a surface weather map. When the isobars are packed together and the winds are between 35 and 64 knots, the tropical depression becomes a **tropical storm**. The tropical storm is classified as a *hurricane* only when its winds exceed 64 knots (74 mi/hr).

Figure 20.5 shows four tropical systems in various stages of development. Moving from east to west, we see a weak tropical disturbance (an easterly wave) crossing over Panama. Further west, a

tropical depression is organizing around a developing center with winds less than 25 knots. In a few days, this system will develop into hurricane Gilma. Further west is hurricane Fico, which we investigated in an earlier section. The swirling band of clouds to the north is Emilia; once a hurricane (but now with winds less than 40 knots), it is rapidly weakening over colder water.

Instead of taking the same path as Emilia, hurricane Fico took a more westerly trajectory. Because it remained over warm water, it was one of the longest lasting eastern Pacific hurricanes ever. It passed about 370 km (230 mi) south of the big island of Hawaii, where its high winds produced waves that pounded the shore, causing some $100,000 in damage to private homes.* It then swung around Hawaii and headed northward, eventually dissipating over the North Pacific more than two weeks after forming off the coast of Mexico.

Hurricane Movement Figure 20.6 shows where most hurricanes are born and the general direction in which they move. Notice that they form over tropical oceans, except in the South Atlantic and in the eastern South Pacific. Presumably, the surface water temperatures are too cold in these areas for their development. It is also possible that the unfavorable location of the ITCZ during the Southern Hemisphere's warm season discourages their development.

Hurricanes that form over the North Pacific and North Atlantic are steered by easterly winds and move west or northwestward at about 10 knots for a week or so. Gradually, they swing poleward around the subtropical high, and when they move far enough north, they become caught in the westerly flow, which curves them to the north or northeast. In the middle latitudes, the hurricane's forward speed normally increases, sometimes to more than 50 knots. The actual path of a hurricane may vary considerably. Some take erratic paths and make odd turns that occasionally catch weather forecasters by surprise. There have been many instances where a storm heading directly for land

---

*The damage caused by hurricane Fico was minor compared to that inflicted on Hawaii by hurricane Iwo during November, 1982. Iwo lashed part of Hawaii with 100-knot winds and huge surf, causing at least $230 million in damages.

Tropical storm
Emilia

Hurricane
Fico

Tropical
depression

Tropical disturbance

Fig. 20.5   Satellite picture taken on July 11, 1978, showing four tropical systems in different stages of development.

suddenly veered away and spared the region from almost certain disaster.

As we saw in an earlier section, many hurricanes form off the coast of Mexico over the North Pacific. In fact, this area usually spawns about seven hurricanes each year, which is slightly more than the yearly average of six storms born over the tropical North Atlantic. Eastern North Pacific hurricanes normally move westward, away from the coast, and so little is heard about them. When one does move northwestward, it normally weakens rapidly over the cool water of the North Pacific. Occasionally, however, one will curve northward or even northeastward and slam into Mexico, causing destructive flooding. Hurricane Tico left 25,000 people homeless and caused an estimated $66 million in property damage after passing over Mazatlán, Mexico, in October, 1983. The remains of Tico even produced record rains and flooding in Texas and Oklahoma. Even less frequently, a hurricane will stray far enough north to bring summer rains to Southern California and Arizona.

Hurricanes that form over the tropical North Atlantic also move westward or northwestward on a collision course with Central or North America. Most hurricanes, however, swing away from land and move northward, parallel to the coasts of the United States. A few storms, perhaps three per year, move inland, bringing with them high winds, huge waves, and torrential rain that may last for days.

FOCUS ON A SPECIAL TOPIC

How Do Hurricanes Compare with Middle Latitude Storms?

By now, it should be apparent that a hurricane is much different from the mid-latitude cyclone that we discussed in Chapter 17. A hurricane derives its energy from the warm water and the latent heat of condensation, whereas the mid-latitude storm derives its energy from horizontal temperature contrasts. The vertical structure of a hurricane is such that its central column of air is warm from the surface upward; consequently, hurricanes are called *warm core lows*. A hurricane weakens with height, and the area of low pressure at the surface may actually become an area of high pressure above 12 km (40,000 ft). Mid-latitude cyclones, on the other hand, usually intensify with increasing height and a cold upper-level low or trough exists to the west of the surface system. A hurricane usually contains an eye where the air is sinking, while mid-latitude cyclones are characterized by centers of rising air.

Further contrasts can be seen on a surface weather map. Figure 1 shows hurricane Allen over the Gulf of Mexico and a mid-latitude storm north of New England. Around the hurricane, the isobars are more circular, the pressure gradient is much steeper, and the winds are stronger. The hurricane has no fronts and is smaller (although

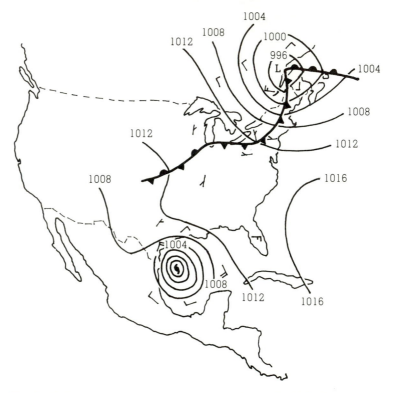

Fig. 1 Surface weather map for the morning of August 9, 1980, showing hurricane Allen over the Gulf of Mexico and a middle latitude storm system north of New England.

Allen is larger than most hurricanes). There are similarities between the two systems: Both are areas of surface low pressure, with winds moving more or less counterclockwise about their respective centers.

Even though hurricanes weaken rapidly as they move in-land, their counterclockwise circulation may draw in air with contrasting properties. If the hurricane links with an upper-level trough, it may actually become a mid-latitude cyclone. This is what happened to hurricane Hazel in 1955 and to Agnes in 1972.

A hurricane moving northward over the Atlantic will normally survive as a hurricane for a longer time than will its counterpart at the same latitude over the Pacific. The reason is, of course, that the surface water of the Atlantic is much warmer. (See Chapter 15.)

Destruction and Warning When a hurricane is approaching from the east, its highest winds are usually on its north (poleward) side. The reason for this phenomenon is that the winds that push the storm along add to the winds on the north side and subtract from the winds on the

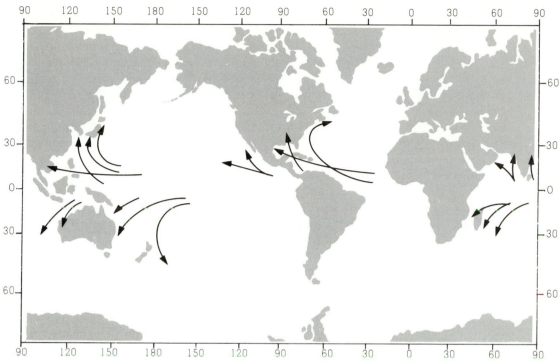

Fig. 20.6 Regions where hurricanes form, and the typical paths hurricanes take.

south (equator) side. Hence, a hurricane with 110-knot winds moving westward at 10 knots will have 120-knot winds on its north side and only 100-knot winds on its south side. The high winds of the storm generate large waves, sometimes 10 to 15 m high. These waves move outward, away from the storm, in the form of *swells* that carry the storm's energy to distant beaches. Consequently, the effects of the storm may be felt days before the hurricane arrives.

Although the hurricane's high winds inflict a great deal of damage, it is the huge waves, high seas, and flooding that cause most of the destruction. The flooding is due, in part, to winds pushing water onto the shore and to the heavy rains, which may exceed 63 cm (25 in.) in 24 hours. Flooding is also aided by the low pressure of the storm. The region of low pressure allows the ocean level to rise (perhaps half a meter), much like a soft drink rises up a straw as air is withdrawn. The combined effect of high water (which is usually well above the high-tide level) and high winds produces the **storm surge**—an abnormal rise of several meters in the ocean level—that inundates

low-lying areas and turns beachfront homes into piles of splinters. The storm surge is particularly damaging when it coincides with normal high tides.

Considerable damage may also occur from hurricane-spawned tornadoes. The exact mechanism by which these tornadoes form is not yet known; however, studies suggest that surface topography may play a role by initiating the convergence (and, hence, rising) of surface air. Recent studies suggest that swathlike areas of extreme damage once attributed to tornadoes may actually be due to downbursts associated with the large thunderstorms around the eye wall. In addition, scientists while studying hurricane Debby with Doppler radar in 1982, observed a mesocyclone (Chapter 19) in the southern part of the developing eye wall.

With the aid of ship reports, satellites, radar, and reconnaissance aircraft, the location and intensity of hurricanes are pinpointed and their movements carefully monitored. When a hurricane poses a direct threat to an area, a **hurricane watch** is issued, if possible several days before the storm arrives. When it appears that the storm

Fig. 20.7 Hurricane Allen in the Gulf of Mexico on August 8, 1980. This large storm with its spiral cloud bands occupies most of the Gulf. Compare Allen's size with the smaller hurricane positioned off the west coast of Mexico.

will strike an area within 24 hours, a **hurricane warning** is then issued. The warning is designed to give residents ample time to secure property and, if necessary, to evacuate the area.

Ample warning by the National Weather Service probably saved the lives of many people as hurricane Allen moved onshore* along the south Texas coast during the morning of August 10, 1980 (Fig. 20.7). The storm formed over the warm, tropical Atlantic and moved westward on a rampage through the Caribbean, where it killed almost 300 people and caused extensive damage. After raking the Yucatán Peninsula with 150-knot winds, Allen howled into the warm Gulf of Mexico. It

reintensified and its winds increased to 160 knots. Gale-force winds reached outward for 320 km (200 mi) north of its center. As it approached the south Texas coast, it was one of the greatest storms to ever enter that area. The central pressure of the storm dropped to a low of 899 mb (26.55 in.), making it the second mightiest Atlantic hurricane on record—only the 1935 Labor Day storm that hit the Florida Keys with a pressure of 892 mb (26.35 in.) was stronger. But its path became wobbly and it stalled offshore just long enough to lose much of its intensity. It moved sluggishly inland on the morning of August 10.

Once over land, it quickly became a tropical storm with peak winds of less than 50 knots. Before weakening, however, Allen hammered the south Texas coast with winds of 100 knots. It

*The position along a coast where the center of a hurricane passes from ocean to land is termed *landfall*.

Modifying Hurricanes

Because of the potential destruction and loss of lives that hurricanes can inflict, attempts have been made to reduce their winds by seeding them with silver iodide. The idea is to seed the clouds just outside the eye wall with just enough artificial ice nuclei so that the latent heat given off will stimulate cloud growth in this area of the storm. These clouds, which grow at the expense of the eye wall thunderstorms, actually form a new eye wall farther away from the hurricane's center. As the storm center widens, its pressure gradient should weaken, which may cause its spiraling winds to decrease in speed. During project Storm Fury, a joint effort of NOAA and the U.S. Navy, several hurricanes were seeded by aircraft. In 1963, shortly after hurricane Beulah was seeded with silver iodide, surface pressures in the eye began to rise and the region of maximum winds moved away from the storm's center. Even more encouraging results were obtained from the multiple seeding of hurricane Debbie in 1969. After one day of seeding, Debbie showed a 30 percent reduction in maximum winds. However, the question remains: Would the winds have lowered naturally had the storm not been seeded? Although showing promise, there are still uncertainties about the effectiveness of seeding hurricanes in an attempt to reduce their winds.

spawned several tornadoes and swamped the area with high tides and flooding. In Corpus Christi, the storm felled trees and power lines; streets were flooded and marinas were littered with the wreckage of small boats. In all of this, only a few people lost their lives, as 200,000 were evacuated from the Texas coast many hours before the storm struck. Further inland, the remains of the storm dumped heavy rains over a wide area of southern Texas. This was a blessing for some because it marked a break in the summer-long drought; for others, it meant despair as they fled from swollen rivers that had overflowed their banks.

A costlier hurricane (in terms of loss of life and total damage) occurred on August 18, 1983, when Hurricane Alicia slammed into the Texas coast just southwest of Galveston. With winds of 100 knots, a central pressure of 962 mb, and a storm surge of at least 3 m (10 ft), Alicia uprooted trees, knocked out skyscraper windows, tossed hundreds of mobil homes about like cardboard boxes, and flooded an extensive area. Total damage from Alicia was estimated at $2 billion, and 21 persons lost their lives.

In recent years, the hurricane death toll in the United States has averaged between 50 and 100 persons. This relatively low total is partly due to the advanced warning provided by the National Weather Service and to the fact that few really intense storms have reached land during the past 20 years. However, as the population density continues to increase in vulnerable coastal areas, the potential for a hurricane-caused disaster continues to increase also.

Hurricane Camille (1969) stands out as one of the most intense and devastating hurricanes to reach the coastline of the United States in recent years. With a central pressure of 905 mb, tempestuous winds reaching 160 knots, and a storm surge more than 7 m (23 ft) above the normal high-tide level, Camille unleashed its fury on Mississippi, destroying thousands of buildings. (See Fig. 20.8.) During its rampage, it caused an estimated $1.5 billion in property damage and took more than 200 lives. But Camille does not even come close to the costliest storm on record—that dubious distinction goes to hurricane Frederic. In early September, 1979, Frederic struck the Gulf Coast of Mississippi and Alabama with heavy rains and maximum sustained winds of 115 knots. The total damage in the United States attributed to this storm was nearly $2.3 billion. Luckily, the death toll was only 5 people. In June, 1972, hurricane Agnes caused over $2 billion in damage to Florida and to the northeastern United States. The damage was caused by flooding due to heavy rainfall that was estimated to be more than 28 trillion gallons! Agnes' death toll was 122.

Fig. 20.8 Destruction caused by hurricane Camille was widespread along the Gulf Coast in August, 1969. This is the remains of a 32-unit apartment building in Gulfport, Mississippi.

Before the era of satellites and radar, catastrophic losses of life had occurred. In 1900, about 6000 people lost their lives when a hurricane slammed into Galveston, Texas, with a huge storm surge. Most of the deaths occurred in the low-lying coastal regions as flood waters pushed inland. In October, 1893, nearly 2000 people perished on the Gulf Coast of Louisiana as a giant storm surge swept that region. Spectacular losses are not confined to the Gulf Coast. Nearly 1000 people lost their lives in Charleston, South Carolina, during August of the same year. But these statistics are small compared to the more than 300,000 lives taken as a killer cyclone and storm surge ravaged the coast of Bangladesh with flood waters in 1970. Unfortunately, in this region the potential for a repeat of this type of disaster remains high, as many people live along the relatively low, wide flood plain that slopes outward to the Bay. And, historically, this region is in a path frequently taken by tropical cyclones.

In an effort to estimate the possible damage a hurricane's sustained winds and storm surge could do to a coastal area, the *Saffir-Simpson scale* was developed (see Table 20.1). The scale numbers are based on actual conditions at some time during the life of the storm. As the hurricane intensifies or weakens, the scale number is reassessed accordingly.

Naming Hurricanes In reading the previous sections, you noticed that hurricanes were as-

Table 20.1 Saffir-Simpson Hurricane Damage-Potential Scale

SCALE NUMBER (CATEGORY)	CENTRAL PRESSURE		WINDS		STORM SURGE		DAMAGE
	millibars	inches	mi/hr	knots	ft	m	
1	≥980*	≥28.94	74–95	64–82	4–5	~1.5	damage mainly to trees, shrubbery, and unanchored mobile homes
2	965–979	28.50–28.91	96–110	83–95	6–8	~2.0–2.5	some trees blown down; major damage to exposed mobile homes; some damage to roofs of buildings
3	945–964	27.91–28.47	111–130	96–113	9–12	~2.5–4.0	foliage removed from trees; large trees blown down; mobile homes destroyed; some structural damage to small buildings
4	920–944	27.17–27.88	131–155	114–135	13–18	~4.0–5.5	all signs blown down; extensive damage to roofs, windows, and doors; complete destruction of mobile homes; flooding inland as far as 10 km (6 mi); major damage to lower floors of structures near shore
5	<920	<27.17	>155	>135	>18	>5.5	severe damage to windows and doors; extensive damage to roofs of homes and industrial buildings; small buildings overturned and blown away; major damage to lower floors of all structures less than 4.5 m (15 ft) above sea level within 500 m of shore

*Symbol > means greater than; < means less than; ≥ means equal to or greater than; ~ means approximately equal to.

signed names. Before this practice was started, hurricanes were identified according to their latitude and longitude. This method was confusing, especially when two or more storms were present over the same ocean. To reduce the confusion, hurricanes were identified by letters of the alphabet. During World War II, names like Able and Baker were used. (These names correspond to the radio code words associated with each letter of the alphabet.) This method also seemed cumbersome so, beginning in 1953, the National Weather Service began using female names to identify hurricanes. The list of names for each year was in alphabetical order, so that the name of the season's first storm began with the letter *A*, the second with *B*, and so on.

From 1953 to 1977, only female names were used. However, beginning in 1978, hurricanes in the eastern Pacific were alternately assigned female and male names (recall Gilma and Fico). This practice was started for North Atlantic hurricanes in 1979. Table 20.2 gives the proposed list of

Table 20.2 Names for Hurricanes

ATLANTIC HURRICANE NAMES				EASTERN NORTH PACIFIC HURRICANE NAMES			
1986	1987	1988	1989	1986	1987	1988	1989
Allen	Arlene	Alberto	Alicia	Agatha	Adrian	Aletta	Adolph
Bonnie	Bret	Beryl	Barry	Blas	Beatriz	Bud	Barbara
Charley	Cindy	Chris	Chantal	Celia	Calvin	Carlotta	Cosme
Danielle	Dennis	Debby	Dean	Darby	Dora	Daniel	Dalilia
Earl	Emily	Ernesto	Erin	Estelle	Eugene	Emilia	Erick
Frances	Floyd	Florence	Felix	Frank	Fernando	Fabio	Flossie
Georges	Gert	Gilbert	Gabrielle	Georgette	Greg	Gilma	Gil
Hermine	Harvey	Helene	Hugo	Howard	Hilary	Hector	Henriette
Ivan	Irene	Isaac	Iris	Isis	Irwin	Iva	Ismael
Jeanne	Jose	Joan	Jerry	Javier	Jova	John	Juliette
Karl	Katrina	Keith	Karen	Kay	Knut	Kristy	Kiko
Lisa	Lenny	Leslie	Luis	Lester	Lidia	Lane	Lorena
Mitch	Maria	Michael	Marilyn	Madeline	Max	Miriam	Manuel
Nicole	Nate	Nadine	Noel	Newton	Norma	Norman	Narda
Otto	Ophelia	Oscar	Opal	Orlene	Otis	Olivia	Octave
Paula	Philippe	Patty	Pablo	Paine	Pilar	Paul	Priscilla
Richard	Rita	Rafael	Roxanne	Roslyn	Ramon	Rosa	Raymond
Shary	Stan	Sandy	Sebastien	Seymour	Selma	Sergio	Sonia
Tomas	Tammy	Tony	Tanya	Tina	Todd	Tara	Tico
Virginie	Vince	Valerie	Van	Virgil	Veronica	Vicente	Velma
Walter	Wilma	William	Wendy	Winifred	Wiley	Willa	Winnie

names for both North Atlantic and eastern Pacific hurricanes.

Summary

Hurricanes are tropical cyclones with winds that exceed 64 knots (74 mi/hr) and blow counterclockwise about their centers. A hurricane consists of a mass of organized thunderstorms that spiral in toward the extreme low pressure of the storm's eye. The most intense thunderstorms, the heaviest rain, and the highest winds occur outside the eye, in the region known as the eye wall. In the eye itself, the air is warm, winds are light, and skies may be broken or overcast.

Hurricanes are born over warm tropical waters where surface winds converge and thunderstorms become organized. Convergence may occur along the ITCZ, on the eastern side of an easterly wave, or along a front that has moved into the tropics from higher latitudes. If the disturbance becomes more organized, it becomes a tropical depression. If central pressures drop and surface winds increase, the depression becomes a tropical storm. Some tropical storms continue to deepen into full-fledged hurricanes.

The easterly winds in the tropics usually steer hurricanes westward. Most storms then gradually swing northwestward around the subtropical high to the north. If the storm moves into middle latitudes, the prevailing westerlies steer it northeastward. Because hurricanes derive their energy from the warm surface water and from the latent heat of condensation, they tend to dissipate rapidly when they move over cold water or over a large mass of land.

Although the high winds of a hurricane can inflict a great deal of damage, it is the huge waves and the flooding associated with the storm surge that cause the most destruction. The Saffir-Simpson hurricane scale was developed to estimate the potential destruction that a hurricane can cause.

Questions for Review

1. What are easterly waves and how do they generally move?

2. Why are streamlines, rather than isobars, used on surface weather maps in the tropics?

3. In your own words, what is a hurricane?

4. Where do hurricanes derive their energy?

5. Describe the horizontal and vertical structure of a hurricane.

6. What conditions at the surface and aloft are most conducive for the formation of a hurricane?

7. What factors tend to weaken hurricanes?

8. Distinguish among a tropical disturbance, a tropical depression, a tropical storm, and a hurricane.

9. In what ways is a hurricane different from a mid-latitude cyclone? In what ways are these two systems similar?

10. Why do most hurricanes move westward over tropical waters?

11. If the high winds of a hurricane are not responsible for inflicting the most damage, what is?

12. Explain how a storm surge forms.

13. Why have hurricanes been seeded with silver iodide?

14. Give two reasons why hurricanes are more likely to strike New Jersey than Oregon.

Questions for Thought

1. Why are hurricanes more apt to form in October than in May?

2. Would it be possible for a hurricane to form in the tropical North Atlantic or North Pacific during December? Explain.

3. Would the winds of a hurricane decrease more quickly as the storm moves over cooler water or over warmer land? Explain.

4. Explain why the surface water temperature of the ocean is usually cooler after the passage of a hurricane.

5. Suppose, in the North Atlantic, an eastward-moving ocean vessel is directly in the path of a westward-moving hurricane. What would be the ship's wisest course—to veer to the north of the storm or to the south of the storm? Explain.

6. Suppose this year five tropical storms develop into full-fledged hurricanes over the North Atlantic Ocean. Would the name of the third hurricane begin with the letter ''C''? Explain.

Problems and Exercises

1. A hurricane just off the coast of northern Florida is moving northeastward, parallel to the eastern seaboard. Suppose that you live in North Carolina along the coast:

(a) How will the surface winds in your area change direction as the hurricane's center passes due east of you? Illustrate your answer by making a sketch of the hurricane's movement and the wind flow around it.

(b) If the hurricane passes east of you, the strongest winds would most likely be blowing from which direction? Explain your answer. (Assume that the storm does not weaken as it moves northeastward.)

(c) The lowest sea level pressure would most likely occur with which wind direction? Explain.

2. Use the Saffir-Simpson hurricane scale (Table 20.1, p. 429) and the text material to determine the category of each of the following hurricanes:

(a) hurricane Fico in Fig. 20.2, p. 419.

(b) hurricane Allen in the Gulf of Mexico on August 9, 1980

(c) hurricane Alicia before landfall on August 18, 1983

(d) hurricane Camille before landfall in 1969

The cool, summer climate near the top of the Minarets in the Sierra Nevada allows winter snow to survive in sheltered coves and valleys. Afternoon thunderstorms that build along the windward slopes produce ample summer rains. However, just a few hundred kilometers to the east and a few thousand meters lower in elevation, the climate is hot (with afternoon summer temperatures hovering near 40°C or 104°F) and dry (as the region is in the rain shadow of this mountain range). (Photo: T. Ansel Toney)

CHAPTER 21

Global Climate

Climate affects nearly everything profoundly. It influences our housing, our clothing, the shape of landscapes, agriculture, and so on. Entire civilizations have flourished in favorable climates and moved away from or perished in unfavorable ones. We learned at the beginning of this text that climate is the average of the day-to-day weather over a long duration. But climate is much more than this, for it includes daily and seasonal extremes of weather in a particular region.

When we speak of climate, we must be careful to specify the region we are talking about. For example, the Chamber of Commerce of a rural town may boast that its community has mild winters with air temperatures seldom below freezing. This may be true several meters above the ground in an instrument shelter, but near the ground the temperature may drop below freezing on many winter nights. This small climatic region near or on the ground is referred to as a **microclimate**. Because a much greater extreme in daily air temperatures exists near the ground than several meters above, the microclimate for small plants is far more harsh than the thermometer in an instrument shelter would indicate.

When we examine the climate of a small area of the earth's surface, we are looking at the **mesoclimate**. The size of the area may range from a few acres to several square kilometers. Mesoclimate includes regions such as forests, valleys, beaches, and towns. The climate of a much larger area, such as a state or a country, is called **macroclimate**. The climate extending over the entire earth is often referred to as **global climate**.

In this chapter, we will concentrate on the larger scales of climate. We will begin with the factors that regulate global climate, then we will look at the different types of climate. Next we will see how climates are classified and, finally, discuss

Contents

some relationships between climate and human comfort.

A World with Many Climates

The world is rich in climatic types. From the teeming tropical jungles to the frigid polar "wastelands" there seems to be an almost endless variety of climatic regions. The factors that produce the climate in a particular place, the *climatic controls*, are the same that produce our day-to-day weather. Briefly, the controls are the (1) intensity of sunshine and its variation with latitude; (2) distribution of land and water; (3) ocean currents; (4) prevailing winds; (5) positions of high- and low-pressure areas; (6) mountain barriers; and (7) altitude.

We can ascertain the effect these controls have on climate by observing the global patterns of two weather elements—temperature and precipitation.

Global Temperatures Figure 21.1 shows mean annual temperatures for the world. To eliminate

the distorting effect of topography, the temperatures are corrected to sea level. Notice that in both hemispheres the isotherms are oriented east-west, reflecting the fact that locations at the same latitude receive nearly the same amount of solar energy. In addition, the annual solar heat that each latitude receives decreases from low to high latitude; hence, annual temperatures tend to decrease from equatorial toward polar regions.

The bending of the isotherms along the coastal margins is due in part to the unequal heating and cooling properties of land and water, and to ocean currents and upwelling. For example, along the west coast of North and South America, ocean currents transport cool water equatorward. In addition to this, the wind in both regions blows toward the equator, parallel to the coast. This situation favors upwelling of cold water (Chapter 15), which cools the coastal margins. In the area of the eastern North Atlantic Ocean (north of 40°N), the poleward bending of the isotherms is due to the Gulf Stream and the North Atlantic Drift, which carry warm water northward.

The highest mean temperatures occur in the

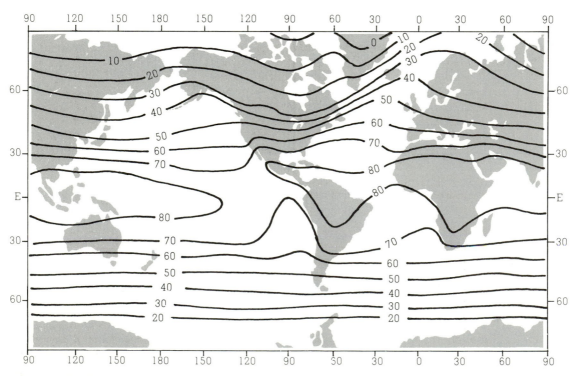

Fig. 21.1 Average annual sea level temperatures throughout the world (°F).

The Hottest and Coldest Places on Earth

Most people are aware of the extreme heat that exists during the summer in the desert southwest of the United States. But how hot does it get there? On July 10, 1913, at an elevation of 54 m (178 ft) below sea level, Greenland Ranch in Death Valley, California, reported the highest temperature ever observed in North America: 57°C, or 134°F. Here, air temperatures are persistently hot throughout the summer, with the average maximum for July being 47°C (116°F). During the summer of 1917, there was an incredible period of 43 consecutive days when the maximum temperature reached 49°C (120°F) or higher.

Probably the hottest urban area in the United States is Yuma, Arizona. Located along the California–Arizona border, Yuma's high temperature during July averages 42°C (108°F). In 1937, the high reached 38°C (100°F) or more for 101 consecutive days.

In a more humid climate, the maximum temperature rarely climbs above 41°C (106°F). However, during the record heat wave of 1936, the air temperature reached 49°C (121°F) near Alton, Kansas. And during the heat wave of 1983, which destroyed about $7 billion in crops and increased the nation's air-conditioning bill by an estimated $1 billion, Fayetteville reported North Carolina's all-time record high temperature when the mercury hit 43°C (110°F). During the same year, West Germany recorded its hottest day ever when the thermometer at Ihringen reached 40°C (104°F).

Table 1 Some Record High Temperatures Throughout the World

LOCATION (LATITUDE)	RECORD HIGH TEMPERATURE		RECORD FOR:	DATE
	(°C)	(°F)		
El Azizia, Libya (32°N)	58	136	The world	September 13, 1922
Death Valley, Calif. (36°N)	57	134	Western Hemisphere	July 10, 1913
Tirat Tsvi, Israel (32°N)	54	129	Asia	June 21, 1942
Cloncurry, Queensland (21°S)	53	128	Australia	January 16, 1889
Seville, Spain (37°N)	50	122	Europe	August 4, 1881
Rivadavia, Argentina (35°S)	49	120	South America	December 11, 1905
Midale, Saskatchewan (49°N)	45	113	Canada	July 5, 1937
Fort Yukon, Alaska (66°N)	38	100	Alaska	June 27, 1915
Pahala, Hawaii (19°N)	38	100	Hawaii	April 27, 1931
Esparanza, Antarctica (63°S)	14	58	Antarctica	October 20, 1956

These readings, however, do not hold a candle to the hottest place in the world. That distinction probably belongs to Dallol, Ethiopia. Dallol is located south of the Red Sea, near latitude 12°N, in the hot, dry Danakil Depression. A prospecting company kept weather records at Dallol from 1960 to 1966. During this time, the average daily maximum temperature exceeded 38°C (100°F) every month of the year, except during December and January, when the average maximum lowered to 37°C (98°F) and 36°C (97°F), respectively. On many days, the air temperature exceeded 49°C (120°F). The average annual temperature for the six years at Dallol was 34°C (94°F). In comparison, the average annual temperature in Yuma is 23°C (74°F) and at Death Valley, 24°C (76°F). The highest temperature reading on earth (under standard conditions) occurred about 3500 km northeast of Dallol at El Azizia, Libya (32°N), when, on September 13, 1922, the temperature reached a scorching 58°F (136°F). (Table 1 gives record high temperatures throughout the world.)

One of the coldest spots in the United States is International Falls, Minnesota, where the average temperature for January is −16°C (3°F). Located about 400 km (250 mi) to the south, Minneapolis–St. Paul, with an average temperature of −9°C (16°F) for the three winter months, is the coldest major urban area in the nation. For duration of extreme cold, Minneapolis reported 186 consecutive hours of temperatures below −18°C (0°F) during the winter of 1911–1912. Within the 48 adjacent states, however, the record for the longest duration of severe cold belongs to Langdon, North Dakota, where the thermometer remained below −18°C (0°F) for 41 consecutive days during the winter of 1936. The official record for the lowest temperature in the 48 adjacent states belongs to Rogers Pass, Montana, where on the morning of January 20, 1954, the mercury dropped to −57°C (−70°F). The lowest official temperature for Alaska, −62°C (−80°F), occurred at Prospect Creek on January 23, 1971.

The coldest areas in North America are found in the Yukon and Northwest Territory of Canada. Resolute, Canada (latitude 75°N), has an average temperature of −32°C (−26°F) for the month of January.

The coldest winters in the Northern Hemisphere are found in the interior of Siberia and Greenland. For example, the average January temperature in Yakutsk, USSR (latitude 62°N), is −43°C (−46°F). There, the mean temperature for the entire year is a bitter cold −11°C (12°F). At Eismitte, Greenland, the average temperature for February (the coldest month) is −47°C (−53°F), with the mean

Table 2   Some Record Low Temperatures Throughout the World

LOCATION (LATITUDE)	RECORD LOW TEMPERATURE (°C)	(°F)	RECORD FOR:	DATE
Vostok, Antarctica (72°S)	−88	−127	The world	August 24, 1960
Verkhoyansk, USSR (67°N)	−68	−90	Northern Hemisphere	February 7, 1892
Northice, Greenland (72°N)	−66	−87	Greenland	January 9, 1954
Snag, Yukon (62°N)	−63	−81	North America	February 3, 1947
Prospect Creek, Alaska (66°N)	−62	−80	Alaska	January 23, 1971
Rogers Pass, Montana (47°N)	−57	−70	United States (excluding Alaska)	January 20, 1954
Sarmiento, Argentina (34°S)	−33	−27	South America	June 1, 1907
Ifrane, Morocco (33°N)	−24	−11	Africa	February 11, 1935
Charlotte Pass, Australia (36°S)	−22	−8	Australia	July 22, 1949
Mt. Haleakala, Hawaii (20°N)	−10	14	Hawaii	January 2, 1961

annual temperature being a frigid −30°C (−22°F). Even though these temperatures are extremely low, they do not come close to the coldest area of the world: the Antarctic.

At the geographical South Pole, nearly 2800 m (9200 ft) above sea level, where the Amundsen-Scott scientific station has been keeping records for more than 25 years, the average temperature for the month of July (winter) is −59°C (−74°F) and the mean annual temperature is −49°C (−57°F). The lowest temperature ever recorded there (−83°C, or −117°F) occurred under clear skies with a

light wind on the morning of June 23, 1983. Cold as it was, it was not the record low for the world. That belongs to the Soviet station at Vostok, Antarctica (latitude 72°S), where the temperature plummeted to −88°C (−127°F) on August 24, 1960. (See Table 2 for record low temperatures throughout the world.)

subtropical deserts of the Northern Hemisphere. Here, the subsiding air associated with the subtropical anticyclones produces generally clear skies and low humidity. In summer, the high sun beating down upon a relatively barren landscape produces scorching heat.

The lowest mean temperatures occur in the Antarctic. During part of the year, the sun is below the horizon; when it is above the horizon, it is low in the sky and its rays do not effectively warm the surface. Consequently, the land remains snow- and ice-covered year-round. The snow and ice reflect perhaps 80 percent of the sunlight that reaches the surface. Much of the unreflected solar energy is used to transform the ice and snow into water vapor. The relatively dry air and the Antarctic's high elevation permit rapid radiational cooling during the dark winter months, producing extremely cold surface air.

Global Precipitation Figure 21.2 shows the worldwide general pattern of annual precipitation, which varies from place to place. There are, however, certain regions that stand out as being wet or dry. For example, equatorial regions are typically wet, while the subtropics and the polar regions are relatively dry. The global distribution of precipitation is closely tied to the general circulation of the atmosphere (Chapter 15) and to the distribution of mountain ranges and high plateaus.

Figure 21.3 shows in simplified form how the general circulation influences the north-to-south distribution of precipitation to be expected on a uniformly water-covered earth. Precipitation is most abundant where the air rises; least abundant where it sinks. Hence, one expects a lot of precipitation in the tropics and along the polar front, and little near subtropical highs and at the poles. Let's look at this in more detail.

In tropical regions, the trade winds converge along the Intertropical Convergence Zone (ITCZ), producing rising air, towering clouds, and heavy precipitation all year long. Poleward of the equator, near latitude 30°, the sinking air of the subtropical highs produces a "dry belt" around the globe. The Sahara Desert of North Africa is in this region. Here, annual rainfall is exceedingly light and varies considerably from year to year. Because the major wind belts and pressure systems shift with the season—northward in July and southward in Jan-

uary—the area between the rainy tropics and the dry subtropics is influenced by both the ITCZ and the subtropical highs.

In polar regions, the cold air can hold little moisture, so there is little precipitation. Winter storms drop light, powdery snow that remains on the ground for a long time because of the low evaporation rates. In summer, a ridge of high pressure tends to block storm systems that would otherwise travel into the area; hence, precipitation in polar regions is meager in all seasons.

There are exceptions to this idealized pattern. For example, in middle latitudes the migrating position of the subtropical anticyclones also has an effect on the west-to-east distribution of precipitation. The sinking air associated with these systems is more strongly developed on their eastern sides. Hence, the air along the eastern side of an anticyclone tends to be more stable; it is also drier, as cooler air moves equatorward because of the circulating winds around these systems. In addition, along coastlines, cold upwelling water cools the surface air even more, adding to the air's stability. Consequently, in summer, when the Pacific high moves to a position centered off the California coast, a strong, stable subsidence inversion forms above coastal regions. With the strong inversion and the fact that the anticyclone tends to steer storms to the north, central and southern California experience little, if any, rainfall during the summer months.

On the western side of subtropical highs, the air is less stable and more moist, as warmer air moves poleward. In summer, over the North Atlantic, the Bermuda high pumps moist, tropical air northward from the Gulf of Mexico into the eastern two-thirds of the United States. The humid air is conditionally unstable to begin with, and by the time it moves over the heated ground, it becomes even more unstable. If conditions are right, the moist air will rise and condense into cumulus clouds, which may build into towering thunderstorms.

In winter, the subtropical North Pacific high moves south, allowing storms traveling across the ocean to penetrate the western states, bringing much needed rainfall to California after a long, dry summer. The Bermuda high also moves south in winter. Across much of the United States, intense winter storms develop and travel eastward, frequently dumping heavy precipitation as they go.

Fig. 21.2 Annual global pattern of precipitation. (Numbers are in inches.)

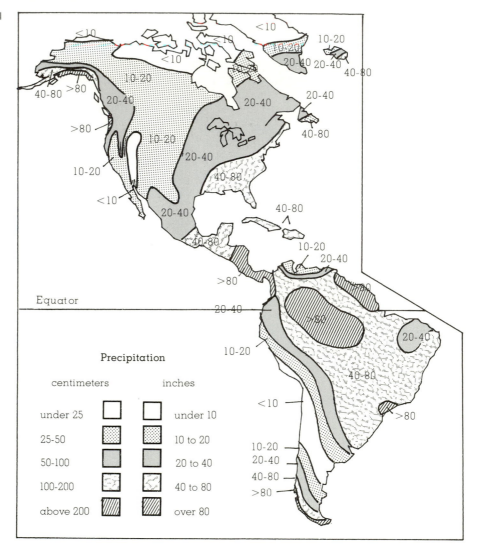

Usually, however, the heaviest precipitation is concentrated in the eastern states, as moisture from the Gulf of Mexico moves northward ahead of these systems. Therefore, cities on the plains typically receive more rainfall in summer, those on the west coast have maximum precipitation in winter, while cities in the midwest and east usually have abundant precipitation all year long. The contrast in seasonal precipitation among a west coast city (San Francisco), a central plains city (Kansas City), and an eastern city (Baltimore) is shown in Fig. 21.4.

Mountain ranges disrupt the idealized pattern of global precipitation (1) by promoting convection (because their slopes are warmer than the surrounding air) and (2) by forcing air to rise along

their windward slopes (*orographic uplift*). In fact, most of the "rainiest" places in the world are located on the windward side of mountains. For example, situated at the base of the Olympic Mountains in the northwestern part of Washington State, the Hoh river valley receives an average 380 cm (150 in.) of precipitation.

As air descends and warms along the leeward side of a mountain range, there is less likelihood of clouds and precipitation. Such regions are called a **rain shadow** and are often deserts. On the eastern side of the Olympic Mountains in Washington State—only about 100 km (62 mi) from the Hoh rain forest—the mean annual precipitation is less than 43 cm (17 in.) and irrigation is necessary to grow certain crops. Figure 21.5 shows a classic

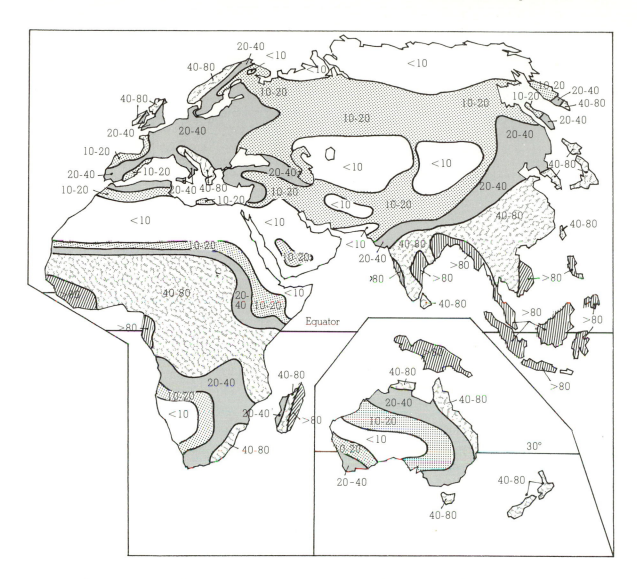

example of a topography that produces several rain shadow effects.

Snowfalls tend to be heavier where cool, moist air rises along the windward slopes of mountains. One of the snowiest places in North America is located at the Paradise Ranger Station in Mt. Rainier National Park, Washington. Situated at an elevation of 1646 m (5400 ft) above sea level, this station receives an average 1575 cm (620 in.) of snow annually. However, a record 2850 cm (1122 in.) was received during the winter of 1971–1972. (Table 21.1 gives some precipitation records throughout the world.)

Classifying Climates The climatic controls interact to produce such a wide array of different

climates that no two places experience exactly the same climate. However, the similarity of climates within a given area allows us to divide the earth into climatic regions.

The Ancient Greeks By considering temperature and worldwide sunshine distribution, the ancient Greeks categorized the world into three climatic regions:

1. A low-latitude *torrid zone*; bounded by the northern and southern limit of the sun's vertical rays (23½°N and 23½°S); here, the noon sun is always high, day and night are of nearly equal length, and it is warm year-round

2. A high-latitude *polar zone*; bounded by the Arctic or Antarctic Circle; cold all year long

Fig. 21.3 A vertical cross section along a line running north to south illustrates the main global regions of rising and sinking air and how each region influences precipitation.

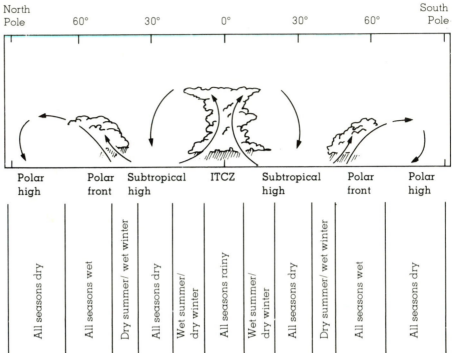

North Pole 60° 30° 0° 30° 60° South Pole

Polar high | Polar front | Subtropical high | ITCZ | Subtropical high | Polar front | Polar high

All seasons dry | All seasons wet | Dry summer/ wet winter | All seasons dry | Wet summer/ dry winter | All seasons rainy | Wet summer/ dry winter | All seasons dry | Dry summer/ wet winter | All seasons wet | All seasons dry

Fig. 21.4 Variation in annual precipitation for three Northern Hemisphere cities.

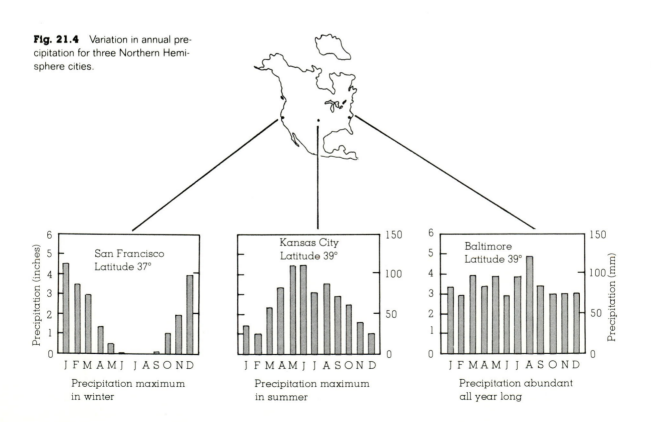

San Francisco Latitude 37°
Precipitation maximum in winter

Kansas City Latitude 39°
Precipitation maximum in summer

Baltimore Latitude 39°
Precipitation abundant all year long

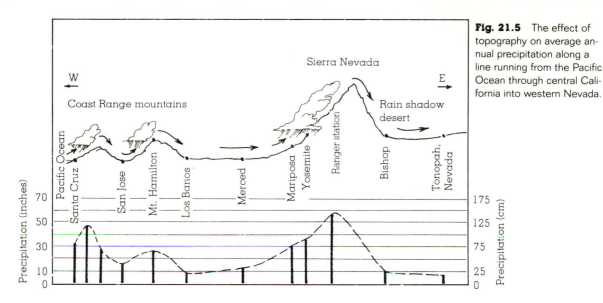

Fig. 21.5 The effect of topography on average annual precipitation along a line running from the Pacific Ocean through central California into western Nevada.

due to long periods of winter darkness and a low summer sun

3. A *temperate zone*; sandwiched between the other two zones; has distinct summer and winter, so exhibits characteristics of both extremes.

Such a sunlight, or temperature-based, climatic scheme is, of course, far too simplistic. It excludes precipitation, so there is no way to differentiate between wet and dry regions. The best classification of climates would take into account as many meteorological factors as can possibly be obtained.

The Köppen System   A widely used classification of world climates based on the annual and monthly averages of temperature and precipitation was devised by the famous German scientist Waldimir Köppen (1846–1940). Initially published in 1918, the original system has since been modified and refined. Faced with the lack of adequate observing stations throughout the world, Köppen related the distribution and type of native vegetation to the various climates. In this way, climatic boundaries could be approximated where no climatological data were available.

Köppen's scheme employs five major climatic types labeled: (A) tropical moist climates; (B) dry climates; (C) moist climates with mild winters; (D) moist climates with severe winters; and (E) polar climates. Each group contains major subregions describing special regional characteristics, such

as seasonal changes in temperature and precipitation. In mountainous country, where rapid changes in elevation bring about sharp changes in climatic type, delineating the climatic regions is impossible. These regions are designated by the letter H for highland climates.

Köppen's system has been criticized primarily because his boundaries (which relate vegetation to monthly temperature and precipitation values) do not correspond to the natural boundaries of each climatic zone. In addition, the Köppen system implies that there is a sharp boundary between climatic zones when in reality there is a gradual transition.

The Köppen system has been revised several times, most notably by the German climatologist Rudolf Geiger, who worked with Köppen on amending the climatic boundaries of certain regions. A popular modification of the Köppen system was developed by the American climatologist Glenn T. Trewartha, who redefined some of the climatic types and altered the climatic world map. Figure 21.6 shows the world distribution of the major climatic types based on a modified Köppen system.

Thornthwaite's System   To correct some of the Köppen deficiencies, the American climatologist C. Warren Thornthwaite (1899–1963) devised a new classification system in the early 1930s. Both systems utilized temperature and precipitation

Table 21.1 Some Precipitation Records Throughout the World

LOCATION (LATITUDE)	AMOUNT (cm)	(inches)	REMARKS	DATE
Cherrapunji, India	2647	1042	Greatest one-year rainfall total	1861
Mt. Waialeale, Hawaii	1168	460	World's greatest annual average rainfall	
Cherrapunji, India	930	366	Greatest one-month rainfall total	July 1861
Belouve, La-Réunion Island	135	53	Greatest 12-hour rainfall total	February 28, 1964
Alvin, Texas	109	43	Greatest 24-hour rainfall total in U.S.	July 25, 1979
Holt, Missouri	30	12	Greatest 42-minute rainfall total	June 22, 1947
Unionville, Maryland	3	1.2	Greatest one-minute rainfall total	July 4, 1956
Bataques, Mexico	3	1.2	Lowest annual average rainfall in Northern Hemisphere	
Arica, Chile	0.08	0.03	Lowest annual average rainfall in the world	
Bagdad, California	0.0	0.0	Longest period without measurable precipitation in U.S. (993 days)	August 1909 to May 1912
Paradise Ranger Station, Mt. Rainier, Washington	2850	1122	Greatest annual snowfall in U.S.	1971–1972
Tamarack, California	991	390	Greatest snowfall in one month	January 1911
Mt. Shasta Ski Bowl, California	480	189	Greatest snowfall in a single storm	February 13–19, 1959
Silverlake, Boulder Co., Colorado	193	76	Greatest snowfall in 24 hours	April 14–15, 1921

measurements and both related natural vegetation to climate. However, to emphasize the importance of precipitation (P) and evaporation (E) on plant growth, Thornthwaite developed a **P/E ratio**, which is essentially monthly precipitation divided by monthly evaporation. The annual sum of the P/E ratios gives the **P/E index**. Using this index, the Thornthwaite system defines five major humidity provinces and their characteristic vegetations: rain forest, forest, grasslands, steppe, and desert.

To better describe the available moisture for plant growth, Thornthwaite proposed a new classification system in 1948 and slightly revised it in 1955. His new scheme emphasized the concept of **potential evapotranspiration*** (PE), which is the amount of moisture that would be lost from the soil and vegetation if the moisture were available.

*Evapotranspiration refers to the evaporation from soil and transpiration of plants.

Thornthwaite incorporated potential evapotranspiration into a moisture index that depends essentially on the differences between precipitation and PE. The index is high in moist climates and negative in arid climates. An index of 0 marks the boundary between wet and dry climates.

Climates of North America   Figure 21.7 shows the generalized distribution of the major climatic regions of North America based mainly on the work of Köppen.* We will first examine the climates in low latitudes and then move northward. Bear in mind that each climatic region has many subregions of local climatic differences wrought by such factors as topography, elevation, and large bodies of water. Remember, too, that boundaries of climatic regions represent gradual transitions. Thus, the major climatic characteristics of a given region are best observed away from its periphery.

Tropical Moist Climates (Group A)

Characteristics: year-round warm temperatures (all months have a mean temperature above 18°C or 64°F); abundant rainfall (typical annual average, exceeds 150 cm, or 59 in.)

Extent: northward from the equator into Mexico and southern Florida (Fig. 21.7)

Types (based on seasonal distribution of rainfall): *tropical wet* (Af) and *tropical wet and dry* (Aw)

At low elevations near the equator, high temperatures and abundant yearly rainfall combine to produce a dense, broadleaf, evergreen forest called a *tropical rain forest*. Within this **tropical wet climate** (Af), seasonal temperature variations are small (normally less than 3°C) because the noon sun is always high and the number of daylight hours is relatively constant. However, there is a greater variation in temperature between day (about 32°C) and night (about 22°C) than there is between the warmest and coolest months. This is why people remark that winter comes to the tropics at night.

*The major climatic types with their subdivisions are given in Appendix G.

The weather in a tropical wet climate is monotonous and sultry. Almost everyday, towering cumulus clouds form and produce heavy, localized rainshowers by early afternoon. As evening approaches, the showers usually end and skies clear. Typical annual rainfall totals are greater than 150 cm (59 in.) and, in some cases, especially along the windward side of hills and mountains, the total may exceed 400 cm (157 in.).

The high humidity and cloud cover tend to keep maximum temperatures from reaching extremely high values. In fact, summer afternoon temperatures are normally higher in middle latitudes than here. Nighttime cooling can produce saturation and, hence, a blanket of dew and, occasionally, fog covers the ground.

Poleward of the tropical wet region, total annual rainfall diminishes and there is a gradual transition from the tropical wet climate to the **tropical wet-and-dry climate** (Aw), where a distinct dry season prevails. Even though the annual precipitation is above 150 cm, the dry season, where the monthly rainfall is less than 6 cm (2.4 in.), lasts at least two months. Because tropical rain forests cannot survive this "drought," the jungle gradually gives way to tall, coarse *savanna grass*, scattered with low, drought-resistant deciduous trees. The dry season occurs during the winter (low sun period), when the region is under the influence of the subtropical highs. In summer, the ITCZ moves poleward, bringing with it heavy precipitation, usually in the form of showers.

As in the tropical wet region, the daily range of temperature usually exceeds the annual range, but the climate here is much less monotonous. There is a cool season in winter when the maximum temperature averages between 30° to 32°C (86° to 90°F). At night, the low humidity and clear skies allow for rapid radiational cooling and, by early morning, minimum temperatures drop to 20°C (68°F) or below.

As spring approaches, the noon sun is slightly higher and the more intense sunshine produces greater surface heating and higher temperatures—usually above 32°C (90°F) and occasionally above 38°C (100°F)—creating desert-like conditions. In summer, during the rainy season, the weather turns hot and muggy and resembles that of the tropical wet climate. Although June and July are hot, April and May are hotter, as

Fig. 21.6 Worldwide distribution of climatic regions (after Köppen).

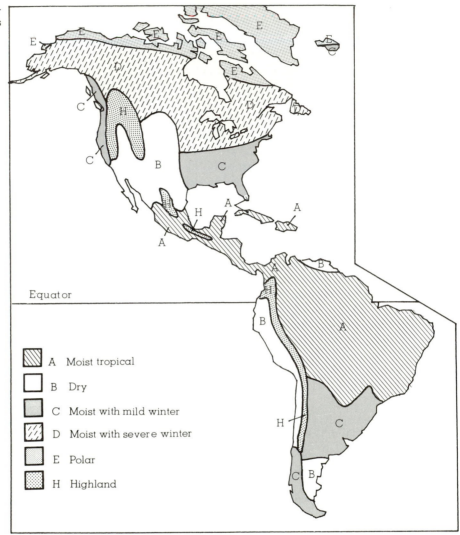

Equator

A   Moist tropical

B   Dry

C   Moist with mild winter

D   Moist with severe winter

E   Polar

H   Highland

more surface heating takes place in the absence of the heavy summertime cloud cover.

Northward of this region, the dry season becomes more severe. When the potential annual water loss through evaporation and transpiration exceeds the annual water gain from precipitation, the climate is described as dry.

Dry Climates (Group B)

Characteristics: deficient precipitation most of the year; potential evaporation and transpiration exceed precipitation

Extent: roughly from northcentral Mexico to southern Canada

Types: *arid* (BW)—the "true desert"—and *semiarid* (BS)

The **arid climate** (BW) extends from northern Mexico into the southern interior of the United States and northward along the leeward slopes of the Sierra Nevada mountains (Fig. 21.7). This region includes both the Sonoran and Mojave deserts and the Great Basin. Precipitation is deficient all year long, with most stations receiving less than 13 cm (5 in.) annually. The precipitation that does fall is spotty. The southern desert region is dry because it is dominated by the subtropical high most of the year, and winter storm systems tend to weaken before they move into this area.

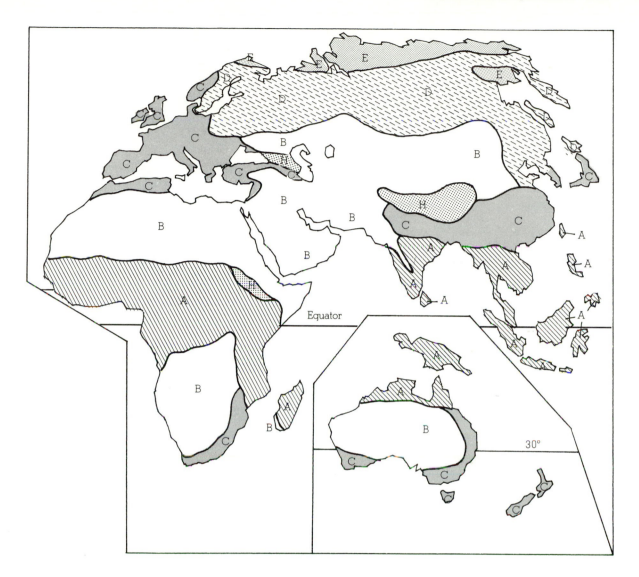

The northern region is in the rain shadow of the Sierra Nevada.

In the southern region, where average annual temperatures are above 18°C (64°F), summers are hot, winters mild. In the northern region, average annual temperatures are markedly cooler (less than 18°C) and winter morning lows often dip well below freezing.

Vegetation throughout the arid land must depend on the infrequent rains. Thus, most of the native plants are *xerophytes*—those capable of surviving prolonged periods of drought. Such vegetation includes the creosote bush and a variety of cacti.

Along the margin of the arid region, where rainfall amounts are greater, the climate gradually changes into **semi-arid** (BS). This region typically has short grass (*steppe*) and scattered low bushes or trees. Notice in Fig. 21.7 that this climatic region includes most of the Great Plains, the southern coastal sections of California, and the northern valleys of the Great Basin. As in the arid region, northern areas experience lower winter temperatures and more frequent snowfalls. Annual precipitation is generally between 20 and 40 cm (8 and 16 in.).

Along the eastern border of the semi-arid region the annual precipitation increases to about 40 cm (16 in.). Here, the climate is classified as humid. Thus, the semi-arid climate marks the tran-

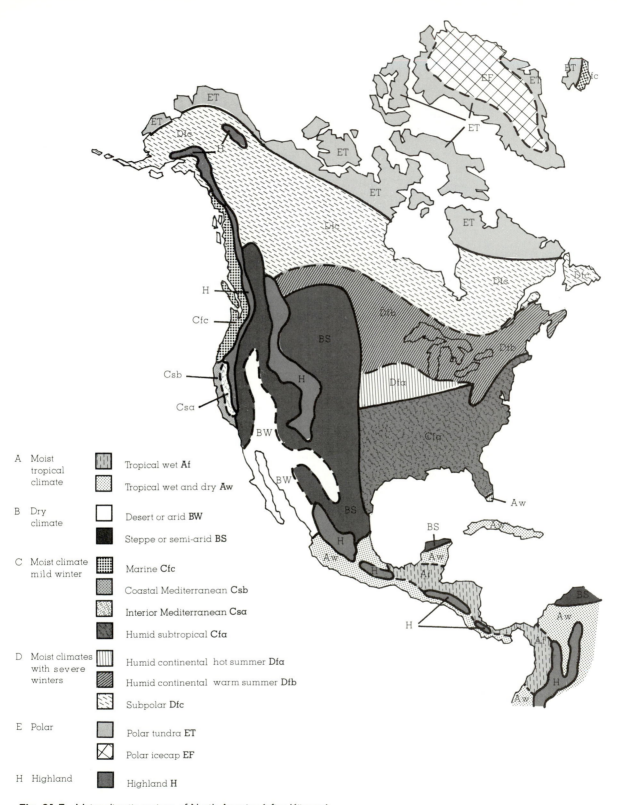

A Moist tropical climate
- Tropical wet **Af**
- Tropical wet and dry **Aw**

B Dry climate
- Desert or arid **BW**
- Steppe or semi-arid **BS**

C Moist climate mild winter
- Marine **Cfc**
- Coastal Mediterranean **Csb**
- Interior Mediterranean **Csa**
- Humid subtropical **Cfa**

D Moist climates with severe winters
- Humid continental hot summer **Dfa**
- Humid continental warm summer **Dfb**
- Subpolar **Dfc**

E Polar
- Polar tundra **ET**
- Polar icecap **EF**

H Highland
- Highland **H**

Fig. 21.7 Major climatic regions of North America (after Köppen).

sition between the arid and the humid climatic regions.

Moist Climates with Mild Winters (Group C)

Characteristics: humid with mild winters (i.e., average temperature of coldest month below 18°C, or 64°F, and above −3°C, or 27°F)

Extent: west coast of North America; southeastern United States

Types: marine (Cfc), *dry-summer subtropical* or *Mediterranean* (Cs), and *humid subtropical* (Cfa)

The Pacific Ocean greatly affects the climate along the west coast of North America. From southeastern Alaska to about central Oregon, the climate is classified as **marine** (Cfc). Because mountains closely parallel the coastline, the marine influence is restricted to narrow belts, where the onshore winds keep summers cool and winters mild, compared to stations located at the same latitude farther inland. In winter, temperatures do drop below freezing when an occasional cold blast of air from the east produces an abnormal cold spell. Snow does fall, but most of the precipitation is in the form of rain. Although adequate in all months, heaviest precipitation occurs in winter. It rains on many days and when it is not raining, skies are usually overcast. The heavy rains produce a dense forest of Douglas fir.

Portland, Oregon, because it has rather dry summers, marks the transition between the marine climate and the dry-summer subtropical climate to the south.

From about the Oregon-California border southward into Baja California, most of the precipitation falls during the winter and spring. The reason for this phenomenon is that the Pacific high, centered off the northern California coast in summer, directs most of the storms to the north. Yearly precipitation amounts range between 35 and 90 cm (14 and 35 in.), with characteristically wet winters. The climate in this region is called **dry-summer subtropical** or **Mediterranean** (Cs) because this climate type is also found along the Mediterranean coast.

Mediterranean climates have cool, wet winters and mild to hot, dry summers. Along the Pacific coastal margins, where upwelling keeps temper-

atures cool all summer long, the climate is called **coastal Mediterranean** (Csb). Here, summer daytime maximum temperatures usually reach about 21°C (70°F), overnight lows often drop below 15°C (59°F). Even though coastal morning fog gives way to afternoon clearing, a strong subsidence inversion usually keeps the cool air (and pollution) near the surface and, as a consequence, skies are often hazy. Precipitation is greatest in the northern region and diminishes sharply as we move south.

In the interior valleys of California, away from the marine influence, summers are much warmer and winters a little cooler. In this **interior Mediterranean climate** (Csa) summer afternoon temperatures usually climb above 35°C (95°F) and occasionally above 40°C (104°F). When the cool, summer marine air penetrates into the area, low temperatures may drop below 15°C (59°F). There is occasional frost in winter, but snow is extremely rare at low elevations. A persistent winter fog is common, especially in the great Central Valley. (See Chapter 8.)

The coastal Mediterranean climate supports tall stands of Douglas fir and coast redwood in the wetter regions to the north. These give way to chaparral with oak and grass in the drier south. The warmer interior valleys support a vegetation cover of grass and annual plants.

In Fig. 21.7 observe that the **humid subtropical climate** (Cfa) occurs in the southeastern United States. This region experiences adequate and fairly well-distributed precipitation throughout the year, with annual averages between 50 and 150 cm (20 and 60 in.). Summers are hot and humid as air flows northward around the western side of the Bermuda high over the Gulf of Mexico and into the region. Afternoon temperatures usually climb above 32°C (90°F) and relative humidities remain high even in the afternoon. Air-mass thunderstorms are frequent during the sultry days of summer. Winters are cool with an occasional outbreak of polar air dropping temperatures well below freezing. Winter precipitation is most often associated with midlatitude storms that sweep through the area. Snow is rare except in the more northern interior.

Generally, coastal regions are somewhat warmer in winter and slightly cooler in summer than the interior regions. The abundant rainfall supports a thick pine forest that becomes mixed with oak at higher latitudes.

Moist Climates with Severe Winters (Group D)

Characteristics: warm summers and cold winters (i.e., average temperature of warmest month exceeds 10°C, or 50°F, and the coldest monthly average drops below −3°C, or 27°F; winters are severe with snowstorms, blustery winds, bitter cold; winter temperatures moderate slightly along northeast coastal margins; climate controlled by North American continent

Extent: north of subtropical climate

Types: *humid continental with hot summers* (Dfa), *humid continental with warm summers* (Dfb), and *subpolar* (Dfc)

In **humid continental climates**, precipitation tends to be adequate and evenly distributed throughout the year. Moving westward toward midcontinent, we find annual precipitation totals dropping, with summers being wetter than winters (Fig. 21.4). In the wetter regions of the United States and southern Canada, typical native vegetation includes forests of spruce, fir, pine, and oak, while the drier regions are populated with tall grasses.

Humid continental climates may be subdivided on the basis of summer temperatures. For example, where summers are long and hot, with the average temperature of the warmest month above 22°C (72°F) and the monthly mean temperature exceeding 10°C (50°F) for at least 4 months, the climate is described as **humid continental with hot summers** (Dfa). Further north, where summers are shorter and the average temperature of the warmest month is below 22°C (72°F) and at least 4 months have average temperatures exceeding 10°C (50°F), the climate is described as **humid continental with warm summers** (Dfb).

When winters are severe and summers short and cool and only 1 to 3 months have mean temperatures exceeding 10°C (50°F), the climate is described as **subpolar** (Dfc). From Fig. 21.7, we can see that this climate occurs in a broad belt across Canada and Alaska. Extremely cold winters coupled with cool summers produce large annual tem-

perature ranges. Precipitation is comparatively light, especially in the interior regions, with most places receiving less than 50 cm (20 in.) annually. A good percentage of the precipitation falls during convective shower activity in summer. The total snowfall is usually not large but the cold air prevents melting, so snow stays on the ground for months at a time. Because of the low temperatures, there is a low annual rate of evaporation that ensures adequate moisture to support the boreal forests of conifers and birches known as *taiga** (Fig. 21.8).

Polar Climates (Group E)

Characteristics: extremely cold winters and summers (i.e., average temperature of the warmest month is below 10°C or 50°F)**

Extent: northern coast of North America; Greenland

Types: *polar tundra* (ET) and *polar ice cap* (EF)

In the **polar tundra** (ET), the average temperature of the warmest month is below 10°C (50°F) but above freezing. Here the ground is permanently frozen to depths of hundreds of meters, a condition known as *permafrost*. Summer weather, however, is just warm enough to thaw out the upper meter or so of soil. Hence, during the summer, the tundra turns swampy and muddy.

Annual precipitation on the tundra is meager, with most stations receiving less than 20 cm (8 in.). In lower latitudes, this would constitute a desert, but in the cold polar regions evaporation rates are very low and moisture remains adequate. Because of the extremely short growing season, tundra vegetation consists of mosses, lichens, dwarf trees, and scattered woody vegetation, fully grown and only several centimeters tall (Fig. 21.9).

When the average temperature for every month drops below freezing, plant growth is impossible and the region is perpetually covered with snow and ice. This climatic type is known as **polar ice**

---

*The term *taiga* is of Russian origin and not only refers to the subpolar forests of North America and Eurasia but also to the climate of the Soviet Union.
**The 10°C summer isotherm is believed to correspond to the poleward limit of tree growth.

Fig. 21.8 Coniferous forests (taiga) such as this occur where winter temperatures are low and precipitation is abundant.

Fig. 21.9 Tundra vegetation in Denali National Park, Alaska. This dry type of tundra is composed mostly of sedges and dwarfed wild flowers that bloom during the brief growing season.

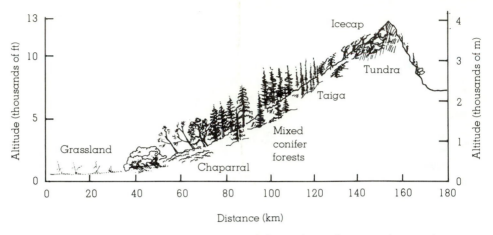

Fig. 21.10 Vertical view of changing vegetation and climate due to elevation in the central Sierra Nevada.

cap (EF). It occupies the interior of Greenland, where the depth of ice in some places measures thousands of meters. In this region, temperatures are never much above freezing and precipitation is extremely meager, with many places receiving less than 10 cm (4 in.) annually. Most precipitation falls as snow during the "warmer" summer.

Highland Climates (Group H)

It is not necessary to visit the polar regions to experience a polar climate. Because temperature decreases with altitude, climatic changes experienced when climbing 300 m (1000 ft) in elevation are about equivalent to horizontal changes experienced when traveling 300 km (186 mi) northward. (This distance is equal to about 3° latitude.) Therefore, when ascending a high mountain one can travel through many climatic regions in a relatively short distance.

Figure 21.10 shows how the climate and vegetation change along the western slopes of the central Sierra Nevada. (See Fig. 21.5 for the precipitation patterns for this region.) Notice that, at the base of the mountains, the climate and vegetation represent semi-arid conditions, while in the foothills the climate becomes Mediterranean and the vegetation changes to chaparral. Higher up, thick fir and pine forests prevail. At still higher elevations, the climate is subpolar and the taiga gives way to dwarf trees and tundra vegetation. Near the summit there are permanent patches of ice and snow, with some small glaciers nestled in protected areas. Hence, in less than 13,000 ver-

tical feet, the climate has changed from semi-arid to polar.

Now that we have looked at climate on a global scale, let's turn our attention to how the climate of a region influences human comfort.

Climate and Human Comfort

As we saw earlier, the climate affects our lives in many ways. Perhaps its greatest effect, however, is on our comfort. In order to survive the cold of winter and the heat of summer, we build homes, heat them, air condition them, insulate them— only to find that when we leave our shelter, we are at the mercy of the climatic environment. Even when we are properly dressed for the weather elements, wind, humidity, and precipitation can make us feel colder or warmer than it really is. In a cold climate, if an individual loses heat too rapidly to the environment, the body temperature may lower and *hypothermia* may set in. In extreme cold, how does an individual maintain a fairly constant inner body temperature? Why is hypothermia more common when the air temperature is above freezing? And, why does the air temperature normally feel higher in a hot, humid climate than in a hot, dry climate? These are some of the questions we will address in the next several sections.

Coping with the Cold The body stabilizes its temperature primarily by converting food into heat (*metabolism*). To maintain a constant temperature, the heat produced and absorbed by the body must be equal to the heat it loses to its surroundings. There is, therefore, a constant exchange of heat—especially at the surface of the skin—between the body and the environment.

One way the body loses heat is by emitting infrared energy. But we not only emit radiant energy, we absorb it as well. Another way the body loses and gains heat is by conduction and convection, which transfer heat to and from the body by air motions. On a cold day, slower-moving "cold" air molecules bounce against the exposed warm skin, gain heat, then dart away, taking some body heat with them. When the air is calm, a thin layer of heated air molecules forms close to the skin, protecting it from the cooler molecules and from the rapid transfer of heat. Thus, in cold weather, when the air is calm, the temperature we perceive—called the **sensible temperature**—is often higher than a thermometer might indicate.

Once the wind starts to blow, the insulating layer of warm molecules is swept away, and heat is rapidly removed from the skin by the constant bombardment of cold air molecules. When all other factors are the same, the faster the wind blows, the greater the heat loss, and the colder we feel. How cold the wind makes us feel is usually expressed as a **wind-chill factor**. The wind-chill chart (Fig. 21.11) translates the ability of the air to take heat away from the body with wind (its cooling power) into an equivalent temperature with no wind. For example, an outside air temperature of −10°C (14°F) with a wind speed of 25 knots (29 mi/hr) produces an equivalent temperature of −34°C (−30°F). This means that a naked body would lose as much heat in one minute in air with a temperature of −10°C and a wind speed of 25 knots as it would in calm air with a temperature of −34°C.* Of course, how cold we feel actually depends on a number of factors, including the fit and type of clothing we wear, and the amount of exposed skin.

High winds, in below freezing air, can remove heat from exposed skin so quickly that the skin may actually freeze and discolor. The freezing of skin, called *frostbite*, usually occurs on the body extremities first because they are the greatest distance from the source of body heat.

*Wind chill is also judged against a wind of 3 knots—the so-called "walking wind speed."

In a cold climate, the moisture content of the air also plays a role in how cold we feel. A cold, damp day often feels colder than a cold "dry" one because moist air conducts heat away from the body better than "dry" air. If exposed skin becomes wet, the removal of body heat is enhanced and the sensible temperature lowers. In fact, in cold, windy, wet weather, a person may lose body heat faster than the body can produce it. This may even occur in relatively mild weather with an air temperature as high as 10°C (50°F). The rapid loss of body heat may lower the body temperature below its normal level and bring on a condition known as **hypothermia**—the rapid, progressive mental and physical collapse that accompanies the lowering of human body temperature.

The first symptom of hypothermia is exhaustion. If exposure continues, judgment and reasoning power begin to disappear. Prolonged exposure, especially at temperatures near or below freezing, produces stupor, collapse, and death when the internal body temperature drops to 26°C (79°F). Most cases of hypothermia occur when the air temperature is between 0°C (32°F) and 10°C (50°F). This may be because many people apparently do not realize that wet clothing in windy weather greatly enhances the loss of body heat, even when the temperature is well above freezing.

In cold weather, heat is more easily dissipated through the skin. To counteract this rapid heat loss, the peripheral blood vessels of the body constrict, cutting off the flow of blood to the outer layers of the skin. In hot weather, the body responds differently.

Some Like It Hot   When the weather turns hot, the body tries to expel heat quickly and efficiently to prevent its internal temperature from rising too high. If the body temperature starts to rise, the brain's *hypothalamus* (a gland that regulates body temperature) activates the body's heat-regulatory mechanisms. The heart rate and blood flow both increase. Blood vessels enlarge, so that more warm blood will circulate closer to the skin's surface, allowing a greater loss of heat to the surroundings.

Even though we lose some heat through breathing and emitting infrared energy, in a warm climate the main source of cooling usually occurs through the process of evaporating perspiration as the hypothalamus calls upon over 10 million sweat

Fig. 21.11 The lines on the chart are equivalent temperatures in °C (left side) and °F (right side). To obtain the equivalent temperature (wind-chill factor), find the intersection of the wind speed and the air temperature. (The broken line corresponds to the example within the text.)

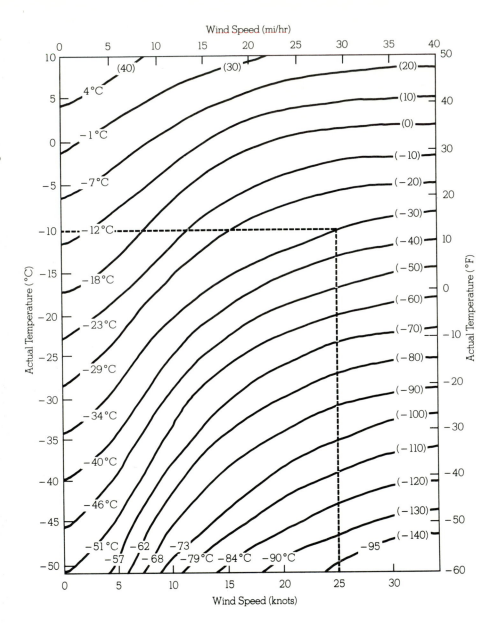

glands to wet the body with as much as two liters per hour. Evaporation removes this moisture and cools the skin.

The *wet-bulb temperature*—because it represents the lowest temperature that can be reached by evaporating water into the air—is a good measure of how cool the skin can become. When the wet-bulb temperature is low, rapid evaporation takes place at the skin's surface. As the wet-bulb temperature rises, less perspiration evaporates, and the skin temperature rises. If the wet-bulb temperature rises above the temperature of the skin, no evaporation occurs, and the body temperature can rise rapidly. Fortunately, most of the time, the wet-bulb temperature is considerably below the skin's temperature.

As perspiration evaporates, the rapid loss of water can produce problems. For example, the loss of salt from the blood and tissues can result in a chemical imbalance that may cause painful *heat cramps*. When water loss exceeds 5 percent of a person's body weight, intense thirst develops as the body temperature and pulse rate increase. These conditions can upset the body's temperature regulator, causing an insufficient supply of blood to the brain, which may result in fatigue, headache, nausea, or even fainting. This condition is commonly known as *heat exhaustion*. A more serious problem—**heatstroke**—occurs when the body temperature rises beyond the critical point of 41°C (106°F). At this temperature, there is complete failure of the circulatory functions. If the body temperature continues to rise, death may follow.

In hot weather, how hot we feel depends not only on the air temperature, but also on the relative humidity, as a temperature of 38°C (100°F) feels much different on a humid summer day in Virginia than on a "dry" summer day in Arizona. On a hot day, when the relative humidity is low, the air is holding a small percent of its capacity for water vapor. Hence, perspiration evaporates quickly into the air, cooling the skin, and we feel cooler than the thermometer indicates. However, when the air temperature and relative humidity are both high, the air is holding close to its capacity for water vapor. Consequently, less perspiration evaporates, less cooling takes place, and we feel warmer than we did in air with a similar temperature, but lower relative humidity. Because a high relative humidity retards evaporation and is mainly responsible for the high sensible temperature, the expression, "It's not the heat, it's the humidity" is common in hot, muggy weather.

In the United States alone, mortality estimates range up to 20,000 for the number of deaths directly related to excessive heat and humidity (from the mid 1930s through the early 1980s). Approximately 1,250 deaths were attributed to the heat wave of 1980. In an effort to draw attention to a serious weather-related health hazard, several indices have been devised.

An index that relates both air temperature and relative humidity to how hot it feels is the **humiture**. (See Table 21.2.) The index works best during the warmest part of the day, and it can be used as a measure of heat stress on persons exercising outdoors.*

*The humiture, developed by George Winterling, combines air temperature and actual vapor pressure in the following formula: $T_h = T + (e - 21)$, where T_h is the humiture (°F), T is the air temperature (°F), and e is the actual vapor pressure of air in millibars.

Table 21.2   The Humiture Index*

AIR TEMPERATURE (°F)	RELATIVE HUMIDITY (%) 10	20	30	40	50	60	70	80	90	100
104 (Uncomfortably warm)	90	98	106	112	120					
102	87	95	102	108	115	123				
100 (Pleasant)	85	92	99	105	111	118				
98	82	90	96	101	107	114				
96	79	87	93	98	103	110	116			
94	77	84	90	95	100	105	111			
92	75	82	87	91	96	101	106	112		
90	73	79	84	88	93	97	102	108		
88	70	76	81	85	89	93	98	102	108	
86	68	73	78	82	86	90	94	98	103	
84	66	71	76	79	83	86	90	94	98	103
82	63	68	73	76	80	83	86	90	93	97
80	62	66	70	73	77	80	83	86	90	93
78 (Crisp, invigorating)		63	67	70	74	76	79	82	85	89
76		60	63	67	70	73	76	79	82	85

(Upper-right region: Unbearably hot)

*A measure of how hot it feels. To obtain the humiture, find the intersection of the air temperature and relative humidity.

An index called the **heat index** (HI), implemented by the National Weather Service during the summer of 1984, combines air temperature with relative humidity to determine an *apparent temperature*—what the air temperature "feels like" to the average person for various combinations of air temperature and relative humidity. For example, in Fig. 21.12, an air temperature of 38°C (100°F) and a relative humidity of 60 percent produce an apparent temperature of 54°C (130°F). Heatstroke or sunstroke is imminent when the index reaches this level.

Even in air with a similar temperature and relative humidity, some people feel the heat before others. This may be due to many factors, including body size. Larger people have less surface area per pound than smaller people. (This makes sense when we realize that we would wind up with less skin by peeling one five-pound potato as compared to peeling five pounds of many smaller potatoes.) The amount of heat that a body can dissipate depends largely on that body's surface area. (Animals that live in the desert are usually slender and have more exposed surface area per body weight.) Consequently, fat people tend to lose body heat more slowly than thin people and, as a result, are the first to feel uncomfortable in hot weather. In cold weather, thinner people are often the first to feel chilly.

Summary

In this chapter, we have examined global temperature and precipitation patterns, as well as the

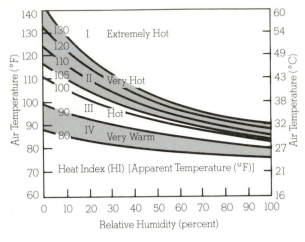

Category	Apparent Temperature (°F)	Heat Syndrome
I	130° or higher	Heatstroke or sunstroke *imminent*
II	105°–130°	Sunstroke, heat cramps, or heat exhaustion *likely*, heatstroke *possible* with prolonged exposure and physical activity.
III	90°–105°	Sunstroke, heat cramps and heat exhaustion *possible* with prolonged exposure and physical activity
IV	80°–90°	Fatigue *possible* with prolonged exposure and physical activity.

Fig. 21.12 The Heat Index (HI). To calculate the apparent temperature, find the intersection of the air temperature and the relative humidity.

various climatic regions throughout the world. Tropical climates are found in low latitudes, where the noon sun is always high, day and night are of nearly equal length, every month is warm, and no real winter season exists. Some of the rainiest places in the world exist in the tropics, especially where warm, humid air rises upslope along mountain ranges.

Dry climates prevail where potential evaporation and transpiration exceed precipitation. Some deserts, such as the Sahara, are mainly the result of sinking air associated with the subtropical highs, while others, due to the rain shadow effect, are found on the leeward side of mountains. Many deserts form in response to both of these effects.

Middle latitudes are characterized by a distinct winter and summer season. Winters tend to be milder in lower latitudes and more severe in higher latitudes. Along the east coast of North America, summers are hot and humid, as air from the Gulf of Mexico sweeps northward, often rising and condensing into afternoon thunderstorms. Along the west coast of North America, summers are drier, as the combination of a cool ocean and sinking air tends to inhibit the formation of towering cumulus clouds.

In the middle of a large continent such as North America, summers are usually wetter than winters. Winter temperatures are generally lower than those experienced in coastal regions. As one moves northward, summers become shorter and winters longer and colder. Polar climates prevail at high latitudes, where winters are severe and there is no real summer. When ascending a high mountain, one can travel through many climatic zones in a relatively short distance.

How cold or warm we feel in a particular climate depends to some extent on the wind and humidity. Wind tends to enhance evaporation and remove body heat. Consequently, wind makes us feel cooler than it really is. In a humid climate, when the relative humidity is high, less perspiration evaporates from the skin, less cooling takes place, and we feel warmer than we would in a climate where the air temperature is the same, but the relative humidity is lower.

Questions for Review

1. Distinguish among microclimate, mesoclimate, and global climate.

2. Describe as many factors as you can that influence the global pattern of precipitation.

3. Describe how a rain shadow desert forms.

4. Explain why, in North America, precipitation typically is a maximum along the west coast in winter, a maximum on the Central Plains in summer, and fairly evenly distributed between summer and winter along the east coast.

5. What constitutes a dry climate?

6. On a blank map of the world, roughly outline where Köppen's major climatic regions are located.

7. What two factors did Thornthwaite emphasize in his first climatic classification system? Why did he revise this system to include the concept of potential evapotranspiration?

8. Define the following terms and then describe in which climatic region each would be observed: tropical rain forest, xerophytes, steppe, taiga, and tundra.

9. What are the controlling factors that produce the following climatic regions: (a) tropical wet and dry; (b) Mediterranean; (c) marine; (d) humid subtropical; (e) subpolar; (f) polar ice cap.

10. Explain why summer afternoons often feel warmer in a hot, humid climate than they do in a hot, "dry" climate with the same air temperature.

Questions for Thought

1. Why do cities directly east of the Rockies (such as Denver, Colorado) receive much more precipitation than cities east of the Sierra Nevada (such as Reno and Lovelock, Nevada)?

2. What climatic controls affect the climate in your area?

3. According to Köppen's climatic system (Fig. 21.6, p. 444), what major climatic type is most abundant (a) in North America, (b) in South America, (c) throughout the world?

4. According to Köppen's system of classifying climates, there are no D climates in the Southern Hemisphere. Why not?

5. In a cold climate, hypothermia can be a problem. Explain why most cases occur in rainstorms rather than in snowstorms.

Problems and Exercises

1. On a map of North America, draw the approximate boundaries for the five major climatic regions. Briefly describe the climate of each region.

Table 21.3 Average Conditions for July

CITY	CLIMATE	AVERAGE DAILY MAXIMUM TEMPERATURE (°F)	AVERAGE MINIMUM RELATIVE HUMIDITY (%)
Denver, Colo.	Semi-arid (BSk)	88	36
Seattle, Wash.	Marine (Csb)	76	50
San Diego, Cal.	Semi-arid (BSk)	77	67
Minneapolis, Minn.	Humid continental (Dfa)	84	54
New Orleans, La.	Humid subtropical (Cfa)	90	62
Phoenix, Ariz.	Arid (BWh)	104	20
New York, N.Y.	Humid subtropical (Cfa)	86	52
Kansas City, Mo.	Humid subtropical (Cfa)	92	52

2. Suppose a city has the mean annual precipitation and temperature given in the table below. Based on Köppen's climatic types, how would this climate be classified? On the map of North America (Fig. 21.7, p. 446), approximately where would this city be located? What type of vegetation would you expect to see there?

3. What is the wind-chill factor when the air temperature is −30°C (−22°F) and the wind speed is 15 knots (17 mi/hr)?

4. (a) What would be the humiture and how would an individual normally feel under the average July conditions you see in Table 21.3?
(b) Use the HI index on p. 454 to determine the apparent temperature for each of the cities listed in Table 21.3.

	Jan.	Feb.	Mar.	Apr.	May	June	July	Aug.	Sept.	Oct.	Nov.	Dec.	Year
Temperature (°F)	40	42	50	60	68	77	80	79	73	62	49	42	60
Precipitation (in.)	4.9	4.2	5.3	3.7	3.8	3.2	4.0	3.3	2.7	2.5	3.4	4.1	45

Ten thousand years ago, additional water from precipitation and melting glaciers formed giant Lake Bonneville—a lake that covered over 50,000 square kilometers of northwestern Utah. Since then, the climate has changed. Evaporation in this semiarid region has gradually lowered the lake until, today, most of the landscape is a flat, salt-covered bed that occasionally becomes dotted with puddles after a summer rain shower. (Photo by author)

CHAPTER 22

Climatic Modification and Change

The climate is always changing. Evidence shows that climate has changed in the past and nothing suggests that it will not continue to do so. Some climatic changes are small, and go practically unnoticed from one year to the next. Others, such as a persistent drought or a delay in the annual monsoon rains, can adversely affect the lives of millions. Even small changes can have an adverse effect when averaged over many years, as grasslands once used for grazing gradually become uninhabited deserts.

In this chapter, we will investigate not only how the global climate has changed, but also some of the theories as to why it has changed. First, however, we will examine climate modifications on a smaller scale—the climate of the urban environment.

The Modification of the Urban Climate

Climatic change can take place on a small scale as green pastures and forests gradually give way to concrete and asphalt. As the urban environment grows, its climate differs from that of the region around it. Sometimes the difference is striking, as when city nights are warmer than the nights of the outlying rural areas. Other times, the difference is subtle, as when a layer of smoke and haze covers the city.

Urban Heat Islands   Probably the most widely studied element of the urban climate is air temperature. For more than 100 years, it has been known that cities are generally warmer than rural areas. This region of city warmth is called the **urban heat island**.

Contents

This warming is attributed to industrial and urban development. In rural areas, a large part of the incoming solar energy is used to evaporate water from vegetation and soil. However, in cities, where less vegetation and exposed soil exists, the majority of the sun's energy is absorbed by urban structures and asphalt. Hence, during warm daylight hours, less evaporative cooling in cities allows surface temperatures to rise higher than in rural areas. Additional city heat is given off by vehicles and factories, as well as by industrial and domestic heating and cooling units.

At night, the solar energy (stored as vast quantities of heat in city buildings and roads) is slowly released into the city air. This slow release of heat tends to keep city temperatures higher than those of the faster cooling rural areas.

On clear, still nights when the heat island is pronounced, a small, thermal low-pressure area forms over the city. Sometimes a light breeze—

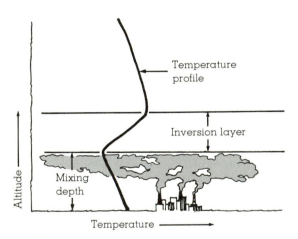

Fig. 22.1 The inversion layer prevents pollutants from escaping into the air above it. If the inversion lowers, the mixing depth decreases and the pollutants are concentrated within a smaller volume.

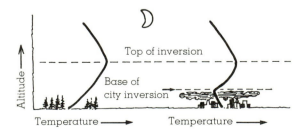

Fig. 22.2 Nighttime temperature profile over a city and a rural area.

called a **country breeze**—blows from the countryside into the city. If there are major industrial areas along the city's outskirts, pollutants are carried into the heart of town, where they tend to concentrate. This is especially true if an inversion inhibits vertical mixing and dispersion.

Air Pollution and the Urban Climate When the atmosphere is unstable, convection currents mix the air and disperse pollution vertically. However, when an inversion exists above a city, such as the one shown in Fig. 22.1, mixing only occurs from the surface up to the base of the inversion. The stable inversion prevents mixing and, therefore, pollutants cannot escape into the air above it. Consequently, the inversion acts like a blanket on the air below. The low-lying unstable mixing layer extending upward from the surface to the base of the inversion is called the **mixing depth**. If the inversion should rise, the mixing depth would increase and the pollutants are dispersed; if the inversion lowers, the mixing depth decreases and the pollutants are more concentrated.

On sunny days, as the ground warms, the layer of mixing usually increases as the inversion is slowly eroded away by rising thermals. At night, as the wind dies down and the air cools and becomes more stable, the mixing layer decreases, and the concentration of pollutants tends to increase.

During the afternoon, the mixing layer over the city is similar to that over the nearby rural area. However, at night, the extra warmth of the city produces a shallow unstable layer near the ground (Fig. 22.2). The depth of the nighttime city mixing layer varies but typically it averages between 100 and 300 m. Pollutants emitted from low-level sources, such as home heating units, tend to concentrate in this shallow air layer, often making it unhealthy to breathe.

When a deep anticyclone moves over a region and stalls, conditions may become ripe for an episode of severe air pollution. At the surface, winds are usually light. Aloft, sinking air warms by compression, producing clear skies and a strong subsidence inversion that suppresses vertical mixing. Near the surface, the weak winds and a shallow mixing layer are unable to adequately disperse the pollutants. This leads to a large concentration of pollution and a condition known as atmospheric stagnation. Some of the worst air pollution disasters (such as the one in the valley city of Donora, Pennsylvania, where in 1948 seventeen persons died within fourteen hours) have occurred when this situation exists for several days to a week or more.

Pollution can influence the climate of a city directly. Certain particles reflect solar radiation, thereby reducing the sunlight that reaches the surface. Some particles serve as nuclei upon which water and ice form. Water vapor condenses onto these particles when the relative humidity is as low as 70 percent, forming haze that greatly reduces visibility. Moreover, the added nuclei increase the frequency of city fog.

Studies suggest that precipitation may be greater in cities than in the surrounding countryside. This may be due partly to the added nuclei and partly to the city heat, which reduces air stability and enhances rising air motions. The rising air aids in forming clouds and thunderstorms. This helps explain why both tend to be more frequent over cities. Table 22.1 summarizes the environmental influence of cities by contrasting the urban environment with the rural.

Climate Modification Downwind of Urban Areas

Apparently, cities can affect the climate and weather downwind from them. For example, La Porte, Indiana, which is located about 50 km downwind of south Chicago industries, has experienced annual precipitation increases of between 30 and 40 percent since 1925. Because this rise closely followed the increase in steel production, it was suggested that the phenomenon was due to the additional emission of particles or moisture (or both) by industries to the west of La Porte.

A study conducted in St. Louis, Missouri, indicated that the average annual precipitation downwind from this city increased by about 10 percent. These increases closely followed industrial development upwind. This study also demonstrated that precipitation amounts were significantly greater on weekdays (when pollution emissions were higher) than on weekends (when pollution emissions were lower). Corroborative findings have been reported for Paris, France, and for other cities as well.

Table 22.1 Contrast of the Urban and Rural Environment*

CONSTITUENT	URBAN AREA (CONTRASTED TO RURAL AREA)
Pollution	higher
Sunlight reaching the surface	lower
Temperature	higher
Relative humidity	lower
Visibility	lower
Fog (frequency)	higher
Wind speed	lower
Precipitation	higher
Cloudiness	higher
Thunderstorm (frequency)	higher

*Values are omitted because they vary greatly depending upon city size, type of industry, and season of the year.

Acid Deposition Air pollution emitted from industrial areas, especially products of combustion, such as oxides of sulfur and nitrogen, can be carried many kilometers downwind. Either these particles and gases slowly settle to the ground in dry form (*dry deposition*) or they are removed from the air during the formation of cloud particles and then carried to the ground in rain and snow (*wet deposition*). **Acid rain** and *acid precipitation* are common terms used to describe wet deposition, while **acid deposition** encompasses both dry and wet acidic substances. How, then, do these substances become acidic?

Emissions of sulfur dioxide and oxides of nitrogen may settle on the local landscape, where they transform into acids as they interact with water, especially during the formation of dew or frost. The remaining airborne particles may transform into tiny dilute drops of sulfuric acid (H_2SO_4) and nitric acid (HNO_3) during a complex series of chemical reactions involving sunlight, water vapor, and other gases. These acid particles may then fall slowly to earth, or they may adhere to cloud droplets or to fog droplets, producing *acid fog*. They may even act as nuclei on which the cloud droplets begin to grow. When precipitation occurs in the cloud, it carries the acids to the ground. Because of this, precipitation is becoming increasingly acidic in many parts of the world, especially downwind of major industrial areas.

Although recent studies suggest that acid precipitation may be nearly worldwide in distribution,

The Pollutants

Carbon monoxide (CO): a major pollutant of city air; colorless, odorless, poison gas; forms during incomplete combustion of carbon-containing fuel; over 75 percent in urban areas comes from road vehicles.

Estimates are that over 100 million tons of CO are spewed into the air annually over the United States alone. Fortunately, it is quickly removed from the atmosphere by microorganisms in the soil. But CO is dangerous even in small amounts. Because it cannot be seen or smelled, it can kill without warning. Your cells obtain oxygen through a pigment in your blood called *hemoglobin*. Hemoglobin picks up oxygen from the lungs, combines with it, and carries it throughout your body. Unfortunately, human hemoglobin prefers carbon monoxide to oxygen (O_2). In fact, the attraction between hemoglobin and CO is 300 times as strong as that between hemoglobin and O_2. If there is much CO in the air, your brain will soon be starved of oxygen and headache, fatigue,

drowsiness and even death may result.*

Sulfur dioxide (SO_2): colorless gas; comes primarily from the burning of sulfur-containing fuels (such as coal and oil, the fossil fuels); primary sources include power plants, heating devices, smelters, and petroleum refineries.

Sulfur dioxide readily oxidizes to form **sulfur trioxide (SO_3)** and, in moist air, **sulfuric acid (H_2SO_4)**. High concentrations produce serious respiratory problems, such as bronchitis and emphysema. Sulfur dioxide also has an adverse effect on plant life. In an attempt to reduce SO_2 emissions, certain industries are using fuel with a low-sulfur content.

*Should you become trapped in your car during a snowstorm and you have your engine and heater running to keep warm, roll down the windows just a little. This will allow any carbon monoxide that may have entered the car through leaks in the exhaust system to escape.

Nitrogen oxides: gases that form when some of the nitrogen in the air reacts with oxygen during the high-temperature combustion of fuel; two primary nitrogen pollutants in cities are **nitrogen dioxide (NO_2)** and **nitric oxide (NO)**.

Although NO_2 and NO are produced by natural bacterial action, their concentration in urban environments is between 10 and 100 times greater than in nonurban areas. In moist air, nitrogen dioxide reacts with water vapor to form corrosive nitric acid. Concentrations of NO_2 may be sufficiently high in cities to color the sky a reddish brown. The primary sources of nitrogen oxides are automobiles, industry, and waste disposal. Nitrogen oxides are removed from the atmosphere by natural processes, including reactions that occur in sunlight. High concentrations irritate the lungs and lower the body's resistance to respiratory infections. Nitrogen oxides are highly reactive gases. They play a key role in the production of

regions noticeably affected are eastern North America, central Europe, and Scandinavia. Sweden contends that most of the sulfur emissions responsible for its acid precipitation are coming from factories in England. In some places, acid precipitation occurs naturally, such as in northern Canada, where natural fires in exposed coal beds produce tremendous quantities of sulfur dioxide.

High concentrations of acid deposition can damage plants and water resources (freshwater ecosystems seem to be particularly sensitive to changes in acidity). Concern centers chiefly on areas where interactions with the soil are unable

to neutralize the acidic inputs. In West Germany, over 200,000 acres of woodland have been seriously damaged by acid deposition. Studies indicate that thousands of lakes in the United States and Canada are so acidified that entire fish populations may have been adversely affected. In an attempt to reduce acidity, lime is being poured into some lakes.

Also, acid deposition is eroding the foundations of structures in many cities throughout the world. In Rome, the acidity of rainfall is beginning to disfigure priceless outdoor fountain sculptures and statues. The estimated annual cost of this dam-

FOCUS ON A SPECIAL TOPIC

ozone and other ingredients of photochemical smog.

Hydrocarbons: individual organic compounds composed of hydrogen and carbon; occur as solids, liquids, and gases at room temperature.

Even though tens of thousands of such compounds are known to exist, **methane** is the most abundant in urban air. An estimated 28 million tons of hydrocarbons are emitted into the air over the United States each year, mostly from gasoline vapors that escape burning in auto engines. Certain hydrocarbons are known to be **carcinogens**—cancer-causing agents. Hydrocarbons react with nitrogen oxides in the presence of sunlight to produce photochemical smog, including ozone (O_3), which is harmful to human health.

Particles: solid particles and liquid droplets small enough to remain suspended in the air; include irritants (such as dust, soil, and smoke) as well as substances that may be highly toxic.

Automobiles and industrial processes emit large quantities of particulate matter into the atmosphere. Particles collected in urban environments include iron, lead, copper, nickel, and carbon. Particle pollution has its most immediate influence on the human respiratory system. Lead particles, most of which enter the air from auto exhaust, are especially dangerous because they are absorbed into the body and accumulate in bone and soft tissues. High concentrations of lead in the human body can cause brain damage, convulsions, and death.

Smog: originally, the combining of smoke and fog; today, the word means the type of smog that forms in large cities, such as Los Angeles.

Because this type of smog forms when chemical reactions take place in the presence of sunlight, it is termed **photochemical smog**. The principal component of photochemical smog is **ozone (O_3)**, a gas that irritates the mucous membranes of the respira-

tory system. Ozone and other oxidants are not emitted into the air directly but form in the atmosphere from reactions involving other pollutants.

In polluted air, hydrocarbons disrupt the reactions because they may react with either oxygen atoms or ozone. Although not everything is known about the reactions that take place among hydrocarbons and other gases, it is believed that these reactions disrupt the normal mechanism for ozone removal, and so the concentration of ozone is able to increase. While this is taking place, the product of a reaction between oxygen and hydrocarbons produces a substance (known as a *hydrocarbon free radical*), which further reacts with O_2 and NO_2 to produce other undesirable pollutants, such as **PAN** (*Peroxyacetyl nitrate*), a pollutant extremely harmful to vegetation and organic compounds. Photochemical smog is the mixture of all of these products.

---

age to building surfaces, monuments, and other structures is more than $2 billion.

Precipitation is naturally acidic. The carbon dioxide occurring naturally in the air dissolves in precipitation, making it slightly acidic with a pH* of 5.6. Consequently, precipitation is considered acidic when its pH is below 5.6. In the northeast-

ern United States, where emissions of sulfur dioxide are primarily responsible for the acid precipitation, typical pH values range between 4.0 and 4.5. (See Fig. 22.3.) But acid precipitation is not confined to the Northeast; the acidity of precipitation has increased rapidly during the past 20 years in the southeastern states too. Further west, rainfall acidity also appears to be on the increase. One area north of Denver reported nearly a sevenfold increase in just 4 years. Along the West Coast, the main cause of acid deposition appears to be the oxides of nitrogen released in automobile exhaust. Acid fog, with a pH of 1.7 in the Los

*The pH scale ranges from 0 to 14, with a value of 7 considered neutral. Values greater than 7 are alkaline and below 7 are acidic. The scale is logarithmic, which means that rain with a pH 3 is 10 times more acidic than rain with pH 4 and 100 times more acidic than rain with pH 5.

Fig. 22.3 Annual average value of pH in precipitation weighted by the amount of precipitation in the United States and Canada for 1980.

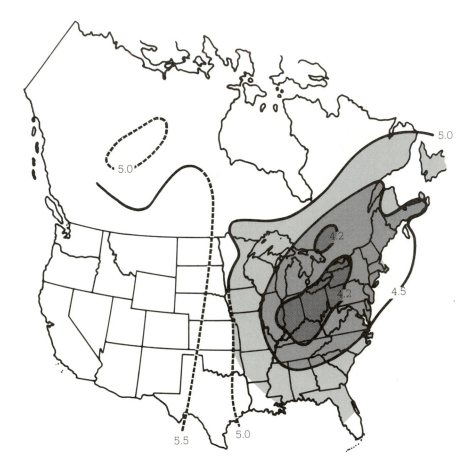

Angeles area, has actually been more acidic than vinegar.

Some scientists believe that if the United States turns more to coal-fired power plants, which are among the leading sources of sulfur oxide emissions, the acid deposition problem will worsen. In fact, there are experts who feel that the acid rain dilemma may be one of the greatest environmental problems facing the world in the near future.

In an attempt to better understand acid deposition, the *Regional Acid Deposition Modeling Project* was developed in 1983. Jointly, the National Center for Atmospheric Research (NCAR) and the Environmental Protection Agency (EPA) have been working together to develop computer models that better describe the many physical and chemical processes contributing to acid deposition. To deal with its acid deposition problem, Canada has recently imposed new pollution control standards and set a goal of reducing industrial air pollution by 50 percent.

Up to this point, we have considered climate modification and change on a relatively small scale. Now, however, we will turn our attention to climatic change over a much broader area—the changing climate of the earth.

The Earth's Changing Climate

We saw earlier in this chapter that the climate of the earth has always been changing. In fact, only about 18,000 years ago the earth was in the grip of a cold spell, with *alpine glaciers* extending their ice fingers down river valleys and huge ice sheets (*continental glaciers*) covering vast areas of North America and Europe. The ice measured several kilometers thick and extended as far south as New York and the Ohio River Valley. At least 4 times during the last 2 million years the glaciers advanced, only to retreat. In the warmer periods between glacier advances, average global temperatures were higher than at present. Hence, we are still in an ice age, but in the comparatively warmer part of it.

Presently, glaciers cover only about 10 percent of the earth's land surface. The total volume of ice over the face of the earth amounts to a little more than 25 million km³. If global temperatures were to rise enough so that all of this ice melted, the level of the ocean would rise about 65 m (213 ft). Imagine the catastrophic results: Many major cities (such as New York, Tokyo, and London) would be inundated. Even a rise in global temperature of several degrees Celsius would be enough to raise sea level by about a meter or so, flooding coastal lowlands.

How has the climate changed, and what evidence is there to suggest that indeed it has changed? These are some of the questions we will address in the next several sections.

Determining Past Climates   The study of the geological evidence left behind by advancing and retreating glaciers is one factor suggesting that global climate has undergone slow but continuous changes. To reconstruct past climates, scientists must examine and then carefully piece together all the available evidence. Unfortunately, the evidence only gives a general understanding of what the climate was like. For example, fossil pollen of a tundra plant collected in a layer of sediment in New England and dated to be 12,000 years old suggests that the climate of that region was much colder than it is today.

Other biological evidence of climatic change comes from the study of annual growth rings of trees, called **dendrochronology**. As a tree grows, it produces a layer of wood cells under its bark. When the tree is cut down, each year's growth appears as a ring. The changes in thickness of the rings indicate climatic changes that may have taken place from one year to the next. The growth of tree rings has been correlated with precipitation and temperature patterns for hundreds of years into the past in various regions of the world.

Still other evidence of global climatic change comes from core samples taken from ocean floor sediments and ice from Greenland. A multiuniversity research project known as CLIMAP (Climate: *long-range investigation mapping and prediction*) studied global climate during the past

million years. Thousands of meters of ocean sediment obtained with a hollow-centered drill were analyzed. The sediment contains the remains of calcium carbonate shells of organisms that once lived near the surface. Because certain organisms live within a narrow range of temperature, the distribution and type of organisms within the sediment indicate the surface water temperature.

In addition, the oxygen-isotope* ratio of these shells provide information about the sequence of glacier advances. For example, the majority of oxygen in sea water is comprised of 8 protons and 8 neutrons in each nucleus, giving an atomic weight of 16. However, about one out of every thousand oxygen atoms contains an extra 2 neutrons, giving it an atomic weight of 18. When ocean water evaporates, the heavy oxygen 18 tends to be left behind. Consequently, during periods of glacier advance, more oxygen 16 is stored on land as ice, and the oceans, which contain less water, have a higher concentration of oxygen 18. Since the shells of marine organisms are constructed from the oxygen atoms existing in ocean water, determining the ratio of oxygen 18 to oxygen 16 within these shells yields information about how the climate may have altered in the past. A higher ratio of oxygen 18 to oxygen 16 in the sediment record suggests a colder climate, while a lower ratio suggests a warmer climate.

Using data such as these, the CLIMAP project was able to reconstruct the earth's surface ocean temperature for various times during the past (Fig. 22.4).

Other data have been used to reconstruct past climates, such as:

1. natural records of lake-bottom sediment and soil deposits
2. the study of pollen in deep ice caves, soil deposits, and sea sediment
3. certain geologic evidence (ancient coal beds, sand dunes, and fossils)
4. documents concerning droughts, floods, and crop yields

Despite all of these data, our knowledge about past climates is still incomplete. Now that we have reviewed *how* the climatologist gains information

*Isotopes are atoms whose nuclei have the same number of protons but different numbers of neutrons.

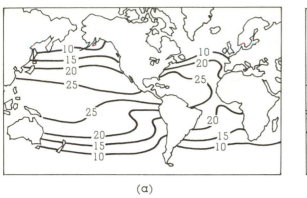 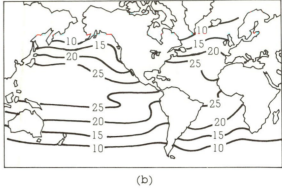

(a)   (b)

Fig. 22.4   (a) Sea surface isotherms (°C) during August 18,000 years ago and (b) during August today.

about the past, let's look at *what* this information reveals.

Climate Through the Ages   Throughout much of the earth's history, the global climate was probably between 8°C and 10°C, warmer than it is today. During most of this time, the polar regions were free of ice. These comparatively warm conditions, however, were interrupted by at least three periods of glaciation. Geologic evidence suggests that one glacial period occurred about 600 million years ago (m.y.a.) and another about 300 m.y.a. The most recent one—the **Pleistocene epoch** or, simply, the **Ice Age**—began about 2 m.y.a. Let's summarize the climatic conditions that led up to the Pleistocene.

About 65 m.y.a., the earth was warmer than it is now; polar ice caps did not exist. Beginning about 55 m.y.a., the earth entered a long cooling trend. After millions of years, polar ice appeared. As average temperatures continued to lower, the ice grew thicker, and by about 10 m.y.a. a deep blanket of ice covered the Antarctic. Meanwhile, snow and ice began to accumulate in high mountain valleys of the Northern Hemisphere, and alpine, or valley glaciers, soon appeared.

About 2 m.y.a., continental glaciers appeared in the Northern Hemisphere, marking the beginning of the Pleistocene epoch. The Pleistocene, however, was not a period of continuous glaciation but a time when glaciers alternately advanced and retreated (melted back) over large portions of North America and Europe. Between the glacial advances were warmer periods called *interglacial periods*, which lasted for 10,000 years or more.

The North American glaciers reached their maximum thickness and extent about 18,000–22,000 years ago (y.a.). At that time, the sea level was perhaps 85 m (280 ft) lower than it is now. The lower sea level exposed vast areas of land, such as the *Bering land bridge* (a strip of land that connected Siberia to Alaska), which allowed human migration from Asia to North America.

The ice began to retreat about 14,000 y.a., and by about 10,000 y.a. the continental ice sheets over North America were gone. From about 7000–5000 y.a. the climate was between 2°C and 3°C warmer than at present. Because this period favored the development of plants, it is called the **climatic optimum**. Then, about 5000 y.a., a cooling trend set in, during which extensive alpine glaciers returned, but not continental ice sheets.

Climate During the Last 1000 Years   About 1000 y.a., the Northern Hemisphere was relatively warm and dry. During this time, vineyards flourished and wine was produced in England, indicating warm, dry summers and the absence of cold springs. It was during this tranquil period of several hundred years that the Vikings colonized Iceland and Greenland.

Sometime around A.D. 1200, the mild climate of western Europe began to show extreme variations. For several hundred years the climate grew stormy. Both great floods and great droughts occurred. Extremely cold winters were followed by relatively warm ones. During the cold spells, the English vineyards and the Viking settlements suffered. Europe experienced several famines during the 1300s.

Around 1400 to 1550, the climate moderated (Fig. 22.5). However, starting in the middle 1550s, the average temperature began to drop. This cooling trend (which continued for almost 300 years) is known as the **Little Ice Age**. During this time, alpine glaciers increased in size and advanced down river canyons. Winters were long and severe, summers short and wet. The vineyards in England vanished and farming became impossible in the more northerly latitudes. Cut off from the rest of the world by an advancing ice pack, the Viking colony in Greenland perished.

During the Little Ice Age, one particular year stands out: 1816. In Europe that year, bad weather contributed to a poor wheat crop, and famine spread across the land. In North America, unusual blasts of cold arctic air moved through Canada and the northeastern United States between May and September. The cold spells brought heavy snow in June and killing frosts in July and August. In the warmer days that followed each cold snap, farmers replanted, only to have another cold outbreak damage the planting. The year 1816 has come to be known as "the year without a summer" or "eighteen hundred and froze-to-death." The unusually cold summer was followed by a bitterly cold winter.

In the late 1800s, the average temperature in the Northern Hemisphere began to rise. From about 1900 until about 1940, the average temperature of the lower atmosphere rose about 0.5°C (<1°F). Following the warmer period, the Northern Hemisphere began to cool by about 0.5°C over the next 25 years or so. In the late 1960s and early 1970s, the cooling trend ended over most of the Northern Hemisphere. During the 1970s and into the 1980s, the average yearly temperature showed considerable fluctuation from year to year and from region to region, with the overall trend pointing to warming (although not all scientists agree on this). The year 1981 was the warmest during the last 100 years, about 0.5°C above the long-term average.

Possible Causes of Climatic Change

Why the earth's climate changes is not totally understood. Many theories attempt to explain the changing climate, but no single theory alone can satisfactorily account for *all* the climatic variations of the geologic past.

Why hasn't the riddle of a fluctuating climate been solved? One major problem facing any comprehensive theory is the intricate interrelationship of the elements involved. For example, if temperature changes, many other elements may be altered as well. The interactions among the atmosphere, the oceans, and the ice are extremely complex and the number of possible interactions among these systems is enormous. No climatic element within the system is isolated from the others. With this in mind, we will first investigate a warmer and a cooler world to see how feedback systems work; then we will consider some of the current theories of climatic change.

Climatic Change and Feedback Mechanisms   Recall (Chapter 3) that the earth-atmosphere system is in a delicate balance between incoming and outgoing energy. If this balance is upset, even slightly, global climate can undergo a series of complicated changes.

Let's assume that the earth-atmosphere system has been disturbed to the point that the earth has entered a slow warming trend. (At this point we need not worry about why—that comes later.) Over the years the temperature slowly rises, and the air's capacity to hold water vapor increases. Water from the ocean rapidly evaporates into the

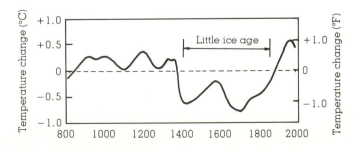

Fig. 22.5   The average temperature variations of eastern Europe for the last 1200 years.

warmer air. The increased quantity of water vapor absorbs more of the earth's infrared energy, thus strengthening the atmosphere (greenhouse) effect. This raises the air temperature even more, which, in turn, further increases the evaporation rate. The atmosphere effect becomes even stronger and the air temperature rises even more. If left unchecked, the temperature would increase until the oceans boiled away. Such a chain reaction is called a *runaway greenhouse effect*. This situation is known as a **positive feedback mechanism** because the initial increase in temperature is reinforced by the other processes that occur.

The earth-atmosphere system has a number of checks and balances that help it readjust itself into a new equilibrium. Hence, a runaway greenhouse effect is not very likely on earth. Helping to counteract the positive feedback mechanisms are **negative feedback mechanisms**—those that tend to weaken the interactions among the variables rather than reinforce them. Let's look at an example of how a negative feedback mechanism might work on a warming planet. As the air warms and becomes more moist, there is a greater likelihood of convection and, perhaps, an increase in global cloudiness. A greater cloud cover reflects a larger percentage of incoming sunlight. Because there is less solar energy to warm the ground, surface temperatures may begin to drop, and a cooling trend begins.

How do feedback systems affect a slow global cooling trend over hundreds or even thousands of years? Lower temperatures would allow for a greater snow cover in middle and high latitudes. Because snow has a high albedo (reflectivity), much of the incident sunlight would be reflected back to space. Less sunlight absorbed at the surface would cause a further drop in temperature. This, in turn, would increase the snow cover, which would lower the temperature even more. This positive feedback mechanism, if left unchecked, would produce a *runaway ice age*. A runaway ice age is not likely because the earth-atmosphere system would respond to the cooling with other feedback mechanisms.

Climate Change and Variations within the Atmosphere We know that sunlight penetrates the atmosphere and warms the surface. The surface, in turn, warms the atmosphere from below

(see Chapter 3). If constituents in the atmosphere either block out some of the incoming sunlight or prevent an additional amount of earth-radiated infrared energy from escaping into space, global climate could be altered.

Climatic Change Due to Increasing Levels of CO_2 We know (see Chapter 2) that carbon dioxide (CO_2) has been increasing steadily in the atmosphere, primarily due to the burning of fossil fuels. However, deforestation may also be adding to this increase as tropical rainforests are removed and replaced with less efficient plants. In 1984, the annual average of CO_2 was about 340 parts per million, and present estimates are that this value may double by the middle of the next century.

Because CO_2 strongly absorbs infrared radiation, it traps heat in the lower atmosphere and plays a major role in the atmosphere (greenhouse) effect. Most numerical model experiments (mathematical models that simulate climate) predict that a doubling of CO_2 will result in a global warming of surface air between 1.5°C and 4.5°C (about 2.5°F to 8°F). There are, however, uncertainties in the models. For example, if deforestation is mainly responsible for the rising levels of CO_2 (rather than the burning of fossil fuels), the increase in CO_2 will occur much more slowly than predicted, and global warming will not be great. Also, the exact effect the oceans will have on the rising levels of CO_2 is uncertain. The oceans are huge storehouses for CO_2. The colder they are, the more CO_2 they can hold; the warmer they are, the less they hold.*

Some climatic models predict that, if average global surface air increases by about 3°C or 4°C, the jet stream might shift from its ''normal'' position, causing a reduction in precipitation over middle latitudes. This, in turn, would put added stress on important agriculture areas, especially those in the western United States that depend greatly on irrigation water from streams and reservoirs. Other consequences might be a rise in global sea level of about half a meter or so, as polar ice begins to melt.

*A few scientists even suggest that the rise in CO_2 levels over the last century is the result of a warmer ocean, rather than the burning of more fossil fuels.

In polar regions, where the warming should be several times greater than in middle and low latitudes, scientists initially turned their attention to ice sheets (especially the west Antarctic ice sheet), to see if shrinkage of the ice might signal a global warming trend due to increasing CO_2 levels. But in polar regions, as elsewhere around the globe, rising temperatures produce complex interactions among temperature, precipitation, and wind patterns. For example, as the average air temperature rises in polar regions, the air's capacity to hold water vapor increases. This, in turn, could lead to added precipitation, which might offset the melting of ice due to higher temperatures.

A few scientists even suggest that increased levels of CO_2 in the atmosphere will have some positive consequences. The higher level of CO_2 will act as a "fertilizer" for some plants, accelerating their growth. Increased plant growth consumes more CO_2, which might retard the increasing rate of CO_2 in the environment. Other scientists feel that the increased plant growth might force some insects to eat more, resulting in a net loss of vegetation. In addition, there is concern that a major increase in CO_2 might upset the balance of nature, with some plant species becoming so dominant that others are eliminated.

Indeed, the interactions between the earth and its atmosphere are complex. Most scientists agree that, because of these complex interactions, more information is needed before definitive answers can be given to the important questions. How much global warming will actually occur? and, Is the recent warming trend due, in fact, to increasing levels of CO_2?

Climatic Change Due to Particles in the Lower Atmosphere   Small particles of dust (especially those less than 5 μm in diameter) spewed into the lower atmosphere from a variety of human activities can remain suspended in the troposphere for several days. (See Fig. 22.6.) The con-

Fig. 22.6 Trapped by an upper-level inversion, a thick blanket of particles hangs over Boulder, Colorado. The effect that particles such as these have on climate is not yet well understood.

Climatic Change Induced by Nuclear War

An increasing number of studies indicate that a nuclear war involving thousands of nuclear detonations might have a disastrous effect on global climate. Sophisticated mathematical models of both large- and small-scale atmospheric circulations even suggest that a medium-scale nuclear war might cause profound climatic changes—in both the Northern and Southern Hemispheres.

Researchers speculate that a nuclear war would raise an enormous pall of thick, sooty smoke from massive fires that would burn for days, even weeks, following an attack. The smoke would drift higher into the atmosphere, where it would be caught in the upper-level westerlies and circle the middle latitudes of the Northern Hemisphere. Unlike soil dust, which mainly scatters and re-flects incoming solar radiation, soot particles readily absorb sunlight. In fact, sooty smoke is a much better absorber of visible solar radiation than infrared radiation. Hence, for several weeks after the war, sunlight would be virtually unable to penetrate the smoke layer, bringing darkness or, at best, twilight at midday. The reduction in solar energy would cause surface air temperatures over land masses to drop well below freezing, even during the summer, resulting in extensive damage to plants and crops. The dark, cold, and gloomy condition that would be brought on by nuclear war is often referred to as **nuclear winter**.

The models predict that the most dramatic drop in temperature would occur over land masses. One model suggests that, in the continental interiors of the Northern Hemisphere, surface air temperatures may fall to as low as $-25°C$ ($-13°F$) within weeks after a large-scale war. The decrease in air temperature would be smaller (perhaps less than $3°C$) over the oceans, where air readily mixes with warm surface water. However, along the coastal margins, where steep temperature gradients exist between land and water, the models suggest that large, intense storms are likely to form.

As the lower troposphere cools, the absorption of solar energy by the smoke particles in the upper troposphere would cause this region to warm. The end result would be a strong, stable temperature inversion extending from the surface up into the higher atmosphere. A strong inversion would lead to a number of adverse effects, such as sup-

stant pouring of such particles into the environment may have an effect on global climate.

The impact of these particles on climate is exceedingly complex and depends upon a number of factors, such as size, shape, color, vertical distribution above the surface, water content of the surface, and surface albedo (reflectivity). For instance, where the particles are bright in relation to the surface below, the amount of sunlight reflected and scattered back to space is likely to increase, causing air temperatures to lower. On the other hand, in regions where the particles are relatively dark compared to the surface below, greater absorption of sunlight by the particles could cause air temperatures to rise.

To further complicate the picture, studies show that, even though certain particles may slightly lower the amount of sunlight reaching the surface, these same particles can selectively absorb out-going infrared radiation from the surface. This, of course, enhances the atmosphere (greenhouse) effect and causes an overall warming of the air. Some scientists even speculate that the pollutant-laden haze layer, which apparently originates from Europe and northern Asia and drifts over the Arctic during the winter, may be responsible for the atmospheric warming in that region. At present, research is still being done, and the net effect of tropospheric dust particles on global climate is not well understood.

Climatic Change Due to Volcanic Eruptions During volcanic eruptions, fine particles of ash and dust (as well as gases) can be ejected into the stratosphere. (See Fig. 22.7.) Scientists agree that the volcanic eruptions having the greatest impact on climate are those rich in sulfur gases. Apparently, volcanic ash falls out

FOCUS ON AN ISSUE

pressing convection, altering precipitation processes, and causing major changes in the general wind patterns. One general circulation model predicts that a large circulating cell would form between the Northern Hemisphere and the Southern Hemisphere. Air aloft, within the cell, would transport smoke into the Southern Hemisphere, where the reduction in sunlight might lead to a lowering of average surface air temperature by 4°C or more. In addition, subfreezing air might extend into tropical latitudes, causing extensive damage to tropical ecosystems.

As the smoke particles gradually settled to earth, sunlight would slowly warm the surface, and air temperatures would rise, perhaps taking 2 years to reach the temperatures of prewar conditions. However, the added sunlight might also lead to additional pollution, as hydrocarbons and oxides of nitrogen (released during the fires) are transformed into photochemical smog. With large quantities of oxides of nitrogen and sulfur in the atmosphere, precipitation might become more acidic (perhaps as low as pH 3). Greater concentrations of atmosphere (greenhouse) effect gasses, such as CO_2 and ozone, might promote a more rapid warming after the smoke had cleared.

If large numbers of nuclear warheads greater than 1 megaton were detonated, the release of oxides of nitrogen into the stratosphere might deplete the ozone shield in the Northern Hemisphere by 50 percent. This, of course, would produce an adverse effect on plants, animals, and humans (see Chapter 4).

For a complete understanding of the impact of a nuclear war on climate, many researchers feel that more realistic models are needed; for example, models that take into account local and seasonal atmospheric variations. Though some of the models are relatively simplistic, the Council of the American Meteorological Society thoroughly recognizes the inevitable and widespread devastation that nuclear war would bring to the atmosphere and to life itself. In 1983, the Council called on the nations of the world to take whatever steps are necessary, such as the adoption of appropriate treaties, to prevent the use of nuclear weapons and avoid the catastrophe of nuclear war.

---

of the stratosphere fairly quickly and does not have a long-lasting effect on climate. However, the sulfur gases, over a period of about 2 months, react with water vapor in the presence of sunlight to produce tiny bright sulfuric acid particles that gradually join together and grow in size, forming a dense haze layer. As heavier particles fall out of the stratosphere, new particles form. And so the haze layer may reside in the stratosphere for years, absorbing and reflecting back to space a portion of the sun's incoming energy. This reduction in sunlight may cause a decrease in average global temperature near the surface.

The largest volcanic eruption of this century, in terms of its sulfur-rich veil, was that of El Chichón in Mexico during April, 1982. The rising cloud from the volcano entered the stratosphere and spread westward. By the end of the month, it had circled the globe. For major eruptions such as this one, mathematical models predict that average hemispheric temperatures can drop by about 0.3°C to 0.5°C for from one to three years. The exact amount of cooling due to El Chichón is difficult to determine, however, because average hemispheric temperatures vary slightly from year to year and because one of the largest El Niño events of this century was occurring at the same time (see Chapter 15). Whether the unusual position of the jet stream (which pushed record cold air across the United States and unseasonably warm air over western Europe during December, 1983) was due to the cooling brought on by El Chichón or by the oceanic warming of El Niño (or a combination of the two) is uncertain at this time. It is interesting to note, however, that the cold year of 1816 (''the year without a summer'') followed the massive eruption in 1815 of Mount Tambora in Indonesia.

In an attempt to correlate sulfur-rich volcanic

Fig. 22.7 Volcanic eruptions rich in sulfur may have an effect on climate. As sulfur gases in the stratosphere transform into tiny bright sulfuric acid particles, they prevent a portion of the sun's energy from reaching the surface.

eruptions with long-term trends in global climate, scientists are measuring the acidity of annual ice layers in Greenland and Antarctica. Generally, the greater the concentration of sulfuric acid particles in the atmosphere, the greater the acidity of the ice layer. Relatively acidic ice has been uncovered from about A.D. 1350 to about 1700, a time that corresponds to the Little Ice Age. Such findings suggest that sulfur-rich volcanic eruptions may have played an important role in triggering this comparatively cool period and, perhaps, other cool periods during the geologic past.

If the Northern Hemisphere were to cool by several degrees Celsius, what immediate effect might this have on weather and climate patterns? Some numerical climate models predict that a cooler Northern Hemisphere would cause a greater contrast in temperature between high and low latitudes that, in turn, would cause the jet stream to move farther south than normal, especially in summer. In order to transfer heat from low to high latitudes, the upper-level flow and the jet stream would circle the world in broad, meandering loops. Well-defined troughs and ridges would form in the flow and remain implanted over a given area for weeks at a time. Vigorous storms would develop in regions near an upper-level trough, while warm and comparatively dry conditions would prevail in areas influenced by a ridge. The total weather pattern would be one of extreme variability, with droughts and floods occurring at the same time in different parts of the world.

Alpine glaciers would slowly advance down river valleys and, in response to this, the sea level might begin to lower. And for crops grown in middle and high latitudes, there would be a greater likelihood of a late spring freeze.

Climatic Change and Variations in Solar Output In the past, it was thought that solar energy does not vary by more than a fraction of a percent over many years. However, recent measurements made by sophisticated radiometers aboard satellites suggest that the sun's energy output may vary considerably more than was thought. One measurement made during the early 1980s showed a decrease of 0.1 percent in the total amount of solar energy reaching the earth over an 18-month period. If this trend were to continue over a span of several decades, it could influence global climate. For example, numerical climatic models predict that a change in solar output of only 0.5 percent per century could alter the earth's climate. And, according to one model, a decrease in solar energy of 1 percent per century would lower the earth's average temperature by 1°C. Does this decrease in solar energy indicate a longer-term trend in solar output, or is it a small variation related to solar sunspot activity? The answer is not presently known.

Scientists have long tried to link sunspots and weather changes. Sunspots are huge magnetic storms that show up as cooler (darker) regions on the sun's surface. They occur in cycles, with the number and size reaching a maximum approximately every 11 years. The decrease in solar energy observed during the early 1980s corresponded to a period of maximum sunspots. In addition, measurements made with a solar tele-scope from 1976 to 1980 showed that during this period, as the number and size of sunspots in-creased, the sun's surface cooled slightly by about 6°C. Sunspots, apparently, block some of the sun's energy from escaping. Some researchers feel that this trapped energy will slowly seep out over a period of many years.

During periods of maximum sunspot activity, the sun's magnetic field is strong; during periods of low activity it weakens. Interestingly, the mag-netic field of the sun reverses every 22 years, during a sunspot minimum. Some scientists point to the fact that periodic 22-year droughts on the Great Plains of the United States seem to cor-relate with this 22-year solar cycle. It should also be noted that the sunspot cycle has not always prevailed. Apparently, between 1645 and 1715, during the period known as the **Maunder mini-mum**,* there were few, if any, sunspots. Although scientists are gaining a better understanding of how the sun operates, much remains to be learned about how solar changes can influence weather and climate on earth.

In summary, many theories have been proposed linking solar variations to changes in climate. To date, none have been proven. However, instru-ments aboard satellites and solar telescopes on the earth are monitoring the sun to observe how the sun's energy output may vary. Because many years of data are needed, it may be some time before the actual relationship between solar vari-ations and climatic changes on earth can be determined.

Climatic Change and Variations in the Earth's Orbit A popular theory ascribing cli-matic changes to variations in the earth's orbit is the **Milankovitch theory**, named for the Yugosla-vian astronomer Milutin Milankovitch, who first proposed the idea in the 1930s. The basic prem-ise of this theory is that, as the earth travels through space, three separate cyclic movements combine to produce variations in the amount of solar energy that falls on the earth.

The first cycle deals with changes in the shape (*eccentricity*) of the earth's orbit as the earth

revolves about the sun. Notice in Fig. 22.8 that the earth's orbit changes from being elliptical to being nearly circular. To go from circular to ellip-tical and back to circular takes about 100,000 years. The greater the eccentricity of the orbit, the greater the variation in solar energy received at the top of the atmosphere between the earth's closest and farthest approach to the sun.

Presently, we are in a period of low eccentricity. The earth is closer to the sun in January and farther away in July (see Chapter 5). The differ-ence in distance (which only amounts to about 3 percent) is responsible for a nearly 7 percent in-crease in the solar energy received at the top of the atmosphere from July to January. When the difference in distance is 9 percent (a highly ec-centric orbit), the difference in solar energy re-ceived will be on the order of 20 percent. In addition, the more eccentric orbit will change the length of seasons in each hemisphere by changing the length of time between the vernal and autumnal equinox.

The second cycle takes into account the fact that, as the earth rotates on its axis, it wobbles like a spinning top. This wobble, known as the *precession of the equinox*, occurs in a cycle of about 22,000 years. Presently, the earth is closer to the sun in January and farther away in July. Due to precession, the reverse will be true in about 11,000 years (see Fig. 22.9). In about 22,000 years we will be back to where we are today. This means, of course, that if everything else remains the same, 11,000 years from now seasonal vari-ations in the Northern Hemisphere should be

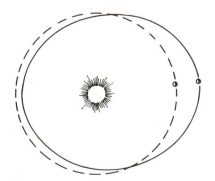

Fig. 22.8 For the earth's orbit to stretch from nearly a cir-cle (dotted line) to an elliptical orbit (solid line) and back again takes nearly 100,000 years. (Diagram is not to scale.)

---

*This period is named after E. W. Maunder, who first discov-ered the low sunspot period sometime in the late 1880s.

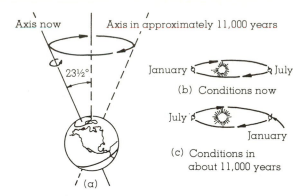

Axis now   Axis in approximately 11,000 years

23½°

January ⟨☀⟩ July
(b) Conditions now

July ⟨☀⟩
            January
(c) Conditions in about 11,000 years

(a)

Fig. 22.9 (a) Like a spinning top, the earth's axis of rotation slowly moves and traces out the path of a cone in space. (b) Presently the earth is closer to the sun in January, when the Northern Hemisphere experiences winter. (c) In about 11,000 years, due to precession, the earth will be closer to the sun in July, when the Northern Hemisphere experiences summer.

greater than at present. The opposite would be true for the Southern Hemisphere.

The third cycle takes about 41,000 years to complete and relates to the changes in tilt (obliquity) of the earth as it orbits the sun. Presently, the earth's orbital tilt is 23½°, but during the 41,000-year cycle the tilt varies from about 22½° to 24½°. The smaller the tilt, the less seasonal variation there is between summer and winter in middle and high latitudes. Thus, winters tend to be milder and summers cooler. During the warmer winters, more snow would probably fall in polar regions due to the air's increased capacity for water vapor. And during the cooler summers, less snow would melt. As a consequence, the periods of smaller tilt would tend to promote the formation of glaciers in high latitudes. In fact, when all of the cycles are taken into account, the present trend should be toward a cooler earth and a new ice age.

In the 1970s, scientists of the CLIMAP project found strong evidence in deep-ocean sediments that variations in climate during the past several hundred thousand years were closely associated with the Milankovitch cycles. Thus, the Milankovitch theory may explain the advance and retreat of ice over periods of 10,000 to 100,000 years. But what caused the Ice Age to begin in the first place? And why have periods of glaciation been so infrequent during geologic time? The Milan-

kovitch theory does not attempt to answer these questions.

Climatic Change and Surface Modifications   During the geologic past, the earth's surface has undergone extensive modifications. One involves the slow shifting of the continents and the ocean floors. This motion is explained in the widely acclaimed **theory of plate tectonics** (formerly called the *theory of continental drift*). According to this theory, the earth's outer shell is composed of huge plates that fit together like pieces of a jigsaw puzzle. The plates, which slide over a partially molten zone below them, move in relation to one another. Continents are embedded in the plates and move along like luggage riding piggyback on a conveyer belt. The rate of motion is extremely slow, only a few centimeters per year.

Besides providing insights into many geological processes, plate tectonics also explains past climates. For example, we find glacial features near sea level in Africa today. This suggests that the area underwent a period of glaciation hundreds of millions of years ago. Were temperatures at low elevations near the equator ever cold enough to produce ice sheets? Probably not. The ice sheets formed when this land mass was located at a much higher latitude. Over the many millions of years since then, the land has slowly moved to its present position. Along the same line, we can see how the fossil remains of tropical vegetation can be found under layers of ice in polar regions today.

According to plate tectonics, the now existing continents were at one time joined together in a single huge continent, which broke apart. Its pieces slowly moved across the face of the earth, thus changing the distribution of continents and ocean basins. Some scientists feel that, when land masses are concentrated in middle and high latitudes, ice sheets are more likely to form. During these times, there is a greater likelihood for the development of the positive feedback mechanism mentioned earlier.

The various arrangements of the continents may also affect the path of ocean currents. This would alter the transport of heat from low to high latitudes and change both the global wind system and the climate in middle and high latitudes.

There are other surface modifications to consider when dealing with climatic changes. For ex-

FOCUS ON A SPECIAL TOPIC

The Sahel—An Example of Climatic Variability and Human Existence

The Sahel is in North Africa, located between about 14° and 18°N latitude. (See Fig. 1.) Bounded on the north by the dry Sahara and on the south by the grasslands of the Sudan, the Sahel is a semi-arid region of variable rainfall. Precipitation totals may exceed 50 cm (20 in.) in the southern portion while in the north, rainfall is scanty. Yearly rainfall amounts are also variable as a year with adequate rainfall can be followed by a dry one.

During the winter, the Sahel is dry but, as summer approaches, the intertropical convergence zone (ITCZ) with its rain usually moves into the region. The inhabitants of the Sahel are mostly nomadic people who migrate to find grazing land for their cattle and goats. In the early and middle 1960s, adequate rainfall led to improved pasture lands; herds grew larger and so did the population. However, in 1968, the annual rains did not reach as far north as usual, marking the beginning of a series of dry years and a severe drought.

Rain fell in 1969, but the totals were far below those of the favorable years in the mid-1960s. The decrease in rainfall along with overgrazing turned thousands of square kilometers of pasture into barren wasteland. By 1973, when the severe drought reached its climax, rainfall totals were 50 percent of the long-term averages and perhaps 50 percent of the cattle and goats had died. The Sahara desert had migrated southward into the northern fringes of the region, and a great famine had

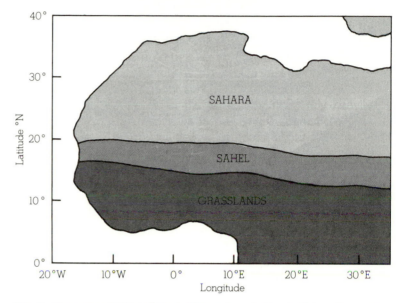

Fig. 1   The semiarid Sahel of North Africa is bounded by the Sahara Desert to the north and grasslands to the south.

taken the lives of more than 100,000 people. Many more of the 2 million or so inhabitants would have perished had it not been for massive outside aid.

The rains returned in 1974 and 1975, but were still 15 to 20 percent below normal. Drought, or at least relatively dry conditions, prevailed in the Sahel from 1976 into the 1980s, with rainfall totals in 1976 and 1977 being comparable to those of 1972 and 1973.

Studies suggest that the wetter years of the 1960s appear to be due to the northward displacement of the ITCZ. The drought, however, appears to be independent of the position of the ITCZ and to be more related to the intensity of rain that falls during the so-called rainy season. Some scientists feel that the lack

of intense rain is due to a *biogeophysical feedback mechanism* wherein less rainfall and reduced vegetation cover modify the surface and promote a positive feedback relationship: Surface changes act to reduce convective activity, which in turn promotes or reinforces the dry conditions. As an example, when the vegetation is removed from the surface, the surface albedo increases and the surface temperature drops. Also, the removal of vegetation reduces the soil moisture and the supply of latent heat to the atmosphere.

Is this a long-term fluctuation in climate, or will the more favorable climate of the early and middle 1960s soon return? Will the Sahara continue to migrate southward? At present, we have no answers.

ample, when large areas of open water are covered with ice, the amount of sensible and latent heat transported to the atmosphere is reduced. In addition, the ice allows snow (a better reflector) to accumulate on top of it. These conditions could lead to even lower temperatures.

Mountain building might play a role in altering climate. For instance, a north-south-trending mountain chain disrupts the flow of air over it. This could lead to changes in the general circulation pattern. Mountains also promote precipitation on their windward slopes. The higher one climbs, the colder it becomes; hence, mountains help to reduce snowmelt in summer.

Another factor that might contribute to the advance and retreat of glaciers is the effect the massive ice sheets have on the land beneath them. For example, as a mass of ice grows thicker, its surface rises to high altitudes, where temperatures are lower and snow is likely to accumulate even faster. The added weight of the ice and snow pushes down on the land beneath it, causing it to

sink. As the ice begins to melt, some of its weight is removed, and the underlying rock slowly rises in response to the lighter load. This lifts the remaining ice to higher altitudes, causing the cycle to begin again.

Many of the theories we have so far described cause climatic changes that span millions of years. Modification of the earth's surface may be taking place right now, modification that could potentially influence the immediate climate in certain areas, especially the microclimate. For example, the overgrazing of grasslands in semiarid regions can increase the surface albedo and cause an increase in desert conditions (a process known as **desertification**). Similar changes in albedo result from the deforestation of large areas.

A report issued in 1984 by the United Nations Environment Program stated that a total of 8.6 billion acres of the world's range and cropland (an area approximately the size of North and South America combined) is affected by desertification. The report goes on to say that each year about 52 million acres are reduced to a state of near or complete uselessness. The main cause is overgrazing, although overcultivation, poor irrigation practices, and deforestation also play a role. The effect this will have on climate, as surface albedos increase and more dust is swept into the air, is uncertain. (However, for a look at how a modified surface is influencing the inhabitants of a region in Africa, read the focus section on the Sahel.)

Fig. 22.10 Altering the microclimate. Notice in the picture that the leaves are still on the tree near the streetlight. Apparently, this sodium vapor lamp emits enough warmth and light during the night to trick the leaves into behaving as if it were September rather than the middle of November.

Summary

In this chapter, we considered some of the many ways climate can be modified and changed. First, we investigated climate modification on a relatively small scale as we examined the urban environment and saw that cities tend to be warmer and more polluted than rural areas. We also saw that industrial areas can modify weather and climate downwind of them. Oxides of sulfur and nitrogen may be swept into the air, where they may transform into acids that fall to the surface as acid precipitation. Acid deposition shows no national boundaries, as the pollution of one country becomes the acid rain of another.

Next, we investigated the changing climate on a global scale and found that the earth's climate has changed considerably during the geologic past. Some of the evidence for a changing climate comes from tree rings (dendrochronology), chemical analysis of oxygen isotopes in fossil shells, and geologic evidence left behind by advancing and retreating glaciers. The evidence from these suggest that, throughout much of the geologic past, the earth was much warmer than it is today. There were cooler periods, however, during which glaciers advanced over large sections of North America and Europe.

We examined some of the possible causes of climatic change, noting that the problem is extremely complex, as a change in one variable in the climatic system almost immediately changes other variables. One theory on climate change proposes that the earth will warm by several degrees Celsius over the next century, due to increasing levels of CO_2 enhancing the atmosphere (greenhouse) effect. Another theory suggests that certain cooler periods in the geologic past may have been caused by volcanic eruptions rich in sulfur. Still another theory postulates that climatic variations on earth might be due to variations in the sun's energy output.

One theory believed to be important in explaining the advance and retreat of glaciers over periods of 10,000 to 100,000 years is the Milankovitch theory. This suggests that the ice ages are the result of small variations in the tilt of the earth's axis and in the geometry of the earth's orbit around the sun.

Even today, modification of the earth's surface may be altering the climate, as overgrazing and deforestation are changing the earth's surface albedo and, in certain regions, rendering the land useless.

Finally, we learned that recent studies indicate that the Northern Hemisphere is in a slight warming trend. Will the climate slowly continue to warm, or will it warm at an accelerated rate due to increasing levels of CO_2? Will the temperature level off, then slowly drop? Will the Northern Hemisphere enter a cooler period, possibly brought on by a slight decrease in the amount of solar energy reaching the earth's surface? Unfortunately, at present, we have no answers to these important questions—only time will tell.

Questions for Review

1. Describe how the climate of a large city usually compares with the climate of surrounding rural areas.

2. What are the main causes of the urban heat island?

3. What causes the "country breeze"? Why is it usually better developed at night than during the day? Would it be best developed in summer or in winter? Explain.

4. Describe how the mixing layer over a city usually changes during the course of a 24-hour day. As it changes, how does it affect the concentration of pollution near the surface?

5. Explain why most severe air pollution episodes are usually associated with high-pressure areas.

6. List the major pollutants and their primary sources.

7. What factors are believed responsible for the increase in precipitation in and downwind of cities?

8. Why is acid deposition considered a serious problem in many regions of the world? How does precipitation become acidic?

9. Describe some of the methods scientists use to determine past climatic conditions.

10. How does the overall climate of the world today compare with the so-called "normal" climate throughout earth's history?

11. Define each of the following: Pleistocene epoch, interglacial period, Bering land bridge, climatic optimum, Little Ice Age.

12. Is a runaway greenhouse effect a positive or negative feedback mechanism? Explain.

13. List as many ways as you can how changes within the earth's atmosphere could alter global climate.

14. Explain how solar variations might influence global climate.

15. Describe the Milankovitch theory of climatic change by explaining how each of the three cycles alters the amount of solar energy reaching the earth.

16. How does the theory of plate tectonics explain the fossil remains of tropical vegetation beneath layers of ice in polar regions today?

17. Describe the biogeophysical feedback mechanism that may be responsible for the lack of intense rain in the African Sahel.

Questions for Thought

1. Generally, when the mixing depth increases, pollutants are dispersed throughout a greater thickness of air. Why, then, do high concentrations of ozone tend to form when the mixing layer is fairly deep?

2. What surface and upper-air conditions lead to atmospheric stagnation?

3. About 18,000 years ago, when glaciation was at a maximum, was global precipitation greater or less than at present? Explain your reasoning.

4. Explain how an increase in the concentration of atmospheric particles could either raise or lower global temperatures.

5. Are ice ages in the Northern Hemisphere more likely when:

(a) the tilt of the earth is at a maximum or a minimum?

(b) the sun is closest to the earth during summer in the Northern Hemisphere, or during winter?

Explain your reasoning for both (a) and (b).

Problems and Exercises

1. If the annual precipitation near Hudson Bay (latitude 55°N) is 38 cm (15 in.) per year, calculate how long it would take snow falling on this region to reach a thickness of 3000 m (about 10,000 ft). (Assume that all the precipitation falls as snow, that there is no melting during the summer, and that the annual precipitation remains constant. To account for compaction of the snow, use a water equivalent of 5 to 1.)

2. Develop a feedback system where increasing levels of CO_2 will lead to lower global temperatures. Within your system, use both positive and negative feedback mechanisms. Devise your system so that it will alternate in a cycle from global cooling to global warming.

APPENDIX A

Units, Conversions, Abbreviations, and Constants

Length

1 kilometer (km)	=	1000 meters (m)
	=	3281 feet (ft)
	=	0.62 mile (mi)
1 mile (mi)	=	5280 feet (ft)
	=	1609 meters (m)
	=	1.61 kilometers (km)
1 centimeter (cm)	=	0.39 inch (in.)
	=	0.01 meter (m)
	=	10 millimeters (mm)
1 inch (in.)	=	2.54 cm
	=	0.08 ft
1 meter (m)	=	100 cm
	=	3.28 ft
	=	39.37 in.
1 micrometer (μm)	=	0.0001 cm
	=	0.000001 m
1 degree latitude	=	111 km
	=	60 nautical mi
	=	69 statute mi

Area

1 square centimeter (cm²)	=	0.15 in.²
1 square inch (in.²)	=	6.45 cm²
1 square meter (m²)	=	10.76 ft²
1 square foot (ft²)	=	0.09 m²

Volume

1 cubic centimeter (cm³)	=	0.06 in.³
1 cubic inch (in.³)	=	16.39 cm³
1 liter (l)	=	1000 cm³
	=	0.264 gallon (gal) U.S.

Speed

1 knot	=	1 nautical mi/hr
	=	1.15 statute mi/hr
	=	0.51 m/sec
	=	1.85 km/hr
1 mile per hour (mi/hr)	=	0.87 knot
	=	0.45 m/sec
	=	1.61 km/hr
1 kilometer per hour (km/hr)	=	0.54 knot
	=	0.62 mi/hr
	=	0.28 m/sec
1 meter per second (m/sec)	=	1.94 knots
	=	2.24 mi/hr
	=	3.60 km/hr

Force

1 dyne	=	1 gram centimeter per second per second
	=	2.2481×10^{-6} pound (lb)
1 newton (N)	=	1 kilogram meter per second per second
	=	10^5 dyne
	=	0.2248 lb

Mass

1 gram (g)	= 0.035 ounce
	= 0.002 lb
1 kilogram (kg)	= 1000 g
	= 2.2 lb

Energy

1 erg	= 1 dyne cm
	= 2.388×10^{-8} cal
1 joule (J)	= 0.239 cal
	= 10^7 erg
1 calorie (cal)	= 4.186 J
	= 4.186×10^7 erg

Pressure

1 millibar (mb)	= 1000 dynes/cm²
	= 0.75 millimeter of mercury (mm Hg)
	= 0.02953 inch of mercury (in. Hg)
	= 0.01450 pound per square inch (lb/in.²)
	= 100 pascals (Pa)
1 standard atmosphere	= 1013.25 mb
	= 760 mm Hg
	= 29.92 in. Hg
	= 14.7 lb/in.²
1 inch of mercury	= 33.865 mb
1 millimeter of mercury	= 1.3332 mb
1 pascal	= 0.01 mb
	= 1 N/m²

Power

1 watt (W)	= 1 J/sec
	= 14.3353 cal/min
1 cal/min	= 0.06973 W
1 horse power (hp)	= 746 W

Temperature °C = 5/9 (°F −32)

To convert degrees Fahrenheit (°F) to degrees Celsius (°C): Subtract 32 degrees from °F, then divide by 1.8

To convert degrees Celsius (°C) to degrees Fahrenheit (°F): Multiply °C by 1.8, then add 32 degrees

To convert degrees Celsius (°C) to Kelvins (K): Add 273 to Celsius temperature, as

$$K = °C + 273.$$

Powers of Ten

Prefix

nano	one-billionth	= 10^{-9} = 0.000000001
micro	one-millionth	= 10^{-6} = 0.000001
milli	one-thousandth	= 10^{-3} = 0.001
centi	one-hundredth	= 10^{-2} = 0.01
deci	one-tenth	= 10^{-1} = 0.1
hecto	one hundred	= 10^2 = 100
kilo	one thousand	= 10^3 = 1000
mega	one million	= 10^6 = 1,000,000
giga	one billion	= 10^9 = 1,000,000,000

Table A.1 SI Units and Their Symbols

QUANTITY	NAME	UNITS	SYMBOL
length	meter	m	m
mass	kilogram	kg	kg
time	second	sec	sec
temperature	kelvin	K	K
density	kilogram per cubic meter	kg/m³	kg/m³
speed	meter per second	m/sec	m/sec
force	newton	m · kg/sec²	N
pressure	pascal	N/m²	Pa
energy	joule	N · m	J
power	watt	J/sec	W

Table A.2   Some Useful Equations and Constants

NAME	EQUATION	CONSTANT VALUE
gas law (equation of state)	$p = \rho RT$	R = 287 J/kg · K (SI) R = 2.87 × 10⁶ erg/g · K
Stefan-Boltzmann law	$R = \sigma T^4$	σ = 5.67 × 10⁻⁸ W/m² · K⁴ (SI) σ = 5.67 × 10⁻⁵ erg/ cm² · K⁴ · sec
Wien's law	$\lambda_{max} = \dfrac{w}{T}$	w = 0.2897 μm K
solar constant		1376 W/m² (SI) 1.97 cal/cm²/min
Coriolis parameter	$2\Omega \sin\phi$	Ω = 7.29 × 10⁻⁵ radian*/ sec
hydrostatic equation	$\dfrac{\Delta p}{\Delta z} = -\rho g$	g = 9.8 m/sec²

*2π radians equal 360°

Weather Symbols and the Station Model

Simplified Surface-Station Model

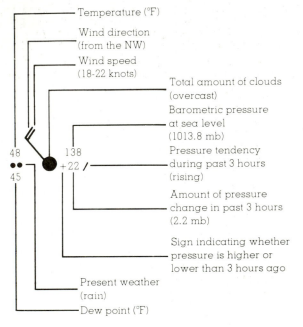

- Temperature (°F)
- Wind direction (from the NW)
- Wind speed (18–22 knots)
- Total amount of clouds (overcast)
- Barometric pressure at sea level (1013.8 mb)
- Pressure tendency during past 3 hours (rising)
- Amount of pressure change in past 3 hours (2.2 mb)
- Sign indicating whether pressure is higher or lower than 3 hours ago
- Present weather (rain)
- Dew point (°F)

Common Weather Symbols

Light rain	Rain shower
Moderate rain	Snow shower
Heavy rain	Showers of hail
Light snow	Drifting or blowing snow
Moderate snow	Dust storm
Heavy snow	Fog
Light drizzle	Haze
Ice pellets (sleet)	Smoke
Freezing rain	Thunderstorm
Freezing drizzle	Hurricane

Upper-Air Model (500 mb)

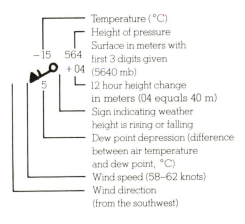

- Temperature (°C)
- Height of pressure Surface in meters with first 3 digits given (5640 mb)
- 12 hour height change in meters (04 equals 40 m)
- Sign indicating weather height is rising or falling
- Dew point depression (difference between air temperature and dew point, °C)
- Wind speed (58–62 knots)
- Wind direction (from the southwest)

Front Symbols

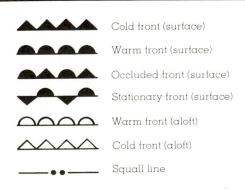

- Cold front (surface)
- Warm front (surface)
- Occluded front (surface)
- Stationary front (surface)
- Warm front (aloft)
- Cold front (aloft)
- Squall line

Total Sky Cover

○ No clouds

◑ Less than one-tenth or one-tenth

◔ Two tenths or three-tenths

◕ Four-tenths

◐ Five-tenths

◒ Six-tenths

◕ Seven-tenths or eight-tenths

◍ Nine-tenths or overcast with openings

● Completely overcast

⊗ Sky obscured

Pressure Tendency

 Rising, then falling

 Rising, then steady; or rising, then rising more slowly

 Rising steadily, or unsteadily

 Falling or steady, then rising: or rising, then rising more quickly

} Barometer now higher than 3 hours ago

 Steady, same as 3 hours ago

 Falling, then rising, same or lower than 3 hours ago

 Falling, then steady; or falling, then falling more slowly

} Barometer now lower than 3 hours ago

 Falling steadily, or unsteadily

Steady or rising, then falling; or falling, then falling more quickly

Wind Entries

	Miles (Statute) Per Hour	Knots	Kilometers Per Hour
◎	Calm	Calm	Calm
—	1-2	1-2	1-3
	3-8	3-7	4-13
	9-14	8-12	14-19
	15-20	13-17	20-32
	21-25	18-22	33-40
	26-31	23-27	41-50
	32-37	28-32	51-60
	38-43	33-37	61-69
	44-49	38-42	70-79
	50-54	43-47	80-87
	55-60	48-52	88-96
	61-66	53-57	97-106
	67-71	58-62	107-114
	72-77	63-67	115-124
	78-83	68-72	125-143
	84-89	73-77	135-143
	119-123	103-107	144-198

APPENDIX C

Changing GMT to Local Time

The system of time used in meteorology is Greenwich Mean Time (GMT), which is also known as Zulu (Z) Time. This is the time measured on the prime meridian (0° longitude) in Greenwich, England. Because Eastern Standard Time (EST) in the United States is 5 hours slower than GMT, to convert from GMT to EST simply requires subtracting 5 hours from GMT. Conversely, to change EST to GMT entails adding 5 hours to EST. Figure C.1 shows how to convert to GMT in various time zones of the United States. Since in meteorology the time is given on a 24-hour clock, Table C.1 shows the relationship between the familiar two 12-hour periods of A.M. and P.M. and the 24-hour system.

Table C.1 Conversion of A.M. and P.M. Time System to 24-Hour System

TIME	24-HR SYSTEM TIME	TIME	24-HR SYSTEM TIME	TIME	24-HR SYSTEM TIME	TIME	24-HR SYSTEM TIME
A.M.		A.M		P.M.		P.M.	
12:00 (midnight)	0000	6:00	0600	12:00 (noon)	1200	6:00	1800
1:00	0100	7:00	0700	1:00	1300	7:00	1900
2:00	0200	8:00	0800	2:00	1400	8:00	2000
3:00	0300	9:00	0900	3:00	1500	9:00	2100
4:00	0400	10:00	1000	4:00	1600	10:00	2200
5:00	0500	11:00	1100	5:00	1700	11:00	2300

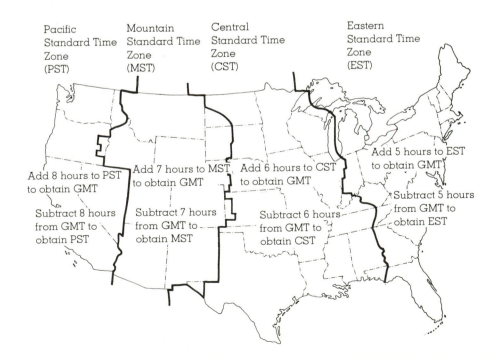

APPENDIX D

Humidity and Dew-Point Tables

To obtain the dew point (or relative humidity), simply read down the temperature column and then over to the wet-bulb depression. For example, in Table D.1 a temperature of 10°C with a wet-bulb depression of 3°C produces a dew-point temperature of 4°C.

Table D.1 Dew-Point Temperature (°C)

AIR (DRY-BULB) TEMPERATURE (°C)	WET-BULB DEPRESSION (DRY-BULB TEMPERATURE MINUS WET-BULB TEMPERATURE) (°C)																
	0.5	1.0	1.5	2.0	2.5	3.0	3.5	4.0	4.5	5.0	7.5	10.0	12.5	15.0	17.5	20.0
−20	−25	−33														
−17.5	−21	−27	−38													
−15	−19	−23	−28													
−12.5	−15	−18	−22	−29												
−10	−12	−14	−18	−21	−27	−36										
−7.5	−9	−11	−14	−17	−20	−26	−34									
−5	−7	−8	−10	−13	−16	−19	−24	−31								
−2.5	−4	−6	−7	−9	−11	−14	−17	−22	−28	−41						
0	−1	−3	−4	−6	−8	−10	−12	−15	−19	−24						
2.5	1	0	−1	−3	−4	−6	−8	−10	−13	−16						
5	4	3	2	0	−1	−3	−4	−6	−8	−10	−48					
7.5	6	6	4	3	2	1	−1	−2	−4	−6	−22					
10	9	8	7	6	5	4	2	1	0	−2	−13					
12.5	12	11	10	9	8	7	6	4	3	2	−7	−28				
15	14	13	12	12	11	10	9	8	7	5	−2	−14				
17.5	17	16	15	14	13	12	12	11	10	8	2	−7	−35			
20	19	18	18	17	16	15	14	14	13	12	6	−1	−15			
22.5	22	21	20	20	19	18	17	16	16	15	10	3	−6	−38		
25	24	24	23	22	21	21	20	19	18	18	13	7	0	−14		
27.5	27	26	26	25	24	23	23	22	21	20	16	11	5	−5	−32	
30	29	29	28	27	27	26	25	25	24	23	19	14	9	2	−11	
32.5	32	31	31	30	29	29	28	27	26	26	22	18	13	7	−2	
35	34	34	33	32	32	31	31	30	29	28	25	21	16	11	4	
37.5	37	36	36	35	34	34	33	32	32	31	28	24	20	15	9	0
40	39	39	38	38	37	36	36	35	34	34	30	27	23	18	13	6
42.5	42	41	41	40	40	39	38	38	37	36	33	30	26	22	17	11
45	44	44	43	43	42	42	41	40	40	39	36	33	29	25	21	15
47.5	47	46	46	45	45	44	44	43	42	42	39	35	32	28	24	19
50	49	49	48	48	47	47	46	45	45	44	41	38	35	31	28	23

Table D.2 Relative Humidity (Percent)

AIR (DRY-BULB) TEMPERATURE (°C)	WET-BULB DEPRESSION (DRY-BULB TEMPERATURE MINUS WET-BULB TEMPERATURE) (°C)																	
	0.5	1.0	1.5	2.0	2.5	3.0	3.5	4.0	4.5	5.0	7.5	10.0	12.5	15.0	17.5	20.0	22.5	25.0
−20	70	41	11															
−17.5	75	51	26	2														
−15	79	58	38	18														
−12.5	82	65	47	30	13													
−10	85	69	54	39	24	10												
−7.5	87	73	60	48	35	22	10											
−5	88	77	66	54	43	32	21	11										
−2.5	90	80	70	60	50	42	37	22	12	3								
0	91	82	73	65	56	47	39	31	23	15								
2.5	92	84	76	68	61	53	46	38	31	24								
5	93	86	78	71	65	58	51	45	38	32	1							
7.5	93	87	80	74	68	62	56	50	44	38	11							
10	94	88	82	76	71	65	60	54	49	44	19							
12.5	94	89	84	78	73	68	63	58	53	48	25	4						
15	95	90	85	80	75	70	66	61	57	52	31	12						
17.5	95	90	86	81	77	72	68	64	60	55	36	18	2					
20	95	91	87	82	78	74	70	66	62	58	40	24	8					
22.5	96	92	87	83	80	76	72	68	64	61	44	28	14	1				
25	96	92	88	84	81	77	73	70	66	63	47	32	19	7				
27.5	96	92	89	85	82	78	75	71	68	65	50	36	23	12	1			
30	96	93	89	86	82	79	76	73	70	67	52	39	27	16	6			
32.5	97	93	90	86	83	80	77	74	71	68	54	42	30	20	11	1		
35	97	93	90	87	84	81	78	75	72	69	56	44	33	23	14	6		
37.5	97	94	91	87	85	82	79	76	73	70	58	46	36	26	18	10	3	
40	97	94	91	88	85	82	79	77	74	72	59	48	38	29	21	13	6	
42.5	97	94	91	88	86	83	80	78	75	72	61	50	40	31	23	16	9	2
45	97	94	91	89	86	83	81	78	76	73	62	51	42	33	26	18	12	6
47.5	97	94	92	89	86	84	81	79	76	74	63	53	44	35	28	21	15	9
50	97	95	92	89	87	84	82	79	77	75	64	54	45	37	30	23	17	11

Table D.3 Dew-Point Temperature (°F)

	WET-BULB DEPRESSION (DRY-BULB TEMPERATURE MINUS WET-BULB TEMPERATURE) (°F)																							
AIR (DRY-BULB) TEMP. (°F)	1	2	3	4	5	6	7	8	9	10	11	12	13	14	15	16	17	18	19	20	25	30	35	40
0	-7	-20																						
5	-1	-9	-24																					
10	5	-2	-10	-27																				
15	11	6	0	-9	-26																			
20	16	12	8	2	-7	-21																		
25	22	19	15	10	5	-3	-15	-51																
30	27	25	21	18	14	8	2	-7	-25															
35	33	30	28	25	21	17	13	7	0	-11														
40	38	35	33	30	28	25	21	18	13	7	-1	-14												
45	43	41	38	36	34	31	28	25	22	18	13	7	-1	-14										
50	48	46	44	42	40	37	34	32	29	26	22	18	13	8	0	-13								
55	53	51	50	48	45	43	41	38	36	33	30	27	24	20	15	9	1	-12						
60	58	57	55	53	51	49	47	45	43	40	38	35	32	29	25	21	17	11	4		-8			
65	63	62	60	59	57	55	53	51	49	47	45	42	40	37	34	31	27	24	19		14			
70	69	67	65	64	62	61	59	57	55	53	51	49	47	44	42	39	36	33	30		26	-11		
75	74	72	71	69	68	66	64	63	61	59	57	55	54	51	49	47	44	42	39		36	15		
80	79	77	76	74	73	71	70	68	66	64	62	60	59	56	54	52	50	47			44	28	-7	
85	84	82	81	80	78	77	75	73	72	70	68	66	64	62	61	59	57	54			52	39	19	
90	89	87	86	85	83	82	80	79	77	76	74	73	72	70	69	67	65	63	61		59	48	32	24
95	94	93	91	90	89	87	86	85	83	81	80	79	78	76	74	73	71	70	68		66	56	43	
100	99	98	98	96	95	94	93	91	91	89	87	86	85	83	82	80	79	77	76	74	72	63	52	37
105	104	103	103	101	100	99	98	96	95	94	93	91	90	89	87	86	84	83	82	80	78	70	61	48
110	109	108	108	106	105	104	103	102	100	99	98	97	95	94	93	91	90	89	87	86	84	77	68	57
115	114	113	113	112	110	109	108	107	106	104	103	102	101	99	98	97	96	94	93	92	90	83	75	65
120	119	118	118	117	115	114	113	112	111	110	108	107	106	105	104	102	101	100	98	97	96	89	81	73

Table D.4 Relative Humidity (Percent)

WET-BULB DEPRESSION (DRY-BULB TEMPERATURE MINUS WET-BULB TEMPERATURE) (°F)

AIR (DRY-BULB) TEMPERATURE (°F)	1	2	3	4	5	6	7	8	9	10	11	12	13	14	15	16	17	18	19	20	25	30	35	40
0	67	33	1																					
5	73	46	20																					
10	78	56	34	13																				
15	82	64	46	29	11																			
20	85	70	55	40	26	12																		
25	87	74	62	49	37	25	13	1																
30	89	78	67	56	46	36	26	16	6															
35	91	81	72	63	54	45	36	27	19	10	2													
40	92	83	75	68	60	52	45	37	29	22	15	7												
45	93	86	78	71	64	57	51	44	38	31	25	18	12	6										
50	93	87	80	74	67	61	55	49	43	38	32	27	21	16	10	5								
55	94	88	82	76	70	65	59	54	49	43	38	33	28	23	19	14	9	5						
60	94	89	83	78	73	68	63	58	53	48	43	39	34	30	26	21	17	13	9	5				
65	95	90	85	80	75	70	66	61	56	52	48	44	39	35	31	27	24	20	16	12				
70	95	90	86	81	77	72	68	64	59	55	51	48	44	40	36	33	29	25	22	19	3			
75	96	91	86	82	78	74	70	66	62	58	54	51	47	44	40	37	34	30	27	24	9			
80	96	91	87	83	79	75	72	68	64	61	57	54	50	47	44	41	38	35	32	29	15	3		
85	96	92	88	84	80	76	73	69	66	62	59	56	52	49	46	43	41	38	35	32	20	8		
90	96	92	89	85	81	78	74	71	68	65	61	58	55	52	49	47	44	41	39	36	24	13	3	
95	96	93	89	85	82	79	75	72	69	66	63	60	57	54	51	49	46	43	41	38	27	17	7	1
100	96	93	89	86	83	80	77	73	70	68	65	62	59	56	54	51	49	46	44	41	30	21	12	4
105	97	93	90	87	83	80	77	74	71	69	66	63	60	58	55	53	50	48	46	43	33	23	15	7
110	97	93	90	87	84	81	78	75	73	70	67	65	62	60	57	55	52	50	48	46	36	26	18	11
115	97	94	91	88	85	82	79	76	74	71	68	66	63	61	58	56	54	52	49	47	37	28	21	13
120	97	94	91	88	85	82	80	77	74	72	69	67	65	62	60	58	55	53	51	49	40	31	23	17

Instant Weather Forecast Chart

This chart is a guide to forecasting the weather. It is applicable to most of the United States, especially the eastern two-thirds. It works best during the fall, winter, and spring, when the weather systems are most active.

Table E.1 Instant Weather Forecast Chart

SEA LEVEL PRESSURE (mb)	PRESSURE TENDENCY	SURFACE WIND DIRECTION	SKY CONDITION	24-HR WEATHER FORECAST (See Weather Forecast Code)
1023 or higher	rising, steady, or falling	any direction	clear, high clouds, Cu	1, 18 (in winter, 14)
1022 to 1016	rising or steady	SW, W, NW, N	clear, high clouds or Cu	1, 18
	falling	SW, S, SE	clear, high clouds	1, 3, 17, 5
		SW, S, SE	middle or low clouds	6, 17
		E, NE	middle or low clouds	6, 14
			clear or high clouds	3, 5, 14
1015 to 1009	rising	SW, W, NW, N	clear	1, 14
			overcast	2, 16
			precipitation	11, 2, 16
	falling	any direction	clear	3, 17 (dry climate summer, 1, 15)
		SW, S, SE	high clouds	3, 17, 5
		SW, S, SE	middle or low clouds	7
	falling	E, NE	middle or low clouds	7, 12, 14
		SE, E, NE	overcast, precipitation	9
		S, SW	overcast, precipitation	10, 13
1008 or below		SW, W, NW, N	clear	1, 12
		SW, W, NW, N	overcast	2, 12, 16
	rising	SW, W, NW, N	overcast with precipitation	11, 12, 16
		NE	overcast	4, 12, 13, 14
	rising	NE	overcast with precipitation	11, 12, 13, 14
		SW, S, SE	clear	3, 6, 8, 12, 15
		SW, S, SE	overcast	7, 8, 12, 13
	falling	SW, S, SE	overcast with precipitation	8, 10, 12, 13, 16
		N	overcast	4, 14
		E, NE	overcast	7, 12, 14
		E, NE	overcast with precipitation	8, 9, 12, 13

Weather Forecast Code

1 = clear or scattered clouds

2 = clearing

3 = increasing clouds

4 = continued overcast

5 = precipitation possible within 24 hours

6 = precipitation possible within 12 hours

7 = precipitation possible within 8 hours

8 = possible period of heavy precipitation

9 = precipitation continuing

10 = precipitation ending within 12 hours

11 = precipitation ending within 6 hours

12 = windy

13 = possible wind shift to W, NW, or N

14 = continued cool or cold

15 = continued mild or warm

16 = turning colder

17 = slowly rising temperatures

18 = little temperature change

Simplified Adiabatic Charts

The adiabatic chart is a valuable tool for anyone who studies the atmosphere. The chart itself is a graph that shows how various atmospheric elements change with altitude. (See Fig. F.5.) At first glance, the chart appears complicated because of its many lines. We will, therefore, construct these lines on the chart step by step.

Figure F.1 shows horizontal lines of pressure decreasing with altitude, and vertical lines of temperature in °C increasing toward the right. The height values on the far right are approximate elevations that have been computed assuming that the air temperature decreases at a standard rate of 6.5°C per km.

In Fig. F.2, the slanted heavy dark lines are called **dry adiabats**. They show how the air temperature would change inside a rising or descending air parcel. Suppose, for example, that an unsaturated air parcel at the surface (pressure 1013 mb) with a temperature of 10°C rises and cools at the dry adiabatic rate (10°C per km). What would be the parcel temperature at a pressure of 900 mb? To find out, simply follow the dry adiabat from the surface temperature of 10°C up to where it crosses the 900-mb line. Answer: 0°C. If the same parcel returns to the surface, follow the dry adiabat back to the surface and read the temperature, 10°C.

On some charts, the dry adiabats are expressed as a potential temperature in Kelvins. The **potential temperature** is the temperature an air parcel would have if it were moved dry adiabatically to a pressure of 1000 mb. Moving parcels to the same level allows them to be observed under identical conditions. Thus, it can be determined which parcels are potentially warmer than others.

The sloping dashed lines in Fig. F.3 are called **moist adiabats**. They show how the air temperature would change inside a rising or descending parcel of saturated air. In other words, they represent the moist adiabatic rate for a rising or sinking saturated air parcel, such as in a cloud (see Chapter 10).

The sloping lines in Fig. F.4 are lines of constant mixing ratio. At any given temperature and pressure, they show how much water vapor the air could hold if it were saturated. At a given dew point, they show how much water vapor the air is actually holding. Hence, given the air temperature and dew point at some level, we can compute the relative humidity of the air. For example, suppose at the surface (pressure 1013 mb) the air temperature and dew point are 29°C and 15°C, respectively. In Fig. F.4, observe that at 29°C the air could hold 26 g of water vapor per kg of dry air, and with a dew point of 15°C the air is actually holding 11 g of water vapor per kg of dry air. This produces a relative humidity of 11/26 × 100%, or 42%.

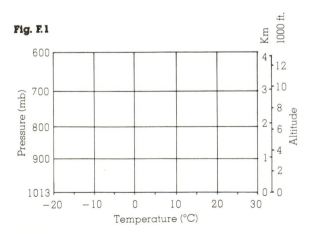

Fig. F.1

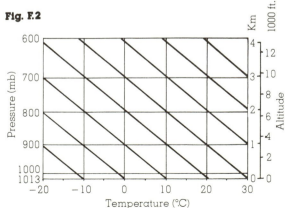

Fig. F.2

Fig. F.3

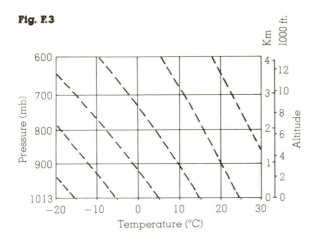

Fig. F.4

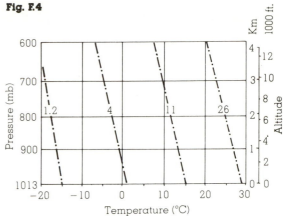

The mixing ratio lines also show how the dew-point temperature changes in a rising or sinking air parcel. If an air parcel with a dew point of 15°C rises from the surface (pressure 1013 mb) up to where the pressure is 700 mb (approximately 3 km), notice in Fig. F.4 that the dew-point temperature inside the parcel would have dropped to a temperature near 9°C.

Figure F.5 shows all of the lines described thus far on a single chart. We have already seen that the chart can be used to obtain graphically a number of atmospheric mathematical relationships. Let's examine one more. Suppose, for example, the surface air temperature is 30°C and the dew point is 21°C. If this air were lifted, at what level would condensation occur and a cloud form?

Assuming that the surface pressure is 1013 mb, follow the mixing ratio line from 21°C up to

where it intersects the dry adiabat that slopes upward from 30°C. In Fig. F.5, notice that the intersection occurs at an altitude near 1 km. This marks the level where the relative humidity is 100% and condensation begins. This level is called the **lifting condensation level** or **LCL.**

Suppose that the cloud continues to build upward from the LCL to a level near 3 km (700 mb). What would be the temperature inside the cloud at this level? From the LCL, follow the moist adiabat up to an altitude of 3 km and read the temperature. (Answer: approximately 10.5°C.)

When data from a radiosonde are plotted on an adiabatic chart, even more information about the atmosphere can be determined, such as atmospheric stability and the possibility of severe weather.

Fig. F.5

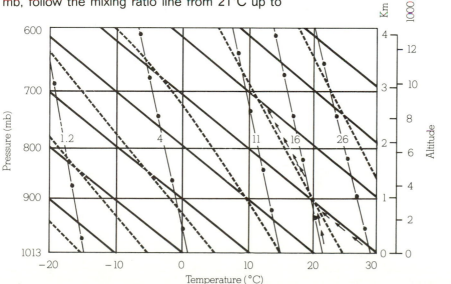

Köppen's Climatic Classification

LETTER SYMBOL			CLIMATIC CHARACTERISTICS	CRITERIA
1st	2nd	3rd		
A			Humid tropical	All months have an average temperature of 18°C (64°F) or higher
	f		Tropical wet (rain forest)	Wet all seasons; all months have at least 6 cm (2.4 in.) of rainfall
	w		Tropical wet and dry (savanna)	Winter dry season; rainfall in driest month is less than 6 cm (2.4 in.) and less than $10 - P/25$ (P is mean annual rainfall in cm)
	m		Tropical monsoon	Short dry season; rainfall in driest month is less than 6 cm (2.4 in.) but equal to or greater than $10 - P/25$
B			Dry	Potential evaporation and transpiration exceed precipitation. The dry/humid boundary is defined by the following formulas: $p = 2t + 28$ when 70% or more of rain falls in warmer 6 months (dry winter) $p = 2t$ when 70% or more of rain falls in cooler 6 months (dry summer) $p = 2t + 14$ when neither half year has 70% or more of rain (p is the mean annual precipitation in cm and t is the mean annual temperature in °C)*
	S		Semiarid (steppe)	The BS/BW boundary is exactly ½ the dry/humid boundary
	W		Arid (desert)	
		h	Hot and dry	Mean annual temperature is 18°C (64°F) or higher
		k	Cool and dry	Mean annual temperature is below 18°C (64°F)
C			Moist with mild winters	Average temperature of coolest month is below 18°C (64°F) and above −3°C (27°F)
	w		Dry winters	Average rainfall of wettest summer month at least 10 times as much as in driest winter month
	s		Dry summers	Average rainfall of driest summer month less than 4 cm (1.6 in.); average rainfall of wettest winter month at least 3 times as much as in driest summer month
	f		Wet all seasons	Criteria for w and s cannot be met
		a	Summers long and hot	Average temperature of warmest month above 22°C (72°F); at least 4 months with average above 10°C (50°F)
		b	Summers long and cool	Average temperature of all months below 22°C (72°F); at least 4 months with average above 10°C (50°F)
		c	Summers short and cool	Average temperature of all months below 22°C (72°F); 1 to 3 months with average above 10°C (50°F)
D			Moist with cold winters	Average temperature of coldest month is −3°C (27°F) or below; average temperature of warmest month is greater than 10°C (50°F)
	w		Dry winters	Same as under C
	s		Dry summers	Same as under C
	f		Wet all seasons	Same as under C
		a	Summers long and hot	Same as under C
		b	Summers long and cool	Same as under C
		c	Summers short and cool	Same as under C
		d	Summers short and cool; winters severe	Average temperature of coldest month is −38°C (−36°F) or below
E			Polar climates	Average temperature of warmest month is below 10°C (50°F)
	T		Tundra	Average temperature of warmest month is greater than 0°C (32°F) but less than 10°C (50°F)
	F		Ice cap	Average temperature of warmest month is 0°C (32°F) or below

*The dry/humid boundary is defined in English units as: $p = 0.44t - 3$ (dry winter); $p = 0.44t - 14$ (dry summer); and, $p = 0.44t - 8.6$ (rainfall evenly distributed). Where p is mean annual rainfall in inches and t is mean annual temperature, °F.

APPENDIX H

Climatic Data for Selected Cities

ALBUQUERQUE, N.M.			**ELEVATION 5311 Ft**		**CLIMATE BSk**		**LAT. 35°N**		**LONG. 106.5°W**			
	Jan.	Feb.	Mar.	April	May	June	July	Aug.	Sept.	Oct.	Nov.	Dec.	Year
t (°F)	35	40	46	56	65	75	78	76	70	58	44	37	57
p (in.)	0.4	0.4	0.5	0.5	0.8	0.6	1.2	1.3	0.9	0.8	0.4	0.5	8.1

ANCHORAGE, ALK.			**ELEVATION 114 Ft**		**CLIMATE Dfc**		**LAT. 61°N**		**LONG. 150°W**			
	Jan.	Feb.	Mar.	April	May	June	July	Aug.	Sept.	Oct.	Nov.	Dec.	Year
t (°F)	12	19	25	37	47	56	58	56	48	36	22	14	36
p (in.)	0.8	0.7	0.5	0.4	0.5	1.0	1.9	2.6	2.5	1.9	1.0	0.9	14.7

ATLANTA, GA.			**ELEVATION 1010 Ft**		**CLIMATE Cfa**		**LAT. 33.5°N**		**LONG. 84.5°W**			
	Jan.	Feb.	Mar.	April	May	June	July	Aug.	Sept.	Oct.	Nov.	Dec.	Year
t (°F)	45	46	51	60	69	77	79	78	73	62	51	45	61
p (in.)	4.4	4.5	5.4	4.5	3.2	3.8	4.7	3.6	3.3	2.4	3.0	4.4	47.1

BALTIMORE, MD.			**ELEVATION 146 Ft**		**CLIMATE Cfa**		**LAT. 39°N**		**LONG. 76.5°W**			
	Jan.	Feb.	Mar.	April	May	June	July	Aug.	Sept.	Oct.	Nov.	Dec.	Year
t (°F)	34	35	43	52	63	72	76	74	68	56	46	36	55
p (in.)	3.7	3.0	3.6	3.7	4.0	3.5	3.9	4.4	3.5	3.4	3.0	3.0	42.6

BARROW, ALK.			**ELEVATION 31 Ft**		**CLIMATE ET**		**LAT. 71°N**		**LONG. 157°W**			
	Jan.	Feb.	Mar.	April	May	June	July	Aug.	Sept.	Oct.	Nov.	Dec.	Year
t (°F)	−16	−18	−15	0	18	33	39	38	30	17	−1	−11	10
p (in.)	0.2	0.2	0.1	0.1	0.1	0.4	0.8	0.9	0.6	0.5	0.2	0.2	4.3

BOSTON, MA.			**ELEVATION 15 Ft**		**CLIMATE Cfa**		**LAT. 42.5°N**		**LONG. 71°W**			
	Jan.	Feb.	Mar.	April	May	June	July	Aug.	Sept.	Oct.	Nov.	Dec.	Year
t (°F)	30	30	38	48	59	68	74	72	65	55	45	33	51
p (in.)	3.9	3.3	4.2	3.8	3.3	3.5	2.9	3.7	3.5	3.1	3.9	3.6	42.8

BUFFALO, N.Y.			**ELEVATION 705 Ft**		**CLIMATE Dfb**		**LAT. 43°N**		**LONG. 78.5°W**			
	Jan.	Feb.	Mar.	April	May	June	July	Aug.	Sept.	Oct.	Nov.	Dec.	Year
t (°F)	24	24	32	44	55	65	70	68	61	51	39	28	47
p (in.)	2.8	2.7	3.2	3.0	3.0	2.5	2.6	3.0	3.1	3.0	3.6	3.0	35.6

CALGARY, CANADA			**ELEVATION 3540 Ft**		**CLIMATE Dfb**		**LAT. 51°N**		**LONG. 114°W**			
	Jan.	Feb.	Mar.	April	May	June	July	Aug.	Sept.	Oct.	Nov.	Dec.	Year
t (°F)	13	17	26	40	50	56	62	60	51	42	28	19	39
p (in.)	0.5	0.5	0.8	1.0	2.3	3.1	2.5	2.3	1.5	0.7	0.7	0.6	16.7

CHICAGO, IL.			**ELEVATION 607 Ft**		**CLIMATE Dfa**		**LAT. 42°N**		**LONG. 88°W**			
	Jan.	Feb.	Mar.	April	May	June	July	Aug.	Sept.	Oct.	Nov.	Dec.	Year
t (°F)	26	28	36	49	60	70	76	74	66	55	40	29	51
p (in.)	1.9	1.6	2.7	3.0	3.7	4.1	3.4	3.2	2.7	2.8	2.2	1.9	33.2

DALLAS, TX.   ELEVATION 481 Ft   CLIMATE Cfa   LAT. 33°N   LONG. 97°W

	Jan.	Feb.	Mar.	April	May	June	July	Aug.	Sept.	Oct.	Nov.	Dec.	Year
t (°F)	46	50	56	65	73	81	85	85	78	68	55	48	66
p (in.)	2.3	2.6	2.8	4.0	4.8	3.2	1.9	1.9	2.8	2.7	2.7	2.7	34.6

DENVER, CO.   ELEVATION 5280 Ft   CLIMATE BSk   LAT. 40°N   LONG. 105°W

	Jan.	Feb.	Mar.	April	May	June	July	Aug.	Sept.	Oct.	Nov.	Dec.	Year
t (°F)	28	32	36	46	56	66	73	72	63	51	38	32	50
p (in.)	0.6	0.7	1.2	2.1	2.7	1.4	1.5	1.3	1.1	1.0	0.7	0.5	14.8

DETROIT, MI.   ELEVATION 619 Ft   CLIMATE Dfa   LAT. 42°N   LONG. 83.5°W

	Jan.	Feb.	Mar.	April	May	June	July	Aug.	Sept.	Oct.	Nov.	Dec.	Year
t (°F)	27	27	35	48	59	70	74	73	65	54	40	30	50
p (in.)	2.0	2.1	2.4	3.0	3.5	2.8	2.8	2.9	2.4	2.6	2.2	2.1	30.9

FARGO, N.D.   ELEVATION 895 Ft   CLIMATE Dfb   LAT. 47°N   LONG. 97°W

	Jan.	Feb.	Mar.	April	May	June	July	Aug.	Sept.	Oct.	Nov.	Dec.	Year
t (°F)	7	11	25	42	55	65	71	69	59	46	28	13	41
p (in.)	0.6	0.7	0.9	1.9	2.2	3.0	2.3	2.7	1.7	1.3	0.9	0.6	18.7

LOS ANGELES, CA.   ELEVATION 97 Ft   CLIMATE BSk   LAT. 34°N   LONG. 118.5°W

	Jan.	Feb.	Mar.	April	May	June	July	Aug.	Sept.	Oct.	Nov.	Dec.	Year
t (°F)	54	55	57	59	62	65	69	69	68	65	61	57	62
p (in.)	2.7	2.9	1.8	1.0	0.1	0.1	0.0	0.0	0.2	0.4	1.1	2.4	12.6

MEMPHIS, TN.   ELEVATION 263 Ft   CLIMATE Cfa   LAT. 35°N   LONG. 90°W

	Jan.	Feb.	Mar.	April	May	June	July	Aug.	Sept.	Oct.	Nov.	Dec.	Year
t (°F)	42	44	52	62	70	78	81	80	74	64	51	43	62
p (in.)	5.6	4.6	5.6	4.8	3.9	3.3	3.2	2.9	2.6	3.3	4.6	5.1	48.4

MIAMI, FL.   ELEVATION 7 Ft   CLIMATE Am   LAT. 26°N   LONG. 80°W

	Jan.	Feb.	Mar.	April	May	June	July	Aug.	Sept.	Oct.	Nov.	Dec.	Year
t (°F)	67	68	70	74	78	81	82	82	81	78	72	68	75
p (in.)	2.0	1.9	2.3	3.9	6.4	7.4	6.8	7.0	9.5	8.2	2.8	1.7	59.8

MINNEAPOLIS—ST. PAUL, MN.   ELEVATION 834 Ft   CLIMATE Dfa   LAT. 45°N   LONG. 93°W

	Jan.	Feb.	Mar.	April	May	June	July	Aug.	Sept.	Oct.	Nov.	Dec.	Year
t (°F)	12	16	27	44	57	67	72	70	60	49	31	18	44
p (in.)	0.7	0.8	1.5	1.8	3.2	4.0	3.3	3.2	2.4	1.6	1.4	0.9	24.8

MONTREAL, CANADA   ELEVATION 187 Ft   CLIMATE Dfa   LAT. 45.5°N   LONG. 73.5°W

	Jan.	Feb.	Mar.	April	May	June	July	Aug.	Sept.	Oct.	Nov.	Dec.	Year
t (°F)	16	18	28	43	57	66	72	70	61	48	36	43	45
p (in.)	3.8	3.0	3.5	2.6	3.1	3.4	3.7	3.5	3.7	3.4	3.5	3.6	40.8

NEW ORLEANS, LA.   ELEVATION 9 Ft   CLIMATE Cfa   LAT. 30°N   LONG. 90°W

	Jan.	Feb.	Mar.	April	May	June	July	Aug.	Sept.	Oct.	Nov.	Dec.	Year
t (°F)	56	58	63	70	76	82	83	83	80	73	62	57	70
p (in.)	4.8	4.2	6.6	5.4	5.4	5.6	7.1	6.4	5.8	3.7	4.0	4.6	63.5

NEW YORK, N.Y.   ELEVATION 13 Ft   CLIMATE Cfa   LAT. 41°N   LONG. 74°W

	Jan.	Feb.	Mar.	April	May	June	July	Aug.	Sept.	Oct.	Nov.	Dec.	Year
t (°F)	32	32	39	49	60	70	76	74	68	58	46	35	53
p (in.)	3.2	2.9	4.2	3.5	3.7	3.4	4.0	5.0	4.2	3.2	3.5	3.2	43.9

OMAHA, NB.   ELEVATION 978 Ft   CLIMATE Dfa   LAT. 41.5°N   LONG. 96°W

	Jan.	Feb.	Mar.	April	May	June	July	Aug.	Sept.	Oct.	Nov.	Dec.	Year
t (°F)	23	27	38	52	63	73	78	76	67	56	39	27	52
p (in.)	0.8	0.9	1.3	2.2	2.8	4.5	3.3	3.1	3.2	1.7	1.3	0.8	25.9

PHOENIX, AZ.   ELEVATION 1117 Ft   CLIMATE BWh   LAT. 33.5°N   LONG. 112°W

	Jan.	Feb.	Mar.	April	May	June	July	Aug.	Sept.	Oct.	Nov.	Dec.	Year
t (°F)	50	54	59	67	75	84	90	88	83	71	58	52	69
p (in.)	0.7	0.8	0.7	0.3	0.1	0.1	0.8	1.1	0.7	0.5	0.5	0.8	7.2

PITTSBURGH, PA.   ELEVATION 1137 Ft   CLIMATE Cfa   LAT. 40.5°N   LONG. 80°W

	Jan.	Feb.	Mar.	April	May	June	July	Aug.	Sept.	Oct.	Nov.	Dec.	Year
t (°F)	29	29	37	49	60	68	72	71	64	53	41	31	50
p (in.)	3.0	2.2	3.3	3.1	3.9	3.8	3.9	3.3	2.5	2.5	2.2	2.4	36.1

PORTLAND, ME.   ELEVATION 43 Ft   CLIMATE Dfb   LAT. 44°N   LONG. 70°W

	Jan.	Feb.	Mar.	April	May	June	July	Aug.	Sept.	Oct.	Nov.	Dec.	Year
t (°F)	22	23	31	42	53	62	68	67	59	49	38	26	45
p (in.)	4.4	3.8	4.3	3.7	3.4	3.2	2.9	2.4	3.5	3.2	4.2	3.8	42.8

PORTLAND, OR.   ELEVATION 30 Ft   CLIMATE Csb   LAT. 45.5°N   LONG. 124°W

	Jan.	Feb.	Mar.	April	May	June	July	Aug.	Sept.	Oct.	Nov.	Dec.	Year
t (°F)	40	44	48	54	59	64	68	68	64	56	47	42	55
p (in.)	5.4	4.9	4.2	2.4	1.9	1.6	0.4	0.6	1.8	3.5	6.0	7.1	39.9

RENO, NV.   ELEVATION 4397 Ft   CLIMATE BSk   LAT. 39.5°N   LONG. 120°W

	Jan.	Feb.	Mar.	April	May	June	July	Aug.	Sept.	Oct.	Nov.	Dec.	Year
t (°F)	31	36	41	48	55	62	70	67	60	51	40	33	50
p (in.)	1.0	1.0	0.7	0.5	0.5	0.4	0.2	0.2	0.2	0.6	0.6	0.9	7.0

ST. LOUIS, MO.   ELEVATION 535 Ft   CLIMATE Cfa   LAT. 38.5°N   LONG. 90°W

	Jan.	Feb.	Mar.	April	May	June	July	Aug.	Sept.	Oct.	Nov.	Dec.	Year
t (°F)	32	35	43	55	64	74	78	77	70	58	44	35	55
p (in.)	2.0	2.0	3.1	3.7	3.7	4.3	3.3	3.0	2.8	2.9	2.6	2.0	35.3

SALT LAKE, UT.   ELEVATION 4220 Ft   CLIMATE BSk   LAT. 41°N   LONG. 112°W

	Jan.	Feb.	Mar.	April	May	June	July	Aug.	Sept.	Oct.	Nov.	Dec.	Year
t (°F)	26	33	41	50	59	67	77	74	64	53	39	32	51
p (in.)	1.2	1.2	1.7	1.8	1.6	0.9	0.6	1.0	0.7	1.3	1.4	1.3	14.7

SAN DIEGO, CA.   ELEVATION 13 Ft   CLIMATE BSk   LAT. 33°N   LONG. 117°W

	Jan.	Feb.	Mar.	April	May	June	July	Aug.	Sept.	Oct.	Nov.	Dec.	Year
t (°F)	55	56	59	62	64	66	70	72	70	66	62	57	63
p (in.)	2.0	2.2	1.6	0.8	0.2	0.1	0.0	0.1	0.2	0.5	0.9	2.0	10.4

SEATTLE, WA.   ELEVATION 387 Ft   CLIMATE Csb   LAT. 47.5°N   LONG. 122.5°W

	Jan.	Feb.	Mar.	April	May	June	July	Aug.	Sept.	Oct.	Nov.	Dec.	Year
t (°F)	38	41	44	49	56	60	65	64	60	52	44	41	51
p (in.)	5.7	4.2	3.8	2.4	1.7	1.6	0.8	1.0	2.0	4.0	5.4	6.3	38.9

ADDITIONAL READING MATERIAL

Periodicals

Selected nontechnical periodicals that contain articles on weather and climate.

Bulletin of the American Meteorological Society. Monthly. The American Meteorological Society, 45 Beacon St., Boston, Mass. 02108.

Meteorological Magazine. Monthly. British Meteorological Office, British Information Services, 845 Third Avenue, New York, New York.

National Weather Digest. Quarterly. National Weather Association, 4400 Stamp Road, Room 404, Marlow Heights, MD 20031. (Deals mainly with weather forecasting.)

NOAA. Bimonthly. Office of Public Affairs, National Oceanic and Atmospheric Administration, Rockville, MD 20852.

Weather. Monthly. Royal Meteorological Society, James Glaisher House, Grenville Place, Bracknell, Berks, United Kingdom.

Weatherwise. Bimonthly. Heldref Publications, 4000 Albermarle St., N.W., Washington, D.C. 20016.

Selected Technical Periodicals

EOS—Transaction of the American Geophysical Union. American Geophysical Union (AGS), Washington, D.C.

Journal of Atmospheric and Oceanic Technology. American Meteorological Society (AMS), Boston, Mass.

Journal of Atmospheric Science. AMS, Boston, Mass.

Journal of Climate and Applied Meteorology. American Meteorological Society (AMS), Boston, Mass.

Journal of Geophysical Research. American Geophysical Union, Washington, D.C.

Monthly Weather Review. AMS, Boston, Mass.

Additional periodicals that frequently contain articles of meteorological interest.

American Scientist. Bimonthly. Sigma Xi, the Scientific Research Society, Inc., New Haven, Conn.

Science. Weekly. American Association for the Advancement of Science, Washington, D.C.

Scientific American. Monthly. Scientific American Inc., New York, New York.

Smithsonian. Monthly. The Smithsonian Association, Washington, D.C.

Books

The titles listed below may be drawn upon for additional information. Many are written at the introductory level. Those that are more advanced are marked with an asterisk.

Anderson, Bette R. *Weather in the West*, American West Publishing Co., Palo Alto, CA, 1975.

Anthes, A. A. *Tropical Cyclones: Their Evolution, Structure, and Effect*, American Meteorological Society, Boston, Mass., 1982.

Bach, W. *Atmospheric Pollution*, McGraw-Hill, New York, 1972.

Battan, Louis J. *Harvesting the Clouds*, Anchor Books, Doubleday and Co., Garden City, N.Y., 1969.

———. *The Nature of Violent Storms*, Anchor Books, Doubleday and Co., Garden City, N.Y., 1961.

*Byers, H. R. *General Meteorology*, McGraw-Hill, New York, 1974.

Craig, R. A. *The Edge of Space*, Anchor Books, Doubleday and Co., Garden City, N.Y., 1968.

Edinger, James G. *Watching for the Wind*, Anchor Books, Doubleday and Co., Garden City, N.Y., 1967.

*Fleagle, Robert G. and Joost A. Businger. *An Introduction to Atmospheric Physics* (2nd ed.), Academic Press, New York, 1980.

Geiger, R. *The Climate Near the Ground*, Harvard University Press, Cambridge, Mass., 1965.

Goody, Richard M. and C. G. Walker. *Atmospheres*, Prentice-Hall, Englewood Cliffs, N.J., 1972.

Greenler, Robert. *Rainbows, Halos and Glories*, Cambridge University Press, New York, 1980.

Hartmann, William K. *Astronomy: The Cosmic Journey* (2nd ed.), Wadsworth, Belmont, CA, 1981.

Hewitt, Paul G. *Conceptual Physics: A New Introduction to Your Environment* (3rd ed.), Little, Brown, Boston, Mass., 1977.

Holdgate, M. F. *A Perspective of Environmental Pollution*, Cambridge University Press, New York, 1979.

*Holton, James R. *An Introduction to Dynamic Meteorology* (2nd ed.), Academic Press, New York, 1979.

Hughes, Patrick. *American Weather Stories*, U.S. Department of Commerce, NOAA, Washington, D.C., 1976.

Imbrie, John, and K. P. Imbrie. *Ice Ages: Solving the Mystery*, Enslow Publishers, Short Hills, N.J., 1979.

Inadvertent Climate Modification. Report of the Study of Man's Impact on Climate (SMIC), The MIT Press, Cambridge, Mass., 1971.

International Cloud Atlas. World Meteorological Organization, Geneva, Switzerland, 1956.

Kellogg, W. W. and M. Mead, Eds. *The Atmosphere: Endangered and Endangering*, U.S. Government Printing Office, Washington, D.C., 1977.

Kotsch, William J. *Weather for the Mariner* (2nd ed.), Naval Institute Press, Annapolis, MD, 1977.

Ladurie, E. L. R. *Times of Feast, Times of Famine: A History of Climate Since the Year 1000*, Doubleday and Co., Garden City, N.Y., 1971.

Landsberg, H. E. *Weather and Health*, Anchor Books, Doubleday and Co., Garden City, N.Y., 1969.

Lowry, W. D. *Weather and Life: An Introduction to Biometeorology*, Oregon State University Book Store, Inc., Corvallis, Oregon, 1968.

*Ludlam, F. H. *Clouds and Storms: The Behavior and Effects of Water in the Atmosphere*, The Pennsylvania State University Press (through AMS), 1980.

Ludlum, D. M. *The Country Journal New England Weather Book*, Houghton Mifflin, Boston, Mass., 1976.

————. *Early American Winters: 1604–1870*, American Meteorological Society, Boston, Mass., 1967.

————. *Weather Record Book*, Weatherwise, Inc., Princeton, N.J., 1971.

Mason, B. J. *Clouds, Rain and Rainmaking* (2nd ed.), Cambridge University Press, New York, 1975.

Mather, J. R. *Climatology: Fundamentals and Applications*, McGraw-Hill, New York, 1974.

Middleton, W. E. K. *The Invention of the Meteorological Instruments*, Johns Hopkins University Press, Baltimore, MD, 1969.

National Academy Press. *Acid Deposition—Atmospheric Processes in Eastern North America*, Washington, D.C., 1983.

————. *Causes and Effects of Stratospheric Ozone Reduction: An Update*, Washington, D.C., 1982.

————. *Changing Climate: A Report of the Carbon Dioxide Assessment Committee*, Washington, D.C., 1983.

————. *Solar Variability, Weather, and Climate*, Washington, D.C., 1982.

National Research Council. *Severe Storms: Prediction, Detection, and Warning*, National Academy of Sciences, Washington, D.C., 1977.

*Palmen, E. and C. W. Newton. *Atmospheric Circulation Systems*, Academic Press, New York, 1969.

Reiter, E. R. *Jet Streams: How Do They Affect Our Weather?* Anchor Books, Doubleday and Co., Garden City, N.Y., 1967.

Roberts, Walter Orr and Henry Landsford. *The Climate Mandate*, W. H. Freeman and Co., San Francisco, 1979.

*Rogers, R. R. *A Short Course in Cloud Physics* (2nd ed.), Pergamon Press, Oxford, England, 1978.

Schaeter, Vincent J. and John A. Day. *A Field Guide to the Atmosphere*, Houghton Mifflin (through AMS), 1981.

Schroeder, Mark S. and Charles C. Buck. *Fire Weather*, U.S. Department of Agriculture, Washington, D.C., 1970.

Simpson, Robert H. and Herbert Riehl. *The Hurricane and Its Impact*, Louisiana State University Press, Baton Rouge, La., 1981.

Stanford, John L. *Tornado: Accounts of Tornadoes in Iowa*, Iowa State University Press, Ames, Iowa, 1977.

U.S. Department of Commerce, NOAA. *Aviation Weather*, U.S. Department of Agriculture, Washington, D.C., 1975.

*Wallace, J. M. and P. V. Hobbs. *Atmospheric Science: An Introductory Survey*, Academic Press, New York, 1977.

GLOSSARY

Absolute humidity   The mass of water vapor in a given volume of air. It represents the density of water vapor in the air.

Absolute zero   A temperature reading of $-273°C$, $-460°F$, or 0K. Theoretically, there is no molecular motion at this temperature.

Absolutely stable air   An atmospheric condition that exists when the environmental lapse rate is less than the moist adiabatic rate.

Absolutely unstable air   An atmospheric condition that exists when the environmental lapse rate is greater than the dry adiabatic rate.

Accretion   The growth of a precipitation particle by the collision of an ice crystal or snowflake with a supercooled liquid droplet that freezes upon impact.

Acid deposition   The depositing of acidic particles (usually sulfuric acid and nitric acid) at the earth's surface. Acid deposition occurs in dry form (*dry deposition*) or wet form (*wet deposition*). Acid rain and acid precipitation often denote wet deposition. (*See* acid rain.)

Acid rain   Cloud droplets or raindrops combining with gaseous pollutants, such as oxides of sulfur and nitrogen, to make falling rain (or snow) acidic—pH less than 5.6. If fog droplets combine with such pollutants it becomes *acid fog*.

Actual vapor pressure   *See* Vapor pressure.

Adiabatic process   A process that takes place without a transfer of heat between the system (such as an air parcel) and its surroundings. In an adiabatic process compression always results in warming, and expansion results in cooling.

Advection   The horizontal transfer of any atmospheric property by the wind.

Advection fog   Occurs when warm, moist air moves over a cold surface and the air cools to below its dew point.

Aerovane   A wind instrument that indicates or records both wind speed and wind direction.

Air glow   A faint glow of light emitted by excited gases in the upper atmosphere. Air glow is much fainter than the aurora.

Air mass   A large body of air that has similar horizontal temperature and moisture characteristics.

Air-mass thunderstorm   A thunderstorm produced by local convection within an unstable air mass.

Air mass weather   A persistent type of weather that may last for several days (up to a week or more). It occurs when an area comes under the influence of a particular air mass.

Air parcel   *See* Parcel of air.

Air pressure (atmospheric pressure)   The pressure exerted by the weight of air above a given point. Usually expressed in millibars or inches of mercury.

Aitken nuclei   *See* Condensation nuclei.

Albedo   The percent of radiation returning from a surface compared to that which strikes it.

Aleutian low   The subpolar low-pressure area that is centered near the Aleutian Islands on charts that show mean sea level pressure.

Altimeter   An instrument that indicates the altitude of an object above a fixed level. Pressure altimeters use an aneroid barometer with a scale graduated in altitude instead of pressure.

Altocumulus   A middle cloud, usually white or gray. Often occurs in layers or patches with wavy, rounded masses or rolls.

Altocumulus castellanus   An altocumulus showing vertical development. Individual cloud elements have towerlike tops, often in the shape of tiny castles.

Altostratus   A middle cloud composed of gray or bluish sheets or layers of uniform appearance. In the thinner regions, the sun or moon usually appears dimly visible.

Analogue method of forecasting   A forecast made by comparison of past large-scale synoptic weather patterns that resemble a given (usually current) situation in its essential characteristics.

Analysis   The drawing and interpretation of the patterns of various weather elements on a surface or upper-air chart.

Anemometer   An instrument designed to measure wind speed.

Aneroid barometer   An instrument designed to measure atmospheric pressure. It contains no liquid.

Angular momentum   The product of an object's mass, speed, and radial distance of rotation.

Annual range of temperature   The difference between the warmest and coldest months at any given location.

Anticyclone   An area of high pressure around which the wind blows clockwise in the Northern Hemisphere and counterclockwise in the Southern Hemisphere.

Arcus cloud   *See* Roll cloud.

Atmosphere effect   The warming of an atmosphere by its absorbing and reemitting infrared radiation while allowing shortwave radiation to pass on through. The gases mainly responsible for the earth's atmosphere effect are water vapor and carbon dioxide. Also called the *greenhouse effect*.

Atmospheric stagnation   A condition of light winds and poor vertical mixing that can lead to a high concentration of pollutants. Air stagnations are most often associated with fair weather, an inversion, and the sinking air of a high-pressure area.

Atmospheric window   The wavelength range between 8 and 11 μm in which little absorption of infrared radiation takes place.

Aurora   Glowing light display in the nighttime sky caused by excited gases in the upper atmosphere giving off light. In the Northern Hemisphere it is called the *aurora borealis* (northern lights); in the Southern Hemisphere, the *aurora australis* (southern lights).

Autumnal equinox The equinox at which the sun approaches the Southern Hemisphere and passes directly over the equator. Occurs around September 23.

Back-door cold front A cold front moving south or southwest along the Atlantic seaboard of the United States.

Backing wind A wind that changes direction in a counterclockwise sense (e.g., north to northwest to west).

Baroclinic The state of the atmosphere where surfaces of constant pressure intersect surfaces of constant density. On an isobaric chart, isotherms cross the contour lines and temperature advection exists.

Baroclinic instability A type of instability arising from a meridional (north to south) temperature gradient, a strong vertical wind speed shear, temperature advection, and divergence in the flow aloft. Many mid-latitude cyclones develop as a result of this instability.

Barograph A recording barometer.

Barometer An instrument that measures atmospheric pressure. The two most common barometers are the *mercury barometer* and the *aneroid barometer*.

Barotropic A condition in the atmosphere where surfaces of constant density parallel surfaces of constant pressure.

Bermuda high *See* Subtropical high.

Billow clouds Broad, nearly parallel lines of clouds oriented at right angles to the wind.

Bimetallic thermometer A temperature-measuring device usually consisting of two dissimilar metals that expand and contract differentially as the temperature changes.

Black body A hypothetical object that absorbs all of the radiation that strikes it. It also emits radiation at a maximum rate for its given temperature.

Blizzard A severe weather condition characterized by low temperatures and strong winds (greater than 32 mi/hr) bearing a great amount of snow. When these conditions continue after the falling snow has ended, it is termed a *ground blizzard*.

Boulder winds Fast-flowing, local downslope winds that may attain speeds of 100 knots or more. They are especially strong along the eastern foothills of the Rocky Mountains near Boulder, Colorado.

Buys-Ballot's law A law describing the relationship between the wind direction and the pressure distribution. In the Northern Hemisphere, if you stand with your back to the wind, then turn clockwise about 30°, lower pressure will be to your left. In the Southern Hemisphere, stand with your back to the wind, then turn counterclockwise about 30°, lower pressure will be to your right.

California current The ocean current that flows southward along the west coast of the United States from about Washington to Baja California.

California norther A strong, dry, northerly wind that blows in late spring, summer, and early fall in northern and central California. Its warmth and dryness is due to downslope compressional heating.

Cap cloud *See* Pileus cloud.

Carbon dioxide (CO_2) A colorless, odorless gas whose concentration is about 0.034 percent (340 ppm) in a volume of air near sea level. It is a selective absorber of infrared radiation and, consequently, it is important in the earth's atmosphere (greenhouse) effect. Solid CO_2 is called *dry ice*.

Ceiling The height of the lowest layer of clouds when the weather reports describe the sky as broken or overcast.

Ceiling balloon A small balloon used to determine the height of the cloud base. The height is computed from the balloon's ascent rate and the time required for its disappearance into the cloud.

Ceilometer An instrument that automatically records cloud height.

Centripetal acceleration The inward-directed acceleration on a particle moving in a curved path.

Centripetal force The radial force required to keep an object moving in a circular path. It is directed towards the center of that curved path.

Chinook A warm, dry wind on the eastern side of the Rocky Mountains. In the Alps, this wind is called a *Foehn*.

Chinook wall cloud A bank of clouds over the Rocky Mountains that signifies the approach of a chinook.

Cirrocumulus A high cloud that appears as a white patch of cloud without shadows. It consists of very small elements in the form of grains or ripples.

Cirrostratus A high cloud appearing as a whitish veil that may totally cover the sky. Often produces halo phenomena.

Cirrus A high cloud composed of ice crystals in the form of thin, white, featherlike clouds in patches, filaments, or narrow bands.

Clear air turbulence (CAT) Turbulence encountered by aircraft flying through cloudless skies. Thermals, wind shear, and jet streams can each be a factor in producing CAT.

Clear ice A layer of ice that appears transparent because of its homogeneous structure and small number and size of air pockets.

Climate The accumulation of daily and seasonal weather events over a long period of time.

Climatic optimum A period in geological history (about 7000 to 5000 years ago) when temperatures were warmer than at present.

Climatological forecast A weather forecast, usually a month or more in the future, which is based upon the climate of a region rather than upon current weather conditions.

Cloudburst Any sudden and heavy rain shower.

Cloud seeding The introduction of artificial substances (usually silver iodide or dry ice) into a cloud for the purpose of either modifying its development or increasing its precipitation.

Cloud streets Lines or rows of cumuliform clouds.

Coalescence The merging of cloud droplets into a single larger droplet.

Cold Clouds *See* Supercooled cloud.

Cold fog *See* Supercooled cloud.

Cold front A transition zone where a cold air mass advances and replaces a warm air mass.

Cold occlusion *See* Occluded front.

Cold wave A rapid fall in temperature within 24 hours that often requires increased protection for agriculture, industry, commerce, and human activities.

Comma cloud A band of organized cumuliform clouds that looks like a comma on a satellite photograph.

Computer enhancement A process where the temperatures of radiating surfaces are assigned different shades of gray on an infrared picture. This allows specific features to be more clearly delineated.

Condensation   The process by which water vapor becomes a liquid.

Condensation nuclei   Tiny particles upon whose surfaces condensation of water vapor begins in the atmosphere. Small nuclei less than 0.2 μm in radius are called *Aitken nuclei*; those with radii between 0.2 and 1 μm are *large nuclei*, while *giant nuclei* have radii larger than 1 μm.

Conditionally unstable air   An atmospheric condition that exists when the environmental lapse rate is less than the dry adiabatic rate but greater than the moist adiabatic rate. Also called *conditional instability.*

Conduction   The transfer of heat by molecular activity from one substance to another, or through a substance. Transfer is always from warmer to colder regions.

Constant-height chart (constant-level chart)   A chart showing variables, such as pressure, temperature, and wind, at a specific altitude above sea level. Variation in horizontal pressure is depicted by isobars. The most common constant-height chart is the surface chart, which is also called the *sea level chart.*

Constant pressure chart (isobaric chart)   A chart showing variables, such as temperature and wind, on a constant pressure surface. Variations in height are usually shown by lines of equal height (contour lines).

Contact nucleation (freezing)   The process by which contact with a nucleus such as an ice crystal causes super-cooled liquid droplets to change into ice.

Contour line   A line that connects points of equal elevation above a reference level, most often sea level.

Contrail (condensation trail)   A cloudlike streamer frequently seen forming behind aircraft flying in clear, cold, humid air.

Convection   Motions in a fluid that result in the transport and mixing of the fluid's properties. In meteorology, convection usually refers to atmospheric motions that are predominantly vertical, such as rising air currents due to surface heating. The rising of heated surface air and the sinking of cooler air aloft is often called *free convection.* (Compare with *forced convection.*)

Convective condensation level (CCL)   The level above the surface marking the base of a cumuliform cloud that is forming due to surface heating and rising thermals.

Convective instability   Instability arising in the atmosphere when a column of air exhibits warm, moist, nearly saturated air near the surface and cold, dry air aloft. When the lower part of the layer is lifted and saturation occurs, it becomes unstable.

Convergence   An atmospheric condition that exists when the winds cause a horizontal net inflow of air into a specified region.

Cooling degree-day   A form of degree-day used in estimating the amount of energy necessary to reduce the effective temperature of warm air. A cooling degree-day is a day on which the average temperature is one degree above a desired base temperature.

Coriolis force   An apparent force observed on any free-moving object in a rotating system. On the earth this deflective force results from the earth's rotation and causes moving particles (including the wind) to deflect to the right in the Northern Hemisphere and to the left in the Southern Hemisphere.

Corona (optic)   A series of colored rings concentrically surrounding the disk of the sun or moon. Smaller than the halo, the corona is caused by the diffraction of light around small water droplets of uniform size.

Country breeze   A light breeze that blows into a city from the surrounding countryside. It is best observed on clear nights when the urban heat island is most pronounced.

Crepuscular rays   Alternating light and dark bands of light that appear to fan out from the sun's position, usually at twilight.

Cumulonimbus   An exceptionally dense and vertically developed cloud, often with a top in the shape of an anvil. The cloud is frequently accompanied by heavy showers, lightning, thunder, and sometimes hail. It is also known as a *thunderstorm cloud.*

Cumulus   A cloud in the form of individual, detached domes or towers that are usually dense and well defined. It has a flat base with a bulging upper part that often resembles cauliflower. Cumulus clouds of fair weather are called *cumulus humilis.* Those that exhibit much vertical growth are called *cumulus congestus* or *towering cumulus.*

Cumulus stage   The initial stage in the development of an air-mass thunderstorm in which rising, warm, humid air develops into a cumulus cloud.

Curvature effect   In cloud physics, as cloud droplets decrease in size, they exhibit a greater surface curvature that causes a more rapid rate of evaporation.

Cutoff low   A cold upper-level low that has become displaced out of the basic westerly flow and lies to the south of this flow.

Cyclogenesis   The development or strengthening of middle latitude (extratropical) cyclones.

Cyclone   An area of low pressure around which the winds blow counterclockwise in the Northern Hemisphere and clockwise in the Southern Hemisphere.

Daily range of temperature   The difference between the maximum and minimun temperatures for any given day.

Dart leader   The discharge of electrons that proceeds intermittently toward the ground along the same ionized channel taken by the initial lightning stroke.

Degassing   The release of gases dissolved in hot, molten rock. Also called *outgassing.*

Dendrochronology   The analysis of the annual growth rings of trees as a means of interpreting past climatic conditions.

Density   The ratio of the mass of a substance to the volume occupied by it.

Deposition   A process that occurs in subfreezing air when water vapor changes directly to ice without becoming a liquid first. (Also called *sublimation* in meteorology.)

Deposition nuclei   Tiny particles (ice nuclei) upon which an ice crystal may grow by the process of deposition. (Also called *sublimation nuclei.*)

Desertification A general increase in the desert conditions of a region.

Desert pavement An arrangement of pebbles and large stones that remains behind as finer dust and sand particles are blown away by the wind.

Dew Water that has condensed onto objects near the ground when their temperatures have fallen below the dew point of the surface air.

Dew point (dew-point temperature) The temperature to which air must be cooled (at constant pressure and constant water vapor content) for saturation to occur. When the dew point falls below freezing it is called the *frost point*.

Diffraction The bending of light around objects, such as cloud and fog droplets, producing fringes of light and dark or colored bands.

Dispersion The separation of white light into its different component wavelengths.

Dissipating stage The final stage in the development of an air mass thunderstorm when downdrafts exist throughout the cumulonimbus cloud.

Divergence An atmospheric condition that exists when the winds cause a horizontal net outflow of air from a specific region.

Doldrums The region near the equator that is characterized by low pressure and light, shifting winds.

Doppler radar A radar that determines the velocity of falling precipitation either toward or away from the radar unit using the *Doppler shift*.

Doppler shift (effect) The change in the frequency of waves that occurs when the emitter or the observer is moving toward or away from the other.

Downburst A severe localized downdraft that can be experienced beneath a severe thunderstorm. (Compare Microburst.)

Drizzle Small drops between 0.2 and 0.5 mm in diameter that fall slowly and reduce visibility more than light rain.

Drought A period of abnormally dry weather sufficiently long enough to cause serious effects on agriculture and other activities in the affected area.

Dry adiabatic rate The rate of change of temperature in a rising or descending unsaturated air parcel. The rate of adiabatic cooling or warming is 10°C per 1000 m (5.5°F per 1000 ft).

Dry line A boundary that separates warm, dry air from warm, moist air. It usually represents a zone of instability along which thunderstorms form.

Dust devil (or whirlwind) A small but rapidly rotating wind made visible by the dust, sand, and debris it picks up from the surface. It develops best on clear, dry, hot afternoons.

Easterly wave A migratory wavelike disturbance in the tropical easterlies. Easterly waves occasionally intensify into tropical cyclones.

Eddy A small volume of air (or any fluid) that behaves differently from the larger flow in which it exists.

Eddy viscosity The internal friction produced by turbulent flow.

Ekman spiral An idealized description of the way the wind-driven ocean currents vary with depth. In the atmosphere it represents the way the winds vary from the surface up through the friction layer.

Electrical resistance thermometers Thermometers that use electrical conducting wires (thermistors) whose electrical resistance changes with the temperature. They are used in radiosondes.

Electromagnetic waves *See* Radiant energy.

El Niño An extensive ocean warming that begins along the coast of Peru and Ecuador. Major El Niño events occur once every 5 to 10 years as a current of nutrient-poor tropical water moves southward along the west coast of South America.

Embryo In cloud physics, a tiny ice crystal that grows in size and becomes an ice nucleus.

Entrainment The mixing of environmental air into a pre-existing air current or cloud so that the environmental air becomes part of the current or cloud.

Environmental lapse rate The rate of decrease of temperature with elevation. It is most often measured with a radiosonde.

Equilibrium vapor pressure The necessary vapor pressure around liquid water that allows the water to remain in equilibrium with its environment. Also called *saturation vapor pressure*.

Evaporation The process by which a liquid changes into a gas.

Evaporation fog Fog produced when sufficient water vapor is added to the air by evaporation. The two common types are *steam fog*, which forms when cold air moves over warm water, and *frontal fog*, which forms as warm raindrops evaporate in a cool air mass.

Exosphere The outermost portion of the atmosphere.

Extratropical cyclone A cyclonic storm that most often forms along a front in middle and high latitudes. Also called a *middle latitude storm*, a *depression*, and a *low*. It is not a tropical storm or hurricane.

Eye A region in the center of a hurricane (tropical storm) where the winds are light and skies are clear to partly cloudy.

Eye wall A wall of dense thunderstorms that surrounds the eye of a hurricane.

Fall streaks Falling ice crystals that evaporate before reaching the ground.

Fall wind A strong, cold katabatic wind that blows downslope off snow-covered plateaus.

Fata Morgana A complex mirage that is characterized by objects being distorted in such a way as to appear as castlelike features.

Feedback mechanism A process whereby an initial change in an atmospheric process will tend to either reinforce the process (*positive feedback*) or weaken the process (*negative feedback*).

Ferrel cell The name given to the middle-latitude cell in the 3-cell model of the general circulation.

Fetch The distance that the wind travels over open water.

Foehn *See* Chinook.

Fog A cloud with its base at the earth's surface. It reduces visibility to below 1 km.

Forced convection On a small scale, a form of mechanical stirring taking place when twisting eddies of air are able to mix hot surface air with the cooler air above. On a larger scale, it can be induced by the lifting of warm air along a front (*frontal uplift*) or along a topographic barrier (*orographic uplift*).

Free convection *See* Convection.

Freeze A condition occurring over a widespread area when the surface air temperature remains below freezing for a sufficient time to damage certain agricultural crops. A freeze most often occurs as cold air is advected into a region, causing freezing conditions to exist in a deep layer of surface air. Also called *advection frost*.

Freezing nuclei Any particle that has a shape similar to that of an ice crystal and allows rapid freezing of supercooled water. Such particles include certain clay minerals, meteoric dust, and ice crystals themselves.

Freezing rain and freezing drizzle Rain or drizzle that falls in liquid form and then freezes upon striking a cold object or ground. Both can produce a coating of ice on objects which is called *glaze*.

Friction layer The atmospheric layer near the surface usually extending up to about 1 km (3300 ft) where the wind is influenced by friction of the earth's surface and objects on it.

Front The transition zone between two distinct air masses.

Frontal fog *See* Evaporation fog.

Frontal thunderstorms Thunderstorms that form in response to forced convection (forced lifting) along a front. Most go through a cycle similar to those of air-mass thunderstorms.

Frontal wave A wavelike deformation along a front in the lower levels of the atmosphere. Those that develop into storms are termed *unstable waves*, while those that do not are called *stable waves*.

Frontogenesis The formation, strengthening, or regeneration of a front.

Frontolysis The weakening or dissipation of a front.

Frost (also called hoarfrost) A covering of ice produced by deposition (sublimation) on exposed surfaces when the air temperature falls below the frost point (the dew point is below freezing).

Frostbite The partial freezing of exposed parts of the body, causing injury to the skin and sometimes to deeper tissues.

Frost point *See* Dew point.

Frozen dew The transformation of liquid dew into tiny beads of ice when the air temperature drops below freezing.

Funnel cloud A rotating conelike cloud that extends downward from the base of a thunderstorm. When it reaches the surface it is called a *tornado*.

Galaxy A huge assembly of stars (between millions and hundreds of millions) held together by gravity.

Gas law The thermodynamic law applied to a perfect gas that relates the pressure of the gas to its density and absolute temperature.

General circulation of the atmosphere Large-scale atmospheric motions over the entire earth.

Geostationary satellite A satellite that orbits the earth at the same rate that the earth rotates and thus remains over a fixed place above the equator.

Geostrophic wind A theoretical horizontal wind blowing in a straight path, parallel to the isobars or contours, at a con-

stant speed. The geostrophic wind results when the Coriolis force exactly balances the horizontal pressure gradient force.

Giant nuclei *See* Condensation nuclei.

Glaciated cloud A cloud or portion of a cloud where only ice crystals exist.

Global scale The largest scale of atmospheric motion. Also called the *planetary scale*.

Glory Colored rings that appear around the shadow of an object.

Gradient wind A theoretical wind that blows parallel to curved isobars or contours.

Graupel *See* Snow pellets.

Green flash A small, green color that occasionally appears on the upper part of the sun as it rises or sets.

Greenhouse effect *See* Atmosphere effect.

Ground fog *See* Radiation fog.

Growing degree-day A form of the degree-day used as a guide for crop planting and for estimating crop maturity dates.

Gulf Stream A warm, swift, narrow ocean current flowing along the east coast of the United States.

Gust front A boundary that separates a cold downdraft of a thunderstorm from warm, humid surface air. On the surface its passage resembles that of a cold front.

Gyre A large circular, surface ocean current pattern.

Haboob A dust or sandstorm that forms as cold downdrafts from a thunderstorm turbulently lift dust and sand into the air.

Hadley cell A thermal circulation proposed by George Hadley to explain the movement of the trade winds. It consists of rising air near the equator and sinking air near 30° latitude.

Hailstones Transparent or partially opaque particles of ice that range in size from that of a pea to that of golf balls.

Halos Rings or arcs that encircle the sun or moon when seen through an ice crystal cloud or a sky filled with falling ice crystals. Halos are produced by refraction of light.

Haze Fine dry or wet dust or salt particles dispersed through a portion of the atmosphere. Individually these are not visible but cumulatively they will diminish visibility.

Heat A form of energy transferred between systems by virtue of their temperature differences.

Heat capacity The ratio of the heat absorbed (or released) by a system to the corresponding temperature rise (or fall).

Heat index (HI) An index that combines air temperature and relative humidity to determine an apparent temperature—how hot it actually feels.

Heating degree-day A form of the degree-day used as an index for fuel consumption.

Heat lightning Distant lightning that illuminates the sky but is too far away for its thunder to be heard.

Heatstroke A physical condition induced by a person's overexposure to high air temperatures, especially when accompanied by high humidity.

Heiligenschein A faint white ring surrounding the shadow of an observer's head on a dew-covered lawn.

Heterosphere The region of the atmosphere above about 85 km where the composition of the air varies with height.

High *See* Anticyclone.

High inversion fog A fog that lifts above the surface but does not completely dissipate because of a strong inversion (usually subsidence) that exists above the fog layer.

Homosphere The region of the atmosphere below about 85 km where the composition of the air remains fairly constant.

Horse latitudes The belt of latitude at about 30° to 35° where winds are predominantly light and weather is hot and dry.

Humidity A general term that refers to the air's water vapor content. (*See* Relative humidity.)

Humiture An index that relates air temperature and relative humidity to how hot it feels.

Hurricane A severe tropical cyclone having winds in excess of 64 knots (74 mi/hr).

Hydrologic cycle A model that illustrates the movement and exchange of water among the earth, atmosphere, and oceans.

Hydrostatic equilibrium The state of the atmosphere when there is a balance between the vertical pressure gradient force and the downward pull of gravity.

Hygrometer An instrument designed to measure the air's water vapor content. The sensing part of the instrument can be hair (*hair hygrometer*), a plate coated with carbon (*electrical hygrometer*), or an infrared sensor (*infrared hygrometer*).

Hydrophobic The ability to resist the condensation of water vapor. Usually used to describe "water-repelling" condensation nuclei.

Hygroscopic The ability to accelerate the condensation of water vapor. Usually used to describe "water-seeking" condensation nuclei.

Hypothermia The deterioration in one's mental and physical condition brought on by a rapid lowering of human body temperature.

Hypoxia A condition experienced by humans when the brain does not receive sufficient oxygen.

Ice age *See* Pleistocene epoch.

Ice crystal process A process that produces precipitation. The process involves tiny ice crystals in a supercooled cloud growing larger at the expense of the surrounding liquid droplets. Also called the *Bergeron-Findeisen process.*

Ice fog A type of fog composed of tiny suspended ice particles that forms at very low temperatures.

Icelandic low The subpolar low-pressure area that is centered near Iceland on charts that show mean sea level pressure.

Ice nuclei Particles that act as nuclei for the formation of ice crystals in the atmosphere.

Ice pellets *See* Sleet.

Indian summer An unseasonably warm spell with clear skies near the middle of autumn. Usually follows a substantial period of cool weather.

Inferior mirage *See* Mirage.

Infrared radiation Electromagnetic radiation with wavelengths between about 0.7 and 1000 μm. This radiation is longer than visible radiation but shorter than microwave radiation.

Infrared radiometer An instrument designed to measure the intensity of infrared radiation emitted by an object. Also called *infrared sensor.*

Insolation The *in*coming *sol*ar radi*ation* that reaches the earth and the atmosphere.

Instrument shelter A boxlike wooden structure designed to protect weather instruments from direct sunshine and precipitation.

Interglacial period A time interval of relatively mild climate during the Ice Age when continental ice sheets were absent or limited in extent to Greenland and the Antarctic.

Intertropical convergence zone (ITCZ) The boundary zone separating the northeast trade winds of the Northern Hemisphere from the southeast trade winds of the Southern Hemisphere.

Inversion An increase in air temperature with height.

Ion An electrically charged atom, molecule, or particle.

Ionosphere An electrified region of the upper atmosphere where fairly large concentrations of ions and free electrons exist.

Iridescence Brilliant spots or borders of colors, most often red and green, observed in clouds up to about 30° from the sun.

Isallobar A line of equal change in atmospheric pressure during a specified time interval.

Isobar A line connecting points of equal pressure.

Isobaric surface A surface along which the atmospheric pressure is everywhere equal. (*See* Constant pressure chart.)

Isotach A line connecting points of equal wind speed.

Isotherm A line connecting points of equal temperature.

Jet maximum *See* Jet streak.

Jet streak A region of high wind speed that moves through the axis of a jet stream. Also called *jet maximum.*

Jet stream Relatively strong winds concentrated within a narrow band in the atmosphere.

Katabatic wind Any wind blowing downslope. Usually cold.

Kelvin A unit of temperature. A Kelvin is denoted by K and 1 K equals 1°C. Zero Kelvin is absolute zero, or −273°C.

Kelvin scale A temperature scale with zero degrees equal to the theoretical temperature at which all molecular motion ceases. Also called the *absolute scale.* The units are sometimes called "degrees Kelvin"; however, the correct SI terminology is "Kelvins," abbreviated K.

Kinetic energy The energy within a body that is a result of its motion.

Kirchhoff's law A law that states: good absorbers of a given wavelength of radiation are also good emitters of that wavelength.

Knot A unit of speed equal to 1 nautical mile per hour. 1 knot equals 1.15 mi/hr.

Lake breeze A wind blowing onshore from the surface of a lake.

Lake-effect snows Localized snowstorms that form on the downwind side of a lake. Such storms are common in late fall and early winter near the Great Lakes as cold, dry air picks up moisture and warmth from the unfrozen bodies of water.

Laminar flow A nonturbulent flow in which the fluid moves smoothly in parallel layers or sheets.

Land breeze A coastal breeze that blows from land to sea, usually at night.

Lapse rate The rate at which an atmospheric variable (usually temperature) decreases with height. (*See* Environmental lapse rate.)

Large nuclei   *See* Condensation nuclei.

Latent heat   The heat that is either released or absorbed by a unit mass of a substance when it undergoes a change of state, such as during evaporation, condensation, or sublimation.

Leeside low   Storm systems (extratropical cyclones) that form on the downwind (lee) side of a mountain chain. In the United States leeside lows frequently form on the eastern side of the Rockies and Sierra Nevadas.

Leeward side   The side of an object away from the direction in which the wind is blowing.

Lenticular cloud   A cloud in the shape of a lens.

Lidar   An instrument that uses a laser to generate intense pulses that are reflected from atmospheric particles of dust and smoke. Lidars have been used to determine the amount of particles in the atmosphere as well as particle movement that has been converted into wind speed. Lidar means *light detection and ranging.*

Lifting condensation level (LCL)   The level at which a parcel of air when lifted dry adiabatically would become saturated.

Lightning   A visible electrical discharge produced by thunderstorms.

Little Ice Age   The period from about 1550 to 1850 when average global temperatures were lower, and alpine glaciers increased in size and advanced down mountain canyons.

Local winds   Winds that tend to blow over a relatively small area. Often due to regional effects, such as mountain barriers, large bodies of water, local pressure differences, and other influences.

Longwave radiation   A term most often used to describe the infrared energy emitted by the earth and the atmosphere.

Long waves in the westerlies   A wave in the major belt of westerlies characterized by a long length (thousands of kilometers) and significant amplitude. Also called *Rossby waves.*

Low   *See* Extratropical cyclone.

Low-level jet streams   Jet streams that typically form near the earth's surface below an altitude of about 2 km and usually attain speeds of less than 60 knots.

Macroclimate   The general climate of a large area, such as a country.

Macroscale   The normal meteorological synoptic scale for obtaining weather information. It can cover an area ranging from the size of a continent to the entire globe.

Magnetic storm   A worldwide disturbance of the earth's magnetic field caused by solar disturbances.

Magnetosphere   The region around the earth in which the earth's magnetic field plays a dominant part in controlling the physical processes that take place.

Mammatus clouds   Clouds that look like pouches hanging from the underside of a cloud.

Maritime air   Moist air whose characteristics were developed over an extensive body of water.

Mature thunderstorm   The second stage in the three-stage

cycle of an air-mass thunderstorm. This stage is characterized by heavy showers, lightning, thunder, and violent vertical motions inside cumulonimbus clouds.

Maunder minimum   A period from about 1645 to 1715 when few, if any, sunspots were observed.

Mean annual temperature   The average temperature at any given location for the entire year.

Mechanical turbulence   Turbulent eddy motions caused by obstructions, such as trees, buildings, mountains, and so on.

Meridional flow   A type of atmospheric circulation pattern in which the north-south component of the wind is pronounced.

Mesocyclone   A vertical column of cyclonically rotating air within a severe thunderstorm.

Mesohigh   A relatively small area of high atmospheric pressure that forms beneath a thunderstorm.

Mesoscale   The scale of meteorological phenomena that ranges in size from a few km to about 100 km. It includes local winds, thunderstorms, and tornadoes.

Mesoscale Convective Complex (MCC)   A large organized convective weather system comprised of a number of individual thunderstorms. The size of an MCC can be 1000 times larger than an individual air-mass thunderstorm.

Mesosphere   The atmospheric layer between the stratosphere and the thermosphere. Located at an average elevation between 50 and 80 km above the earth's surface.

Meteorology   The study of the atmosphere and atmospheric phenomena as well as the atmosphere's interaction with the earth's surface, oceans, and life in general.

Microburst   A strong localized downdraft less than 4 km wide that occurs beneath severe thunderstorms. A strong downdraft greater than 4 km across is called a *downburst.*

Microclimate   The climate structure of the air space near the surface of the earth.

Micrometer (μm)   A unit of length equal to one-millionth of a meter.

Microscale   The smallest scale of atmospheric motions.

Middle latitude cyclones   *See* Extratropical cyclone.

Milankovitch theory   A theory proposed by Milutin Milankovitch in the 1930s suggesting that changes in the earth's orbit were responsible for climatic changes and the ice ages.

Millibar (mb)   A unit for expressing atmospheric pressure. Sea level pressure is normally close to 1013 mb.

Mirage   A refraction phenomenon that makes an object appear to be displaced from its true position. When an object appears higher than it actually is, it is called a *superior mirage.* When an object appears lower than it actually is, it is an *inferior mirage.*

Mixed cloud   A cloud containing both water drops and ice crystals.

Mixing depth   The unstable atmospheric layer that extends from the surface up to the base of an inversion. Within this layer the air is well stirred.

Mixing ratio   The ratio of the mass of water vapor in a given volume of air to the mass of dry air.

Moist adiabatic rate   The rate of change of temperature in a rising or descending saturated air parcel. The rate of cooling or warming varies but a common value of 6°C per 1000 m (3.3°F per 1000 ft) is used.

Molecular viscosity   The small-scale internal fluid friction that is due to the random motion of the molecules within a smooth-flowing fluid, such as air.

Molecule A collection of atoms held together by chemical forces.

Monsoon depressions Weak low-pressure areas that tend to form in response to divergence in an upper-level jet stream. The circulation around the low strengthens the monsoon wind system and enhances precipitation during the summer.

Monsoon wind system A wind system that reverses direction between winter and summer. Usually the wind blows from land to sea in winter and from sea to land in summer.

Mountain and valley breeze A local wind system of a mountain valley that blows downhill (*mountain breeze*) at night and uphill (*valley breeze*) during the day.

Multicell storms Thunderstorms in a line, each of which may be in a different stage of development.

Nacreous clouds Clouds of unknown composition that have a soft, pearly luster and that form at altitudes about 25 to 30 km above the earth's surface. They are also called *mother-of-pearl clouds.*

Nebula A cloud of interstellar gas and dust.

Negative feedback mechanism See Feedback mechanism.

Neutral stability (neutrally stable air) An atmospheric condition that exists in dry air when the environmental lapse rate equals the dry adiabatic rate. In saturated air the environmental lapse rate equals the moist adiabatic rate.

Nimbostratus A dark, gray cloud characterized by more or less continuously falling precipitation. It is not accompanied by lightning, thunder, or hail.

Noctilucent clouds Wavy, thin, bluish-white clouds that are best seen at twilight in polar latitudes. They form at altitudes about 80 to 90 km above the surface.

Nocturnal inversion See Radiation inversion.

Northeaster A name given to a strong, steady wind from the northeast that is accompanied by rain and inclement weather. It often develops when a storm system moves northeastward along the coast of North America.

Northern lights See Aurora.

Nuclear fusion The combination of the nuclei of light atoms to form heavier nuclei, with the release of energy.

Nuclear winter The dark, cold, and gloomy conditions that presumably would be brought on by nuclear war.

Numerical weather prediction (NWP) Forecasting the weather based upon the solutions of mathematical equations by high-speed computers.

Occluded front (occlusion) A complex frontal system that ideally forms when a cold front overtakes a warm front. When the air behind the front is colder than the air ahead of it, the front is called a *cold occlusion.* When the air behind the front is milder than the air ahead of it, it is called a *warm occlusion.*

Oceanic front A boundary that separates masses of water with different temperatures and densities.

Offshore breeze A breeze that blows from the land out over the water. Opposite of an onshore breeze.

Omega high A ridge in the middle or upper troposphere that has the shape of a Greek letter omega (Ω).

Onshore breeze A breeze that blows from the water onto the land. Opposite of an offshore breeze.

Orchard heaters Oil heaters placed in orchards that generate heat and promote convective circulations to protect fruit trees from damaging low temperatures. Also called *smudge pots.*

Orographic uplift The lifting of air over a topographic barrier. Clouds that form in this lifting process are called *orographic clouds.*

Overrunning A condition that occurs when air moves up and over another layer of air.

Ozone (O₃) An almost colorless gaseous form of oxygen with an odor similar to weak chlorine. The highest natural concentration is found in the stratosphere.

Pacific high See Subtropical high.

Parcel of air An imaginary small body of air a few meters wide that is used to explain the behavior of air.

Permafrost A layer of soil beneath the earth's surface that remains frozen throughout the year.

Persistence forecast A forecast that the future weather condition will be the same as the present condition.

Photochemical smog See Smog.

Photodissociation The splitting of a molecule by a photon.

Photon A discrete quantity of energy that can be thought of as a packet of electromagnetic radiation traveling at the speed of light.

Photosphere The visible surface of the sun from which most of its energy is emitted.

Pileus cloud A smooth cloud in the form of a cap. Occurs above, or is attached to, the top of a cumuliform cloud.

Planetary scale The largest scale of atmospheric motion. Sometimes called the *global scale.*

Plasma See Solar wind.

Plate tectonics The theory that the earth's surface down to about 100 km is divided into a number of plates that move relative to one another across the surface of the earth. Once referred to as continental drift.

Pleistocene epoch (or Ice Age) The most recent period of extensive continental glaciation that saw large portions of North America and Europe covered with ice. It began about 2 million years ago and ended about 10,000 years ago.

Polar easterlies A shallow body of easterly winds located at high latitudes poleward of the subpolar low.

Polar front A semipermanent, semicontinuous front that separates tropical air masses from polar air masses.

Polar front jet stream The jet stream that is associated with the polar front in middle and high latitudes. It is usually located at altitudes between 9 and 12 km.

Polar front theory A theory developed by a group of Scandinavian meteorologists that explains the formation, development, and overall life history of cyclonic storms that form along the polar front.

Polar orbiting satellite A satellite whose orbit closely parallels the earth's meridian lines and thus crosses the polar regions on each orbit.

Pollutants Any gaseous, chemical, or organic matter that contaminates the atmosphere, soil, or water.

Positive feedback mechanism See Feedback mechanism.

Positive vorticity advection (PVA) A region of positive vorticity usually several hundred kilometers wide on a upper-level chart that moves with the general wind flow. It aids in weather prediction by showing where regions of rising air, clouds, and storms are likely to form.

Potential energy The energy that a body possesses by virtue of its position with respect to other bodies in the field of gravity.

Potential evapotranspiration (PE) That amount of moisture that, if it were available, would be removed from a given land area by evaporation and transpiration.

Potential temperature The temperature that a parcel of dry air would have if it were brought dry adiabatically from its original position to a pressure of 1000 mb.

Precipitation Any form of water particles—liquid or solid—that falls from the atmosphere and reaches the ground.

P/E index (Precipitation-evaporation index) An index that gives the long-range effectiveness of precipitation in promoting plant growth.

P/E ratio (Precipitation-evaporation ratio) An expression devised for the purpose of classifying climates; based on monthly totals of precipitation and evaporation.

Pressure gradient The rate of decrease of pressure per unit of distance. On the same chart, when the isobars are close together, the pressure gradient is steep. When the isobars are far apart, the pressure gradient is weak.

Pressure gradient force The force due to differences in pressure within the atmosphere that causes air to move and, hence, the wind to blow. It is directly proportional to the pressure gradient.

Pressure tendency The rate of change of atmospheric pressure within a specified period of time, most often three hours. Same as *barometric tendency*.

Prevailing wind The wind direction most frequently observed during a given period.

Probability forecast A forecast of the probability of occurrence of one or more of a mutually exclusive set of weather conditions.

Prognostic chart (Prog) A chart showing expected or forecasted conditions, such as pressure patterns, frontal positions, contour height patterns, and so on.

Prominence *See* Solar flare.

Psychrometer An instrument used to measure the water vapor content of the air. It consists of two thermometers (dry bulb and wet bulb). After whirling the instrument, the dew point and relative humidity can be obtained with the aid of tables.

Radar An electronic instrument used to detect objects (such as falling precipitation) by their ability to reflect and scatter microwaves back to a receiver.

Radiant energy (radiation) Energy propagated in the form of electromagnetic waves. These waves do not need molecules to propagate them, and in a vacuum they travel at nearly 300,000 km per sec.

Radiation fog Fog produced over land when radiational cooling reduces the air temperature to or below its dew point. It is also known as *ground fog* and *valley fog*.

Radiation inversion An increase in temperature with height due to radiational cooling of the earth's surface. Also called a *nocturnal inversion*.

Radiometer *See* Infrared radiometer.

Radiosonde A balloon-borne instrument that measures and transmits pressure, temperature, and humidity to a ground-based receiving station.

Rain Precipitation in the form of liquid water drops that have diameters greater than that of drizzle.

Rainbow An arc of concentric colored bands that spans a section of the sky when rain is present and the sun is positioned at the observer's back.

Rain shadow The region on the leeside of a mountain where the precipitation is noticeably less than on the windward side.

Rawinsonde observation A radiosonde observation that includes wind data.

Reflection The process whereby a surface turns back a portion of the radiation that strikes it.

Refraction The bending of light as it passes from one medium to another.

Relative humidity The ratio of the amount of water vapor actually in the air compared to the amount of water vapor the air can hold at that particular temperature and pressure. The ratio of the air's actual vapor pressure to its saturation vapor pressure.

Return stroke The luminous lightning stroke that propagates upward from the earth to the base of a cloud.

Ridge An elongated area of high atmospheric pressure.

Rime ice A white, granular deposit of ice formed by the freezing of water drops when they come in contact with an object.

Riming *See* Accretion.

Roll cloud A dense, roll-shaped cloud attached to the lower front part of the main cloud. It often forms with thunderstorms along the leading edge of a gust front. Also called an *arcus cloud*.

Rossby waves *See* Long waves in the westerlies.

Rotor cloud A turbulent cumuliform type of cloud that forms on the leeward side of large mountain ranges. The air in the cloud rotates about an axis parallel to the range.

Rotors Turbulent eddies that form downwind of a mountain chain, creating hazardous flying conditions.

St. Elmo's fire A bright electric discharge that is projected from objects (usually pointed) when they are in a high electric field, such as during a thunderstorm.

Saltation The bouncing movement of sand and small particles along the surface due to the wind.

Sand dunes A hill or ridge of loose sand shaped by the winds.

Santa Ana wind A warm, dry wind that blows into southern California from the east off the elevated desert plateau. Its warmth is derived from compressional heating.

Saturation (of air) An atmospheric condition whereby the level of water vapor is the maximum possible at the existing temperature and pressure.

Saturation vapor pressure The maximum amount of water vapor necessary to keep moist air in equilibrium with a surface of pure water or ice. It represents the maximum amount of r vapor that the air can hold at any given temperature and ressure. (*See* Equilibrium vapor pressure.)

Savanna   A tropical or subtropical region of grassland and drought-resistant vegetation.

Scales of motion   The hierarchy of atmospheric circulations from tiny gusts to giant storms.

Scattering   The process by which small particles in the atmosphere deflect radiation from its path into different directions.

Scintillation   The apparent twinkling of a star due to its light passing through regions of differing air densities in the atmosphere.

Sea breeze   A coastal local wind that blows from the ocean onto the land. The leading edge of the breeze is termed a *sea breeze front*.

Sea breeze convergence zone   A region where sea breezes, having started in different regions, flow together and converge.

Sea level pressure   The atmospheric pressure at mean sea level.

Selective absorbers   Substances such as water vapor, carbon dioxide, clouds, and snow that absorb radiation only at particular wavelengths.

Sensible temperature   The sensation of temperature that the human body feels in contrast to the actual temperature of the environment as measured with a thermometer.

Severe thunderstorms   Intense thunderstorms capable of producing heavy showers, flash floods, hail, strong and gusty surface winds and tornadoes.

Shear   *See* Wind shear.

Sheet lightning   A fairly bright lightning flash from distant thunderstorms that illuminates a portion of the cloud.

Shelterbelt   A belt of trees or shrubs arranged as a protection against strong winds.

Shortwave (in the atmosphere)   A small wave that moves around longwaves in the same direction as the air flow in the middle and upper troposphere. Shortwaves are also called *shortwave troughs*.

Shortwave radiation   A term most often used to describe the radiant energy emitted from the sun, in the visible and near ultraviolet wavelengths.

Shower   Intermittent precipitation from a cumuliform cloud, usually of short duration but often heavy.

Siberian high   A strong, shallow area of high pressure that forms over Siberia in winter.

Sleet   A type of precipitation consisting of transparent pellets of ice 5 mm or less in diameter. Same as *ice pellets*.

Smog   Originally smog meant a mixture of smoke and fog. Today, smog means air that has restricted visibility due to pollution, or pollution formed in the presence of sunlight—*photochemical smog*.

Smog front (also smoke front)   The leading edge of a sea breeze that is contaminated with smoke or pollutants.

Smudge pots   *See* Orchard heaters.

Snowflake   An aggregate of ice crystals that falls from a cloud.

Snow flurries   Light showers of snow that fall intermittently.

Snow grains   Precipitation in the form of very small, opaque grains of ice. The solid equivalent of drizzle.

Snow pellets   White, opaque, approximately round ice particles between 2 and 5 mm in diameter that form in a cloud either from the sticking together of ice crystals or from the process of accretion.

Snow rollers   A cylindrical spiral of snow shaped somewhat like a child's muff and produced by the wind.

Snow squall (shower)   An intermittent heavy shower of snow that greatly reduces visibility.

Solar constant   The rate at which solar energy is received on a surface at the outer edge of the atmosphere perpendicular to the sun's rays when the earth is at a mean distance from the sun. The value of the solar constant is about two calories per square centimeter per minute or 1376 W/m^2 in the SI system of measurement.

Solar flare   A rapid eruption from the sun's surface that emits high energy radiation and energized charged particles.

Solar wind   An outflow of charged particles from the sun that escapes the sun's outer atmosphere at high speed.

Solute effect   The dissolving of hygroscopic particles, such as salt, in pure water thus reducing the relative humidity required for the onset of condensation.

Sonic boom   A loud explosive-like sound caused by a shock wave eminating from an aircraft (or any object) traveling at or above the speed of sound.

Sounding   An upper-air observation, such as a radiosonde observation.

Source regions   Regions where air masses originate and acquire their properties of temperature and moisture.

Southern oscillation   The reversal of surface air pressure at opposite ends of the tropical Pacific Ocean that occur during El Niño events.

Specific heat   The ratio of the heat absorbed (or released) by the unit mass of the system to the corresponding temperature rise (or fall).

Specific humidity   The ratio of the mass of water vapor in a given parcel to the total mass of air in the parcel.

Spontaneous nucleation (freezing)   The freezing of pure water without the benefit of any nuclei.

Squall line   Any nonfrontal line or band of active thunderstorms.

Stable air   *See* Absolutely stable air.

Standard atmosphere   A hypothetical vertical distribution of atmospheric temperature, pressure, and density in which the air is assumed to obey the gas law and the hydrostatic equation. The lapse rate of temperature in the troposphere is taken as 6.5°C/1000 m or 3.6°F/1000 ft.

Stationary front   A front that is nearly stationary with winds blowing almost parallel and from opposite directions on each side of the front.

Station pressure   The actual air pressure computed at the observing station.

Steady-state forecast   A weather prediction based on the past movement of surface weather systems. It assumes that the systems will move in the same direction and at approximately the same speed as they have been moving. Also called *trend forecasting*.

Steam fog   *See* Evaporation fog.

Steppe   An area of grass-covered, treeless plains that has a semi-arid climate.

Stepped leader   An initial discharge of electrons that proceeds intermittently toward the ground in a series of steps in a cloud-to-ground lightning stroke.

Storm surge   An abnormal rise of the sea along a shore. Primarily due to the winds of a storm, especially a hurricane.

Stratocumulus   A low cloud, predominantly stratiform with low, lumpy, rounded masses, often with blue sky between them.

Stratosphere   The layer of the atmosphere above the troposphere and below the mesosphere (between 10 km and 50 km), generally characterized by an increase in temperature with height.

Stratus   A low, gray cloud layer with a rather uniform base whose precipitation is most commonly drizzle.

Streamline   A line that shows the wind flow pattern.

Sublimation   The process whereby ice changes directly into water vapor without melting. In meteorology, sublimation can also mean the transformation of water vapor into ice. (*See* Deposition.)

Subsidence   The slow sinking of air, usually associated with high-pressure areas.

Subsidence inversion   A temperature inversion produced by the adiabatic warming of a layer of sinking air.

Subtropical front   A zone of temperature transition in the upper troposphere over subtropical latitudes, where warm air carried poleward by the Hadley cell meets the cooler air of the middle latitudes.

Subtropical high   A semipermanent high in the subtropical high-pressure belt centered near 30° latitude. The *Bermuda high* is located over the Atlantic Ocean off the east coast of North America. The *Pacific high* is located off the west coast of North America.

Subtropical jet stream   The jet stream typically found between 20° and 30° latitude at altitudes between 12 and 14 km.

Suction vortices   Small, rapidly rotating whirls perhaps 10 m in diameter that are found within large tornadoes.

Summer solstice   Approximately June 22 in the Northern Hemisphere when the sun is highest in the sky and directly overhead at latitude 23½°N, the Tropic of Cancer.

Sundog   A colored luminous spot produced by refraction of light through ice crystals that appears on either side of the sun. Also called *parhelion*.

Sun pillar   A vertical streak of light extending above (or below) the sun. It is produced by the reflection of sunlight off ice crystals.

Sunspots   Relatively cooler areas on the sun's surface. They represent regions of extremely high magnetic field.

Supercell storm   An enormous severe thunderstorm whose updrafts and downdrafts are nearly in balance, allowing it to maintain itself for several hours. It can produce large hail and tornadoes.

Supercooled cloud (or cloud droplets)   A cloud composed of liquid droplets at temperatures below 0°C (32°F). When the cloud is on the ground it is called *supercooled fog* or *cold fog*.

Superior mirage   *See* Mirage.

Supernova   A tremendous explosion of a massive star.

Supersaturated air   A condition that occurs in the atmosphere when the relative humidity is greater than 100 percent.

Surface inversion   *See* Radiation inversion.

Synoptic scale   The typical weather map scale that shows features such as high- and low-pressure areas and fronts over a distance spanning a continent. Also called the *cyclonic scale*.

Taiga (boreal forest)   The open northern part of the coniferous forest. Taiga also refers to subarctic climate.

Tangent arc   An arc of light tangent to a halo. It forms by refraction of light through ice crystals.

Tcu   An abbreviation sometimes used to denote a towering cumulus cloud (cumulus congestus).

Temperature   The degree of hotness or coldness of a substance as measured by a thermometer. It is also a measure of the average speed or kinetic energy of the atoms and molecules in a substance.

Terminal velocity   The constant speed obtained by a falling object when the upward drag on the object balances the downward force of gravity.

Texas norther   A strong, cold wind from between the northeast and northwest associated with a cold outbreak of polar air that brings a sudden drop in temperature. Sometimes called a *blue norther*.

Theodolite   An instrument used to track the movements of a pilot balloon.

Thermal   A small, rising parcel of warm air produced when the earth's surface is heated unevenly.

Thermal belts   Horizontal zones of vegetation found along hillsides that are primarily the result of vertical temperature variations.

Thermal circulations   Air flow resulting primarily from the heating and cooling of air.

Thermal lows and thermal highs   Areas of low and high pressure that are shallow in vertical extent and are produced primarily by surface temperatures.

Thermal tides   Atmospheric pressure variations due to the uneven heating of the upper atmosphere by the sun.

Thermal turbulence   Turbulent vertical motions that result from surface heating and the subsequent rising and sinking of air.

Thermistor   An electrical resistance device used in the measurement of temperature.

Thermograph   An instrument that measures and records air temperature.

Thermometer   An instrument for measuring temperature.

Thermosphere   The atmospheric layer above the mesosphere (above about 85 km) where the temperature increases rapidly with height.

Thunder   The sound due to rapidly expanding gases along the channel of a lightning discharge.

Thunderstorm   A local storm produced by cumulonimbus clouds. Always accompanied by lightning and thunder.

Tornado   An intense, rotating column of air that protrudes from a cumulonimbus cloud in the shape of a funnel or a rope and touches the ground. (*See* Funnel cloud.)

Tornado vortex signature (TVS)   An image of a tornado on the Doppler radar screen that shows up as a small region of rapidly changing wind speeds inside a mesocyclone.

Tornado warning A warning issued when a tornado has actually been observed either visually or on a radar screen.

Tornado watch A forecast issued to alert the public that tornadoes may develop within a specified area.

Trade wind inversion A temperature inversion frequently found in the subtropics over the eastern portions of the tropical oceans.

Trade winds The winds that occupy most of the tropics and blow from the subtropical highs to the equatorial low.

Transpiration The process by which water in plants is transferred as water vapor to the atmosphere.

Tropical depression A mass of thunderstorms and clouds generally with a cyclonic wind circulation of between 20 and 34 knots.

Tropical disturbance An organized mass of thunderstorms with a slight cyclonic wind circulation of less than 20 knots.

Tropical rain forest A type of forest consisting mainly of lofty trees and a dense undergrowth near the ground.

Tropical storm Organized thunderstorms with a cyclonic wind circulation between 35 and 64 knots.

Tropopause The boundary between the troposphere and the stratosphere.

Tropopause jets Jet streams found near the tropopause, such as the polar front and subtropical jet streams.

Troposphere The layer of the atmosphere extending from the earth's surface up to the tropopause (about 10 km above the ground).

Trough An elongated area of low atmospheric pressure.

Turboclair A fog-clearing technique that uses hot gases from jet engines to heat the air and evaporate warm fog.

Turbulence Any irregular or disturbed flow in the atmosphere that produces gusts and eddies.

Twilight The time immediately before sunrise and after sunset when the sky remains illuminated.

Typhoon A hurricane that forms in the western Pacific Ocean.

Ultraviolet radiation Electromagnetic radiation with wavelengths longer than X-rays but shorter than visible light.

Unstable air See Absolutely unstable air.

Upslope fog Fog formed as moist, stable air flows upward over a topographic barrier.

Upslope precipitation Precipitation that forms due to moist, stable air gradually rising along an elevated plain. Upslope precipitation is common over the western Great Plains, especially east of the Rocky Mountains.

Upwelling The rising of water (usually cold) toward the surface from the deeper regions of a body of water.

Urban heat island The increased air temperatures in urban areas as contrasted to the cooler surrounding rural areas.

Valley breeze See Mountain breeze.

Valley fog See Radiation fog.

Vapor pressure The pressure exerted by the water vapor molecules in a given volume of air.

Veering wind The wind that changes direction in a clockwise sense—north to northeast to east, and so on.

Ventifact A rock that has been cut, shaped or faceted by wind-driven particles.

Vernal equinox The equinox at which the sun approaches the Northern Hemisphere and passes directly over the equator. Occurs around March 20.

Virga Precipitation that falls from a cloud but evaporates before reaching the ground. (See Fall streaks.)

Viscosity The resistance of fluid flow. See Molecular viscosity and Eddy viscosity.

Visibility The greatest distance an observer can see and identify prominent objects.

Vorticity A measure of the spin of a fluid, usually small air parcels. *Absolute vorticity* is the combined vorticity due to the earth's rotation and the vorticity due to the air's circulation relative to the earth. *Relative vorticity* is due to the curving of the air flow and wind shear.

Wall cloud An area of rotating clouds that extends beneath a severe thunderstorm and from which a funnel cloud may appear. Also called a *collar cloud.*

Warm clouds Clouds that form at temperatures above freezing.

Warm-core low A low-pressure area that is warmer at its center than at its periphery. Tropical cyclones exhibit this temperature pattern.

Warm front A front that moves in such a way that warm air replaces cold air.

Warm occlusion See Occluded front.

Warm sector The region of warm air within a wave cyclone that lies between a retreating warm front and an advancing cold front.

Water equivalent The depth of water that would result from the melting of a snow sample. Typically about 10 inches of snow will melt to 1 inch of water, producing a water equivalent of 10 to 1.

Waterspout A column of rotating wind over water that has characteristics of a dust devil and tornado.

Wave cyclone An extratropical cyclone that forms and moves along a front. The circulation of winds about the cyclone tends to produce a wavelike deformation on the front.

Wavelength The distance between successive crests, troughs, or identical parts of a wave.

Weather The condition of the atmosphere at any particular time and place.

Weather types Certain weather patterns categorized into similar groups. Used as an aid in weather prediction.

Wet-bulb depression The difference in degrees between the air temperature (dry-bulb temperature) and the wet-bulb temperature.

Wet-bulb temperature The lowest temperature that can be obtained by evaporating water into the air.

Whirlwinds See Dust devils.

Wind Air in motion relative to the earth's surface.

Wind-chill factor The cooling effect of any combination of temperature and wind, expressed as the loss of body heat. Also called *wind-chill index.*

Wind direction The direction *from which* the wind is blowing.

Wind machines Fans placed in orchards for the purpose of mixing cold surface air with warmer air above.

Wind rose A diagram that shows the percent of time that the wind blows from different directions at a given location over a given time.

Wind shear   The rate of change of wind speed or wind direction over a given distance.

Wind sock   A tapered fabric shaped like a cone that indicates wind direction by pointing away from the wind. Also called a *wind cone*.

Wind vane   An instrument used to indicate wind direction.

Windward side   The side of an object facing into the wind.

Wind waves   Water waves that form due to the flow of air over the water's surface.

Winter solstice   Approximately December 22 in the Northern Hemisphere when the sun is lowest in the sky and directly overhead at latitude 23½°S, the Tropic of Capricorn.

Xerophytes   Drought-resistant vegetation.

Zonal wind   A wind that has a predominate west-to-east component.

INDEX